Survival & Rescue Equipment of World War II:

Army Air Forces and U.S. Navy

VOLUME 3

Emergency Kits

First Aid Kits

Ration Kits

Dustin Clingenpeel

ISBN 978-1-943492-95-4

Book and cover design © Mick J. Prodger

EGP

ELM GROVE PUBLISHING
San Antonio, Texas, USA
www.elmgrovepublishing.com

Contents

Introduction 5

MULTI-PERSON EMERGENCY KITS

Part I. USAAF 11

Part II. U.S. NAVY 45

INDIVIDUAL EMERGENCY KITS

Part I. USAAF 58

Part II. U.S. NAVY 127

FIRST AID KITS 171

EMERGENCY RATIONS 246

Afterword 298

Acknowledgments 300

Author's Note 301

Index 302

"Our aviators and seamen, with confidence that they will fly and sail again tomorrow, dare to face greater odds in the war today. Vision of our services, and ingenuity of industry have provided survival and rescue equipment, which has lessened the hazards, improved the safety, and given our men greater courage. May we never be content with present equipment, but constantly improve it with experience, continued study and cooperative effort."

Admiral Russell R. Waesche,
Commandant of the United States Coast Guard

Introduction

The first great survival story of the war occurred on 16 January 1942, an odyssey first publicized in an article in *Life* magazine in April, 1942 and chronicled in the book *The Raft*, by Robert Trumbull. It was a first hand account of three Naval aviators of Torpedo Squadron 6: Aviation Chief Machinist's Mate Harold Dixon, Tony Pastula AOM2/c, and radioman/gunner Gene Aldrich. Their experience was so remarkable that no discussion of survival at sea would be complete without reference to it. They proved that men could survive for more than a month without any initial supplies of food and water, as they survived 34 days with very little equipment in a rubber life raft. This great survival story taught several basic lessons that the Navy then incorporated into its survival courses and new equipment, and impacted the development of the war.

Their story began in the waning first months of WWII. The aircraft carrier strength of the US Navy was divided between the Atlantic and the Pacific Oceans, with only three carrier air groups in the Pacific Fleet upon the attack on Pearly Harbor on 7 December 1941. By early January 1942, the Pacific Fleet was reinforced with the transfer of the USS *Yorktown* from the Atlantic Fleet. The carrier air group was then the USS *Enterprise*, USS *Lexington*, USS *Saratoga*, and USS *Yorktown*. The Imperial Japanese Navy had ten aircraft carriers of experienced and seasoned air groups, so the significant numerical disadvantage made the situation critical. On 11 January 1942, a significant loss occurred to the Pacific Fleet: the USS *Saratoga* toke a torpedo hit from a Japanese submarine, forcing her to the US West Coast for repairs.

The remaining three carriers were tasked with protecting the volatile supply convoys of the newly created South Pacific ferry route. That had been created by a coalition in the summer of 1941 to enable supplies to reach the Philippines, bypassing Japanese-held territory. Airfields and seaports were established, by agreement with and permission of territories under jurisdiction by the United Kingdom, New Zealand, Australia, the Netherlands, and the Free French. This route was an arc to the southwest from Hawaii-Christmas Island-Samoa Islands-Fiji Islands-New Caledonia and on to Australia. This route would later prove pivotal in the rescue of Dixon, Pastula, and Aldrich.

Admiral Chester W. Nimitz, Pacific Fleet Commander, pressed plans for the first major offensive to be on the Japanese held territories of the Gilbert and Marshall Islands, striking at the threats of those island bases. The USS *Enterprise*, with Carrier Air Group 6, sailed from Pearl Harbor 12 January 1942, rendezvousing with the USS *Yorktown* to first escort a convoy to Samoa. The two carrier task groups would then sail northwest to launch air strikes over the Gilbert and Marshall Islands, scheduled for 1 February. While en route, an epic journey of ingenuity, perseverance, determination, and some luck commenced.

On the morning of 16 January 1942, Harold Dixon went to the ready room of the USS *Enterprise* to prepare for his assigned anti-submarine patrol of the day. There he studied wind and navigational data for the mission. Dixon was an experienced 22 year Naval veteran. This experience would later prove instrumental in the group's survival. When he wasn't on duty, he worked in the ship's educational department,providing instruction on navigation,mathematics,and other technical subjects of airplane structures and electronics.

Information and mission orders in hand, Dixon reached the flight deck for takeoff in a TBD Devastator. The mission would require the aircraft's full extended range, leaving little fuel for their return. On reaching the aircraft, Dixon briefly laid eyes on his two aircrew, Pastula and Aldrich. In the U.S. Navy, assignments for airplanes and aircrew alternated. Often, a pilot had no formal meeting with his crew prior to takeoff. Introductions would take place during the flight. Fully fueled and armed with a bomb on its belly, Dixon, Pastula, and Aldrich took off at 2:15 P.M.

The flight path was predetermined; they encountered moderate weather conditions. At some point during the return leg, a possible navigational error was made. Dixon took some maneuvers to avoid scattered rain squalls but was an astute student of navigation and convinced his calculations were correct. Pastula, the bombardier, was charged with taking wind estimates for Dixon; it's possible he made a calculation error. The other potentiality was that the Enterprise task group shifted off course at some point. The cause may forever be a mystery. What was to be a routine scouting mission evolved to a fight for survival.

When he realized he was in trouble, Dixon evaluated and executed every possible scenario to contact the *Enterprise* by line of sight and radio. With negative results and down to their last ten gallons of gas, the crew knew they were hopelessly lost. While still airborne, Dixon began preparations for a water landing. This would have been a standard operating procedure for all Naval aviators and aircrew. Dixon directed that Pastula and Aldrich gather up all equipment and articles that weren't bolted down to bring aboard the raft. Signaling devices were the responsibility of the gunner's position on the SBD Dauntless or TBD Devastator. Aldrich gathered the signals, most likely the Mark-III Very signal pistol and shells, the first aid kit, and two flashlights. In the TBD and SBD, first aid kits were stored at the gunner's position. The first aid kit would either have been the aircraft first aid package NAF No.213886 or the M-448. In the bombardier's position was the life raft and emergency supplies, stored separately. Pastula prepared this equipment for immediate removal once they hit the water. Dixon, wanting to fully concentrate on making the water landing, handed his plotting board, .45 automatic pistol and loose shells, and a pair of pliers from his pocket to Pastula. It was the plan to bring the parachutes as well. At this point, the steps taken could be viewed as being textbook emergency procedure.

With the sun setting, Dixon made a perfect power stall landing on a fairly calm sea that day. What happened next was a testimony to preparedness. As soon as the plane settled, Dixon sprang from the cockpit and onto the wing and, as an almost instant relay, Pastula handed him the life raft. Before he could get the raft inflated, however, the airplane plunged beneath the waves.

It happened so fast that Aldrich and Pastula had not yet exited their positions and were pulled under. They returned to the surface after engaging their life preservers. During the chaos and struggle, all emergency equipment and parachutes were left behind. In the melee, the plotting board and flashlights were also lost. The consequence was that their survival depended on what they carried on their persons and the equipment provided in the raft itself. This was fundamental to the eventual development of the pararaft and back-pad kit.

Darkness set in and they regretted the loss of the flashlight. Two-cell flashlights were standard aircraft equipment; they were attached to the fuselage by spring clips in each position. With the integration of the life preserver light, they would then be attached to each person to prevent loss.

At about 8:00 A.M. the next day they sighted an airplane bearing in their general direction they determined to be friendly. Dixon estimated that the plane came within a half mile of their position. They began waving their arms and clothes and screaming but were never spotted, watching hopelessly as the plane drifted from their sight. Dixon believed that the route of the aircraft indicated it was searching for them.

Based on an estimate of where they went down, the aircraft was likely USAAF from Christmas or Palmyra islands from the east. These two islands were US-held territories that had airfields as part of the South Pacific ferry route as the first leg from Hawaii. In retrospect, one could imagine the outcome if they had access to any of the types of distress signals available at the time or in the future such as the sea marker, handheld smoke or projectors. Because of the circumstances, Aldrich was unable to retain the signal pistol. Later in the war, pilots and aircrew would have tracer ammunition readily at their disposal as wells as projectors in their back-pad kits or on their person – again highlighting the significance of personal equipment.

Later that morning Dixon inventoried their equipment and found in the supply pocket a signal mirror he described as stainless steel. In all likelihood it was the round chromium type that was standard equipment at the time, easily mistaken for stainless steel. He later practiced using it, but they would never again see another aircraft or surface vessel. During this era technical training was the priority and very little emergency equipment training was provided. However, Dixon's technical training would prove indispensable. He was on-point and prepared throughout. Dixon later talked about that first night when they were confident of their location and assured the plane would return to search for them. He also talked about feeling around the raft that night, wishing he had a flashlight. Certain that a search effort would continue, Dixon would have surely felt his way to the supply pocket for the mirror, having it at the ready come sunrise. This suggested he had no formal training on the equipment and is an example of how proper training, even on menial topics, could fundamentally save lives.

Dixon spent the better part of their first day studying the behavior of the wind and current. He had a strong working knowledge of that part of the Pacific and a vivid image in his head of a map of the islands. Though the plotting board was lost, he happened to have a pencil in his pocket and drew a chart from memory on his life preserver. There, he marked their estimated location. From mental data he determined their drift was southerly. He earmarked a series of islands directly in that path. The most familiar would had been the Samoa Islands, but it is not known what he exactly calculated. The best direct evidence of that was where they did end up, on Danger Island of the Cook Island group, not far from Samoa. (Danger Island is now known as Pukapuka.)

For the next 30 days, Dixon navigated using a small aerial navigator's mileage scale he'd had in his pocket, and his celestial knowledge. To his credit, he manipulated the raft on a discernible course. This influenced several developments. The first, soon after their rescue, was that the BuAer adopted sails as standard equipment on all life rafts. Second, navigational charts became standard. A series of charts was adopted in early 1943 as a standard for life rafts and paper, later cloth, charts were issued as personal equipment. These would include ocean currents and navigational headings.

For the first five days they endured blistering hot sun, lacking any type of protection. All they had were the clothes they wore. The men attempted to keep cool by soaking their clothes in seawater at regular intervals. It wasn't long before they sacrificed some of their clothing for protection of the head, later commenting there was no way to protect the face. Sunburn cream would have been a valuable aid.

They experienced what would later be observed in field trials in 1943 and resulted in the BuAer first adopting headnet M-565, revising it (as the M-565A) to include a neck and face shield. The sail might have provided additional sun protection, since their clothing eventually deteriorated and was lost experiencing severe sunburns.

Every day was a small adventure. At times they struggled to maintain morale. After losing their supplemental supplies when the plane sank, the biggest hurdle was gathering food and water. Each had carried a .45 caliber automatic pistol but only one was retained, with just three clips of ammunition.

On the first day, birds circled the raft. Dixon fired off a full clip trying to gather a meal but to no avail. Aldrich attempted to fish using the cotton line from the supply pocket and a section of spring from the pistol magazine as a hook but he too had no luck. Five days without food and water and morale plummeted. Salvation seemingly appeared on the fifth night when it rained, but they were completely unprepared for that eventuality. The rain subsided as quickly as it began, so they only moderately refreshed themselves with a shower and a quick drink. They attempted to use the oar and supply pockets to gather water, but soon realized that the fresh water was compromised by the saturation of salt in the pockets. They had learned to rinse them first – but by the time they'd rinsed them out, the rain had stopped.

The men encountered rain squalls periodically throughout their voyage, which were enough to sustain a level of hydration. With the introduction of the sail, water collection would had been much easier. The five-quart water bladder for storage and the M-616, with its integral funnel, were added later.

Dixon, Aldrich and Pastula attempted to fish on multiple occasions but failed. As if the sea were teasing them, their raft was almost always surrounded by some sort of sea life. Frustrated, Aldrich switched to a spearing method, grazing one fish on his first attempt, which gave them hope. Though fishing proved wholly unsuccessful, it was this event that promoted the BuAer to adopt a fishing kit in July 1942, designated as the M-552. After their story was circulated in Life magazine, Gifford Pinchot began experimenting on the harvesting of fresh water from fish. The BuAer, USAAF, USCG, and U.S. Navy also began a coordinated effort in development of an apron type fishing kit.

Arguably, a pocketknife was their most valuable item. It was standard equipment, stored in the supply pocket. As per M-3 specifications it was to be a four-bladed Boy Scout type. After the near-miss of spearing a fish, Aldrich was motivated to try again; after all they hadn't eaten for seven days. He leaned over the boat patiently waiting for a victim and, once it came into range, he extended his arm, impaling a perch. The fish was brought aboard and divided into thirds for consumption. As luck would have it, it rained later that day. A meal, a shower and a fresh drink went a long way toward boosting their spirits.

That same day Aldrich shot an albatross that they skinned; they ate its entrails. Aldrich stepped up his game for a bigger bounty, spearing a shark the next day. The shark was brought aboard the raft and processed with the little knife. The men consumed everything that looked edible: the organs, unprocessed fish in its stomach, and the shark's blood, which they drank.

With the success at spearfishing, a spear was added to later developments of fishing kits and as an accessory to the oars. The pocketknife remained standard equipment for life rafts and was integrated into other types of emergency kits, having proved its viability. Fixed-blade knives would soon become a typical staple for Naval aviators as an authorized allowance.

Catching the shark marked a high point in their odyssey. Their health and morale would slowly decline as the days and weeks went by. Dixon attributes their survival to their prior healthy state. All three had participated in daily exercises, maintaining peak physical condition. Indirectly, surviving 34 days with little amounts of food and water proved that an individual's health played a major role in his ability to survive. The importance of physical conditioning was soon introduced into the training curriculum for all aviation cadets.

As the men declined in physical health, the environment began to take its toll. The sea beat them under rough conditions and the sun baked them, cracking and blistering their skin. Several times the boat capsized and each time precious gear was lost. Dixon, Aldrich and Pastula were weak. They barely had the willpower to live, so it was difficult to muster enough strength to bail the raft.

Being able to drink water fairly regularly was a key to their survival. Emergency kits could only carry so much in provisions.

Two innovated methods were later developed by means of a chemical desalting method and a vaporizing still. These methods provided almost an endless means of obtaining water for an indefinite period.

They ate about once a week, enough to provide some sustenance to renew energy levels. At the two-week mark, their hunger began to subside, which, medically, was not a good sign and indicated a significant decline in physical health.

Until the 30th day, the three men were able to manage. Dixon maintained general navigational headings and manipulated the raft on a general course. They had established a basic routine of housekeeping but their physical and mental wellbeing were declining. On the thirty-first day it took a turn for the worst.

That afternoon they encountered heavy winds and a turbulent sea, capsizing their raft once again. This time they lost all their gear and most of their clothing. The three were nearly naked and were also nearly skin and bones. Their situation had become critical; rescue was needed soon.

For the next three days they drifted, clinging to some shred of hope that they were close to a body of land. As part of their routine they assigned shifts of watches, and on the morning of the thirty-fourth day Aldrich was on watch. He mumbled that he saw a beautiful field of corn. Each time the raft reached the crest of the swell he reiterated that he saw green. Finally, Dixon and Pastula took an interest and, with all three waiting to peak the next swell, they all saw an island.

Adrenaline took over at this point. Dixon realized that, with the wind direction, they were going to drift right on by. Their position needed to be adjusted. Mustering every shred of energy they could, they reached over the side and paddled with their hands all day to correct their drift. Their efforts prevailed and they reached the breakers, which threw the men from the raft. Tumbling and struggling for the beach, they crawled on all fours, too weak to stand.

They took refuge in a hut. At last, a sign of civilization. The next morning, Dixon spotted a native walking along the beach. He attracted his attention by the only signaling device they had left, a police whistle. The whistle was a standard accessory for the life raft, seemingly useless upon their initial inventory of their supplies. A child's toy brought salvation.

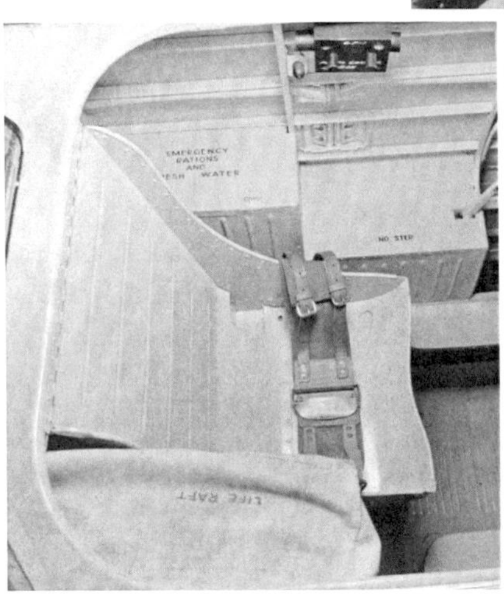

The TBD Devastator entered service in the fall of 1937 as a torpedo bomber manufactured by the Douglas Aircraft Company having a total production of only 130 aircraft. Upon entering WWII, it was considered obsolete slated to be replaced by the TBF Avenger. But the Avenger was not yet suitable for mass production when hostilities began. The TBD had a three-man inline crew; pilot, bombardier/second pilot, and a radioman/gunner at the rear. Automatic inflation bags in the wings were a standard feature for Naval aircraft developed through the 1930's, the TBD was so equipped. These inflation bags had a reputation of only functioning part of the time. The Devastator operated by Dixon should had been equipped with these inflation bags, but they may had been removed. If they had functioned, this would had brought valuable time to finish gathering all the provisions they wanted to retrieve. In the narrative, the aircraft was described to had nosed down quickly enveloping the wings with water creating a rapid decent to the ocean depths. The raft and emergency provisions were stored in the bombardiers' position, the raft in its case to the right, and the emergency rations and water in a compartment to the left. This presents the argument against storing those items separately, having the raft stored tethered to the emergency equipment container proved to be more advantageous. (National Archives)

Natives carried the three survivors to the local governor's plantation where they explained who they were. There they had their first meal in over a week. They soon learned they had landed on Danger Island, the northernmost atoll of the Cook Islands. They had drifted 750 nautical miles from where they ditched in those 34 days. When factoring in swirling currents, they had actually drifted 1,000 miles, an indirect route southward. Here, unable to barely support themselves, they pose with their raft several days after landing on Danger Island. Left to right; Pastula, Dixon, Aldrich. (National Archives)

After nearly a week on the island, Dixon learned there was a radio transmitter. With his technical knowledge, he was able to send a message. Seven days from the date they landed, a seaplane tender, the USS Swan, anchored offshore. The three men were escorted off the island from a float plane to the Swan and returned to Hawaii. L-R; Pastula, Aldrich, Dixon. (National Archives)

Their plight immediately jump-started amazement at their accomplishment. They were celebrated as celebrities. But they had to share the spotlight with the raft itself. Amazingly, it had survived the entire journey, first salvaged from the surf then brought aboard the USS Swan, making its way to Hawaii. It returned to the United States where it was first displayed publicly at a Goodyear Tire and Rubber Company's rally, Rub Out the Axis, in June 1942, Akron, Ohio. Dixon commented that he appreciated the workers at Goodyear for assembling something of such quality. Their MK-IV life raft was manufactured in November of 1937, so was just over four years old when they climbed aboard. It had cradled and brought them home. After its Home Front tour, it earned a place in the Naval Museum at Annapolis. There it was displayed until being placed into storage, hidden for decades. It was later revivified and now has a permanent home on display at the National Naval Aviation Museum in Pensacola, Florida.

MULTI-PERSON EMERGENCY KITS

Part I.
USAAF

World War II survival equipment consisted of a variety of items intended for airmen forced to abandon disabled aircraft either by parachute or crash landing. Survival equipment was therefore divided into two groups: overland and over water. Within those groups, equipment was subdivided into; Personal Aids Kits, -Airplane kits, carried in the aircraft and intended for use after crash landings, -Dropping Kits, transported by airplane to the scene of a disaster.

It is practical to categorize survival equipment items in terms of their intended operational region and function. Survival items were useful for or included: body protection, firearms, food, water, signaling, medical, sanitation, shelter, tools, and travel aids.

At first, development of USAAF emergency kits was based on the premise that all personnel flying outside the continental United States would be provided with a basic parachute emergency kit. This assumption would remain through the duration of the conflict and remained detailed by T/O&E. Over time, supplemental aids were developed: Types E-3, E-6, E-7, E-13 and E-17. These smaller kits were more appropriate for single-seater aircraft pilots with space restrictions and complemented aircrew of larger airplanes.

For large aircraft, Airplane Kits were developed. These kits contained implements and rations to sustain life for extended periods when forced down in isolated territories, and their designers were conscientious of space and weight requirements. The kits contained only items determined to be essential. With the wide expansion of operational regions across the world, multiple kits were developed for different climate zones. For example, airplanes operating over Alaska and Greenland would carry arctic kits; those operating over the Asiatic-Pacific and Africa would carry desert and tropic kits. All large airplanes operating over large bodies of water would carry over water kits. The types of survival equipment formulated in the early stages were developed more from necessity than a plan. As the USAAF began to cover new areas across the world, revised lists of equipment and revised types of kits were made to meet the new demands. The new environments also increased demand, and aerial delivery kits were developed to drop to stranded personnel from searching airplanes. As the war progressed, developments shifted toward the environmentally flexible and the universal.

As early as November 1937, the Air Corps established a requirement for a cross-country emergency kit. Cross-country would encompass areas within and outside of the continental United States. Initially, it was believed that the then-current emergency parachute kit would meet this demand. But it was found unsatisfactory for the purpose. The then-current kits did not provide adequate items if an airplane were forced down. Pending development of a new kit, local commands were directed to assemble improvised kits. Details were to be furnished to the Materiel Division for review of a standardized type. In the subsequent three years, little was done to develop a suitable cross-country kit.

A request submitted by Headquarters, Ladd Field, Fairbanks, Alaska appealed for a kit containing rations and equipment sufficient for personnel in remote sections of Alaska to survive until they could be rescued. The kit was deemed necessary because of the exceedingly rough mountain and swampy terrain, and the minimal transportation facilities. The region offered only air, dog team, boat, a single railroad from Seward to Fairbanks, and a single highway from Valdez to Fairbanks. Other than along these routes, there was very little habitation in inland Alaska. If an airplane were forced down, it would be almost impossible to walk out without assistance. And it was standard operating procedure for all pilots and aircrew to remain with the airplane until found. In many cases it had taken days, sometimes weeks, to locate lost airplanes and crew.

Partially influenced by emergency equipment carried aboard aircraft operated under direction of the Civil Aeronautics Authority and Pacific Alaska Airways, Ladd Field had submitted a proposed Wilderness Subsistence Kit. It consisted of two units: Equipment Kit Wilderness and Ration Kit Wilderness. The suggested ration container included coffee, corned beef, matches, bouillon, canned ham, salt, pepper, flour, dried peaches, navy beans, lard, butter, mess kits, pocket compass, fishing items, flashlight, and preservative oil. The equipment kit contained a hand axe, a 30-06 rifle, ammunition, hunting knife, sleeping bags, canteens, compass, flashlight, rope and fishing supplies. There was some redundancy in supplies, in case one of the kits got lost.

In October 1940, two units were assembled by the Materiel Division and shipped to Ladd Field for experimentation. The Materiel Division tentatively labeled the two units as Kit, Emergency, Type-I and Kit, Wilderness Equipment Type-I. The ration unit now consisted of raisins, bacon, chocolate, tea, sugar, rice, bouillon, corned beef, canned veggies, canned milk, matches, salt, mess kits, and two cooking stoves. The equipment kit included a combination rifle .22 cal. & .410 gauge, ammunition, first aid kit, fishing supplies, camphor, flashlights, compass, candles, whetstone, tin cup, tin pan, frying pan, stew pan, table knife, forks, table spoons, hunting knife, hatchet, fishing supplies, mosquito headnets, gloves, and sun glasses.

Ladd Field's cold-weather experimental station had concluded its field tests in April of 1941, submitting staff recommendations and establishing a tentative standard for operational aircraft in the region. As part of the recommendations, one wilderness equipment kit would be provided per airplane, and one ration kit for every two men. Furthermore, they detailed that sleeping bags and shelter tents be provided for each airplane, proportioned to crew. The last significant suggestion was that one unit consist of a large stove. This project was of extreme importance, and in June of 1941 Ladd had requested authority to make these emergency kits Air Corps items of procurement, storage, and issue.

Development of kits for Alaska was delayed, pending a decision as to whether the items could be procured by the Air Corps. Due to the variety of items that would be included, the kits were organized to include those items normally procured by the Quartermaster Corps, Ordnance Department, and Signal Corps. Requisitions could then be submitted to these arms of service to avoid competition in procurement.

As 1941 ended, wilderness kits for use in remote areas were of paramount importance due to the anticipated reinforcement increasing the volume of air traffic that supplied Alaskan bases. The Air Transport Command and airplanes of the 11th Air Force would fly consistent operations along what was often referred to as the Alaskan route. The principal staging area was Seattle, Washington. From there, aircraft would fly the coastal route along the Canadian coastline to Fairbanks. These aircraft would often fly inland through Edmonton to Fairbanks over predominately uninhabited regions of Canada under certain weather conditions. Air traffic would exponentially increase along the Alaskan route ferrying airplanes to Russia, and increased defensive operations of Dutch Harbor and airfields along the Aleutian Islands.

In 1940, another cold-weather North Atlantic ferrying route had been established by the British and Canadians. After the Lend-Lease Act early in 1941, the United States also assumed an active role. This ferrying route principally consisted of transoceanic hops from Newfoundland to Labrador, to Greenland, to Iceland, and ending in Great Britain. The United States established communication and weather stations and air bases through that area. The North Atlantic route was considered a major hazard and the Alaska kits would be useful.

It took the better part of a year to develop a suitable emergency kit for use in northern regions. The process consisted of obtaining a variety of rations and equipment, assembling the units, and finally shipping them to Alaska for evaluation and testing. With the United States entering the war under considerable pressure, the wilderness kits were released for standard on 4 March 1942 in what would become the first of the "E" type emergency sustenance kits for the USAAF and the first airplane kits.

The first to be detailed was the ration component designated as the Kit, Emergency Sustenance, Type E-1 (Rations) under Specification No. 94-40421. The second unit was designated as Kit, Emergency Sustenance, Type E-2 (Implements) under Specification No. 94-40422. Formal specifications for both kits were drafted and released as standard on 3 June 1942, followed by a limited procurement of 1,000 of each in July. The Liberty Electric Co. was awarded the contracts for the Type E-1, and Cappel MacDonald and Co. for the Type E-2. The Type E-1 was detailed to supply the necessary rations for operations in northern climates, and the basis of issue was one kit for every two men lasting four days. The Type E-2 was intended to supply the necessary implements for use in northern climates. Its issuance was one for each large airplane.

Contents for the Type E-1 were as follows: four cans biscuit (US Army Field Ration, Type-C), four rolls dextrose energy tablets (18 tablets per roll), one box raisins (15 ounce box), one box orange pekoe tea (100 bags), one box cubed sugar, one box rice (one pound), one container salt (five ounces), one can powdered milk, one box dried whole egg (five ounce box), eight bars field ration Type-D (two ounces), two cans corned beef (12 ounces), one can bacon (1--1/2 pounds), four cans dried soup (four ounces each), four cans drinking water AN-W-5A, four cans tomato juice (12 to 16 ounce cans), two boxes waterproof matches, two hunting knives (5" with sheath), two mess kits (M-1932), two mosquito headnets 43G3868, two pairs gloves, two tubes mosquito repellent, four bags tobacco, two forks, two spoons, two pair mukluks, Army sewing kit, two pair ice creepers, two boxes of heat tabs (24 per box), grill for the heat tabs and one Field Manual 31-15 titled Operation in Snow and Extreme Cold.

Components for the Type E-1 were assembled and stored in a container 12" diameter x 20", weighing 29 lbs. when full. The container was a fiber drum referenced as Fiberpak of the Container Corp., Part No. T-751. In the original specification, the drum was to be colored orange; in Amendment-1 of 7 December 1942, the drum was changed to be olive drab. Contract data and nomenclature were stenciled on the outside; contents list in lid. The lid shown would not completely seat due to fiber swelling. This example is of the second procurement, awarded in March 1943 for a quantity of 16,000 to the Liberty Electric Co. under Purchase Order 43-18884, available under stock number 8300-559150. Contract was negotiated in December of 1942. (Robin Bartel Collection)

EMERGENCY RATION TYPE E-1 KIT

Contents:

4 Cans Type C Biscuits
4 Rolls Dextrose Tablets
1 Box Raisins
1 Hundred Tea Bags
1 Box Cube Sugar
1 Box Rice
1 Container Pepper and Salt
1 Can Powdered Milk
1 Box Dried Eggs
8 Bars Chocolate—Ration D
2 Cans Corned Beef Hash
1 Can Bacon
4 Cans Dried Soup
4 Cans Drinking Water
4 Cans Tomato Juice
2 Containers, Matches
2 Hunting Knives
2 Mess Kits
2 Mosquito Head Nets
2 Pair Gloves
2 Mosquito Repellant
4 Bags Tobacco
2 Forks
2 Spoons
2 Pair Mukluks
1 Sewing Kit
2 Pair Ice Creepers
2 Boxes Heat Tabs with Grill
1 Basic Field Manual

(Right) Type E-1 from the July 1942 procurement. Drinking water was not included in the image, likely due to its being a recently adopted A/N item and not yet available. The flat AN-W-5A water can did not yet exist; those detailed by specification were to be the round type. (National Archives)

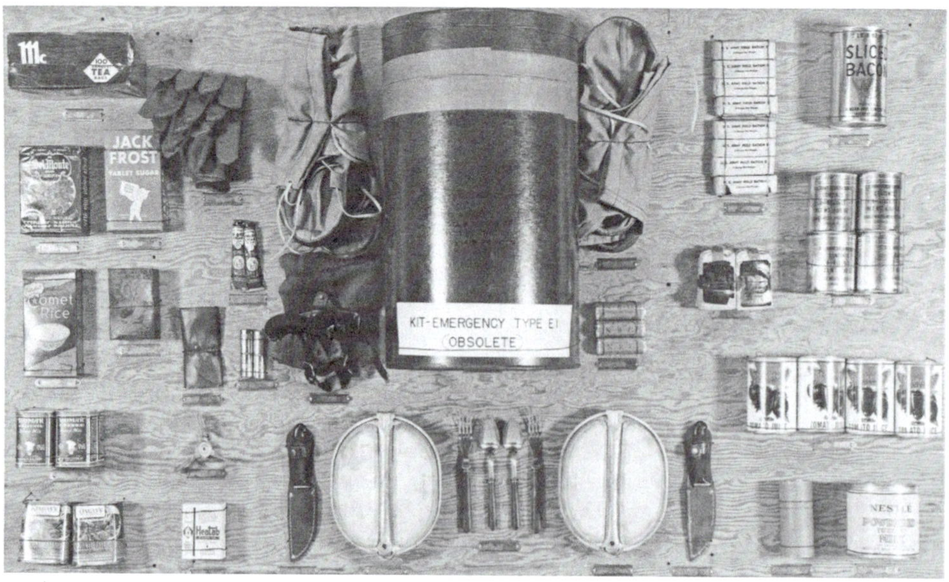

Training aid poster for the Type E-1. Ration complement is that of the QMC Mountain Rations.

Type E-1 from the March 1943 procurement. All components correspond to specification.

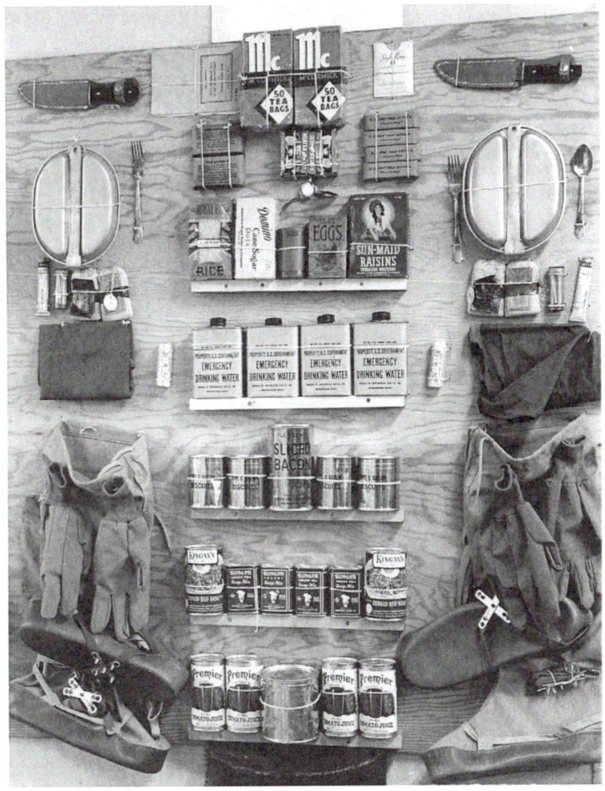

As part of the amendment of 7 December 1942, all foodstuffs were deleted and replaced by two US Army Mountain Rations. The new ration complement should have technically been part of the March 1943 contracts, but as evidenced by the physical example, the contents list provided on the lid reflect those detailed in the original specification. The image is of the proposed revision of the Type E-1 with the mountain ration complement. (National Archives)

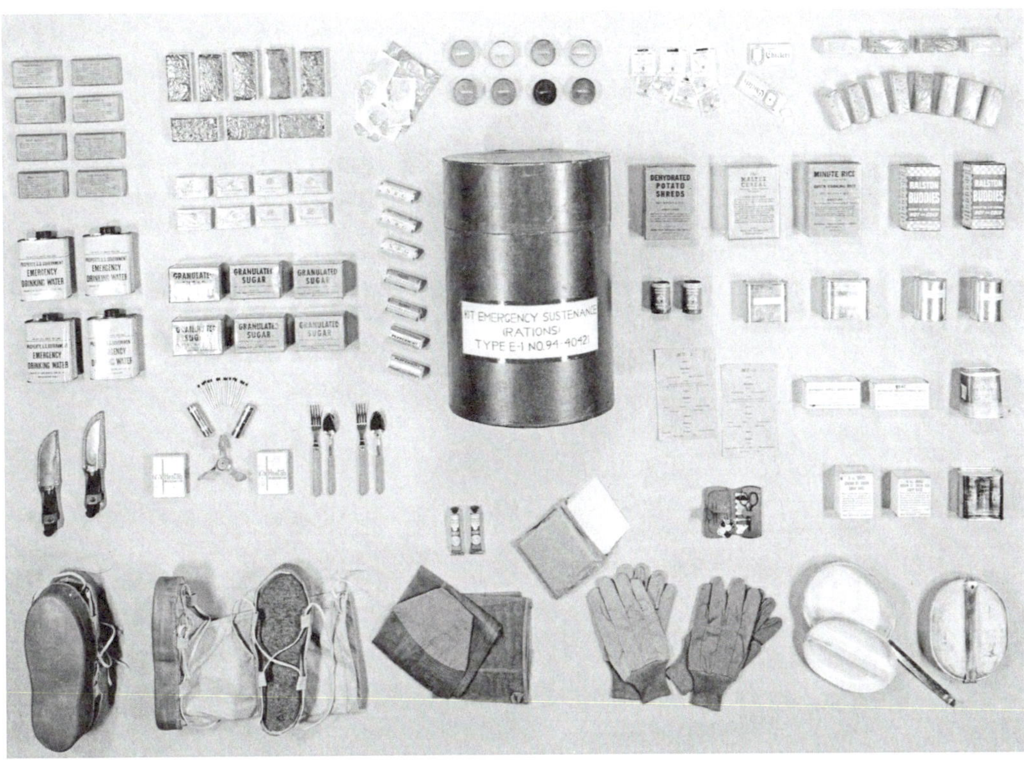

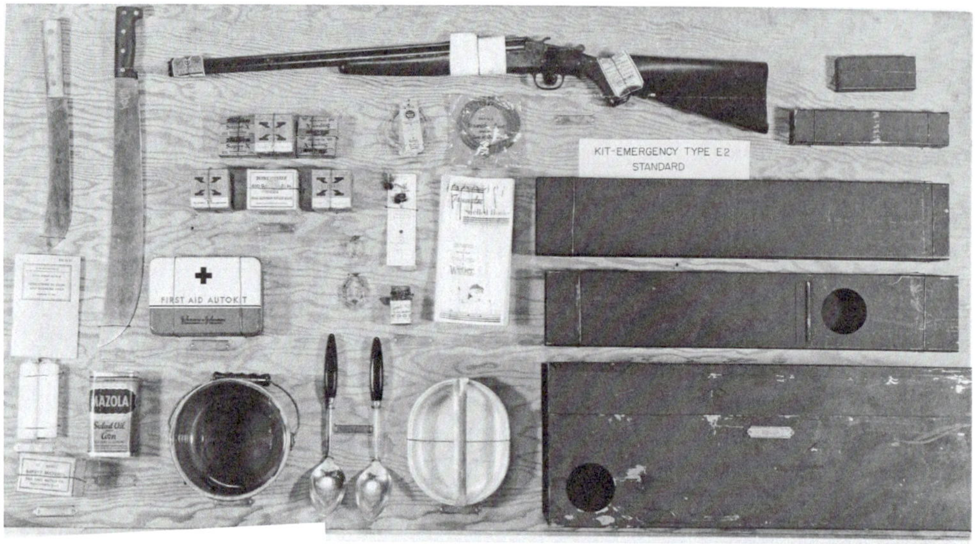

(Right) Type E-2 sustenance kit was to be used in conjunction with Type E-1 containing implements or tools. The Type E-1 was designed to support two men but Type E-2 was intended to be used as a community kit. Image shows original assembly of Type E-2 sustenance kit from the July 1942 procurement. Most of the components of it and the Type E-1 were commercially available, which allowed efficiency for contractors when assembling the units. (National Archives)

One of the 4,000 Type E-2 kits procured in March 1943. Cappel MacDonald and Co. was the original contractor, coinciding with the subsequent contract for the Type E-1 in March 1943; 4,000 Type E-2 kits were negotiated under the same Purchase Order 43-18884, awarded to the Liberty Electric Co.

Working with the guidelines from Ladd Field, the Materiel Command integrated a specially designed stove to be used as the container, referenced under Assembly Dwg. No. 41K6996. The stove comprised of three primary pieces: the stove body, referenced under Dwg. No. 41G7279 and a two-section smokestack, referenced under Dwg. No. 42D7446 and 42D7447. When packaged in the carrying bag, Dwg. No. 42G7206, the unit measured 10 3/4" x 10 3/4" x 28". The components were stored in the stove. When fully assembled, the kit weighed 45 lbs.; available under stock number 8300-559154.

Contents for the Type E-2 were detailed as follows: combination .22 cal. and .410 gauge rifle (Stevens), four boxes .22 cal. long rifle ammunition, three boxes .410 gauge No. 7-1/2 chilled shot ammunition, five boxes .410 gauge slug ammunition (five per box), one container matches (ten boxes), a tube of camphor (five to six ounces), two wax candles (1-1/4" x 5"), a frying pan with handle (M-1932), a stew pan (8" x 6-1/2"), two large spoons, one butcher knife (10"), one pint cooking oil, one machete (18"), three 3-prong fishing hooks, one container salmon eggs, one roll fishing line (25 yards), three 15lb fishing leaders, six snelled flies with No.8 hook (assorted patterns), 24 snelled hooks (assorted sizes), six lead sinkers (1/8 ounce), a first aid kit, a whetstone and the Field Manual 31-15, Operation in Snow and Extreme Cold.

The original specification from 3 June 1942 called for the 10" folding machete 42D7462 but was verbally revised to an 18" version as a tool better suited for clearing brush and defensive purposes. The 42D7462 was a newly adopted tool at the time of standardization.

The Type E-2 was amended on 8 October 1942 to include one hand-energized flashlight and the newly standardized machete, as detailed under Dwg. No. 43E10628. The 43E10628 machete and flashlight are in the above image; the flashlight is the Vizlight model, later replaced to the Daco type. The first aid kit and whetstone were removed from the kit. All large aircraft would carry emergency first aid kits of either the Jungle, Arctic, or Aeronautic types, making the unit in the E-2 redundant.

Cooking assembly on display with pot and M-1932 mess kit

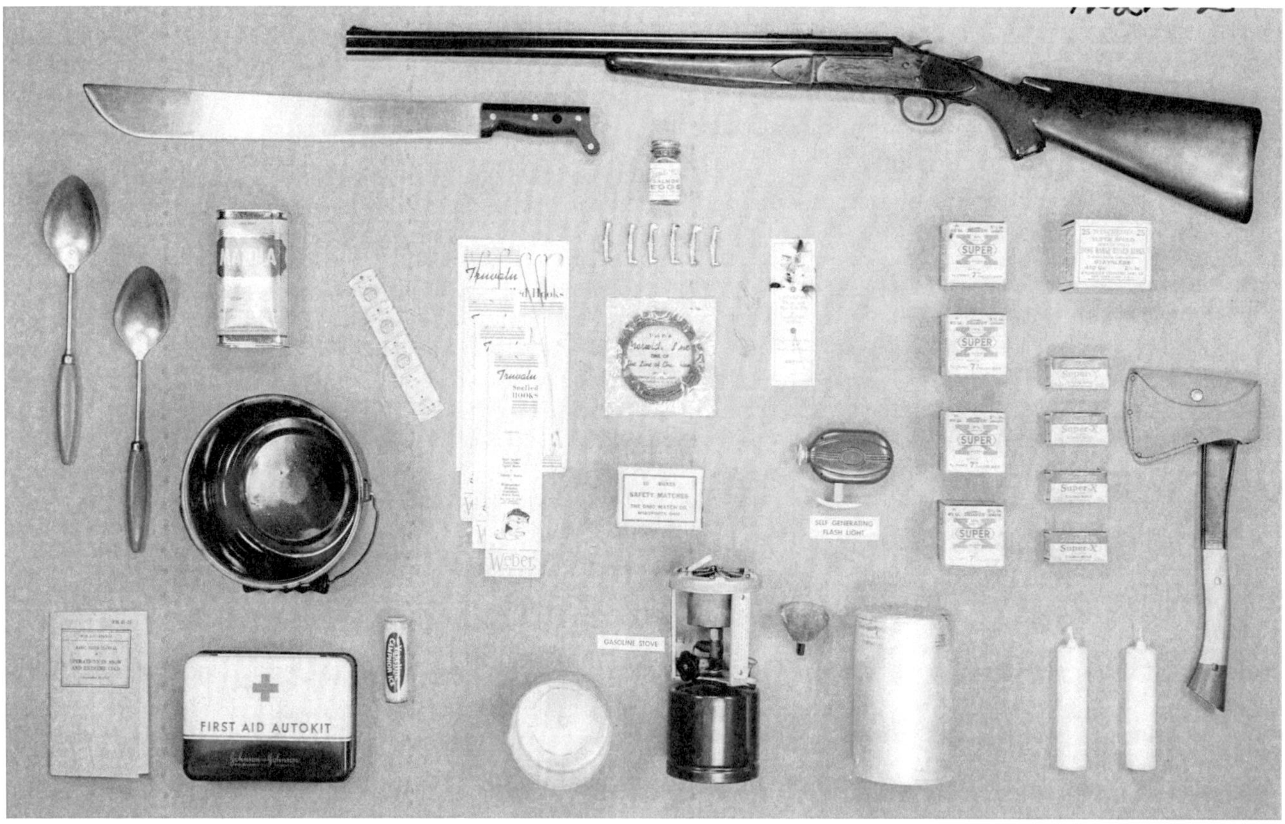

In January 1943, Ladd Field conducted controlled experiments with the Type E-1 and E-2 sustenance kits. Staff recommended adding a hand axe for chopping wood, finding the machete unsatisfactory for the task. Since many areas in Alaska had no adequate combustible wood for the stove, a small gasoline stove was recommended for inclusion. A hand axe and QMC M-1941 field stove were added per Amendment-A 4 June 1943, which also removed the M-1932 mess kit. This assembly retained the first aid kit. 43E10628 machete was recommended to be deleted and replaced by the M-1942 type, as shown here. This image is also included in the training aid poster for the Type E-2. (National Archives)

After the standardization of the Types E-1 and E-2, development progressed rapidly for additional models to meet wartime requirements. In fact, design of emergency kits began to overlap. In one instance, when one kit was adopted as standard, a succeeding type was already in process. The next airplane kit was designed around a Primus stove developed by the QMC for use by mountain troops. The kit was to be used by ground personnel for cooking food under emergency conditions, such as a forced landing in isolated territory, or by special task forces at advance bases. This unit remained under development by Aladdin Industries Inc. throughout 1942. A suitable design was eventually submitted to the Materiel Command in December and released for standard in January 1943 designated as the Kit, Emergency Sustenance, Type E-4 (Cooking). Specifications were drafted and released on 10 March 1943, detailed under Specification No. 94-40455 and Assembly Dwg. No. 43D16144.

The Type E-4 consisted of a sheet metal shroud with three retractable feet referenced under Dwg. No. 43D16145; QMC Stove, Cooking, Gasoline, One Burner, M-1942; cooking pot with flue, Dwg. No. 43D16152; cooking pot (Dwg. Unknown); frying pan, Dwg. No. 43B16154; hot pan holder, Dwg. No. 43B16162; one-gallon fabric gasoline bag with clamp handle, Dwg. No. 43A16157; and a funnel, Dwg. No. 43A16166.

(Below) The shroud was a container and provided wind protection. When fully assembled it weighed 5 lbs. and measured 10" x 10"; the complete assembly was stored in a cotton drawstring bag. With components removed, M-1942 field stove locks into bottom of shroud. Frying pan was also used as a cover and plate; when used as a cover, the handle locked across the top. Two 2 qt. pots served as a double boiler; the lower pot had a flue to permit heating the upper pot. Upper pot included three pivoting feet for stability. Clamp was provided for grasping hot pots. The funnel was used for refueling the field stove. The fuel bag was stored with the unit empty. It was intended to salvage aviation fuel in the event of a forced landing. The accompanying images illustrate the components of the Type E-4. (National Archives)

The complete assembly of the Type E-4 kit (upper pot not pictured). Design was intended to provide a more suitable cooking outfit than the QMC mountain cooking set. Pots, lid, and stove were like those in the mountain cooking set but with added shroud. Each item, except the M-1942 field stove and canvas bag, was stamped with its corresponding USAAF drawing or part number. Kit was never standardized but did undergo one procurement, under P.O. 43-33385 for a quantity of 1,000 units awarded to Cappel MacDonald and Co. Issuance was to be one kit per large airplane for use in a forced landing in isolated territory. When used as emergency cooking kit for special task forces at advance bases, distribution ratio was one kit for every 6-8 men. Available under stock number 8300-559162.

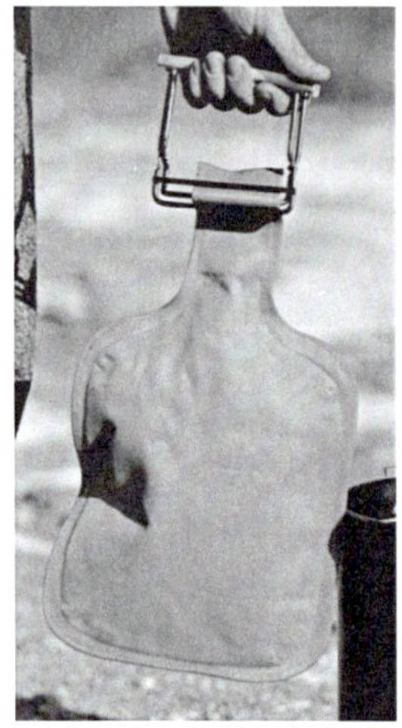

The Chief of the Air Corps noted that, as of September 1941, several missions were performed requiring long over water flights and it was considered essential that the proper type of emergency kit be provided to meet all contingencies. It was further noted that local emergency procurements by Air Corps Depots had to be authorized to supply over water flights to afford some measure of protection for emergency conditions. The articles authorized for procurement include: two gallons of liquid (one gallon water and one gallon tomato juice), six fusee flares, four HC-M8 smoke grenades, one M4 ground projector with three M17 flares, one aluminum sea marker, one oil silk (6 sq. ft. for catching rainwater), one reflector mirror, four yellow skullcaps and 200' of heavy silk cord. Development of an over water kit was prioritized. Two primary needs were being met at this time: anemergency kit for overland and an emergency kit for over water operations.

A floating emergency kit was developed under an OC&R directive in November 1941. On 29 July 1942, specifications had been drafted and released but the standardization was delayed pending the development of a suitable strong and waterproof container. It was anticipated that a suitable container would be available by August, and upon its receipt the emergency kit would be standardized.

The original components included were: eight US Army Field Ration D, eight AN-W-5 cans of water, four HC-M8 smoke grenades, three M17 signals, one M4 ground projector, eight cans tomato juice, two aluminum sea markers, a 6 sq. ft. waterproof cloth, one hunting knife, one flashlight, one signal mirror, one compass, and a first aid kit, as well as 200' of heavy cord. In the interim, the unit underwent several revisions since it was intended to supplement the supplies in life rafts. On 21 October 1942 specifications for the over water kit were prepared with updated components. Multi-person life rafts were also significantly revised. The new over water emergency kit was released for standard on 19 November 1942 as the Kit, Emergency Sustenance, Type E-5 (Over water) under Specification No. 94-40444: adopted as standard on 2 December 1942.

In January 1943 the Type E-5 kit was released for procurement for a quantity of 25,000 units, equally split between the Liberty Electric Co. under P.O. 43-18884 and the S. Buchsbaum Co. under P.O. 43-18886, available under stock number 8300-559166. The Type E-5 was intended for transport in large airplanes mainly over water as a supplement to the life raft.

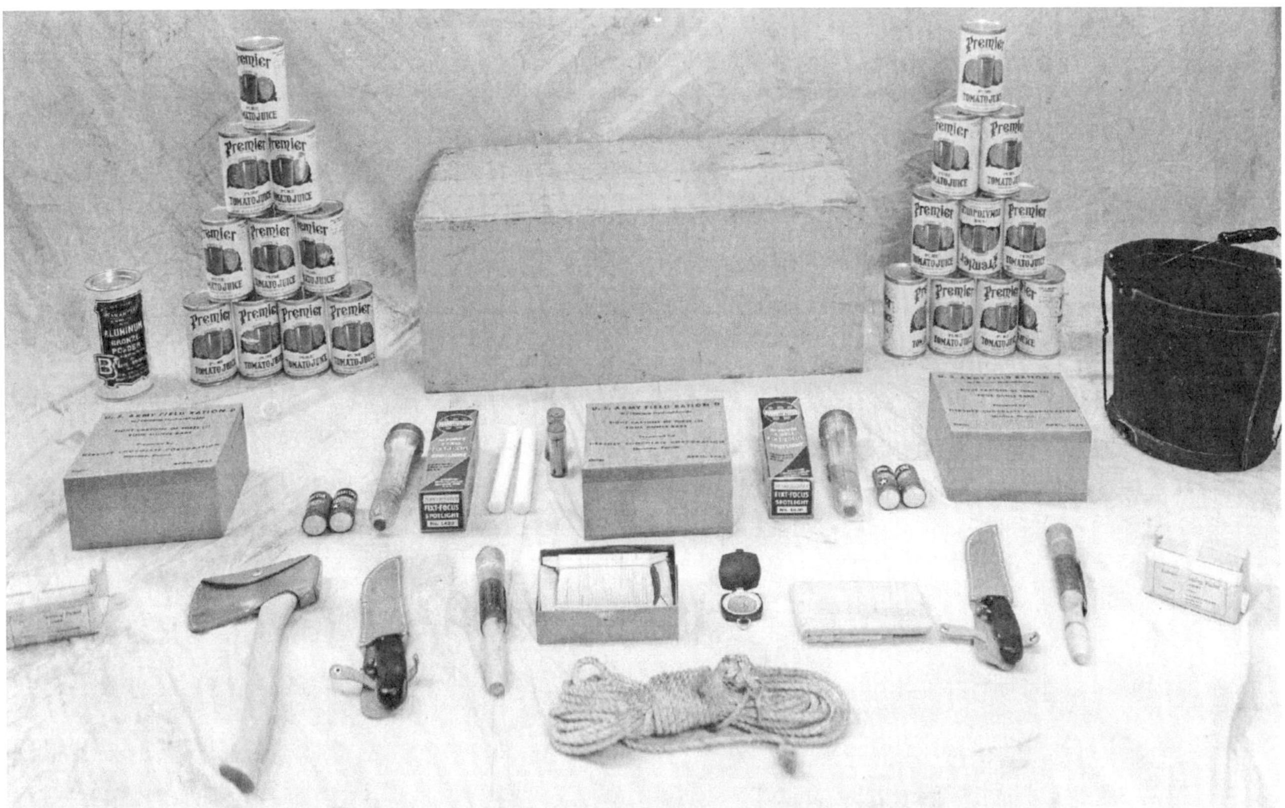

Early concept of the Type E-5 sustenance kit prior to standardization. The container originally chosen was a plywood box found to be unsatisfactory because it seeped water and required tools to open. An easily opened watertight container was required. When released for review, ten cans of tomato juice were deleted, replaced by ten AN-W-5A cans of water. The remaining components pictured were significantly changed after standardization. (National Archives)

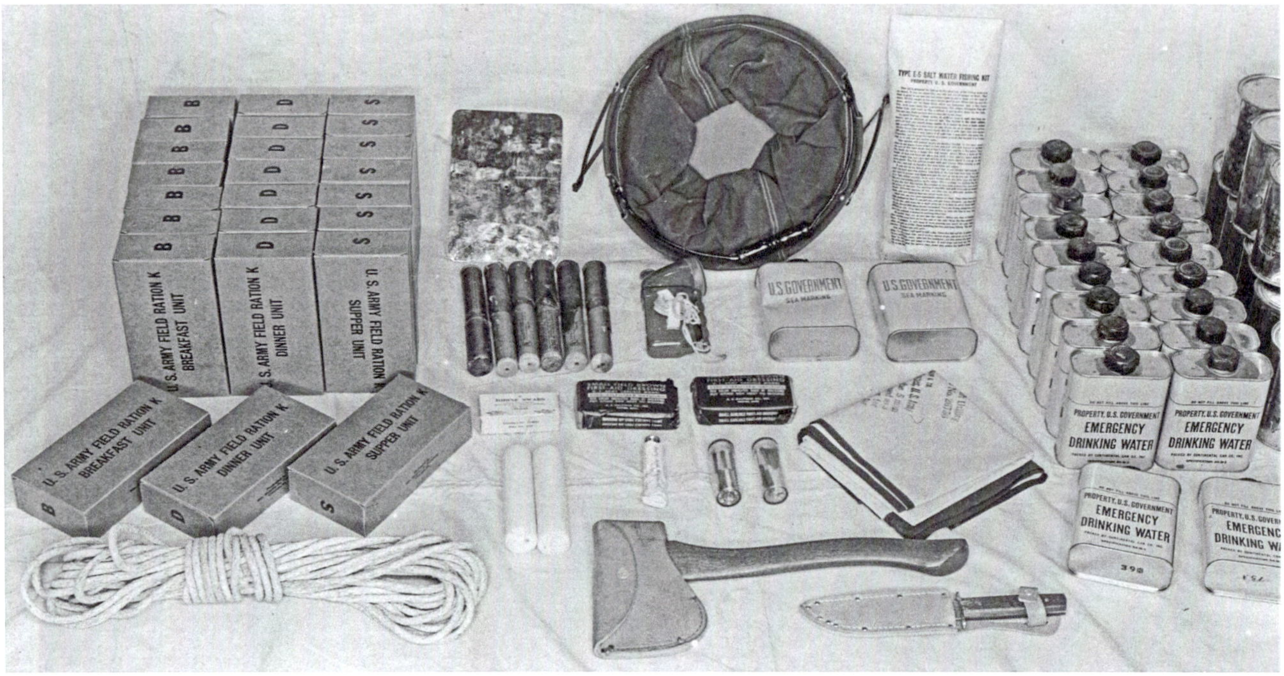

In the revised specification of 2 December 1942, the Type-D rations were replaced by 24 Type-K rations and the remaining tomato juice cans were replaced by ten AN-W-5A (flat) cans of water for total of 30 cans. The remaining components consisted of: two Carlisle packets 92060, box of iodine 91120, tube of boric acid 91010, hand-energized flashlight, collapsible bucket, pocket compass, two match safes with matches 42A6930, hunting knife, reflector mirror (ESM/1 added during later production), hand axe, two candles, package of fishing tackle (Liberty Electric type), six 15-min. fusee flares, paulin Size-77 (Size-18 detailed upon standardization), two cans of sea marker M-528 (revised to AN-S-10 during production) and 40' of ¼" rope. This Type E-5 has updated items, including Type A-9 hand-energized flashlight, 5-min. fusee flares, and camouflaged field brown Carlisle bandages. It retains the early items of a chromium mirror and Size-18 paulin.

Components were stored in a yellow waterproof plywood container secured by five latches. The trunk measures 28" x 14-1/2" x 13-1/2" with contents list stenciled on lid. Some examples by Liberty Electric will have contents in red stenciling.

(Below) On one end of the trunk was a handle serving two functions: aid in removal from the cotton bag and latching point for a lanyard.

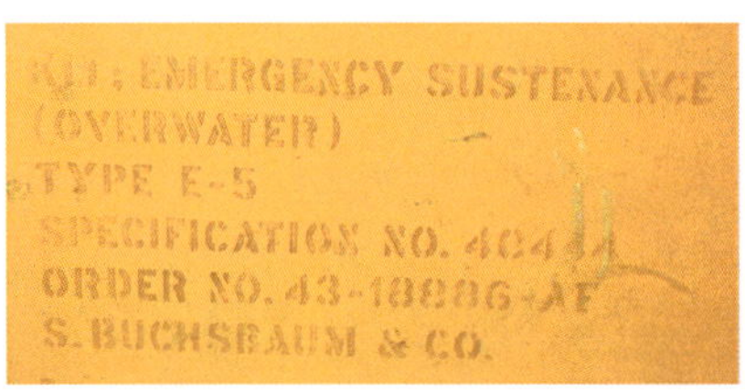

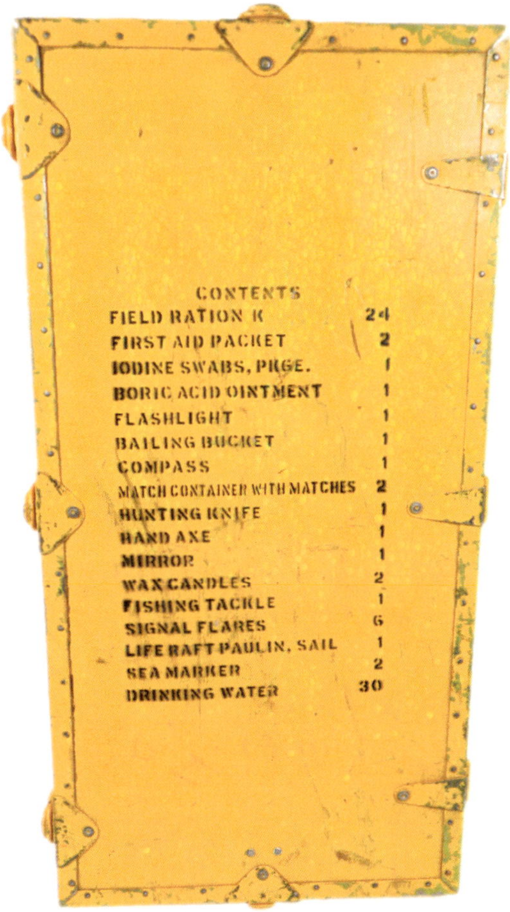

(Above) Type E-5 kit was designed to float, increasing precious raft space. In this wartime image, the Type E-5 kit floats near the life raft connected by a section of rope from the lifeline to the grab handle. (National Archives)

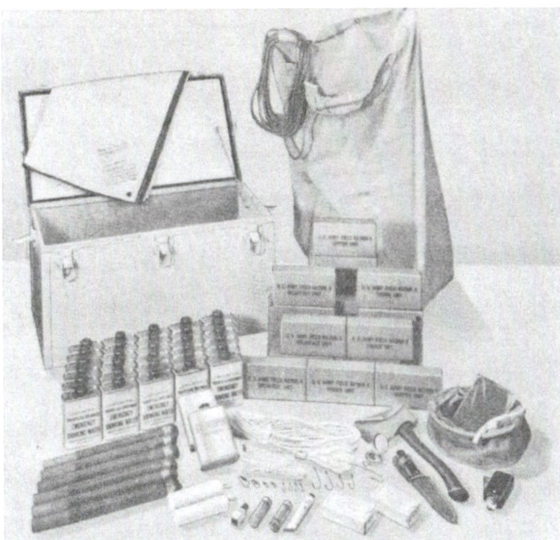

(Above) This assembly included the 15-min. fusee and Vizlight flashlight. Detailed in the last amendment for the Type E-5, 2 April 1943, the Vizlight was replaced by the Daco type.

(Right) The Type E-5 kit remained in demand throughout the war and was one of the more popular airplane kits for bombardment aircraft. The back of a Type E-5 kit can be seen near the waist gunners' positions on a B-24 operating in the SWPA, c.1944. Upon ditching, the gunners would be responsible for the removal of the kit. Likewise, the radio operator would retrieve the SCR-578 emergency transmitter before exiting and boarding the life raft. (National Archives)

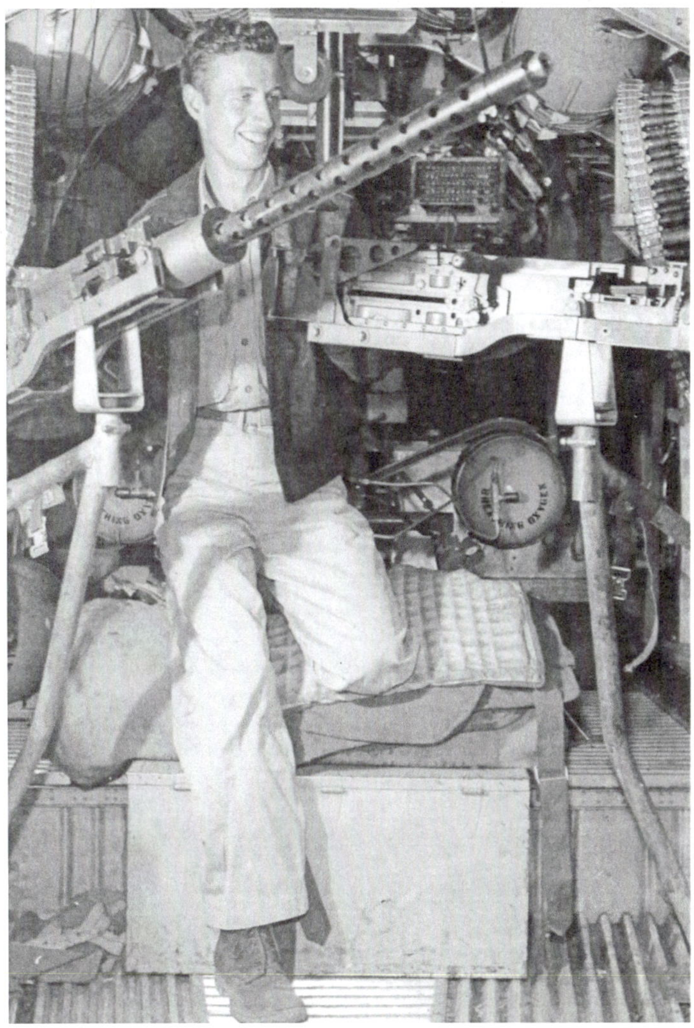

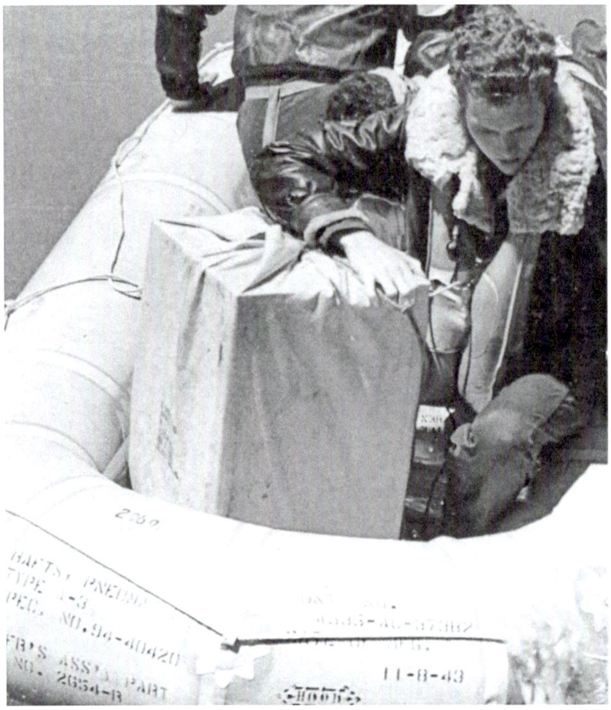

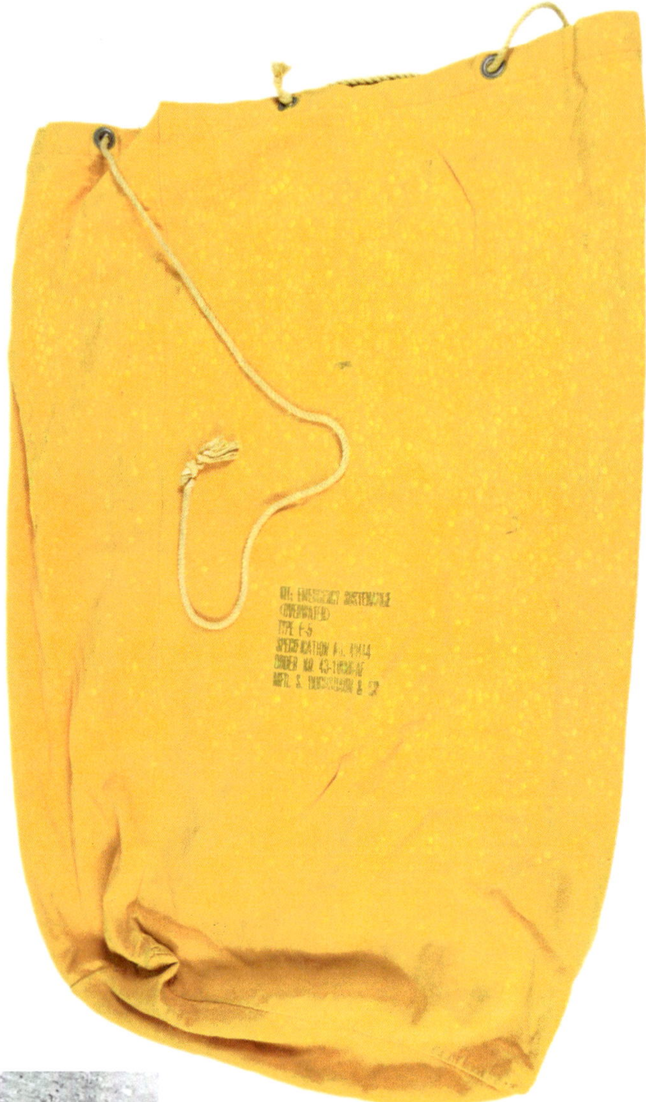

(Above and right) Provided with the unit was a yellow cotton bag for stowing the trunk. The bag included eight zinc grommets along the throat perimeter. The 40' of ¼" rope was part of the bag assembly. The rope was referenced as a trailing rope. After evacuation and inflation of the life raft, the unit would be secured by the rope to the lifeline of the raft, trailing the occupants. The image demonstrates the Type E-5 kit with cotton bag loaded aboard a life raft before field trials in Hawaii, summer 1944. (National Archives)

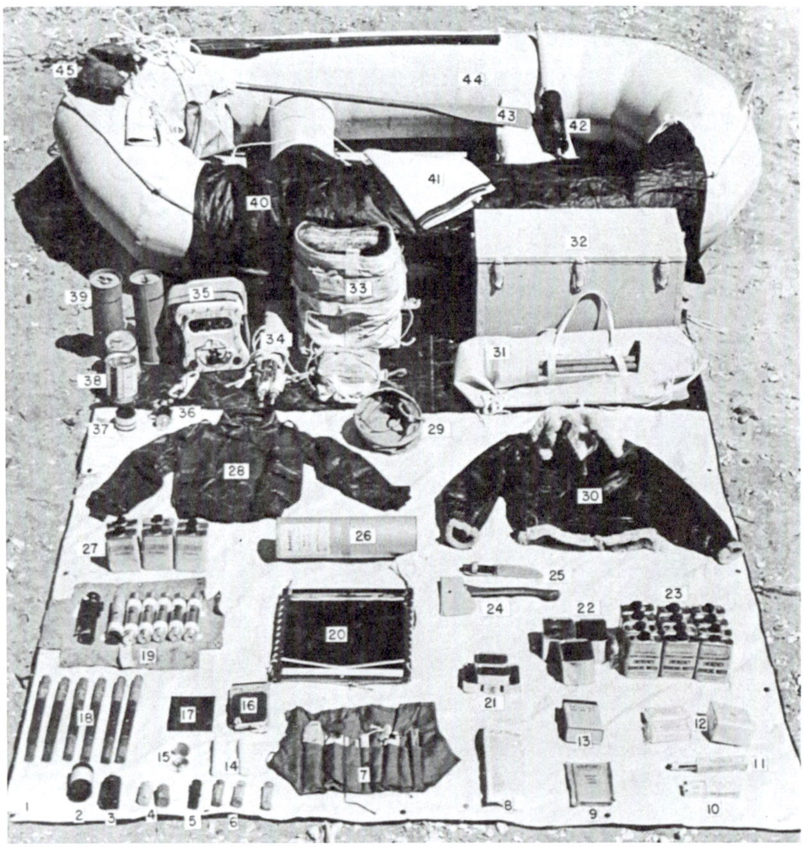

(Left) Representation of complete complement of emergency provisions for flying over water. Large airplanes would be equipped with a standard complement of a life raft with accessories, in this case a Type A-3. Stored in the airplane and intended for use in conjunction with a multi-person life raft would be emergency transmitter SCR-578. Also intended to complement life raft provisions was the Type E-5 sustenance kit. Distributed over the Size-77 paulin was the bulk of accessories of the life raft and Type E-5 kit intermixed. Supplementing the life raft is the Delano solar still, items labeled 20 and 26. (National Archives)

The development of sustenance kits was a necessity, starting with units for operations in northern regions. It was viewed generically as an overland unit. Soon following was a requirement for a similar unit for over water flights. The Types E-1, E-2, and E-5 were the first in those two categories, but the wartime environment required more specialized kits per region of combat operations. The first kit adopted of the subcategory divided by climate zone was a desert and tropical kit. Released for standard on 23 November 1942 was the Kit, Emergency Sustenance, Type E-8 (Desert and Tropical Implements) under Specification No. 94-40491, adopted as standard on 14 January 1943 and available under stock number 8300-559178. This unit contained necessary implements for use in desert and jungle areas. One kit was to be carried aboard large airplanes; it was also authorized for over water operations. The components for the Type E-8 kit included a combination .22 cal. and .410 gauge rifle (Stevens), four boxes .22 cal. long rifle ammunition, three boxes .410 gauge No. 7--1/2 chilled shot ammunition, five boxes .410 gauge slug ammunition (five per box), two Size-18 paulins, two hand-energized flashlights, two machetes 43E10628, one Shurlite Hiway Beacon, one mirror reflector (ESM/1 added during later production), two first aid kits aeronautic 97765, one E-Z Pull sewing kit, one pocket compass, a bar of soap and a tube of boric acid ointment. This implement unit was designed to be used in conjunction with the Type E-9 rations kit, developed under the same concept as the Types E-1 and E-2.

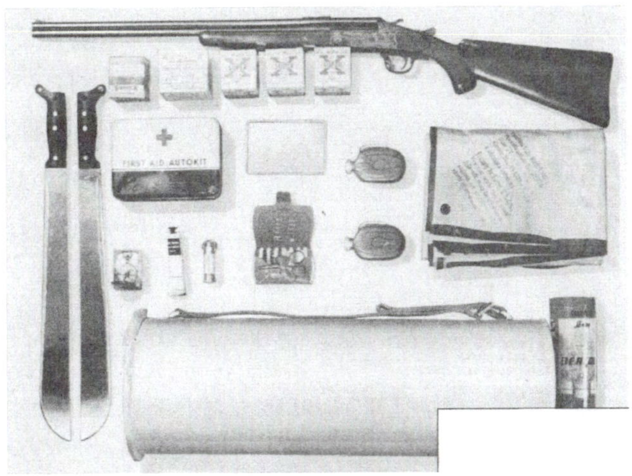

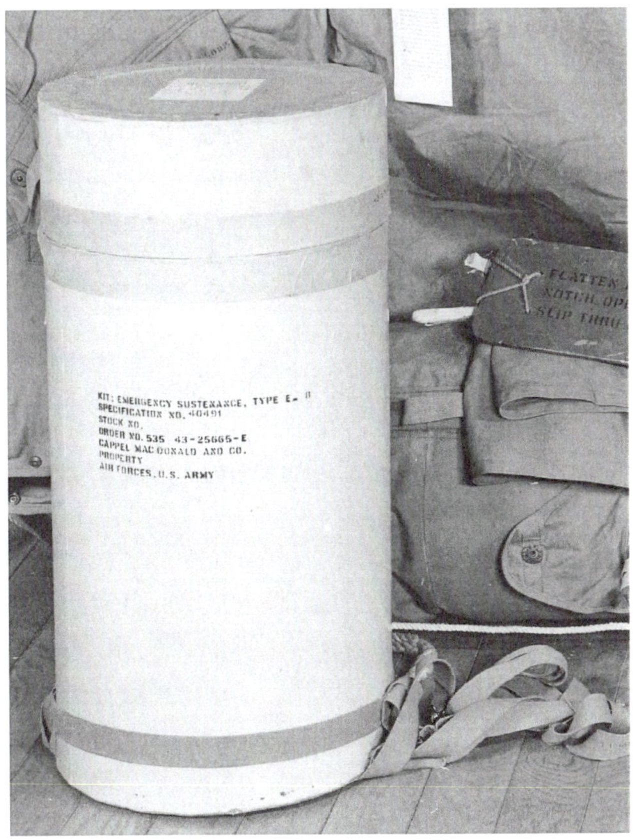

(Above right) Initial components of Type E-8 sustenance kit included two 18" machetes and a commercial auto first aid kit, deleted and replaced later in production. Upon standardization, 1,000 units were procured from the Cappel MacDonald and Co. under P.O. 43-25665. It includes the amended container of a steel drum.

(Right) Components were initially contained in a yellow Fiber-pak drum like that as detailed for the Type E-1 kit, 9" diameter x 26". A web harness was secured to the drum for transporting the unit over land. (National Archives)

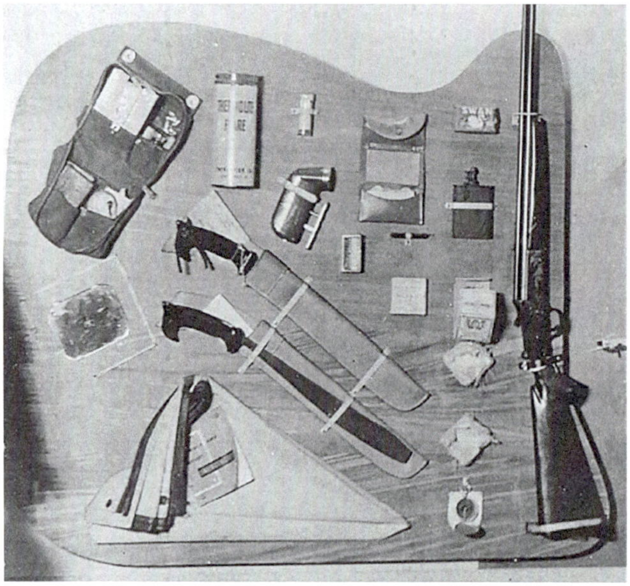

(Above) Type E-8 with components detailed by specification after standardization. Later directives added survival manual Jungle, Desert, Arctic, Ocean Emergencies. Initial draft of specifications details the Fiberpak drum, but a metal drum was provided with shoulder straps. This change occurred under Amendment-A effective 2 July 1943. (National Archives)

Released for standard on 23 November 1942 was the Kit, Emergency Sustenance, Type E-9 (Desert and Tropical Rations) under Specification No. 94-40492, adopted as standard 14 January 1943 and available under stock number 8300-559182. The Type E-9 kit was developed as a streamlined Type E-5, furnishing what were considered the most essential items in an over water emergency. Since it was determined that ration requirements for the ocean and the desert are the same, the Type E-9 was intended for universal use in the desert, jungle, and ocean. It was distributed as one kit for every two men. Upon its anticipated standardization, the Type E-5 kit was placed on a limited standard status on 26 December 1942 based on the premise that it duplicated, to a large extent, the equipment in the life raft and the Type B-4 parachute kit, which had been recently standardized. Upon the standardization of the Type E-9, cancelling the Type E-5 contracts was recommended. But contracts had been negotiated and were already in production. Since the expense had been incurred and the emergency kit was urgently needed in the combat zone, the contract was allowed to proceed. The Type E-9 kit eventually officially superseded the Type E-5. A contract was awarded for the manufacture of 3,000 Type E-9 kits under P.O. 43-25665 to Cappel MacDonald and Co.

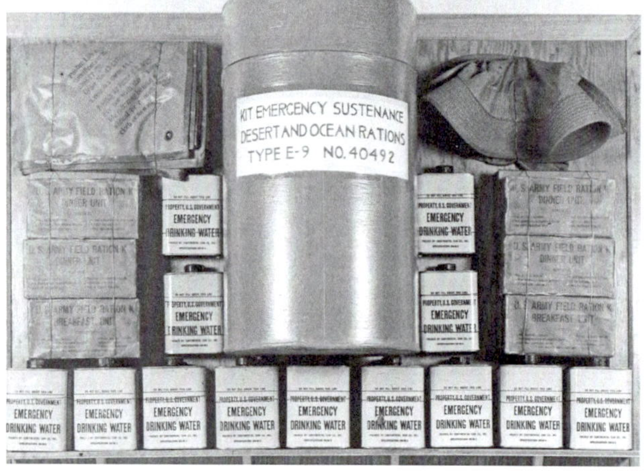

(Above) The essential items for the Type E-9 sustenance kit were two Type-K ration units (one unit consists of a breakfast, dinner, and supper), 22 AN-W-5A (flat) cans of drinking water, one Size-18 paulin and two brown sunhats (commercial brand by R.A. Lloyd Co.). Components were assembled in a cylindrical Fiberpak container 12" diameter x 15-1/2", and weighing 35 lbs. fully assembled. (USAF)

(Left) Type E-9 training aid poster. Types E-1 and E-2 kits were designed for arctic regions; Types E-8 and E-9 were designed for use in desert, jungle, and ocean regions. (USAF)

Developed in conjunction with the Types E-8 and E-9 was the Kit, Emergency Sustenance, Type E-10 (Desert and Tropical Aerial Delivery), released for standard on 23 November 1942 under Specification No. 94-40493. This kit was intended for airdropping the necessary rations and equipment to aircraft personnel stranded in isolated jungle and desert territories by parachute. This kit combined the Types E-8 and E-9. The kits that preceded this unit were classified as airplane kits; this would be the first designated aerial delivery unit. (Any of the previous kits could be airdropped with the assistance of a parachute container but were intended to be stored aboard the airplane.) The three kits, Types E-8, E-9 and E-10, were designed to complement each other, but with a regional emphasis.

Upon their development, operations in North Africa were commencing, which played a role in design. On the other side of the world, aircraft were operating in the Asiatic-Pacific, often surrounded by ocean. Both regions, coincidently, have the same requirements, which the Types E-8 and E-9 would fulfill efficiently. The emphasis behind the Type E-10 were operations in the CBI. The ATC was the aerial lifeline to China, and, at the time, the only means of getting personnel and supplies into China was flying a route known as The Hump, over the Himalayas. An aerial delivery unit was necessary for this region. The Type E-10 kit consisted of two yellow Fiberpak containers 13" diameter x 26" with a combined weight of 150 lbs., assigned stock number 8300-559186. Drums were equipped with harnesses, the same as the Type E-8. For aerial delivery, the Type A-4 aerial delivery container, Specification No. 94-40322, was used. Type E-10 was adopted as standard on 18 January 1943, but no procurements were made until August of 1943 and only for a limited quantity of 300 assemblies awarded to Cappel MacDonald and Co. under P.O. 44-1139. Soon after standardization, procurement was withheld pending a discussion on a new line of superseding sustenance kits. This limited procurement was intended to satisfy immediate requirements in the CBI until new models became available. Upon availability, it was no longer a required item since emphasis in that theater transitioned to providing cache kits along predetermined evasion and escape routes. Air-droppable units were assembled locally by those organizations responsible for rescue.

The components for the Type E-10 include: five Type-K ration units (one unit consists of one breakfast and two dinners), two Type A-2 field shelters (replaced by mountain tent), 20 AN-W-5A cans of water, two brown sunhats (commercial brand R.A. Lloyd Co.), combination .22 cal. and .410 gauge rifle (Stevens), four boxes .22 cal. long rifle ammunition, three boxes .410 gauge No. 7½ chilled shot ammunition, five boxes .410 gauge slug ammunition (5/box), two neckerchiefs, two tubes insect repellent, one Size-77 paulin, two boxes of heat tabs, a grill for the heat tabs, hand-energized flashlight, machete 43E10628, Shurlite Hiway Beacon, mirror reflector (ESM/1 added during production), E-Z Pull sewing kit, pocket compass, bar of soap, Boy Scout knife, tube of sunburn ointment, six tablets of Atabrine, tablets of Benzedrine sulfate, ten tablets of tea, box of matches and 24 vitamin capsules. At procurement, the Type A-2 field shelter was eliminated and replaced by the QMC two-man mountain tent.

Training aid poster for the Type E-10. It shows the revision to include the QMC mountain tent. (National Archives)

Display of components for the Type E-10. For organizational purposes, the Fiberpak drums were numbered. Type A-2 field shelters are included in this assembly. (National Archives)

By the end of 1942, there was a rapid progression in development of multi-person sustenance kits. The Engineering Division was pressed to address the issues by specific request, and theater commands reported their requirements to the Materiel Command attempting to cater to that need. Submitted for standardization on 24 November 1942 was the Kit, Emergency Sustenance, Type E-11 (Marine Survivors) under Specification No. 94-40494. The Type E-11 was developed at the request of the Anti-Submarine Command for specific operations in aiding the rescue of victims of torpedoed vessels and distressed aircraft. This kit was to be carried aboard patrol airplanes.

Type E-11 was adopted as standard on 14 January 1943. It contained: six Type-K ration units (one unit consisted of one breakfast and two dinners), 24 AN-W-5A cans of drinking water, M8 Pyrotechnic Pistol, M11 parachute flares, four 15-min. fusee flares, two match safes, four heat pads, two first aid auto kits, four sea markers AN-S-10, two police whistles, two tin cups, two sponges, reflector mirror, fishing kit (Liberty Electric), Boy Scout knife, two gal. water bag, pocket compass, a Size-77 paulin, two blankets (water-resistant), container of sulfathiazole and six yellow cloth helmets.

Before authorization of procurement, the Anti-Submarine Command required the kit to be used for airdrops, thus needing a suitable container for that task. Under Amendment-A 6 July 1943 a new metal container was adopted for aerial delivery, referenced under Dwg. No. 43G13341. The free-falling container consisted of a sealed lid with four D-rings: two rings allowed the attachment of the Type G-7 sustenance parachute, and two rings were provided for suspension in bomb bays or under the wing of patrol aircraft. The container was colored orange/yellow, measuring 14" diameter x 30" and weighed 110 lbs. when fully equipped.

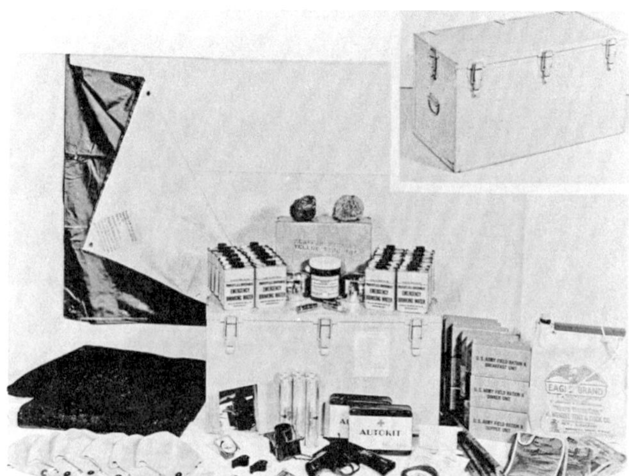

The Type E-5 was used for the base concept, the proposed contents by the Anti-Submarine Command utilizing the trunk from the Type E-5 kit. (National Archives)

(Above) Standardized Type E-11 and its full complement of accessories. (National Archives)

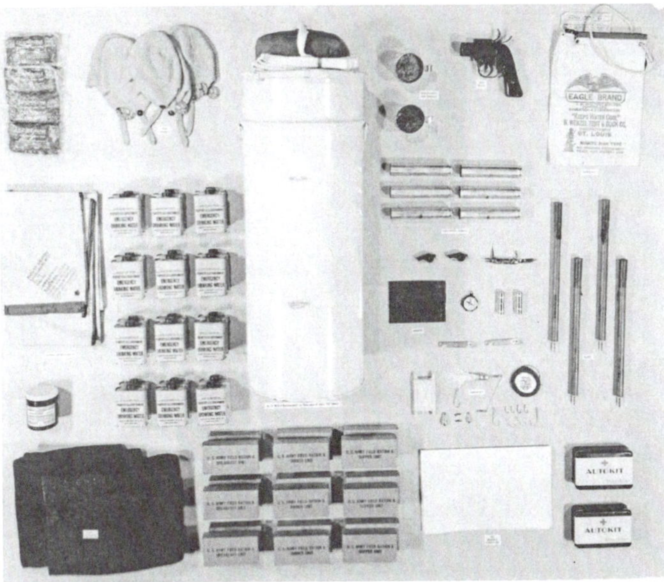

(Above) Type E-11. After the development of the free-falling container, a limited quantity contract was awarded to the Liberty Electric Co. for 300 assemblies, available under stock number 8300-559190. Just like the Type E-10, only a limited amount of funds was released for the procurement pending the development of other sustenance kits. In Amendment-A, 6 July 1943, the Type-K rations were detailed to be replaced by the Life Raft Ration Type-A, the fishing kit was to be the Type-5, M-8 hand projector was to be replaced by the M-9, a hunting knife was added and two first aid kits were to be that of the BuM&S first aid kit for pneumatic rafts S2-1055. It is not clear if these changes were ever applied. (National Archives)

As a result of a directive in October 1942, the Engineering Division began the development of an arctic rescue kit for the Greenland Base Command. The purpose was to supply stranded aircraft personnel on the Greenland Ice Cap and other similar desolate territories with the necessary rations, clothing, and equipment to sustain life at the scene of a forced landing until a guide could be dropped in by parachute to lead the party out on foot. The unit assembled by the Materiel Command was designed to supplement two men for 10 days. By December, experimental units were assembled and sent to the Greenland Base Command for testing. Following the base's approval, the unit was released for standard on 25 January 1943 as the Kit, Emergency Sustenance, Type E-12 (Arctic Aerial Delivery) under Specification No. 94-40471, adopted as standard on 10 March 1943.

The kit consisted of four groups of equipment with a total weight of 175 pounds. Following the basic guidelines established by Ladd Field in 1941, the Type E-12 incorporated the Types E-1 and E-2 sustenance kits, field shelters, and sleeping bags. Arctic clothing and tools were added. Contained and delivered in the Type A-4 aerial delivery container were: one Type E-1 sustenance kit, one Type E-2 sustenance kit, two Type A-2 field shelters, two Type A-2 mattresses, two gallons of gasoline, one M-1941 field stove, one hand rip saw, one Size-77 paulin, two pairs of ski goggles, one ice axe, two pairs of crampons and one entrenching shovel M-1910. The remaining articles would be packaged in three bundles and thrown from the plane in a free-fall. Two bundles consisted of two Type A-2 sleeping bags. The sleeping bags were to remain in their factory fiberboard container and painted yellow. The third bundle consisted of clothing: two reversible ski parkas, two pairs of ski pants, two pairs of trigger finger mittens (outer shells), two pairs of trigger finger mitten inserts, two pairs of wool wristlets, two pairs of cushion soled socks, two pairs of ski socks, two pairs of arctic socks, two wool mufflers and 50' of ¼" rope.

The first procurement was awarded in August 1943 to Cappel MacDonald and Co. for 300 assemblies under contract 33038-AC-145 available under stock number 8300-559194. A subsequent contract was awarded to Cappel MacDonald for an additional 200 assemblies in June of 1944 under contract 20017-AC-1200.

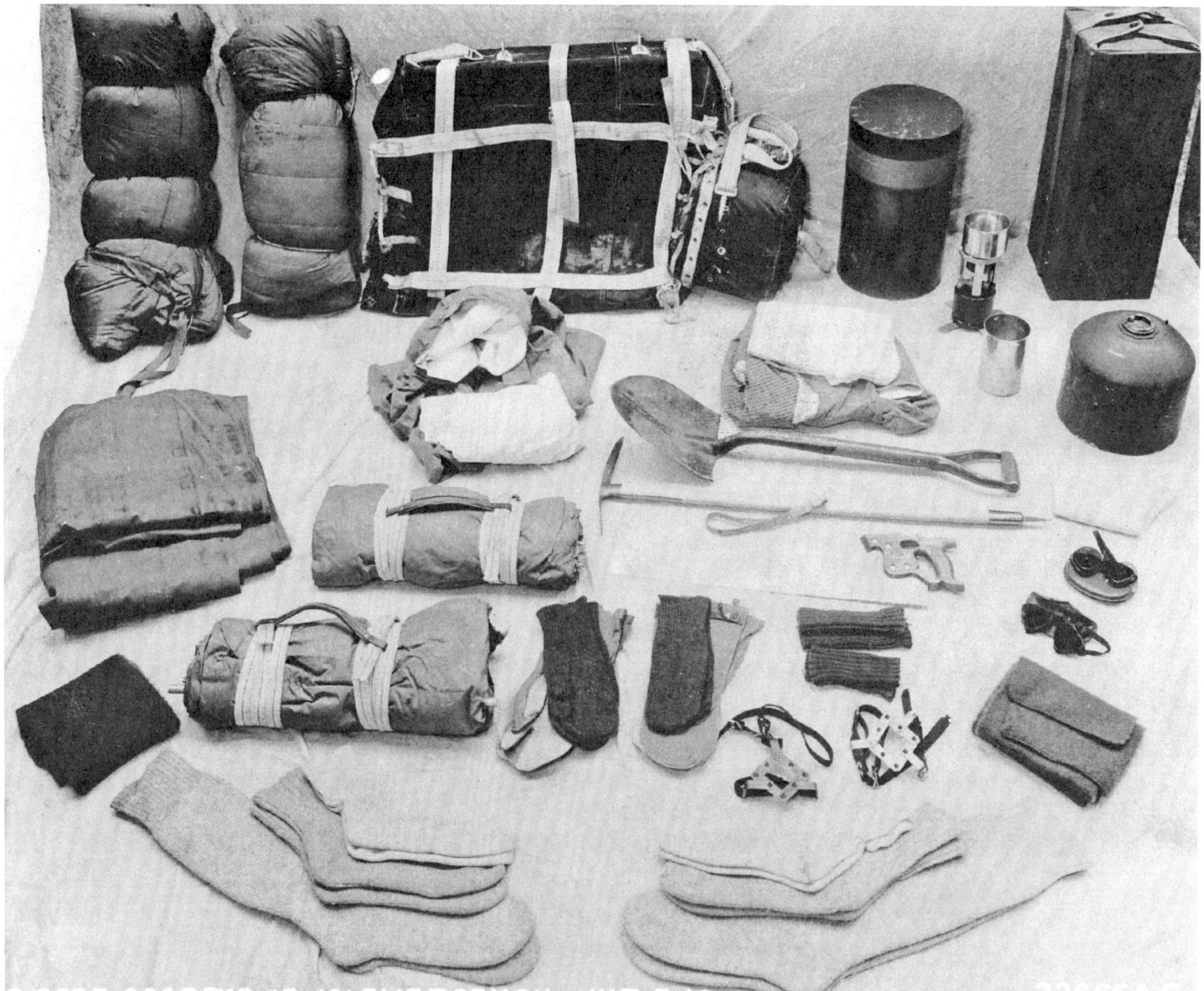

(Opposite page) Complete assembly of the Type E-12 upon standardization. Here, the M-1910 entrenching shovel was substituted by a D-handled spade shovel. Kit was issued to cargo and transport airplanes operating over the arctic; extra kits carried in rescue airplanes. Flights often consisted of a two-man crew, pilot and copilot, ferrying the airplane or delivering supplies. (National Archives)

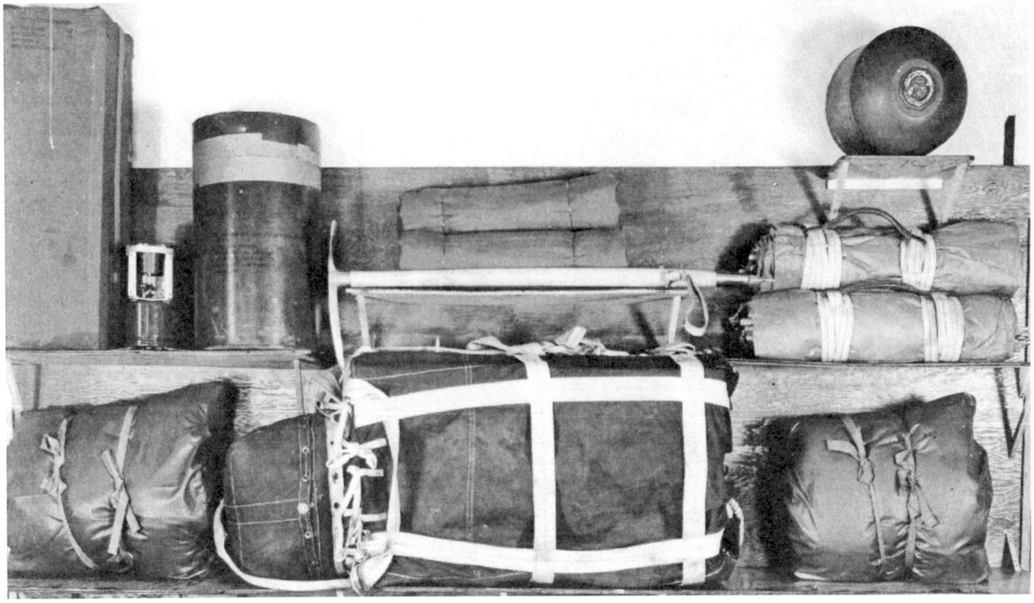

(Left) Multiple views of the Type E-12. Shown is the Type A-4 container assembled and ready for aerial delivery. Sleeping bags are in waterproof outer case. (National Archives)

Specification No. 94-40471 was revised 1 October 1943, substituting the QMC mountain/ski clothing with USAAF standard stock items and eliminated the Type A-2 field shelters. New articles: two mechanics jackets Type D-2, two mechanics trousers Type D-2, two pairs of mechanics gloves Type D-3, two pairs of arctic flying gloves Type A-12, two pairs of rayon glove inserts, two pairs of ski goggles, two pairs of crampons, two-man QMC mountain tent, two arctic sleeping bags Type A-3 with liners, two mattresses Type A-3, two flying helmets Type B-9, and two pairs of mukluks Type A-14 with inserts.

On 13 March 1944, the Type E-12 kit was overhauled to address concerns by theater commands and updated with new items. Type E-1 kit was removed and replaced by the Type E-16, retaining the Type E-2 but with revised components. The assembly now consisted of these additional items: combination .22 cal. and .410 gauge rifle (Stevens), four boxes .22 cal. long rifle ammunition, three boxes .410 gauge No. 7--1/2 chilled shot ammunition, two boxes .410 gauge slug ammunition (5/box), one Size-77 paulin, machete 43E10628 (replaced in April by M-1942 type), hand axe, entrenching shovel M-1910, stove M-1941, two mess kits (M-1932), cooking pot, two Shurlite Hiway Beacon, signal mirror 40653, large spoon, ice axe, hunting knife, engineer wrist compass, hand-energized flashlight Type A-9, two aeronautic first aid kits 97765, fishing kit, two gallons of gasoline, tube of insect repellent, 120' climbing rope, two candles, 3 oz. oil with firearm cleaning supplies and Survival Manual 21W.

The Type E-12 would be expanded multiple times as authorized by local commands. Later additions would include snowshoes, pack boards, QMC field clothing, tool kits, ponchos and replacing the combination rifle with the M1 or M1A1 Carbine.

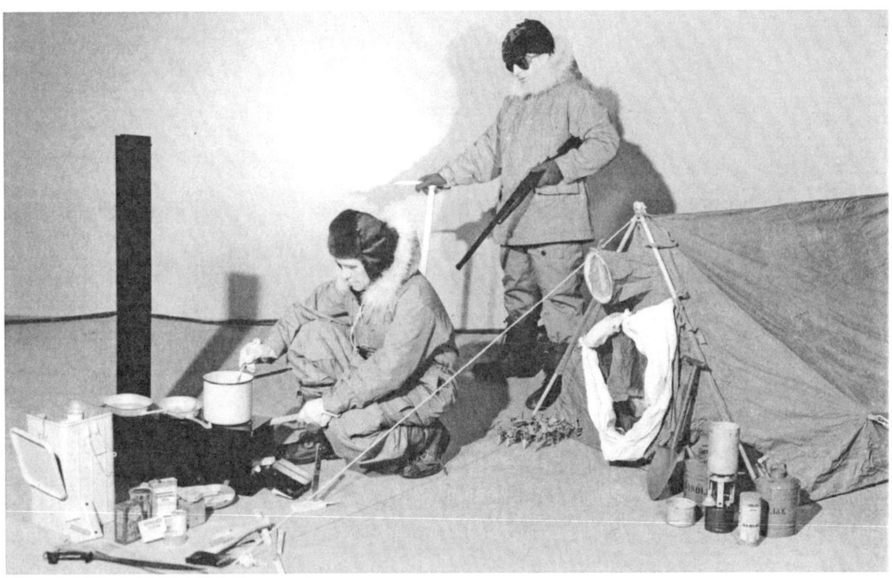

Revised Type E-12 being demonstrated. (National Archives)

Revised Type E-12 with its new complement to include the Type E-16 sustenance kit and USAAF mechanics clothing along with the mountain tent. (National Archives)

Training aid poster for the Type E-12 (revised) sustenance kit. For aerial delivery, clothing articles were bundled with sleeping bags. The Type E-16 kit was a free-falling parachute unit, airdropped separately but in conjunction with the Type A-4 and Two sleeping bag-clothing bundles.

Some factors need to be taken into consideration when reviewing the sustenance kits presented in the photos here. Contractors were only required to meet those Items as detailed by specification, which are often very generic. Examples would be the listing "Axe, Hand, approximately 3-5/8' blade with sheath," and the terminology of "or Equal" that was often used. By that standard, a multiple array of items could be utilized based on the contractor's discretion. When reviewing the limited number of images available, two of the same type of kits may not contain the same exact components. Also, specifications included new items that may not be available for preliminary deliveries of the kits. The components were classified as Government Furnished Equipment or Contractor Furnished Equipment. For GFE, the item was procured by the US Government and supplied to the contractor for assembly in the unit. What was detailed as GFE typically had no or limited variable. For CFE, the items were procured by the contractor from commercial sources meeting the general descriptions or "equal" criteria but having a wide variable of brands and designs.

Upon the adoption of the Type E-12 kit, the logistics of development got slightly out of hand. The kits were becoming too specialized, which led to requisitioning confusion by operational commands, both for the kits themselves and their replenishment. In the spring of 1943, OC&R directed the Equipment Board to investigate the possibility of reducing the number of emergency sustenance kits. It was this directive that held up quantity procurement of several sustenance kits mentioned previously. In May of 1943 the Equipment Board submitted its findings and recommendations. In its review, it recommended the entire line be streamlined into three principal kits: an individual parachute kit, an emergency general purpose kit, and an arctic general purpose kit. All could be carried in or airdropped from an airplane.

In part, the Equipment Board based its recommendations on the established T/O&E for fighter and bombardment squadrons, in that all combat aircrew members would be issued a parachute emergency kit. At the time, the Type B-4 was the standard. Included in this kit were basic provisions: rations, signals, machete, and a poncho. Additionally, the Board was influenced by the proposed idea of a vest type kit. These would be supplemented by the Types E-6 and E-7 sustenance kits. Together they would provide the immediate necessary items in the event of an emergency. If further items were deemed necessary, aerial delivery units were available to supply stranded personnel until rescue. By having only two sustenance kits, duplication and overlap were eliminated, simplifying requisitioning and issuing for operational commands.

Upon review of the report, AAF Headquarters issued a directive in July 1943 outlining four types of kits: a general purpose kit, a desert and tropic aerial delivery kit, an ocean aerial delivery kit, and an arctic aerial delivery kit. These units would supersede all previous types. The Materiel Command immediately began work on a revision for sustenance kits, building on all the recommendations. Type E-12 for arctic requirements was retained and two new sustenance kits were adopted, one for desert and tropic aerial delivery designated as the Type E-14, and one for ocean aerial delivery designated as the Type E-15. The concept of a general purpose unit was met by the adoption of the Type C-1 sustenance vest.

During the development of Types E-14 and E-15 it was determined they did not provide sufficient rations for an extended emergency situation. It was therefore felt practical to retain as standard Types E-1 and E-9 ration kits for replenishment purposes. As part of the redesigning process, sustenance kits would be provided with a free-falling parachute container like that provided with the Type E-11. The Type E-9 kit was revised under Amendment-A, effective 8 April 1944, and it was repackaged in a metal parachute container as detailed under Dwg. No. 44G18525. The components were additionally revised to contain four cans of Life Raft Rations Type-A, one unit Type-K ration (breakfast, supper, dinner), and 20 AN-W-5B cans of drinking water. The unit could then be airdropped, replenishing supplies by aerial delivery to stranded personnel. Upon adoption, a contract was awarded for the assembly of 10,000 Type E-9A sustenance kits to the Standard Molding Corp.

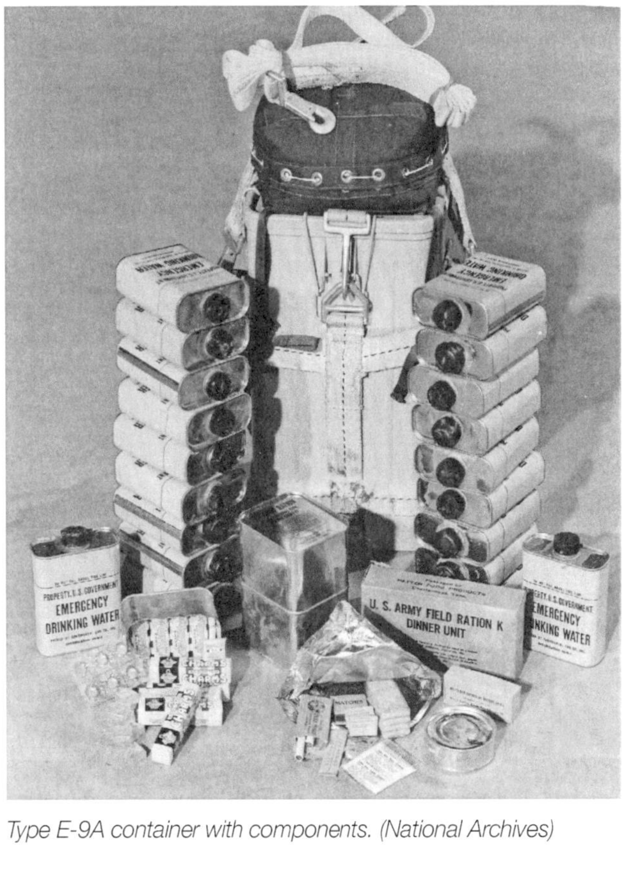

Type E-9A container with components. (National Archives)

The Type E-9A container was orange/yellow, 9" x 9" x 16," available under the same stock number as its predecessor, 8300-559182. A parachute harness was provided for attachment of the Type G-8 sustenance parachute for aerial delivery. Complete assembly was designed to float, weighing 35 lbs. when fully assembled. Lid for Type E-9A was watertight, secured by four wire-framed latches. Chain was provided to prevent loss of the lid. On one side was carrying handle. This unit would eventually be superseded by the Type E-19 sustenance kit. (National Archives)

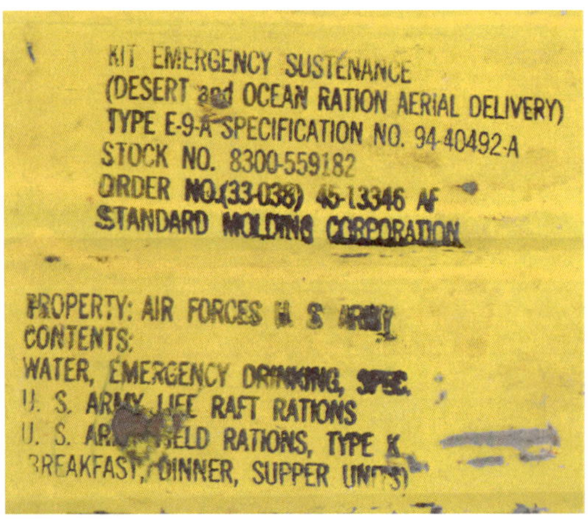

Adopted as standard on 4 November 1943 was the Kit, Emergency Sustenance, Type E-14 (Desert and Tropical Aerial Delivery) under Specification No. 94-40681, available under stock number 8300-559210. This unit combined the best attributes of the Types E-8 and E-10 sustenance kits, which this unit superseded. In line with the directive to make the kits universal to either be carried in the airplane or as an aerial delivery unit, components were packaged in a sealed orange/yellow metal parachute container 12" diameter x 26", detailed under Dwg. No. 44G4558 with Inner Canvas Liner 44G5060. The inner liner was a carrying bag for land transport. The Type G-7 sustenance parachute was used for aerial delivery. Total assembly weighed 80 lbs. 10,600 Type E-14 kits were placed on procurement; contractor and contract unknown.

Type E-14 kit contained accessories necessary for sustaining six men for 10 days in the desert or jungle. Components of original specification include a combination .22 cal. and .410 gauge rifle (Stevens), two boxes .22 cal. long rifle ammunition, four boxes .410 gauge No. 7½ chilled shot ammunition, five boxes .410 gauge slug ammunition (5/box), cleaning brush .45 cal., thong, machete 43E10628, liquid-filled compass, six boxes of safety matches, hand-energized flashlight, twelve 5-min. fusee flares 43B3554, 24 heat tabs, grill for the heat tabs, six sunhats 44G5062, hand axe, fishing kit AN-K-2, Boy Scout knife, two signal mirrors 40653, six red or blue handkerchiefs, six mosquito headnets 43G3868, two tubes of sunburn ointment, survival manual titled Jungle, Desert, Arctic, Ocean Emergencies, 3 oz. can of oil, two first aid kits aeronautic 97765, 50 flannel patches for rifle cleaning, Size-77 paulin, sewing kit, 1½ lbs. bacon, four units Type-K ration (breakfast, supper, dinner), two cans insect repellent, 50' of ¼" rope, package of salt and pepper, vial of salt tablets, bar of soap, 1 lb. cubed sugar, 12 tea tablets, six bags of tobacco with papers, vitamin capsules, 12 AN-W-5A cans of drinking water and two plastic police whistles.

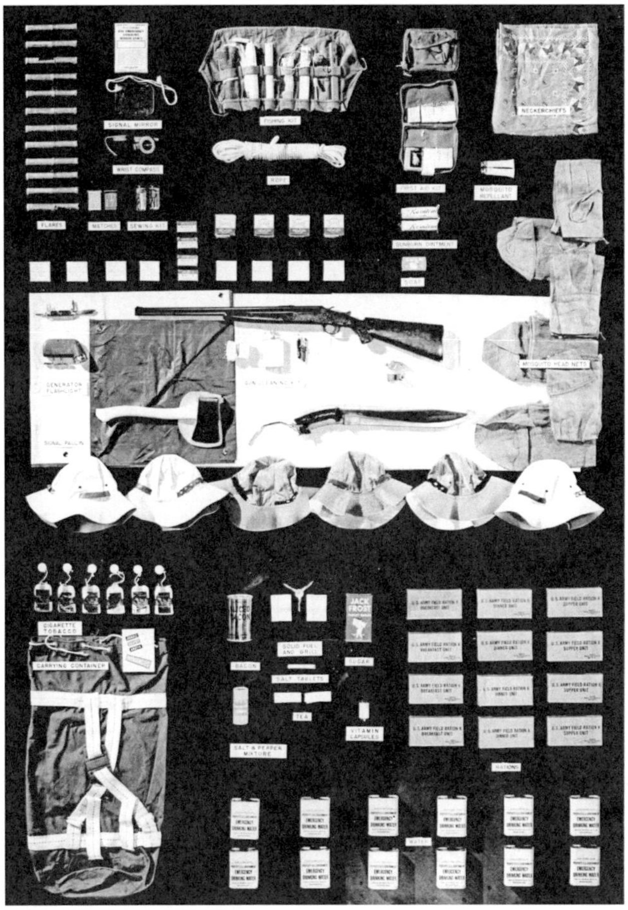

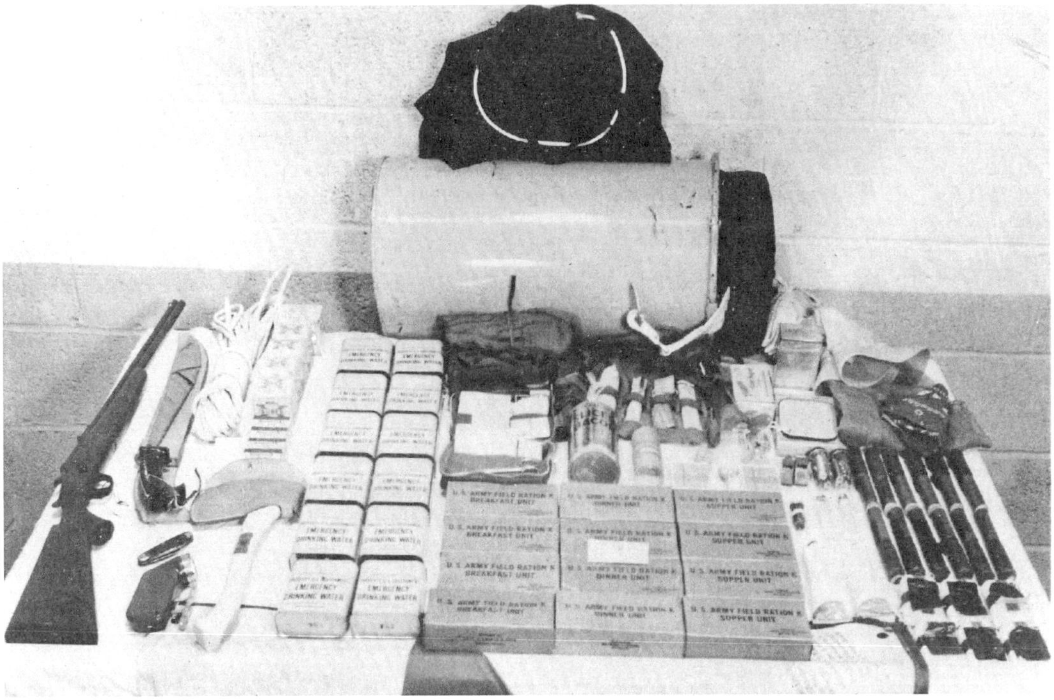

(Above) Exploded view of the Type E-14. (USAF)

(Left) Type E-14 sustenance kit as standardized. Kit contained accessories necessary for sustaining six men for 10 days in desert or jungle. (National Archives)

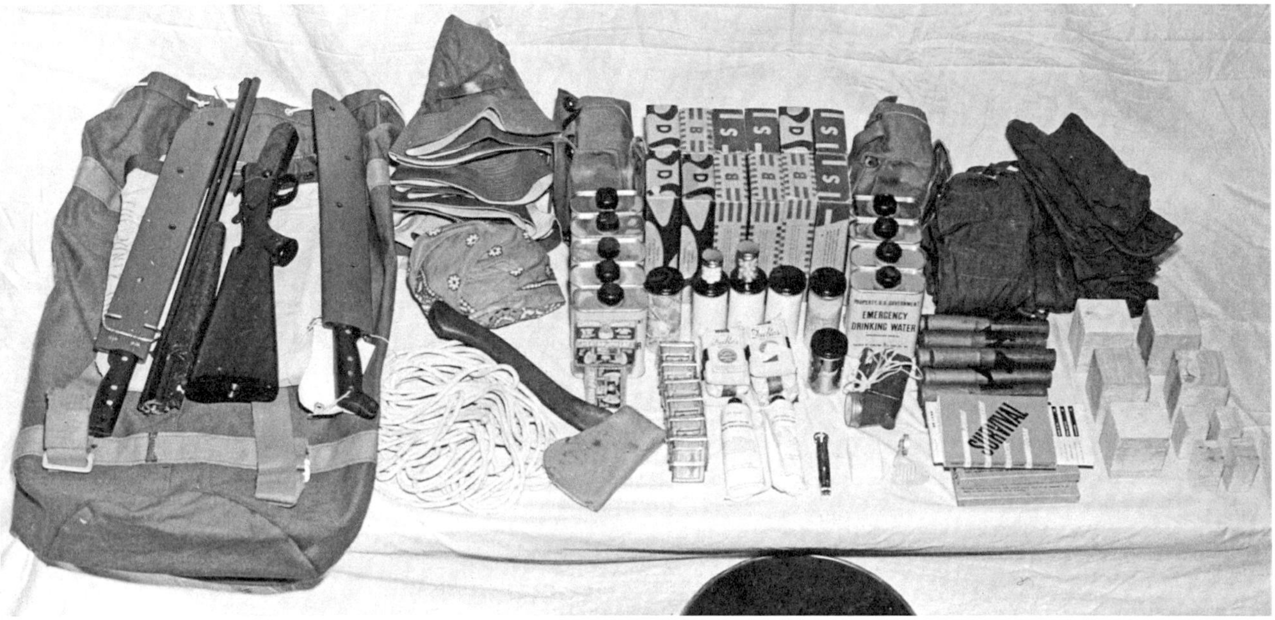

Amended Type E-14 that includes the revised Type-K rations in colored packaging and two 18" machetes with plastic scabbards atop inner canvas bag. Amendment to include two of the new type of machetes occurred 6 May 1944.

Kit, Emergency Sustenance, Type E-15 (Ocean Aerial Delivery) was adopted as standard on 16 October 1943, under Specification No. 94-40682 and available under stock number 8300-559214. This unit combined the best attributes of Types E-5 and E-11 kits, which this unit superseded. Components were packaged in a sealed orange/yellow metal parachute container 14" diameter x 30", detailed under Dwg. No. 43G13341. The Type G-7 sustenance parachute was used for aerial delivery. Assembly weighed 90 lbs. First procurement of 10,600 Type E-15 kits was awarded to Cappel Macdonald and Co. under P.O. 44-7012.

Components for the Type E-15 kit were as follows: two water-resistant blankets, one collapsible bailing bucket, one engineer liquid-filled compass, one match safe 42A6930, one can of sea marker AN-S-10, six 30-minute fusee flares, two solid fuel kits, six pairs of Type D-2 mechanics gloves, six sun hats 44G5062, two aeronautic first aid kits 97765 (revised to life raft type JAN-K-87), two Shurlite Hiway Beacons, two fishing kits AN-K-2, one Boy Scout knife, one signal mirror 40653, six mosquito headnets 43G3868, two tubes of sunburn ointment, four heat pads, one Size-77 paulin, one hand projector M8 (revised to M9), six M11 parachute flares, four Type-K ration units (revised to 12 life raft Type-A), 40' of ¼" cotton rope, a sponge, 12 AN-W-5B cans of drinking water, and one survival manual titled Jungle, Desert, Arctic, Ocean Emergencies (revised to include the survival manual 21W).

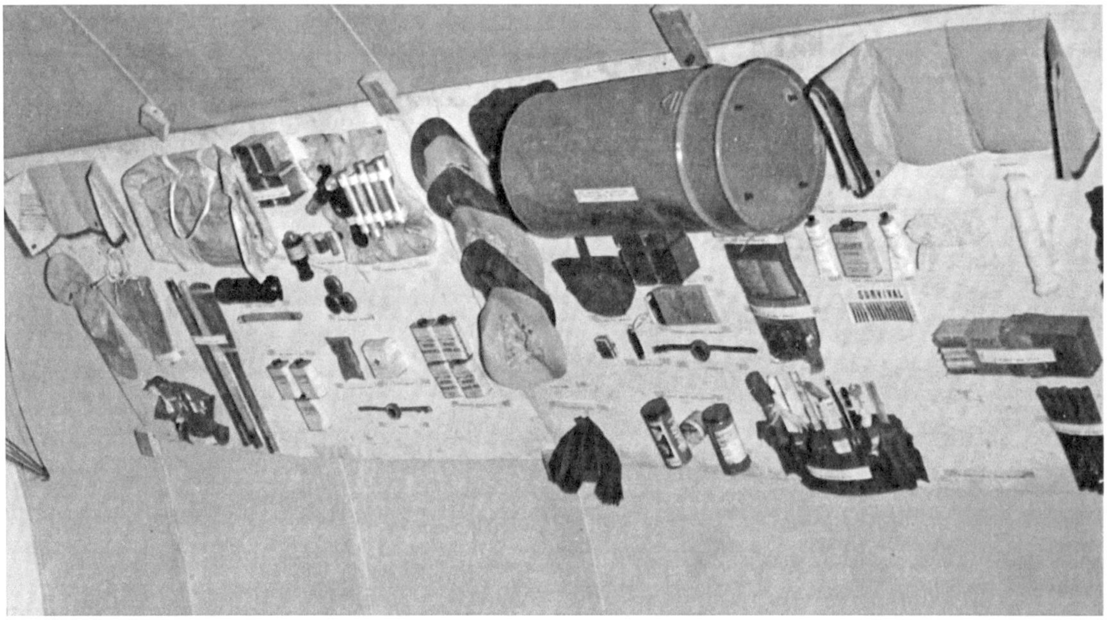

Components for the Type E-15 on a display board of the 500th BG on Saipan, c.1945. (National Archives)

Type E-15 sustenance kit as standardized. Kit was designed to sustain six men for 10 days in a life raft or lifeboat. Like the Type E-5, it was buoyant, allowing conservation of space aboard the life raft. (National Archives)

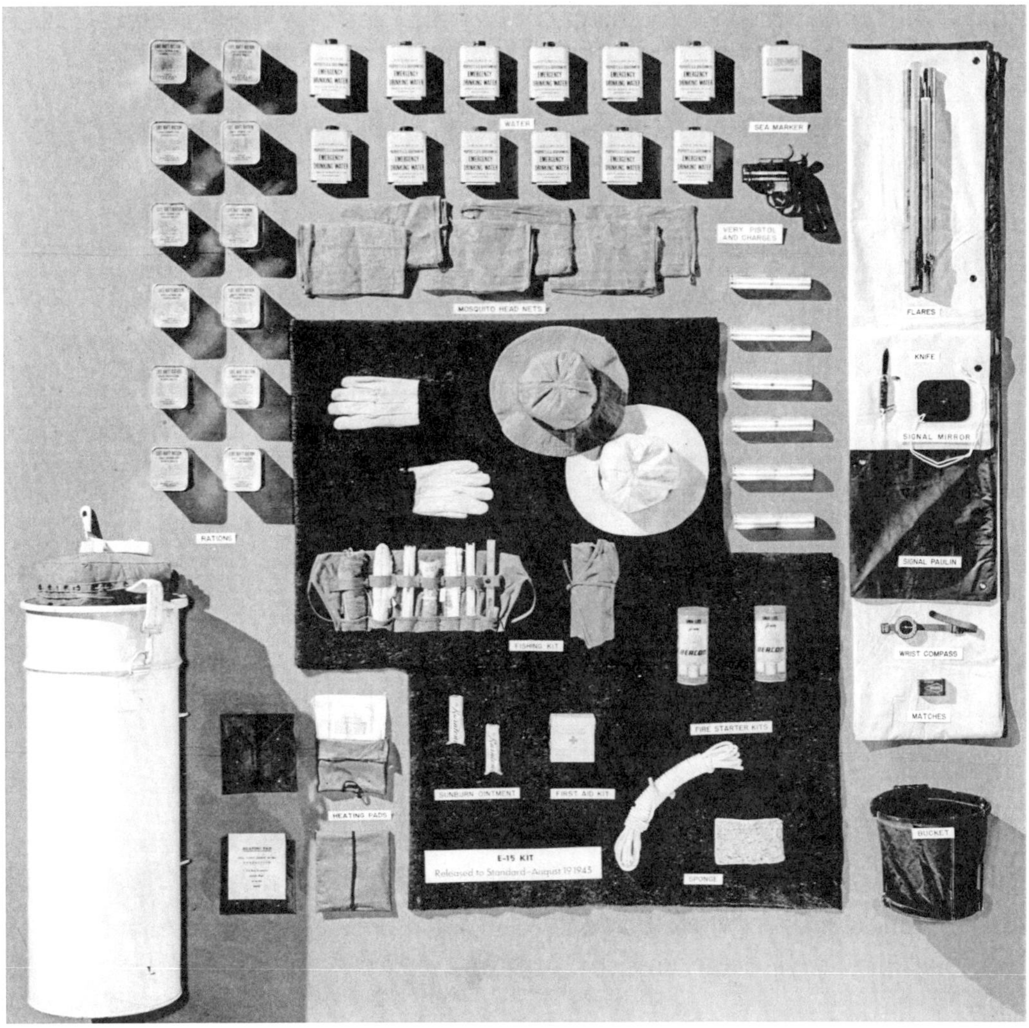

(Left) Exploded view of the Type E-15 sustenance kit with all components. (USAF)

The Type E-15 used the same container as the Type E-11, Dwg. No.43G13341. This example differs in that it is colored Universal Orange. By 1945, the USAAF was transitioning to this color. This unit was part of a procurement from March of 1945 to Exhibit Displays Inc. for an unknown quantity. Likely this contract was eventually cancelled since it would be superseded by the Type E-18. The two features to note are the turn buckles on the lid and crossed out listing of the sun hats. Turn buckles were used to secure the Type G-7 sustenance parachute. By the date of its procurement, the Type C-1 sustenance vest would be the standard for all flying personnel, it inlcudes a sun hat as a standard. It is not clear if there is a correlation to why that line item had been deleted.

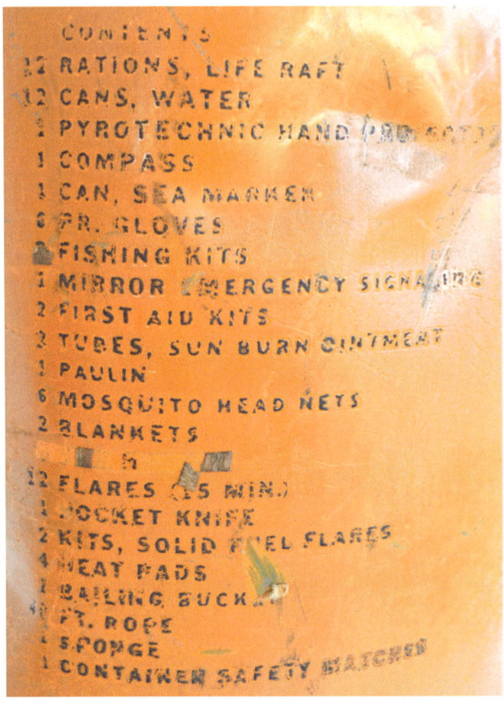

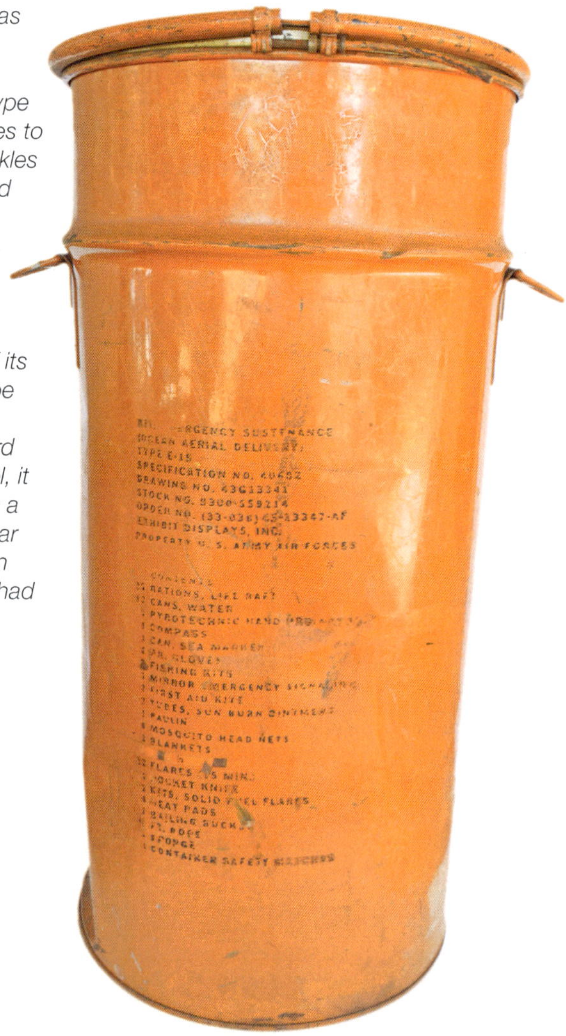

(Above) An E-15 has been airdropped. The Type G-7 parachute included a 12' diameter canopy, the Type G-8 canopy was 8' in diameter. (Left) Te G-7 pack included four tabs on the bottom tray secured by the turn buckles. (USAF)

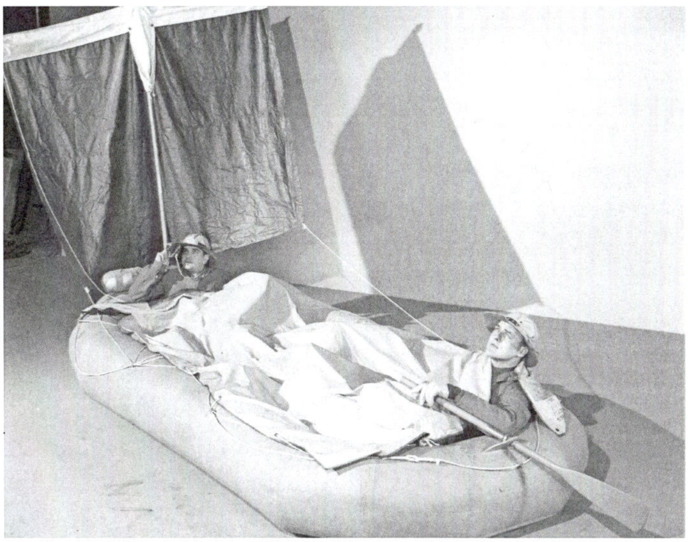

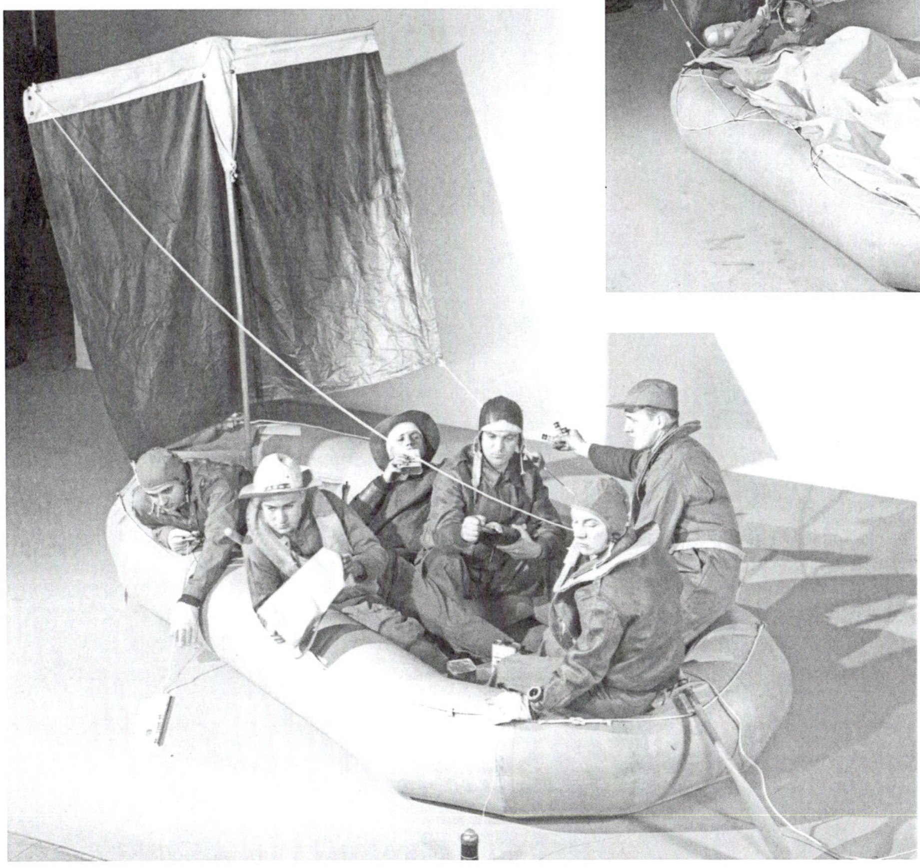

(Left and above) USAAF personnel simulate the use of accessories of the Type A-3 life raft and Type E-15 sustenance kit. (National Archives)

By the end of 1943, multi-person sustenance kits were consolidated for logistical efficiency and had five standardized units: Types E-1, E-9, E-12, E-14, and E-15. All others were placed on limited standard status. The Types E-1 and E-9 were ration replenishment units. After various conversations among the Equipment Laboratory, AAF Board and OC&R, it was concluded that, to economize on procurement and issuing, a combined Type E-1 and E-9 should be developed. Following a directive issued on 13 December 1943 by OC&R, the Materiel Command combined the best attributes of the Types E-1 and E-9, designating it as the Kit, Emergency Sustenance, Type E-16 (Ration Replenishing Aerial Delivery) under Specification No. 94-40753.

Type E-16 was adopted as standard on 4 April 1944, available under stock number 8300-559218.

The container and parachute were the same as used in Type E-9A, but with new components. 10,300 Type E-16 kits were placed on procurement. Components included: three packages of Type-C biscuits, can of bacon, seven 2 oz. fruit bars, three cans of butter, ten packages of Charms candy, seven 2-oz. bars of sweet chocolate, carton of cigarettes, ten packages of cocoa, seven 1-oz. packages of coffee, can of corned beef hash, seven packages of lemon juice, can of meat and rice, two cans of powdered milk, can opener, a box of raisins, a box of salt and pepper mix, two cans of link sausage, four boxes of dehydrated soap, three boxes of sugar and sixteen bags of tea. The complete assembly weighed 35 pounds. For procurement efficiency, all items were standard QMC stock.

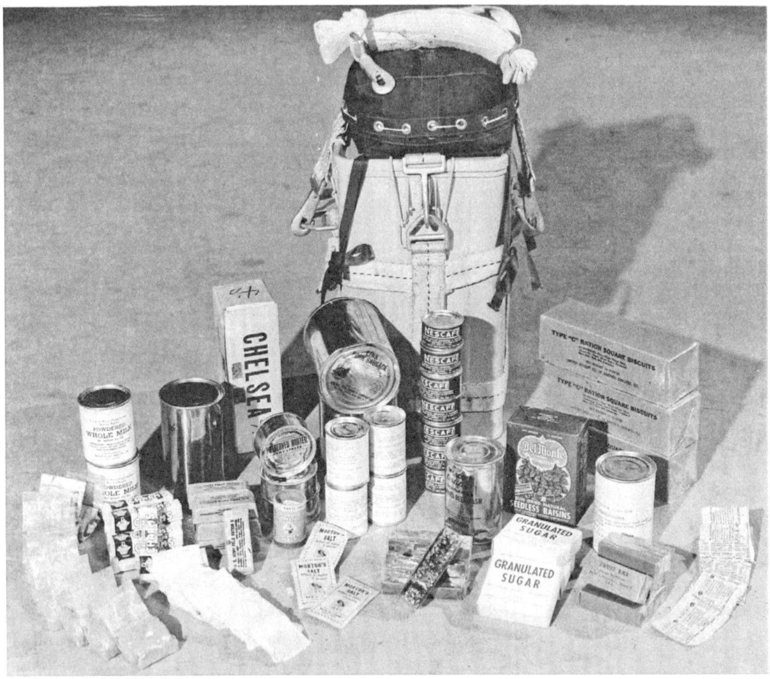

(Left) Type E-16 sustenance kit as standardized.

(Below) This assembly is representative to both the Types E-9A and E-16 since they both used the same container Dwg. No.44G18525 and Type G-8 sustenance parachute. (USAF)

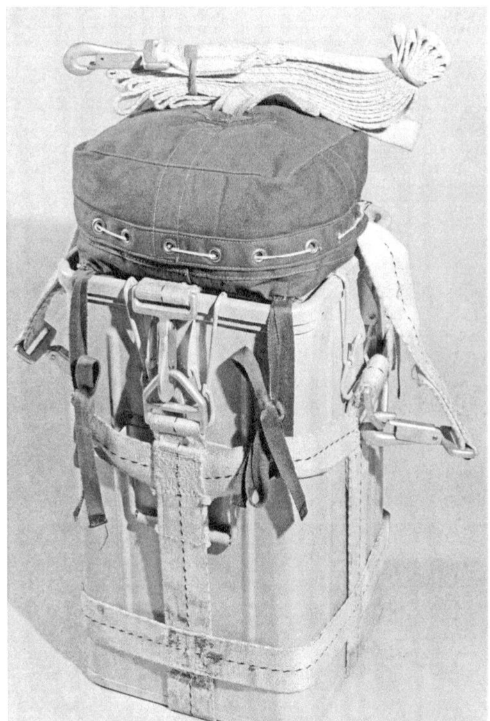

(Left) USAAF personnel review the components of a Type E-16 during a simulation. The Type E-16 superseded the Type E-1 for overland use, with the E-9A standard as an ocean replenishment unit. In the design concept, container and parachute system from the E-9A were used. It, too, would be superseded by the Type E-19. (USAF)

With the adoption of the Type E-16 kit three sustenance kits covered the four climatic zones: Arctic (Type E-12), Jungle (Type E-14), Desert (Type E-14), and Ocean (Type E-15) with the Types E-16 and E-9A available as ration replenishment units, which were considered universal. The standard kits could either be stored aboard the airplane or be airdropped, with each in a free-falling parachute container. All the previous kits remained in inventory and were issued until supplies were exhausted.

In retrospect, the initial standardized kits up to the Type E-12 derived from a necessity. The next generation, Types E-9A, E-14, E-15, and E-16 were based on a plan developed to meet real-time combat theater experiences. Emergency equipment was always in demand and consistently being revised and updated with new items. Never completely satisfied, commands continued to submit recommendations for improvements. As 1944 ended, it became apparent that the volume of the existing kits was not necessary. Multiple factors were considered in this decision, including duration of survival episodes decreasing with the establishment of efficient air-sea rescue organizations, and stranded airmen no longer needing or using the bulk of equipment available before rescue.

The Emergency Rescue Branch and the Personal Equipment Laboratory took a closer look at what was truly necessary. They developed a simplified unit stripped down to the essentials they named the Basic Airplane Kit. What made this unit practical was the adoption and implementation of the Type C-1 sustenance vest. Some items contained within the vest were rations, first aid items, signal devices, protective clothing, fire starter, fishing kit and a sewing kit, which are some of the same items contained in the "E" type sustenance kits. These items were then carried and secured to the user as personal equipment.

To aid in USAAF operations and increase the effectiveness of emergency survival and rescue purposes, on 8 March 1945 OC&R issued a directive to develop a basic airplane sustenance kit. The directive was to provide a suitable standard container for all additional items of emergency equipment to supplement those items available in the Type C-1 sustenance vest and the life raft accessory kit. The Materiel Command's Personal Equipment Laboratory, Emergency Rescue Equipment Branch had already been working on the concept, and by the end of March they had a sample kit assembled. It was to contain items for all areas and be smaller, lighter, and more compact than former kits. As the development process unfolded, two units were created. One container would be filled with items for all areas of the world as a basic implements kit, while another container would be filled with food and water as a basic ration kit. These were intended to be carried in the airplane or for rescue purposes designed to sustain three men.

The concept unit compiled by the Materiel Command consisted of a first aid kit, machete with sheath, three cans .410 gauge 7--1/2 chilled shot, three boxes .22 cal. long rifle, police whistle, three stainless steel cups, 3" file, three mess kit spoons, can of fuel tablets, gun cleaning kit (3oz. oiler, flannel patches, .45 cal. brush, and thong), package of fire starting tabs, sharpening stone, wrist compass, signal mirror 40653, fishing kit 45G9049, three match safes 42A6930, four packages of salt, safety razor in plastic case, tube of sunburn ointment, survival manual 21W, two candles, three toothbrushes, 5 qt. water container, three cans of insect repellent 45B9008, six smoke signals MK-1 Mod-0, six T-49 distress signals, two bars of soap, three pairs of gloves, three pairs of socks, QMC poncho, three sun hats 44G5062, 3 yards. mosquito netting and 10' of copper wire.

Concept Basic Airplane Kit with a new proposed aerial delivery container. (National Archives)

The basic airplane kit remained in development until after VJ-Day. It was not until 20 August 1946 that the kit was adopted as standard as the Kit, Emergency Sustenance, Type E-18 (Basic Airplane Kit) under Specification No. 94-41035. Upon standardization, it superseded Types E-12, E-14, and E-15 sustenance kits. Standardization was delayed pending development of the integrated first aid kit and container. The adopted container was referenced under Dwg. No. 45G18818 measuring 7-1/2" x 7-1/2" x 25-1/2", made from aluminum, designed to float and be used as a cooking container. It was designated as Container Assembly, Emergency Sustenance Kit, Type A-13. D-rings were provided to attach a Type G-8 sustenance parachute.

The components remained unchanged except updating it with the vinylite 5 qt. water bladder Dwg. No. 46G3007-1, wrist compass Type L-1 and Type K-1 flying gloves. Theater commands were authorized to supplement the kit as they saw fit. The Type E-18 would undergo further revisions, eliminating the Stevens rifle and replacing it with the M1A1 carbine. Further amendments would include the addition of the Type A-1 folding machete with scabbard. The Types E-9A and E-16 were soon replaced by arctic and jungle supplemental kits after standardization of Type E-18 using the Type A-13 container. Upon the standardization of the Type E-18, the Type A-13 container was revised and referenced under a new drawing number 46G3207. External configuration changes were applied. The Type E-18 was superseded by the Type D-1 kit in 1950-52.

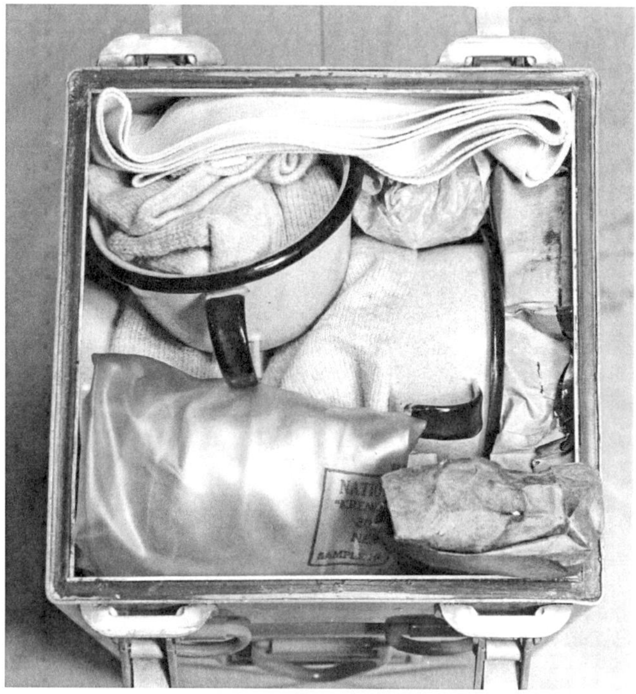

(Above) Components for the adopted Type E-18 sustenance kit packed into the aluminum Type A-13 parachute container. (Below) Spread out for display. (National Archives)

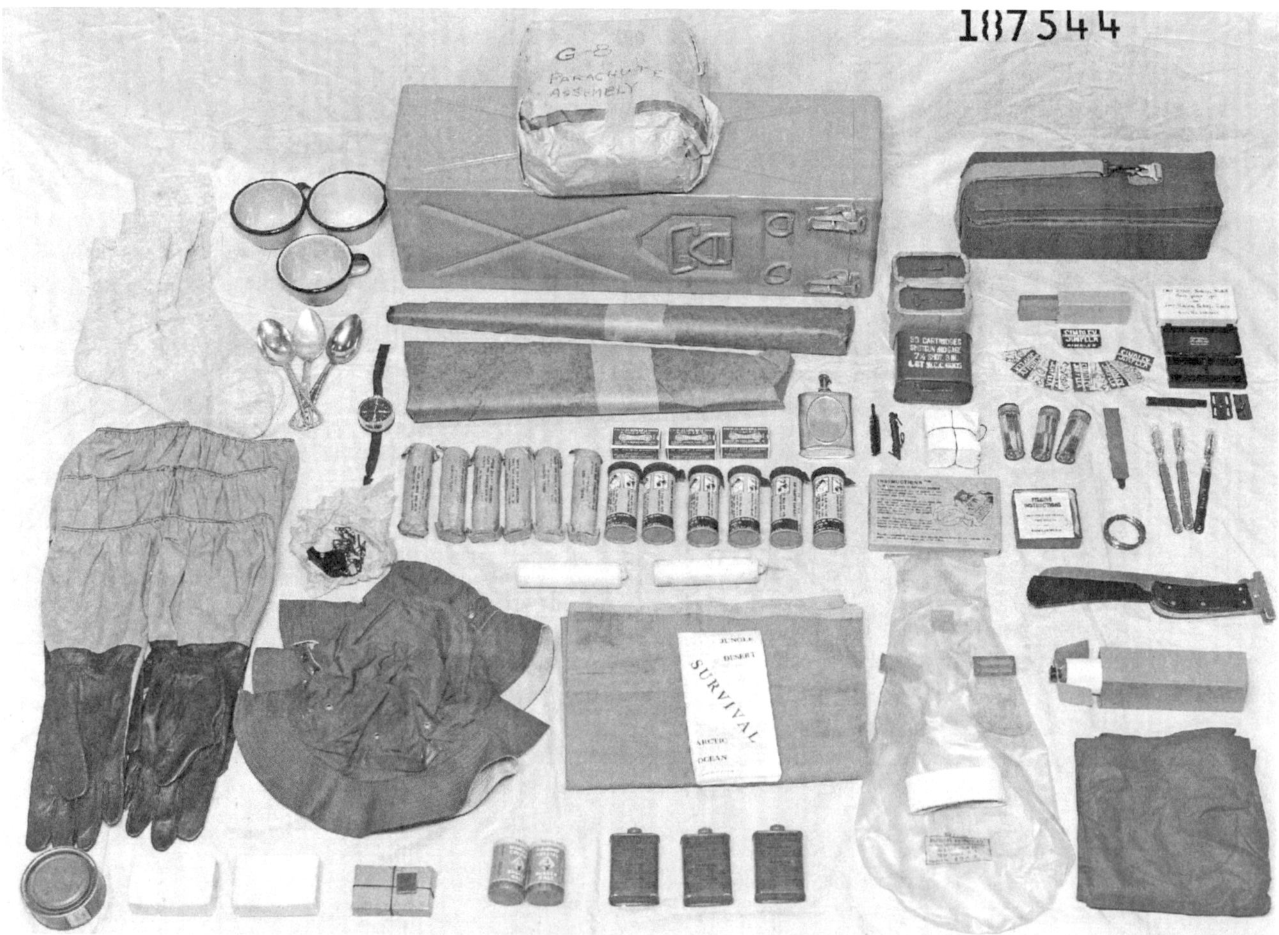

The last sustenance kit developed before VJ-Day was the Kit, Emergency Sustenance, Type E-19 (Rations) under Specification No. 94-41040. Because it was a simpler unit, it was adopted as standard before the Type E-18, 26 October 1945. Like the Types E-1, E-9, E-9A, and E-16, it was issued for ration replenishment using the same container 45G18818 (later, 46G3207) Type A-13. Contents include ten AN-W-5B cans of drinking water, ten cans of Life Raft Ration Type-A, and ten desalting kits Type JJ-1.

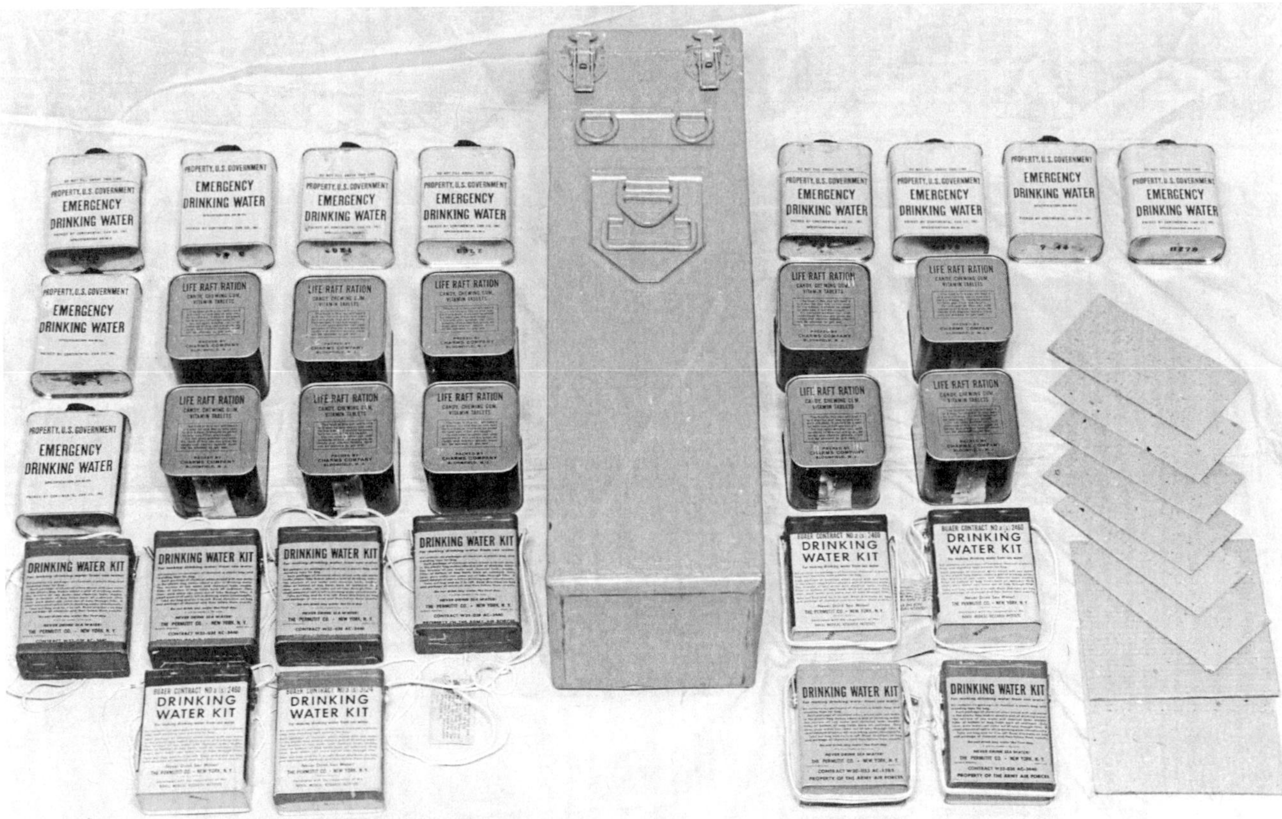

(Above) Type E-19 sustenance kit. In the bottom right of the image are corrugated carboard sheets placed between each item for packing. (National Archives)

(Left and above) After contracts were awarded for sustenance kits, they would not see immediate operational use. As an interim, local commands developed locally fabricated emergency kits. Left, improvised kit using a milk can as the container, filled with items at the local command's discretion and attached to modified standard seat type parachutes. Milk cans were a preferred durable container for storage aboard the airplane for airdropping. (National Archives)

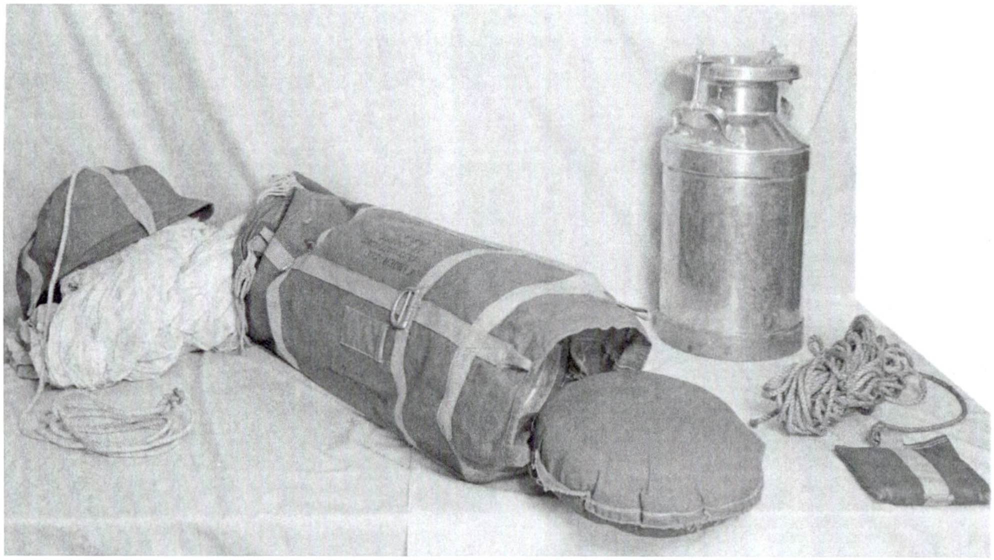

(Left) Standard Type A-3 aerial delivery container is being used for airdropping assembly of milk cans. (National Archives)

(Below left) USAAF procured 10 gal. milk can.

(Below) Unit specifically assembled for airdropping medical supplies to stranded personnel in a modified bomb casing. (National Archives)

(Right) Variety of sealed durable containers for airdropping supplies for air-sea rescue purposes. These were used extensively early in the war due to the nonexistence of aerial delivery sustenance kits. Eventually they would be phased out once standard kits became available in the operational theater. (National Archives)

MULTI-PERSON EMERGENCY KITS

Part II.
U.S. NAVY

Development of multi-person sustenance kits for the Bureau of Aeronautics was simpler than for the USAAF. The BuAer had and would maintain an emphasis on over water operations. Therefore, the development of the life raft met all existing service requirements. The war would shift this emphasis as it did with the USAAF, but in the form of personal emergency kits.

Operational aircraft differed greatly between the two air services. Most USAAF flying personnel operated bombardment airplanes with up to 10-man crews. The BuAer principally operated carrier aircraft occupied by one to three people. With the emphasis on over water and with limited crew capacities, it was unnecessary for the BuAer to develop large sustenance kits for all regions of the world; however, larger kits were not completely disregarded.

The BuAer made great strides in providing improved and more efficient lifesaving equipment and maintaining a consistent communication stream. This was accomplished by fleet requests and recommendations, as well as after action reports by personnel either temporarily lost or rescued after forced water landings. Later, Air Sea Rescue reports were introduced, which largely reported on life raft and emergency equipment performance. The information from these sources was disseminated and addressed. In initial attempts to provide for the fleet, adoption of the pararaft served to orient emergency equipment toward the individual. Soon after came the development of the individual back pad kit for jungle survival. For the war environment, the two primary operational regions were the jungle and ocean; however, the jungle quickly took precedent. The development of personal equipment would remain the primary focus for the flying crew, as it was considered more advantageous to secure emergency equipment with the individual.

The M-3 Type-D life raft was best suited for war requirements, in that it was capable of being air-dropped to survivors. Included in the raft were emergency provisions that were constantly being revised. If survivors were spotted, the Type-D life raft could also be dropped, providing some security and replenishment until rescue was possible.

(Left) MK-IV Type-D life raft being dropped by USCG personnel operating a Grumman G-21 Goose. Patrol aircraft were equipped with the Type-D for rescue purposes. When survivors were spotted, a smoke bomb was dropped to mark the location. After determining wind direction from the smoke, the airplane would make another coordinated pass over the survivors and, on a specific flight path, crew members would drop the life raft upwind. (National Archives).

(Right) Airship crew members of ZP-31 are preparing to drop a MK-VII Type-D life raft to survivors off the coast of California, c.1944. The Type-D was the principle multi-person sustenance kit used by the BuAer. It was a standard stocked item being readily available to all activities. It included all the necessary emergency provisions for survivors. The unit simplified the logistics of having to maintain a large variety of specialized equipment. (National Archives)

With increased combat operations and forced water landings, provisions in the life raft were deemed suboptimal for extended survival. A replenishment kit for survivors was commissioned as a result. However, the Type-D life raft would remain the principal airdrop unit. Working in conjunction with the New England Bedding Co., an aerial delivery sustenance kit was devised and adopted as standard on 30 September 1943. It was designated as Kit, Shipwreck (For Aircraft Use) under NavAer Specification M-594. This kit was designed as a free-falling unit requiring no parachute. All the components were available from standard supply channels.

The assembly was referenced under BuAer Dwg. No. 8731 and was available under stock number R83-K-710400. It weighed 55 pounds. The components included: six tablet rations M-539B, one match safe compass AN-C-101, one fishing kit AN-K-2, two sea markers AN-S-10, eight AN-W-5A (flat) cans drinking water, Boy Scout knife M-575, one signal mirror M-580, six 2" safety pins, adhesive tape, 50' of cotton line, two tins sunburn ointment, one whistle, two woolen blankets, four 1.6 oz. bottles of whiskey, two 12 oz. cans of chicken broth, two 12 oz. cans tomato juice, ten cigarette packs (four per pack with matches), one graduated drinking cup (metal or plastic), two packs chewing gum, a first aid kit S2-1055, a projector kit (MK-IV with 6-red Very's signals) and a sheet of instructions.

The components were secured to a plywood panel in two tiers. Most items were wrapped in waterproof asphalt paper; unit thoroughly padded with wood shavings (excelsior) and cloth. Two woolen blankets provided additional padding around the complete assembly. The extra padding was provided for shock absorption. The outer orange/yellow vinyl-coated container seals the kit, keeping it waterproof. To ensure waterproofing, a zipper closure is on one end to prevent seepage, which, in turn, is sealed with a rubber-cemented strip of the same fabric. The container would be retained for component storage after opening. Unit measured 33"x 15"x 11-1/2". Only one procurement of the M-594 was awarded under contract NOA-923 for 1,000 kits to the New England Bedding Co., 5 June 1943. The M-594 was standardized upon delivery to major supply points after the contract award.

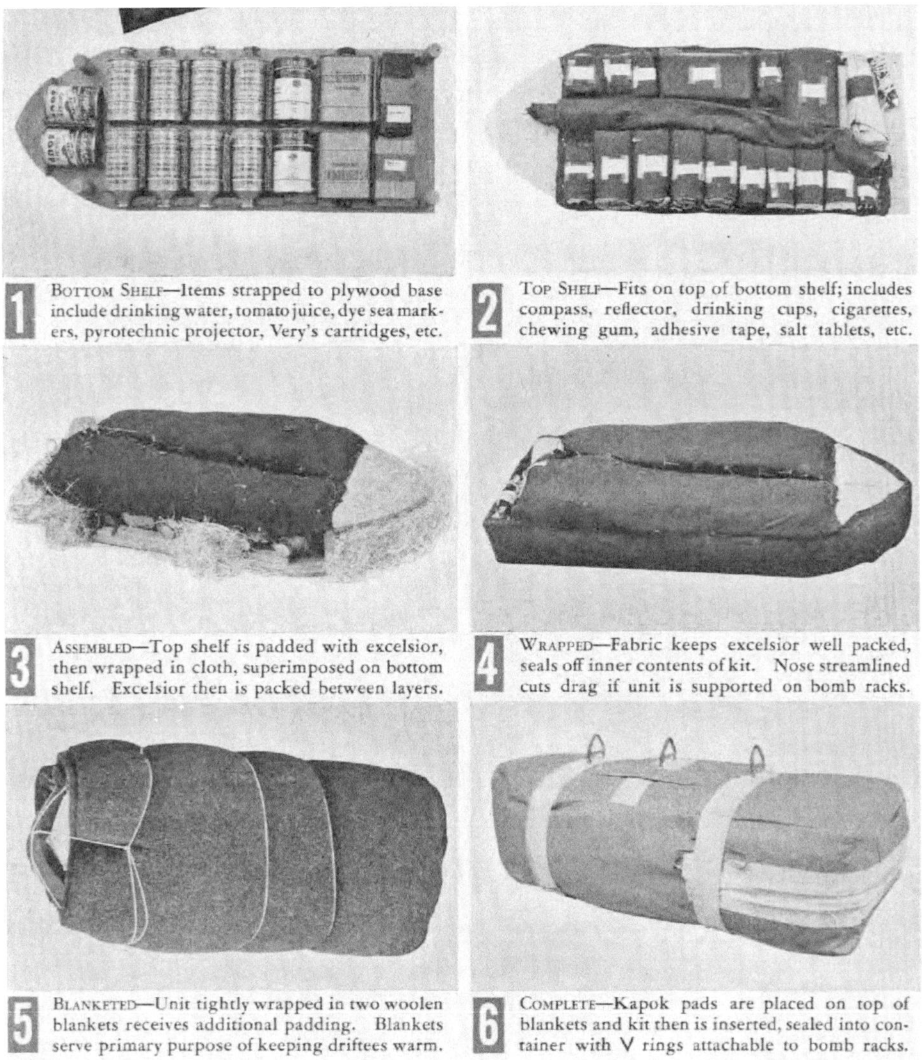

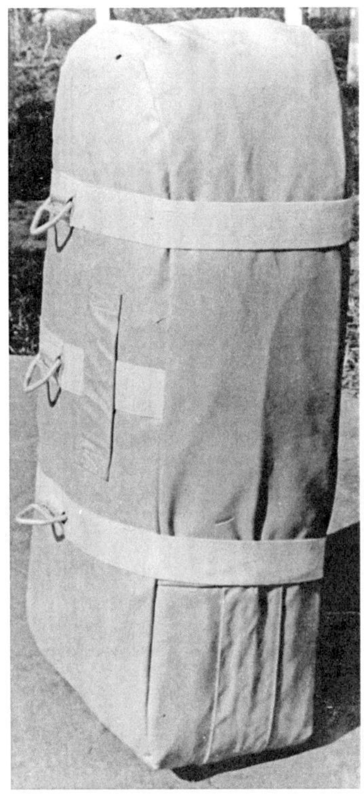

(Above) M-594 shipwreck kit.

(Left) In various states of assembly as illustrated in Naval Aviation News.

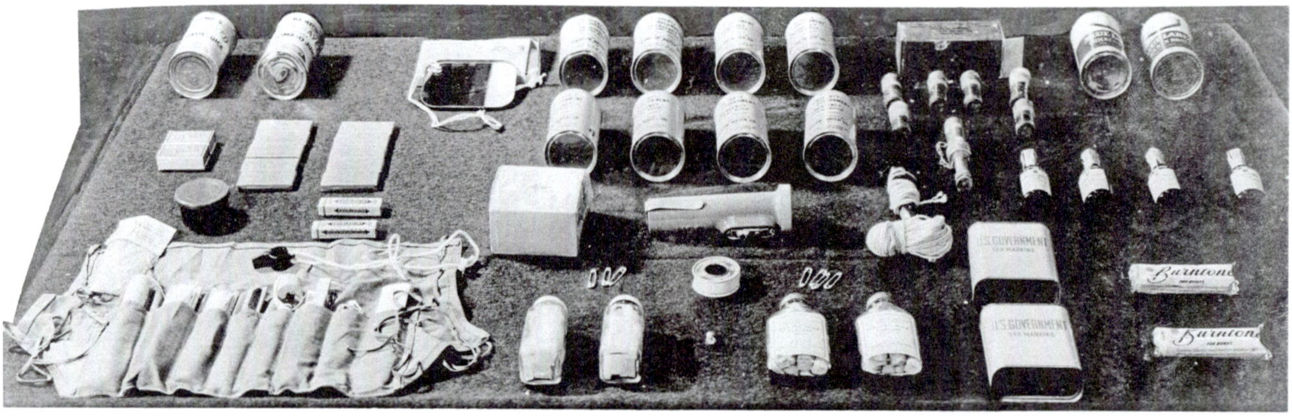

Components of M-594 Shipwreck Kit upon standardization. This specific unit was assembled prior to the availability of the tablet ration and sunburn cream that were provided in a tin. Two tubes of sunburn cream were substituted. Rations included two jars of tablet candies and two jars of malted milk tablets from the ration assembly M-539. (National Archives)

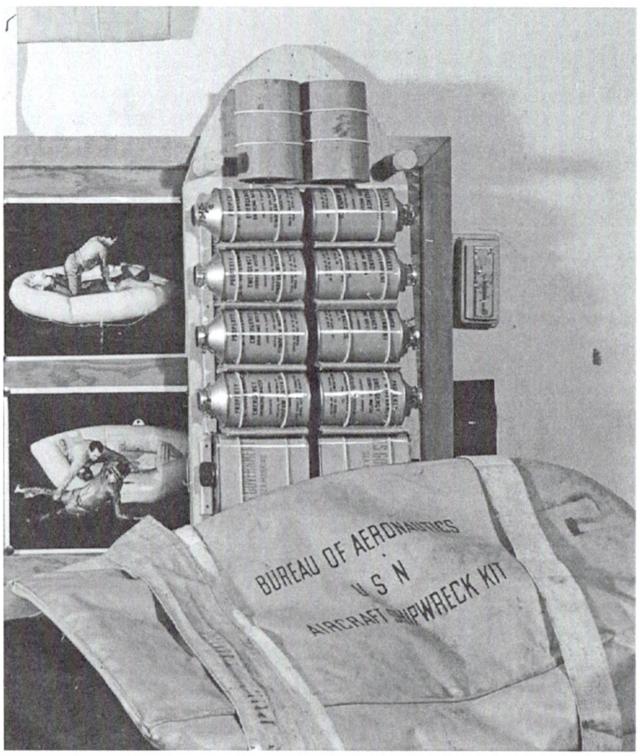

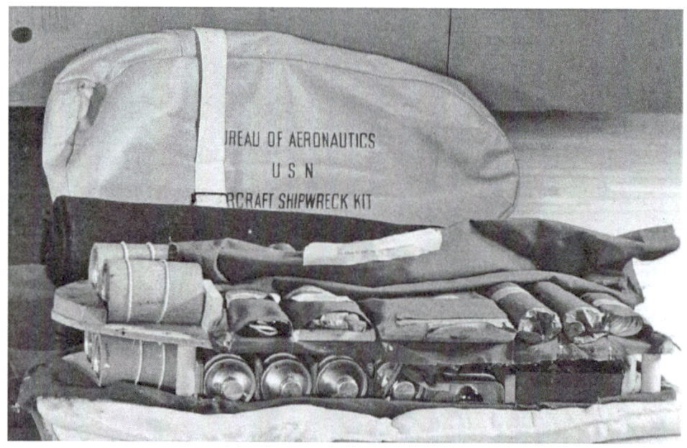

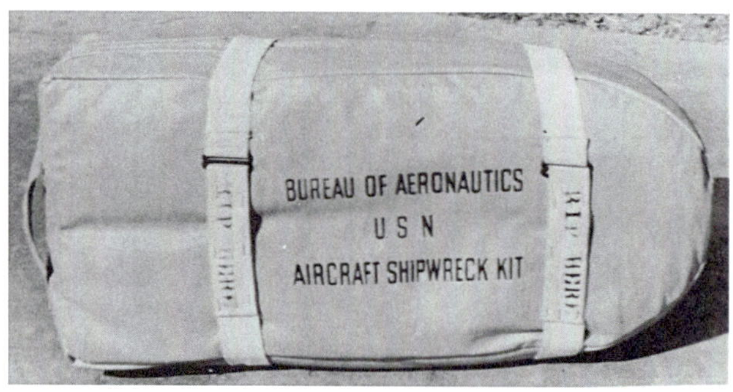

Side profile of the two tiers and how all the components are stowed in the plywood panels: either secured to the panel by cotton string or in pockets. All components were standard stock, whether ASO, BuOrd, or BuM&S. BuAer Dwg. No. 8731 was distributed to the fleet for local assembly. When locally assembled, all the components as detailed in the drawing were requisitioned accordingly, which expedited the use in operational theaters. Though intended for use over water, it could be dropped overland. (National Archives).

(Right) How the assembly appeared before being inserted into the outer container. Jackknife was left exposed near the zipper closure for immedite access after opening. (National Archives)

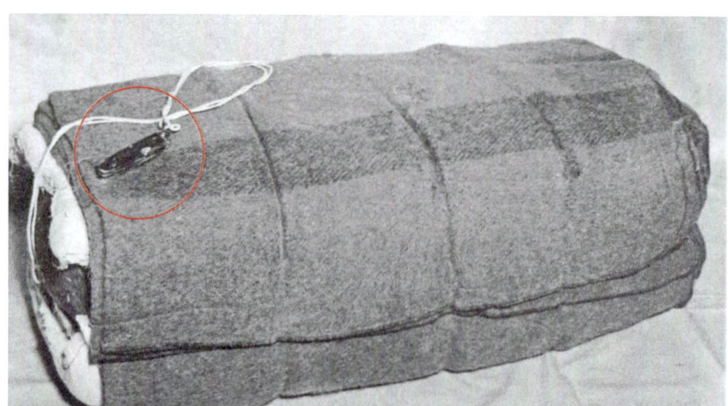

Layout of the M-594A. As part of the directive to eliminate the canned sea marker in all emergency equipment, the M-594 was revised effective 25 March 1944 (Amendment-A) under the new designation as the M-594A. The two sea marker cans AN-S-10 were replaced by six sea marker packets M-566, in addition to the removal of the whiskey and cigarettes. All other articles were maintained from the original specification. In April 1944, contract 288-22029 was awarded for the manufacture of 5,000 M-594A kits; an additional contract, NOA-4228, was awarded in July for 5,000 more. The New England Bedding Co. would be the sole contractor for BuAer shipwreck kits. (NNAM)

(Right) Additional component display of the M-594A. (National Archives)

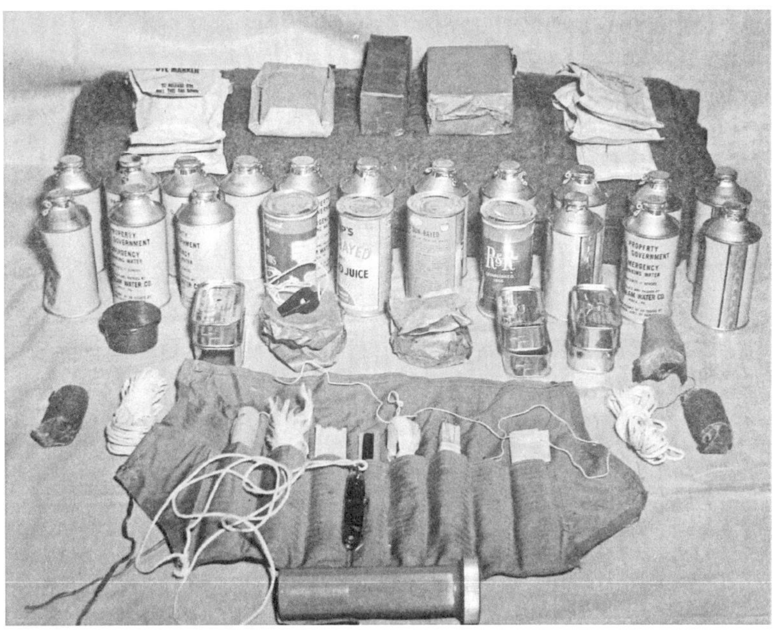

The M-594 was used at the discretion of the requisitioning party, and no applicable technical orders or notes were drafted detailing its issuance. It was felt to be self-explanatory. However, it was intended to be used by patrol and air-sea rescue aircraft, but adaptable enough to be dropped by carrier airplanes. With its adoption, the BuAer had three categories of emergency equipment: personal equipment, which the pilot or aircrew member carries on their person, equipment carried on the airplane, and equipment that could be airdropped.

By the close of 1943 there were multiple development programs in progress to streamline survival and rescue equipment, based on unofficial policy. One of these initiatives combined the pararaft with the back pad kit and increase the pilot's personal equipment. The reasoning was that all flying personnel operating outside the continental United States would be equipped with individual lifesaving aids. As a precaution, all aircraft except VF class would carry an M-3 life raft. In the event of equipment loss, supplies could be airdropped with the M-594 developed to supplement the survivors. In line with this policy, emphasis shifted from survival to rescue and for carrier fleets to become more self-sufficient in air-sea rescue work. At this point, the BuAer began developing a rigid air-droppable boat designated as the AR-8 for PBY aircraft and the AR-10 assembly for aerial delivery from a TBF/TBM Avenger. When airdropping equipment, precision was key because of ocean currents and wind direction. To facilitate survivor recovery of equipment, a train assembly was integrated into the AR-10 which mitigated outside factors such as those ocean and wind currents.

By the summer of 1944, extensive field tests had been conducted, proving train assemblies to be highly effective. Added were the AR-1, AR-2, AR-4, and AR-7 airborne rescue assemblies. For these assemblies to work, the Shipwreck kit had to be modified to facilitate the train. This modification was detailed in technical notes. In turn, specifications were revised, and the modifications implemented by the contractor. Amendment-B of Specification M-594 became effective on 2 December 1944 with the revised title Kit, Shipwreck (For Aircraft Use) Mark-3. The M-594 and M-594A were retroactively designated as the Mark-1 and Mark-2. Mark-3 was detailed in Amendment-B, but no Mark designation for the two previous models.

M-594B fully assembled with components on display. (National Archives)

The M-594B was assigned stock number R83-K-170308-10, specially designed to be an integral part of airborne rescue assemblies. M594 and M-594A shared stock numbers, meaning either of the models could be delivered when requisitioned. Components for the M-594B were significantly revised, increasing the weight to 88 lbs. and altering the dimensions slightly to 31"x 14"x 10".

Contents for the M-594B were revised to include: twelve life preserver CO2 cartridges AN-C-105, four AN-W-5B cans drinking water, twelve tablet rations M-539B, twelve sea markers M-566B, Boy Scout knife M-575, two signal mirrors M-580A, eight desalting kits M-613, two fishing kits M-627, solar still M-663, first aid kit 57K1, flashlight 17F7, 5 qt. water bag Dwg. 8985, six 2" safety pins, adhesive tape, two coils 50' cotton line, four graduated drinking cups R83-C-81525, two tins sunburn ointment, two whistles 74W1, poncho P-61, bailing sponge, two projector kits, woolen blanket, ten handheld smoke signals MK-1 Mod-0, magnetic card compass and a sheet of instructions. Components were secured to two plywood panels stowed facing each other. Padded panels were provided to surround the components. Upon acceptance of Amendment B, a contract was awarded the same month to the New England Bedding Co. for assembly of 2,500 units under contract 288-27741. An additional contract was awarded in February 1945 for assembly of 1,000 more units, under contract NOA-6012.

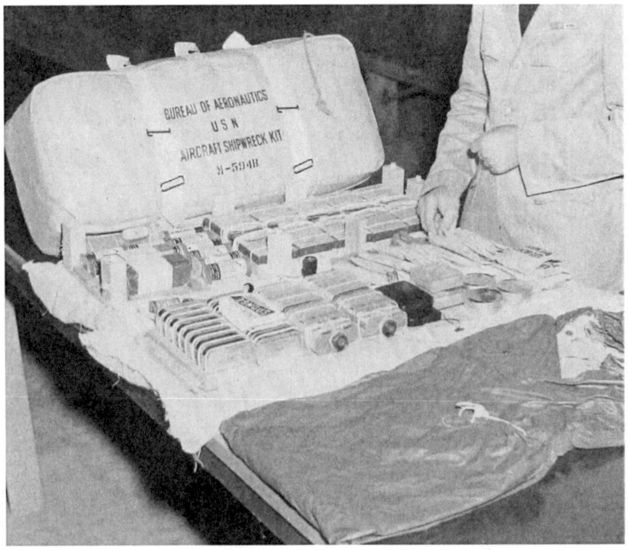

(Top right) M-594B being packed. (National Archives)

(Right) Loaded up and ready to go. (National Archives)

(Below, left and right) Significant external modification to the M-594B was the addition of a rope pocket stowing 105' of floating rope for rescue assembly train. Examples of the air-sea rescue assembly, concept applies to the AR-2, AR-4, or AR-7. Raft bundle had been modified for the assembly as detailed in Technical Note 103-44. These airborne rescue assemblies were designed to be dropped by the TBF/TBM Avenger because of its appropriate bay capacity, although all the airdrop units were adaptable for all carrier aircraft. (National Archives)

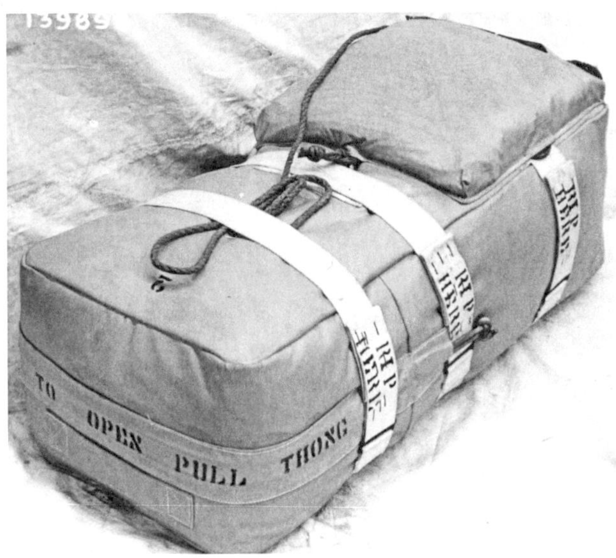

The BuAer took steps to develop equipment in what it viewed as the most practical and efficient way to address all fleet concerns in a universal manner. Often, survivors could not be picked up immediately. Hours or even days may lapse before rescue. Their position would be reported upon the initial sighting and, depending on wind and ocean currents, they may drift miles before being spotted again.

Developed in conjunction with the airborne assemblies were two small supplemental sustenance and signaling units. The first unit developed was oriented toward rescue and designated as the Kit, Signaling, Aircraft Droppable (ADS-1) under NavAer Specification M-697, detailed under BuAer Dwg. No. 9314. Initially, this kit was developed for VH (rescue) squadrons to provide additional signals to survivors, but versatile and compact enough to be used in all carrier aircraft. When immediate rescue was impossible, the ADS-1 could be dropped to assure the survivors had sufficient signaling supplies. Contract NOA-4724 was released in December 1944 for the assembly of 5,000 ADS-1 kits awarded to the New England Bedding Co. Limited quantities became available in March 1945 and were officially adopted as standard on 10 April 1945, available under stock number R83-K-710309-10. ComAirPac directed that, due to limited availability, the initial issuance would be restricted to activities engaged in air-sea rescue operations, with priority given to VH squadrons and fleet units participating in carrier rescue operations in the forward area. When more supplies became available, the ADS-1 was to be supplied to carriers in different quantities: 25 units per CV, 15 per CVL, and 150 per VH squadron.

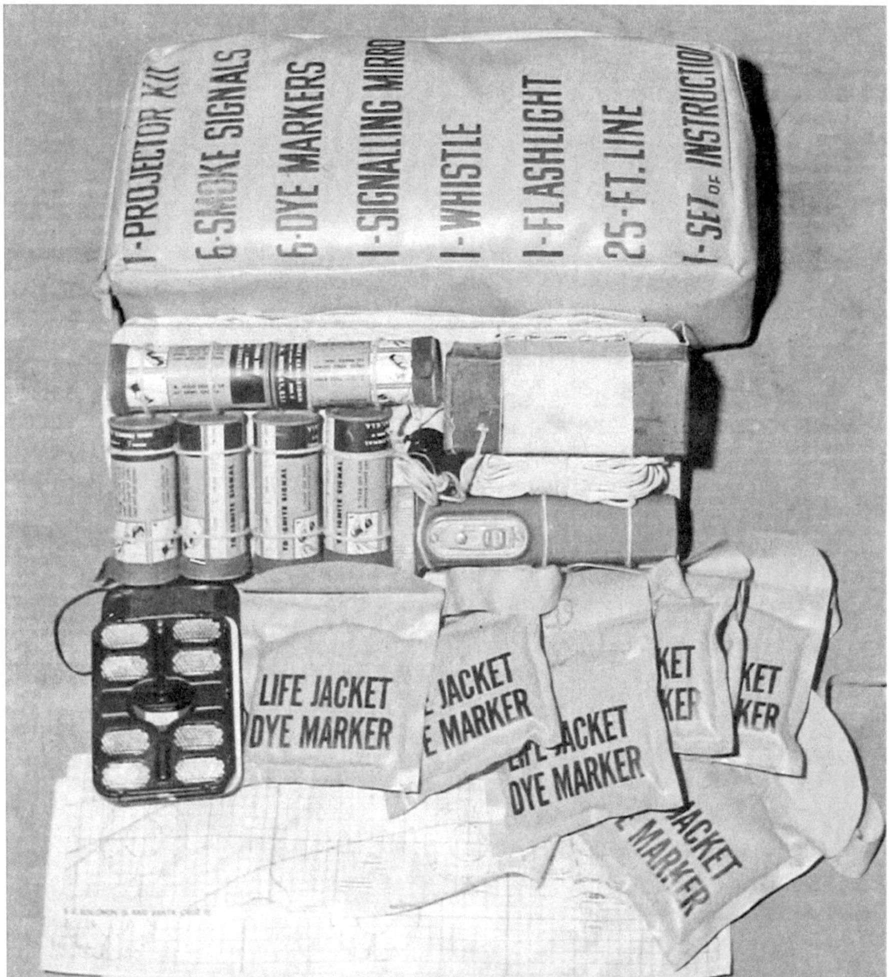

(Above right) Detailed to the fleet in Technical Note 3-45 dated 3 January 1945, the ADS-1 was packed like the shipwreck kit and measured 15"x 7"x 7", weighed 8 lbs. and orange/yellow. Components: pyrotechnic kit, six sea markers M-566B, signal mirror M-580A, flashlight 17F7, six handheld smoke signals MK-1 Mod-0, whistle 74W1 and 25' of cotton line. Images illustrate Delta model A-2060 flashlight, replaced by A-2066 during production. Cloth charts are included in the image. These were not detailed to be included in the technical note nor in specification M-697; charts were issued to the individual as personal equipment. (National Archives)

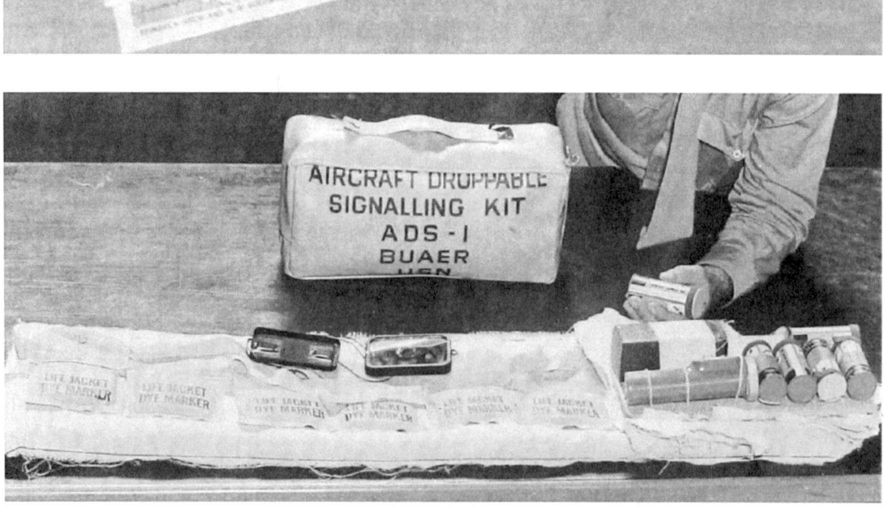

(Right) ADS-1 components and protective padding. (National Archives)

Though rescue was the emphasis, it was critical to develop a unit that contained the maximum amount of drinking water, the most vital element for ocean survival. Second unit was designated as the Kit, Sustenance, Aircraft Droppable (ADR-1) under NavAer Specification M-698, detailed under BuAer Dwg. No. 9315. 5,000 units were procured under contract NOA-4737 in December 1944 from the New England Bedding Co., available under stock number R83-K-710309-40; adopted as standard on 20 April 1945. The ADR-1 was detailed to the fleet in Technical Note 23-45 dated 28 March 1945. Contents: Two AN-W-5B cans drinking water, six tablet rations M-539B, solar still M-663, four desalting kits M-613, 5 qt. water bag Dwg. 8985, Boy Scout knife M-575, 25' cotton line and instruction sheet. Unit measured 8"x 6"x 21".

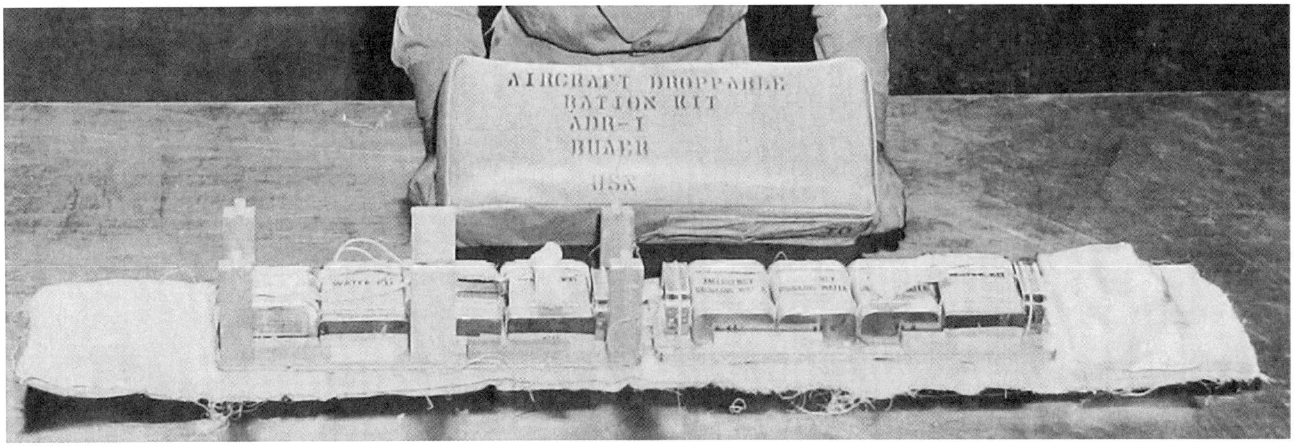

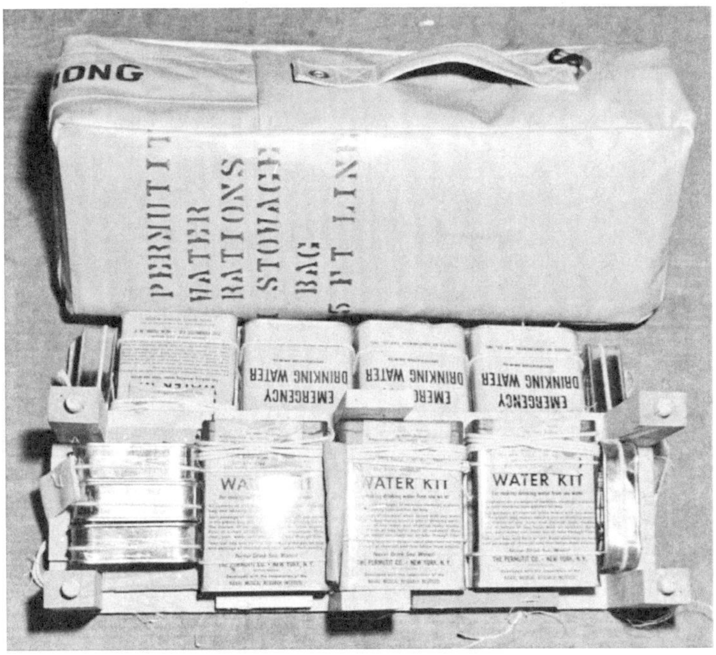

(Above) ADR-1 assembly. (National Archives)

(Left) Each component is secured to the wood panel by a cotton twine. (National Archives)

(Below) F6F Hellcat is dropping rescue gear to aviator in water.

(Below left) All units were designed to float. (National Archives).

(Right) Previous images of the ADR-1 are of preliminary concept, missing the solar still M-663. These kits were not yet available at time of the release of contracts. Shown is configuration illustrating solar still with two drinking water cups nested with the AN-W-5B water cans. The solar still is the updated MK-2 MOD-0.

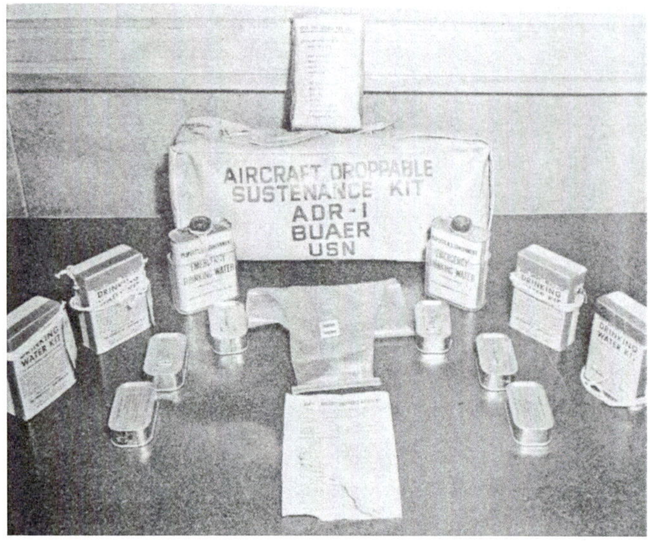

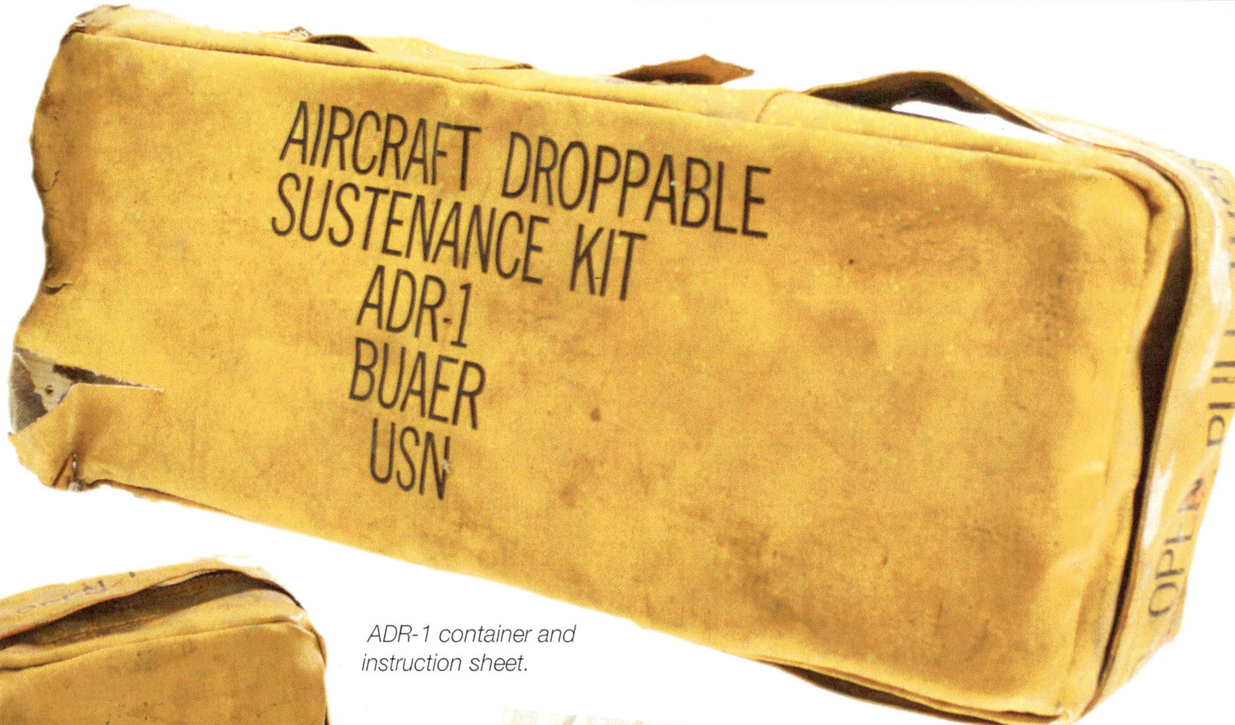

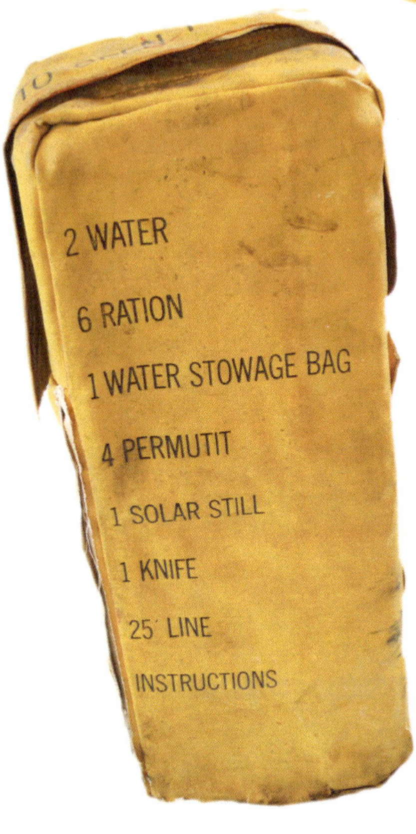

ADR-1 container and instruction sheet.

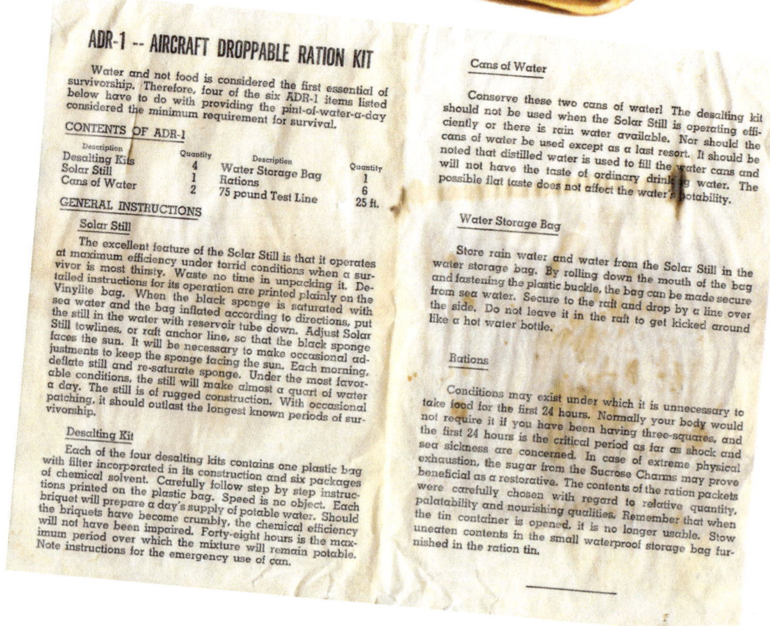

ADR-1 -- AIRCRAFT DROPPABLE RATION KIT

Water and not food is considered the first essential of survivorship. Therefore, four of the six ADR-1 items listed below have to do with providing the pint-of-water-a-day considered the minimum requirement for survival.

CONTENTS OF ADR-1

Description	Quantity	Description	Quantity
Desalting Kits	4	Water Storage Bag	1
Solar Still	1	Rations	6
Cans of Water	2	75 pound Test Line	25 ft.

GENERAL INSTRUCTIONS

Solar Still

The excellent feature of the Solar Still is that it operates at maximum efficiency under torrid conditions when a survivor is most thirsty. Waste no time in unpacking it. Detailed instructions for its operation are printed plainly on the Vinylite bag. When the black sponge is saturated with sea water and the bag inflated according to directions, put the still in the water with reservoir tube down. Adjust Solar Still towlines, or raft anchor line, so that the black sponge faces the sun. It will be necessary to make occasional adjustments to keep the sponge facing the sun. Each morning, deflate still and re-saturate sponge. Under the most favorable conditions, the still will make almost a quart of water a day. The still is of rugged construction. With occasional patching, it should outlast the longest known periods of survivorship.

Desalting Kit

Each of the four desalting kits contains one plastic bag with filter incorporated in its construction and six packages of chemical solvent. Carefully follow step by step instructions printed on the plastic bag. Speed is no object. Each briquet will prepare a day's supply of potable water. Should the briquets have become crumbly, the chemical efficiency will not have been impaired. Forty-eight hours is the maximum period over which the mixture will remain potable. Note instructions for the emergency use of can.

Cans of Water

Conserve these two cans of water! The desalting kit should not be used when the Solar Still is operating efficiently or there is rain water available. Nor should the cans of water be used except as a last resort. It should be noted that distilled water is used to fill the water cans and will not have the taste of ordinary drinking water. The possible flat taste does not affect the water's potability.

Water Storage Bag

Store rain water and water from the Solar Still in the water storage bag. By rolling down the mouth of the bag and fastening the plastic buckle, the bag can be made secure from sea water. Secure to the raft and drop by a line over the side. Do not leave it in the raft to get kicked around like a hot water bottle.

Rations

Conditions may exist under which it is unnecessary to take food for the first 24 hours. Normally your body would not require it if you have been having three-squares, and the first 24 hours is the critical period as far as shock and sea sickness are concerned. In case of extreme physical exhaustion, the sugar from the Sucrose Charms may prove beneficial as a restorative. The contents of the ration packets were carefully chosen with regard to relative quantity, palatability and nourishing qualities. Remember that when the tin container is opened, it is no longer usable. Stow uneaten contents in the small waterproof storage bag furnished in the ration tin.

Martin PBM Mariner airdropping an AR-10 assembly by parachute, May 1945. (National Archives)

On display at the Survival Museum at NAS Pensacola are the principal air-droppable assemblies for the BuAer. The multi-person life raft would be the primary equipment, supplemented by M-594 shipwreck kit, ADS-1 and/or ADR-1 and emergency transmitter SCR-578. Individuals would have the standard allotment of personal equipment. The M-3 Type-D was the primary multi-person sustenance kit used by the BuAer, as a standard stocked item, readily available and containing all necessary emergency provisions. (National Archives)

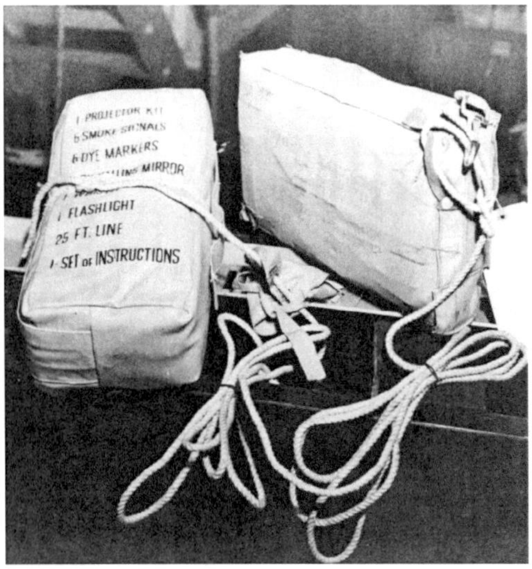

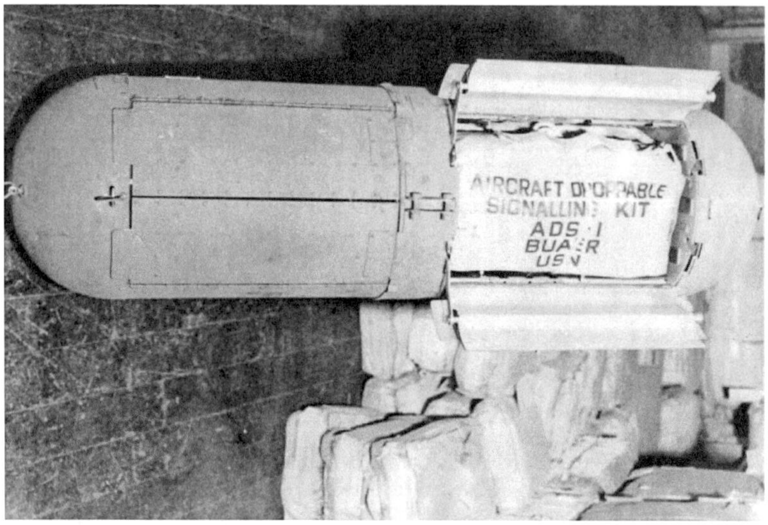

Rescue assembly used on the USS Ticonderoga by VF class aircraft. It comprised of a modified A/N series pararaft and an ADS-1 kit housed in a modified Mark-3 bomb container. Some fighters were assigned to fly Combat Air Patrol (CAP) to lend assistance when needed. They acted as first responders, would call in the emergency, and circle as overwatch until rescue arrived. Fighters from the Ticonderoga assigned air cover could be equipped with the rescue assembly, dropping those provisions to the survivors if necessary. Fleet stations exuded ingenuity when it came to locally fabricating equipment for their needs. Facilities were available on carriers and shore-based CASU units that could fabricate almost anything for any purpose. Almost every aviation unit devised some unique modification to emergency equipment. They were either kept inhouse or shared through bulletins or the Engineering Division. Many survival accounts describe locally fabricated survival equipment, but official images of them are difficult to source. (National Archives)

(Right) Several airdrop containers used to aid survivors at sea: a milk can, cylindrical parachute container, and hollowed bomb. The hollowed bomb pictured was developed at the USCG Elizabeth City, N.C. Air Base and dubbed "Provision Bomb". It was designed to be airdropped to survivors from airplanes on Atlantic patrol and contains a variety of rations and medical items. (National Archives)

For the BuAer, the ocean and jungle were the two principle operational regions. Although they did operate in the arctic and the desert, demand for emergency provisions was not as prevalent for the latter. The Naval Air Transport Service operated worldwide, transporting millions of tons of cargo and thousands of passengers. For combat operations, the Asiatic-Pacific was the predominant region; however, carrier- and land-based units did operate in the Atlantic and Arctic Oceans as well as the Mediterranean Sea. Patrol (VPB) and VF squadrons were based in Britain and in Alaska. Arctic survival was part of training.

(Right) Diorama of arctic survival concepts on display at the Survival Museum in Pensacola. (National Archives)

(Below) Fleet Air Wing-4 aircrew members stationed at NAS Kodiak, Alaska. (National Archives)

To supplement aircrews in Alaska, BuAer requisitioned limited quantities of the Type E-12 arctic sustenance kit from the USAAF. This was deemed expedient. The Stevens 22/410 rifle had been replaced by the M1 Carbine; remaining items illustrated are as detailed in original specification for the Type E-12. (National Archives)

INDIVIDUAL EMERGENCY KITS

Part I.

USAAF

By the early 1930s, the USAAC was primarily operating in three regions outside the continental United States: Alaska, Panama and the Philippines, establishing air routes and air bases that performed various functions. This territory was wild or uninhabited. Airplanes were often hampered by mechanical problems or poor weather, and maps were frequently inadequate or inaccurate. There were many accidents resulting in deaths, revealing some deficiencies.

To adapt to the operational conditions, locally improvised emergency equipment was assembled. The equipment was intended for survival of emergency conditions in the event of a forced landing. To standardize the equipment, the Chief of the Air Corps issued a memorandum on 10 April 1933 for the Materiel Division to study the requirements for an emergency kit to be carried in airplanes on cross-country flights. The kits were to include smoke candles, a hunting knife, compass, waterproof match box, first aid kit, and rations. The Chief of the Air Corps suggested the use of the QMC Reserve Ration.

It was decided to develop the food component first. After that was taken care of, the emergency container would be designed to accommodate all the proposed items. Tentative specifications for the emergency ration and drawings for the new emergency kit were released in March 1934. The cross-country emergency kit was titled Case, Pilots Emergency Kit, detailed under Dwg. No. 34G3416. It was intended to sustain an individual for three days. The container consisted of cotton duck fabric with a kapok cushion that replaced the standard seat pad on parachutes. The basis of issue was to be one kit per member of an airplane engaged in extended flights. On 17 April 1934 action was initiated to fabricate 300 kits, which were assembled at the Fairfield Air Depot, Patterson Field, Fairfield, Ohio. These emergency kits were not contracted items but assembled from items on hand at the air depot. After assembly, they were to be issued by the Operations Officer as Class-13 equipment available under stock number 8300-085000. It was further outlined that they be issued on temporary receipt, that is, the kits would be turned in when the flight was complete. The finalized kit included: one Air Corps, Cross Country, Emergency Ration (three ration packages), a three-cell flashlight, a 5" Marbles hunting knife, pocket compass, waterproof match safe Marbles #18, and an M-1910 canteen with dismounted cover. First aid items were not integrated into the unit due to the recent adoption of aeronautic type first aid kit 99518. These would be provided in the aircraft.

The 34G3416 kit was not an original concept. Operational commands had been experimenting with similar concepts typically stored in the airplane, separate from the pilots and aircrew. After the memorandum of April 1933, the idea of a parachute kit did circulate to various commands. With this concept in mind, the commanding officer at Albrook Field, Panama Canal Zone submitted a report on 26 January 1934 requesting that a parachute pack be developed capable of carrying emergency equipment should a forced jump be necessary. It was reported that 80 percent of the terrain over which air units operate was thick jungle.

The question was whether to remain with the airplane and have the emergency equipment in the luggage compartment or jump and be without the equipment. Considering the hazard of crashing in the jungle, a forced jump with the equipment as part of the parachute assembly was determined more beneficial for the pilot and to increase the chance of rescue. Albrook Field suggested the contents include a bolo machete, a .45 cal. pistol with clips, a first aid kit, compass, and match box. Building from the concept of the 34G3416, the Materiel Division designed a unit around the parameters detailed by Albrook Field.

Drawings for the new kit were completed on 8 October 1934 and titled the Container Assembly-Jungle Emergency Landing, referenced under Assembly Dwg. No. 35D1751. The proposed components by the Materiel Division included a 15" bolo machete with detachable handle, a .45 cal. pistol with 20 rounds of ammunition, first aid kit, pocket compass, waterproof match safe with matches Marbles #18 and an emergency Air Corps ration (two ration packages). These components would be significantly revised in the subsequent months after the release of the drawing. The first aid kit was deleted and replaced by an additional match safe with a quarter-ounce vial of iodine for the treatment of small wounds and water purification and 25 tablets of quinine for malarial treatment. The pocket compass was replaced by the Engineer Corps prismatic type with luminous dial. A mosquito headnet was also added. The Air Corps emergency ration was revised to include two rations and for them to be sealed in a tear-strip tin to help with preservation in extreme climate as well as space conservation. By early 1935, it was recommended that the 35D1751 supersede the 34G3416, but this was not deemed practical, so the latter remained in service.

Deliberation on the design between the Materiel Division and Albrook Field continued through 1935, and they continually made fine adjustments to the unit. One of the more important issues was the design of a suitable machete; the final determination was to adopt a modified commercial model manufactured by Collins & Co., referenced as Model #18, detailed under USAAC Dwg. No. 35B1647. With the machete issue concluded, the final design of the kit was agreed upon. Authority for the fabrication of 300 jungle landing kits was granted on 14 August 1935. The complete assembly.

of the 35D1751 jungle kit was divided among three air depots, which were also responsible for their distribution to three additional regional air depots. The Middletown Air Depot would fabricate 125 for the Panama Air Depot, the Rockwell Air Depot would fabricate 125 for the Philippine Air Depot, and the Scott Field Air Depot would fabricate 50 for the San Antonio Air Depot. The San Antonio Air Depot was the staging area for flights departing to the Panama Canal Zone. Assembly at these three designated depots commenced in January 1936 and continued into that summer, but ration deliveries delayed shipments to overseas stations. Albrook Field reported that 33 kits were available as of 13 July and France Field, Panama Canal Zone reported that 62 were available as of 16 July. Depots could requisition the jungle kit under stock number 8300-127000.

The authority to fabricate 300 jungle kits was based on a requirement for the 1936 fiscal year. The 34G3416 was to be part of a projected combined 600 parachute emergency kits available for issuance. It was calculated that each position on the aircraft would be supplied an emergency kit, with the ratio of two kits per attack, corps, army observation and two-place pursuit plane, four kits per bombardment and long-range observation craft, ten kits per transport plane, and one kit per single-place pursuit plane. As part of the fiscal year projection it was calculated the planes in service would include 291 attack, 269 bombardment, 161 corps and army observation, 55 long-range observation, 260 single-place pursuit, 121 two-place pursuit, and 94 transport aircraft.

Distribution of the 35D1751 was the same as that outlined for the 34G3416, to be by the Operations Officer. Technical Order 13-5-8 dated 30 May 1936 states the jungle kits were to be issued to the Philippine and Panama commands and for flights to the Panama Department, issued for required flights over jungle territory. The 34G3416 was to be issued to meet the requirements for equipment necessary for forced landings in other than jungle territory or temperate zones. At the time, these kits were believed to satisfy USAAC requirements for cross-country flights.

USAAC sergeant demonstrating standard seat type parachute with parachute emergency kit attached, Ladd Field, Alaska, c.1930s. Origins of this kit unknown, but photo illustrates emergency kit like that of the 34G3416 in use. During kit development, conversations occurred as to whether to use snaps or zipper for sealing the kit. Unit illustrates slide fastener. (National Archives)

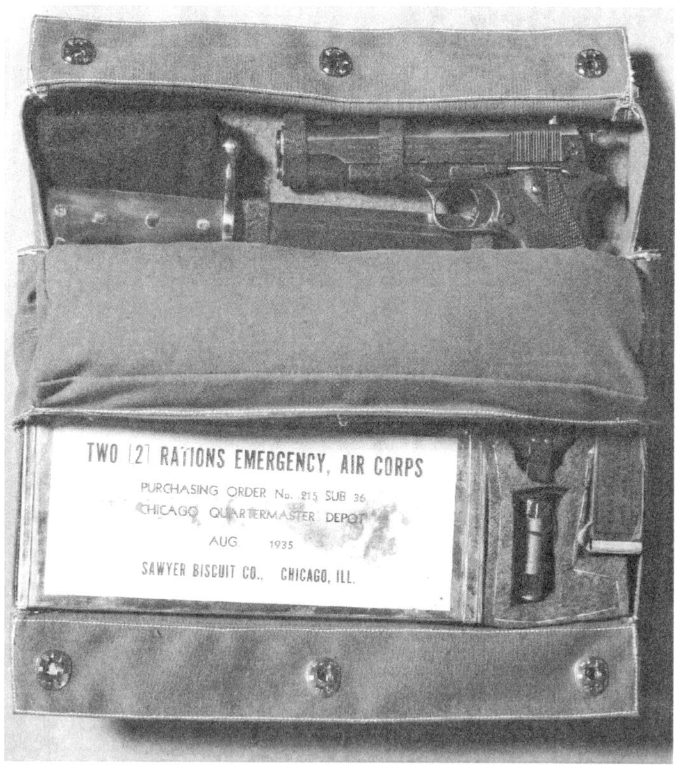

Full assembly of 35D1751 conducted at Middletown Air Depot. Kit was designed to provide essential lifesaving equipment in an emergency jungle landing. Kit was based on the 34G3416 and fabricated using cotton duck fabric with an attached kapok seat cushion. Consisted of two compartments for components. Components were separated in the kit by a felt insert, pockets, and straps. accommodating the following equipment: .45 cal. pistol, 20 .45 cal. cartridges, Air Corps emergency ration (sealed tin of two), two Marbles #18 match safes (one for matches, one for ¼ oz. iodine), engineer prismatic compass, #18 Collins & Co. machete, and mosquito headnet. Tin of 25 quinine tablets was to be included but provision for securing the tin was not made. (USAF)

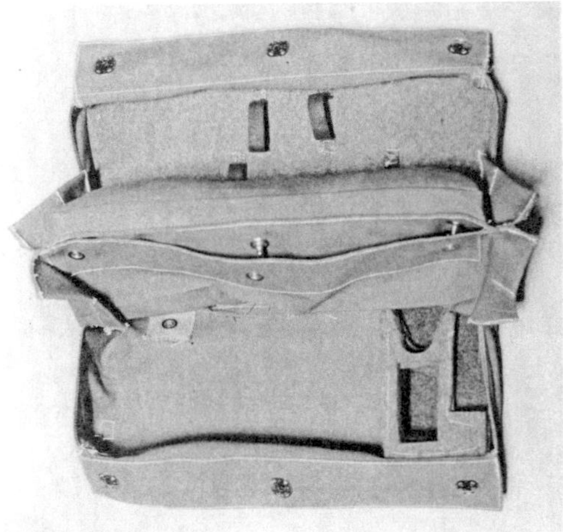

Components removed. Each compartment was accessed seperately. (USAF)

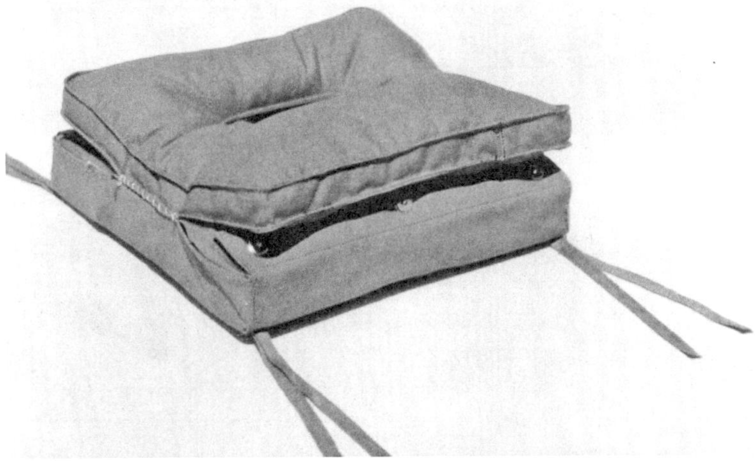

35D1751 with kapok seat cushion. Tie tapes at four corners were provided for attachment to standard seat type parachute; center slot was for parachute harness leg straps. (USAF)

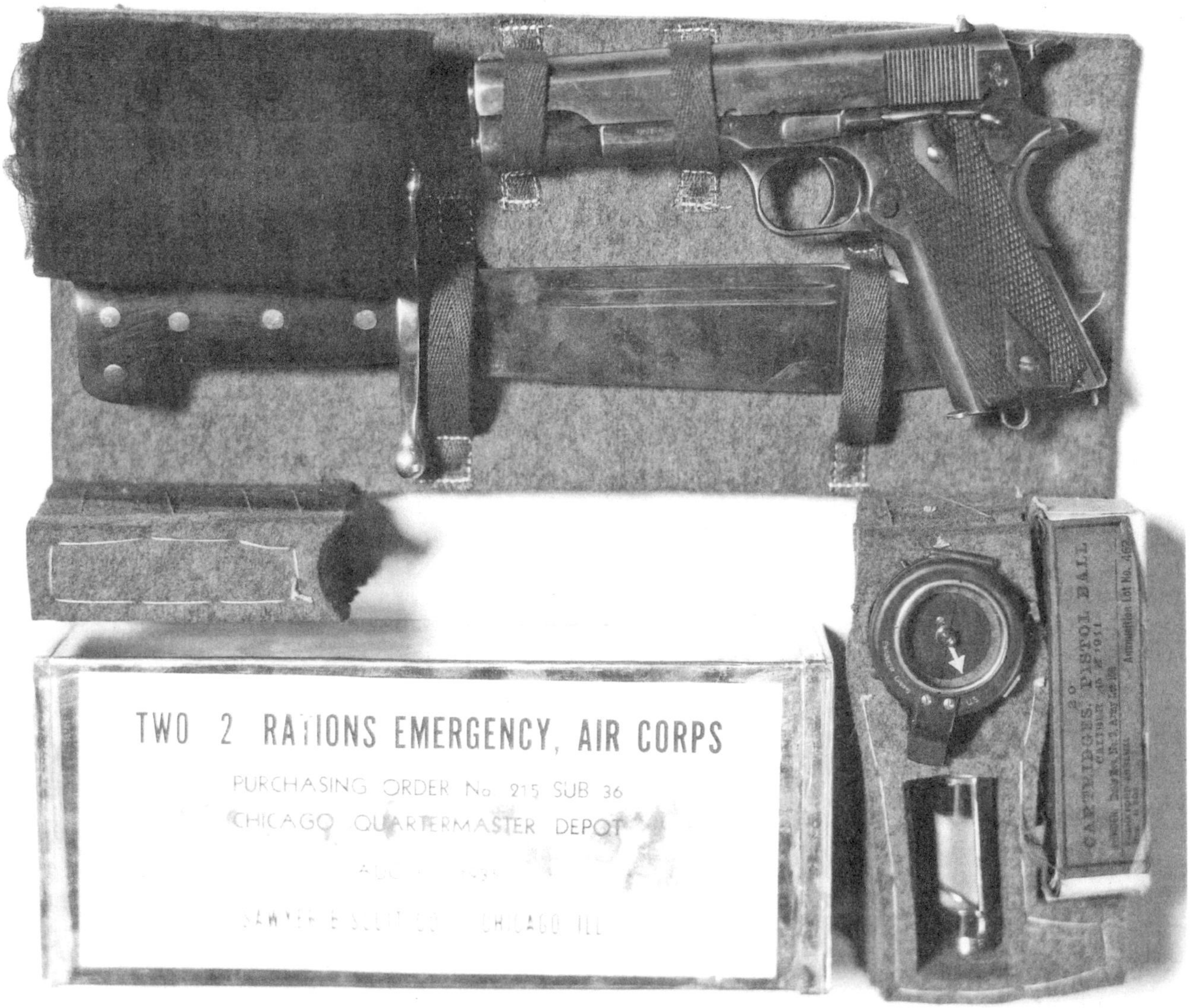

Felt inserts removed with components in their prescribed location. Two emergency rations were packaged in sealed tin. Two types of the #18 machete were used in this kit: those procured by Materiel Division as detailed in Dwg. No. 35B1647, and, when unavailable, air depots assembling the kits made direct procurements from Collins, which were provided with their standard No. 216 handle. The Collins #18 in this kit has the No. 216 handle. (USAF)

As with all development, there was a trial-and-error period, in which items were distinctly different from the conceptual stage, as they were modified for practical application. In July 1936, Albrook Field reported that the limited quantities of the jungle kit had undergone testing. It was found that fewer than half of the pilots could use the kit with any degree of comfort in P-12 pursuit airplanes. The report stated any pilot 5' 10" or more in height, when using the kit, stuck out of the cockpit above the windshield, hampering flying efficiency. Also, pilots with large thighs could not fasten the safety belt when using the kit; with the added bulk made it difficult to enter and exit the cockpit. It was determined by that command that the jungle kit would be worn at the discretion of the individual. France Field reported much the same deficiencies, adding that the jungle kits were completely unsatisfactory as they made it difficult to get out of the front and rear gunners' cockpitsof the B-6 and B-10B bombardment airplanes.

When the Materiel Division evaluated the concerns, they found the bulk was created by the emergency rations. A condensed ration was deemed necessary and the pistol would be worn by the individual. The Materiel Division initiated a study almost immediately after the reports from Albrook and France Fields in August of 1936 to investigate further technical development of an emergency jungle kit. It was believed at this time that there could be a more suitable arrangement for carrying the kit, incorporating the parachute back pad in lieu of the seat cushion. The initial steps were the development of a more suitable ration, and, since problems that may arise differ by territory, only the minimum essential items would be universally included for the territories of Panama and the Philippines. Pilots and aircrew would be authorized to wear the pistol and canteen on their person, principally to supplement the emergency kit. Samples of a revised back type jungle kit with the current emergency ration were fabricated and sent to France and Albrook fields in Panama and Nichols Field in the Philippines for comment and evaluation. By the fall of 1936, it was concluded that the present seat type kit was wholly unsatisfactory: with all stations stating the preference for a kit in the form of a back pad. With exchanges of concepts between the Materiel Division and the commands in Panama and the Philippines, the preliminary concept of a new jungle emergency kit was established. Further development would be delayed pending receipt of a new type of streamlined emergency ration for integration.

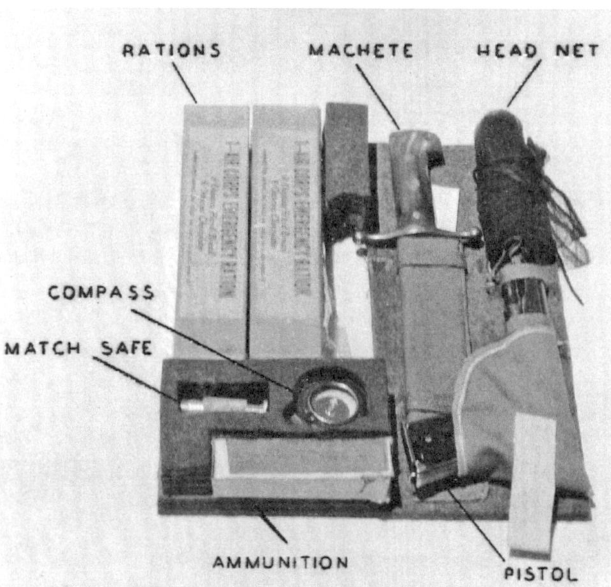

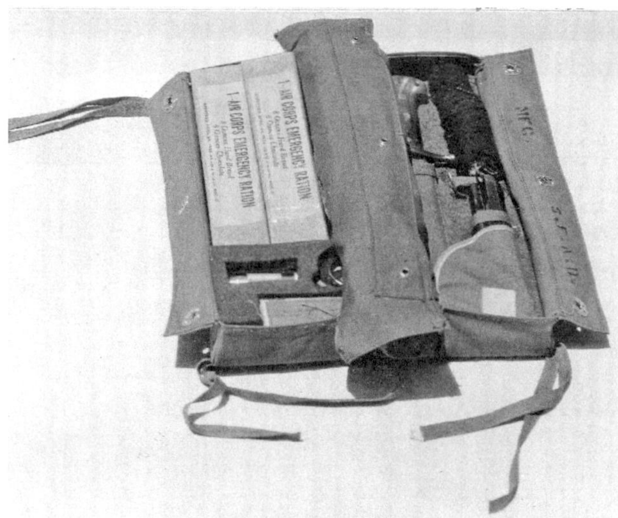

(Above and above right) 35D1751 jungle kit fabricated at Scott Field Air Depot. Rations had been removed from the sealed tin and machete was provided with a canvas scabbard. (National Archives)

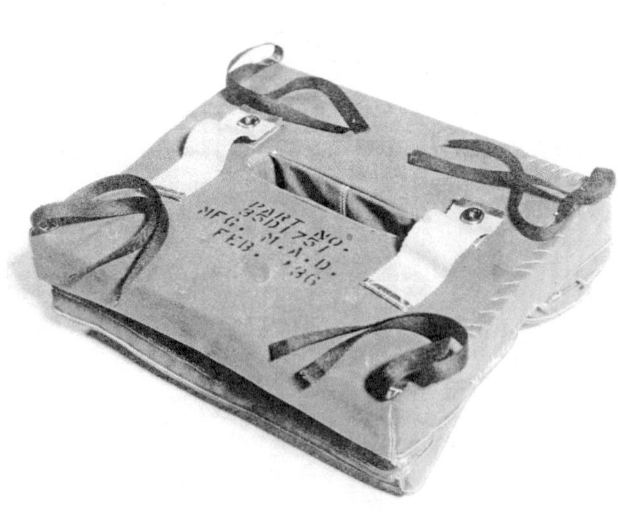

(Right) Underside of fully assembled kit. Only 300 kits were authorized for assembly; however, drawings were available upon request, making it possible more were assembled by authority of local commands. (USAF)

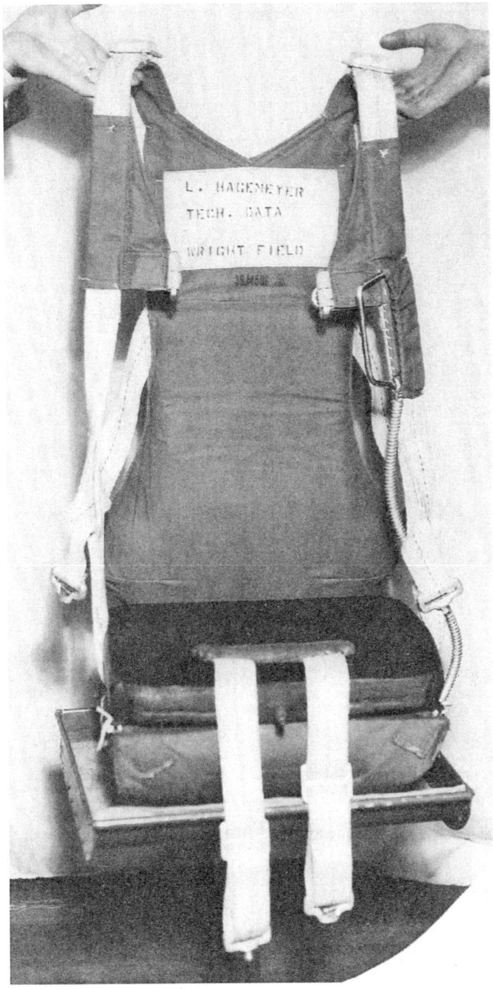

(Above) Experimental back type jungle emergency kit assembled by Materiel Division in 1936 for evaluating reconfigured concept. The standard seat type parachute assembly for referencing purposes; includes a pneumatic seat cushion and sponge rubber back pad, either the seat or back type emergency kits were required to integrate into this assembly. (USAF)

(Below) Mosquito headnet with open drawstring top provided in jungle seat and back type kits were standard QMC type worn over headgear. When front and back were secured to waist belt, netting was away from the face and neck. Two volunteers wearing denim fatigue caps demonstrating proper use of headnet. (National Archives)

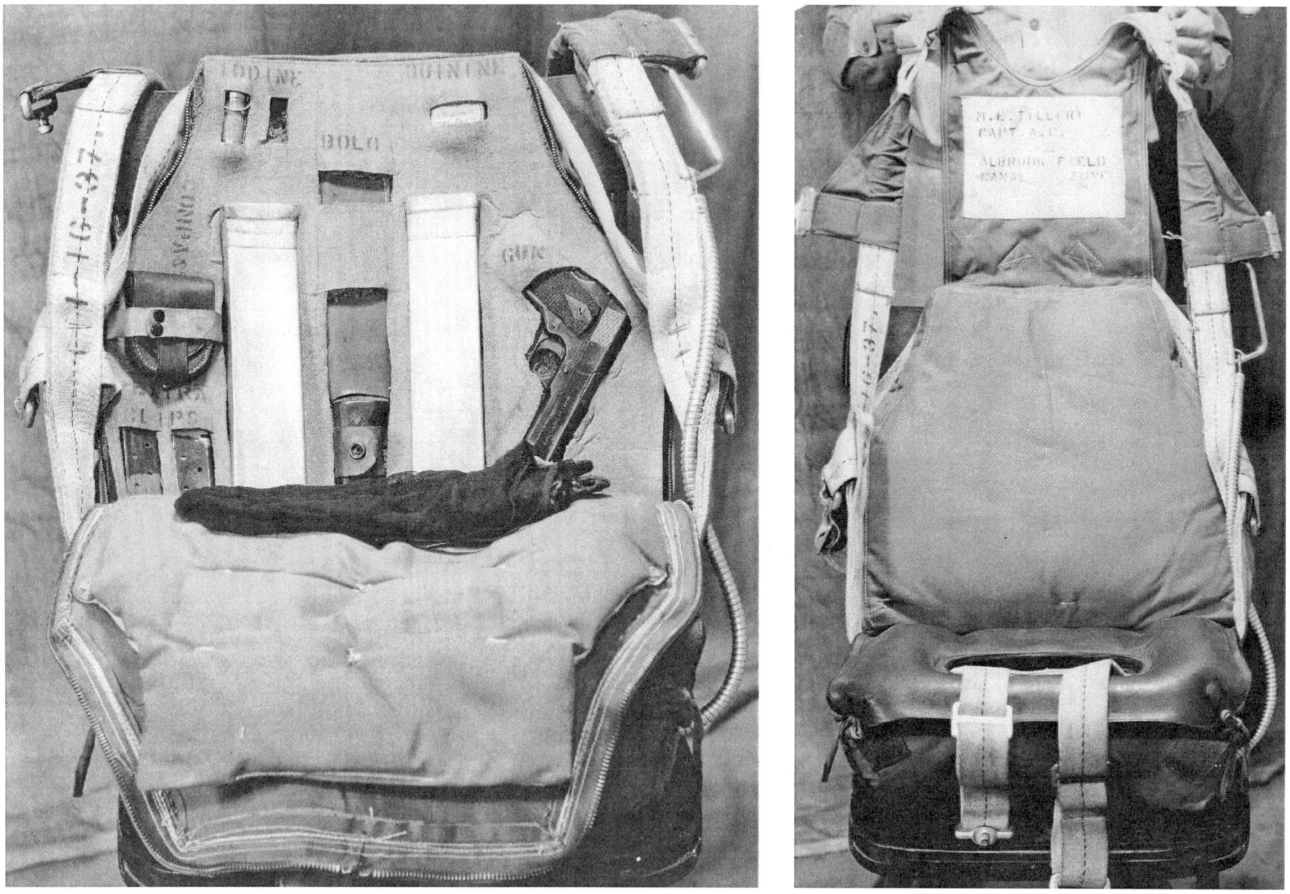

Back type emergency jungle kit fabricated at Albrook Field and engineered from pattern concepts provided by Materiel Division. This unit was field-tested by squadrons operating in Panama through Spring 1937. Components were slightly modified from those in the 35D1751: prismatic compass retains its leather case, the kit now includes two emergency Air Corps rations (seperated sealed tins) and a locally fabricated machete with leather scabbard (a modified standard stock Engineer bolo type). Two prescribed match safes are at upper left, and to the upper right is a round tin for quinine tablets. Configuration preferred over the seat type. (National Archives)

To continue development, the existing emergency ration would need to be amended. The first step was the decision to eliminate the hard bread and to provide only a basic component of sweet chocolate. Working in conjunction with the QMC, which was responsible for ration procurement, it was determined to provide the chocolate in the form of cakes. The Materiel Division developed the appropriate tin container. This development was in its final stages by the spring of 1937 with the complete ration referenced under USAAC Assembly Dwg. No. 38D2666. An immediate expedited procurement was authorized for field testing. With the availability of the service test samples of the new ration, the Materiel Division assembled a limited quantity of experimental jungle back pad kits in two different component arrangements for service testing. Samples were then shipped to Albrook and France Fields in November of 1937 with a suggested 30-day trial period. Albrook Field was the first to submit its final report in January of 1938, stating it had many advantages over the seat type kit and recommended it for standardization. The report additionally stated that it was especially suitable to the type of aircraft now in that command and that which was being contemplated for future delivery. France Field submitted its report in March with comments echoing those of Albrook.

Components of the 35D1751 were transferred to the book type back pad kit, with additions of a fishing kit and new 38D2666 emergency chocolate ration. The complete assembly was determined to contain essentials needed in jungle territory, supplemented by a pistol and a canteen worn by an individual, and aeronautic first aid kit. Medical items of the jungle kit facilitate immediate treatment of injuries (iodine antiseptic ointment for minor wounds, quinine for malaria, and water purification for protection from waterborne diseases). Additional medical items were available in the aeronautic first aid kit 99518. The jungle kit was designed to replace the standard kidney shaped parachute back pad and to assure efficient integration with the parachute harness. It was fabricated in the same dimensions. The top flap incorporated a sponge rubber cushion, with the components secured by a fricyion fit by cutouts in a felt insert.

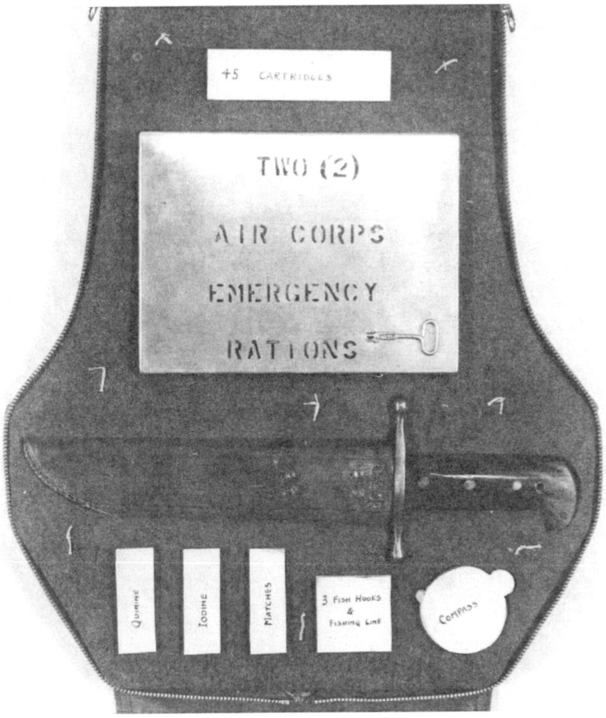

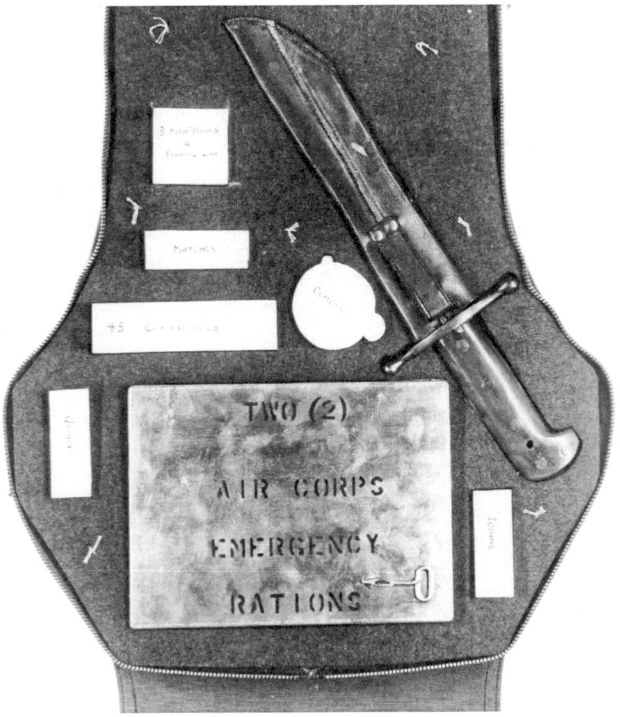

Sample back pad kits submitted for field testing in November of 1937 in two component arrangements. These examples include the machete, Dwg. No. 35B1647. (National Archives)

Based on favorable service test results and recommendations, the Engineering Division drafted drawings for a new jungle emergency back pad kit, assigning the title Container Assembly-Jungle Emergency Back Pad, Dwg. No. 38J5271. The kits were slated to be contracted items from Irving Air Chute Co. and The Switlik Parachute & Equipment Co. Upon delivery, the components were to be assembled at the Middletown Air Depot. Kits were available for requisition under stock number 8300-128000. Components included: machete (Dwg. No. 35B1647), three Marbles #18 match safes 35A5510 for matches, ¼ oz. vial of iodine, and 25 quinine tablets, 20 .45 cal. cartridges, engineer prismatic compass, emergency ration 38D2666, mosquito headnet, and fishing kit (30 ft. fishing line, 3 flies, 6 hooks). Standardization date not known but was available for issue in November 1939, detailed in Air Corps Circular Letter 65-115 and Technical Order 00-30-51. The Circular specified the kit was not authorized for issue in the continental United States and Alaska, but for jungle territory only. It superseded jungle landing kit 35D1751.

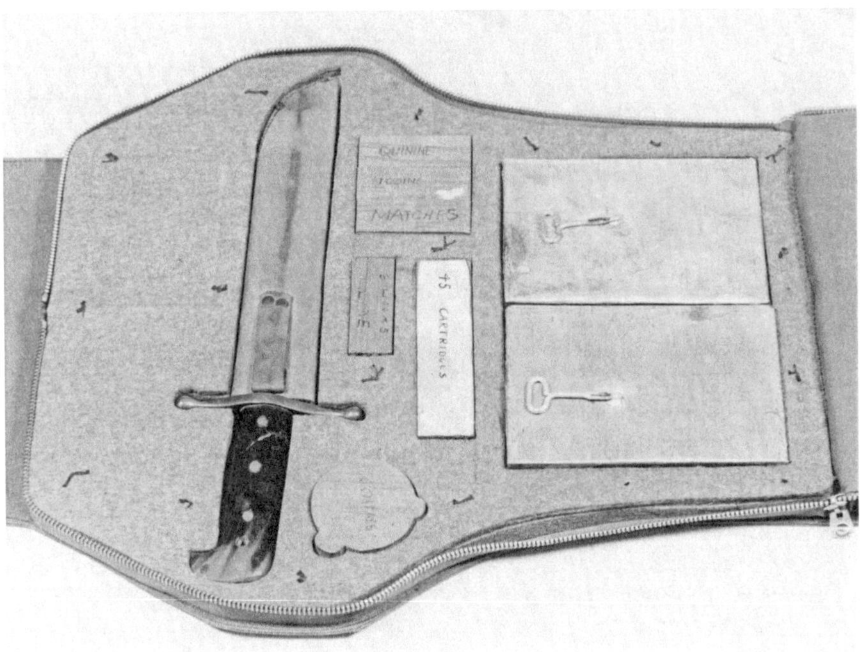

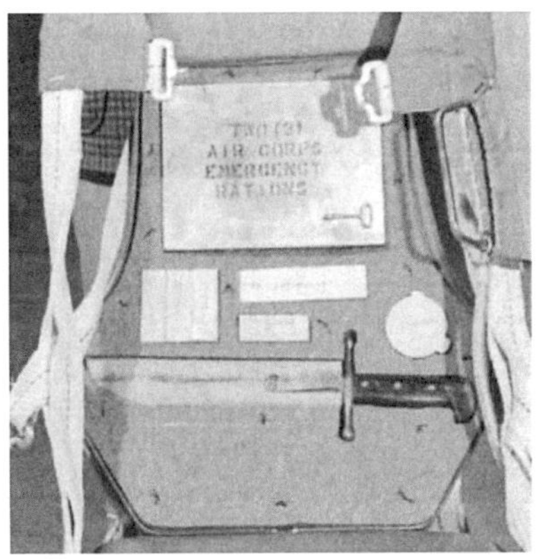

(Above) Personal Equipment Laboratory sample of 38J5271. Note: wooden blocks in four of the inserts. (USAF)

(Left) Arrangment of back pad kit 38J5271. Cutout made to accomodate machete, Dwg. No. 35B1647. Rations are a concept type. (National Archives)

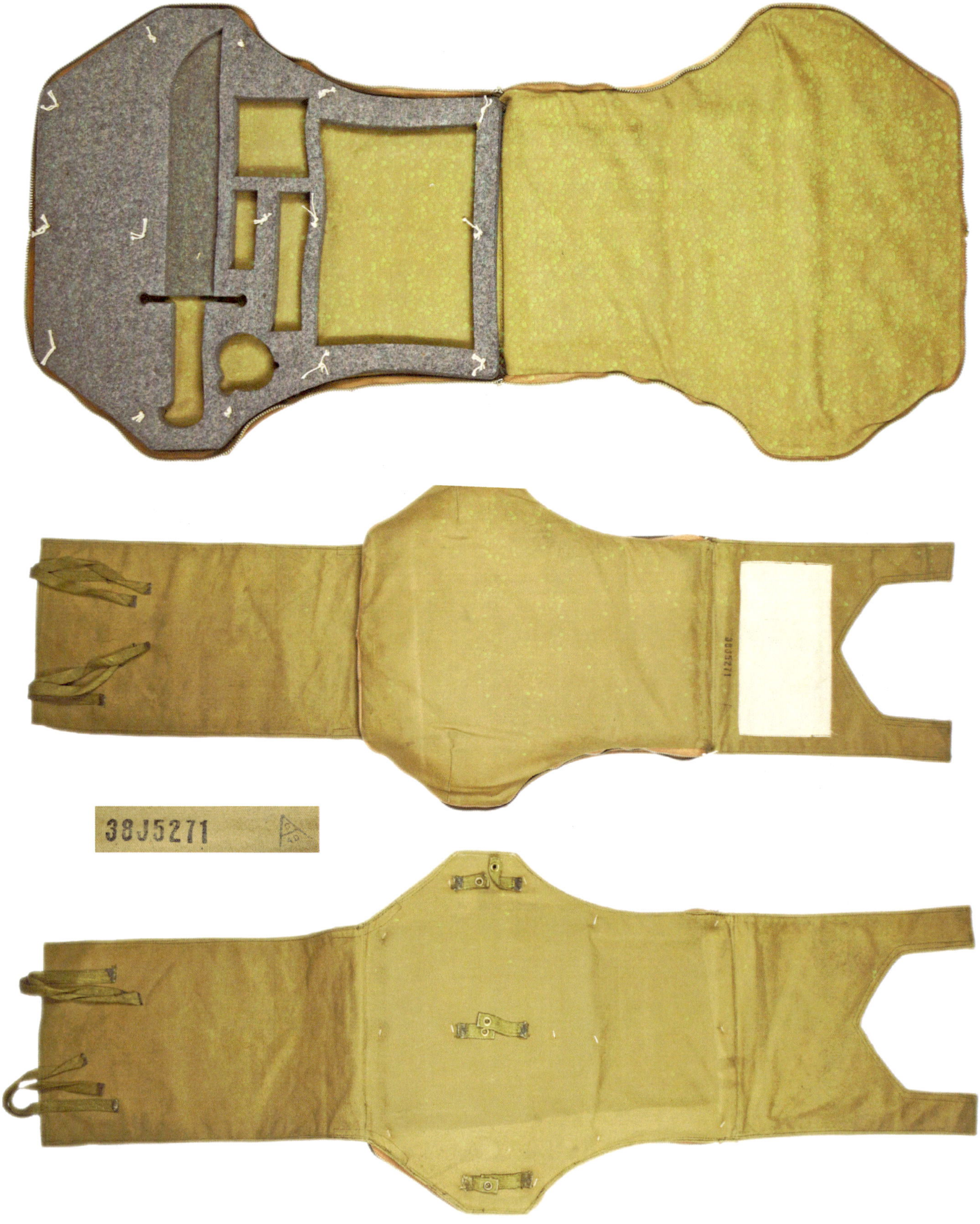

Emergency jungle back pad kit 38J5271. Kit attached to parachute harness like standard back-pad; includes sponge rubber back cushion and felt insert with cutouts for components. Components were placed toward top of kit, allowing more comfort at the small of the back; back cushion was thicker at lower half. Early production kits were stamped only with drawing number and inspector's acceptance stamp. Production quantities are unknown; one confirmed contract, 535-AC-15894, to an unknown contractor.

(Left and far left) Developed for inclusion in the jungle kit was a tin container with dimensions equal to those of match safe referenced as Box, Emergency Container Fish Tackle under Dwg. No. 41A9711. The tin measures 3" x 1" x 3/4", stock number 8300-039930. Fishing tackle tin containing 30 ft. fishing line, three assorted flies, two Size 2 hooks, two hooks Size 4, and two hooks Size 6. Tin had no identifiable markings.

(Right) Waterproof match safe included in both the 35D1751 and 38J5271 emergency kits as required by Air Corps. For procurement, the Marbles #18 match safe was adopted, referenced under USAAC Part No. 35A5510, titled Case-Waterproof Match, available under stock number 8300-088000. Match safes contained matches, vial of iodine and quinine.

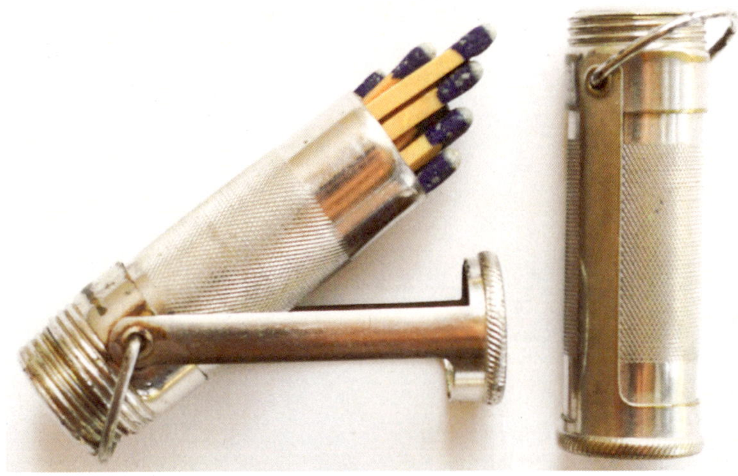

By spring 1941, Middletown Air Depot reported supplies of 35B1647 machetes were exhausted and requested 1900 more. However, a new jungle emergency kit was in development. As an interim, Technical Order 00-30-51 was revised on 3 November 1941, detailing the use of a 5" hunting knife for temporary replacement of depleted stock. With outbreak of war and combat operations in the Asia-Pacific, emergency and temporary procurements of the 38J5271 were authorized in December 1941. Purchase Order 42-7616 was awarded to the Cable Raincoat Co. for their manufacture, quantities unknown. Lacking suitable machetes for inclusion, procurements were awarded for commercial models. Jungle kits, 38J5271, purchased under the emergency order were fabricated according to all drawing details except that the machete cutout was altered to accommodate the #18 Collins & Co. machete with No. 216 handle.

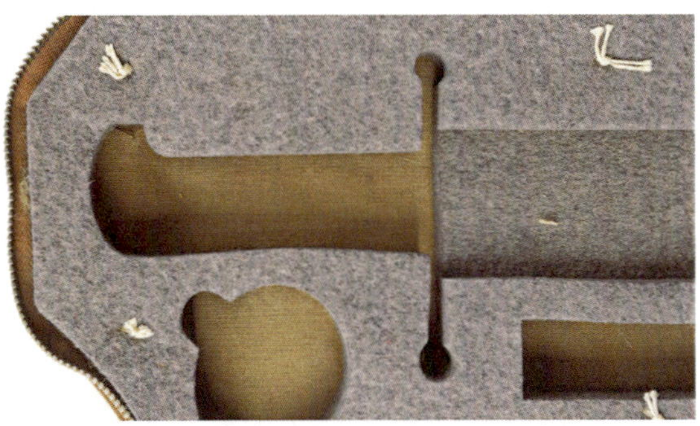

Comparative view between the cutouts for the 9" machete. 35B1647 handle was rounded (top). Emergency procurement was cut for standard No. 216 handle (bottom).

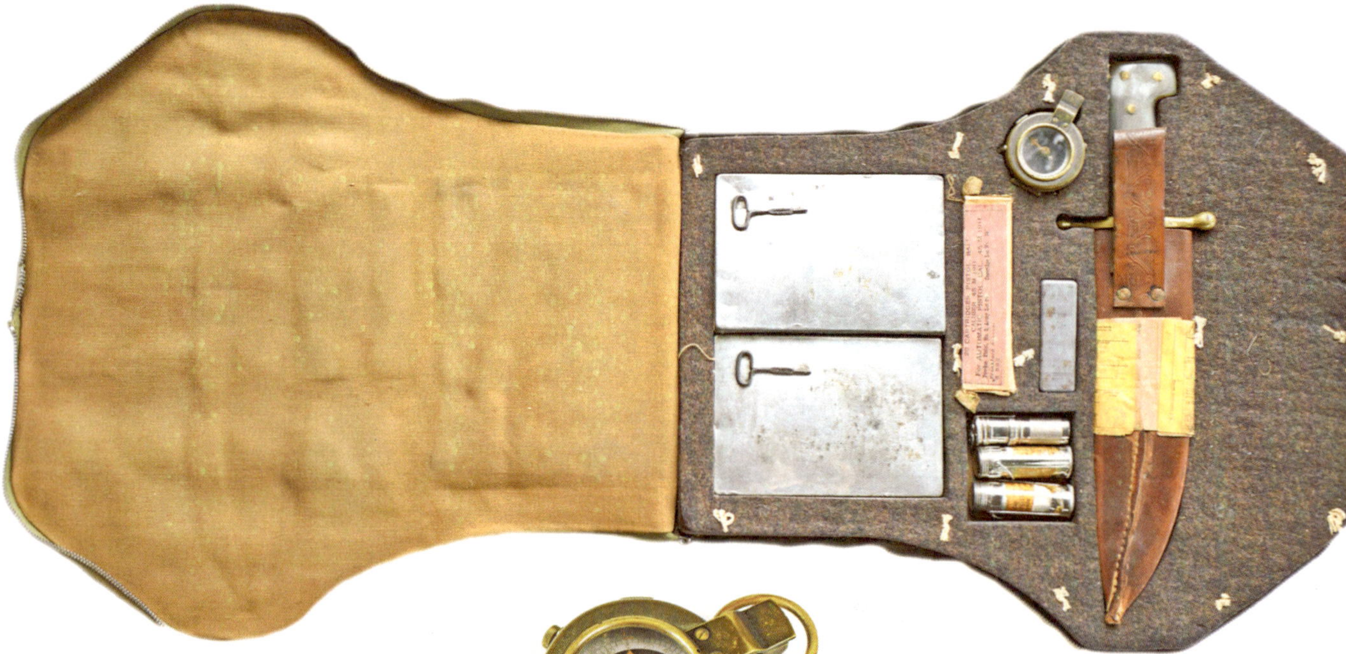

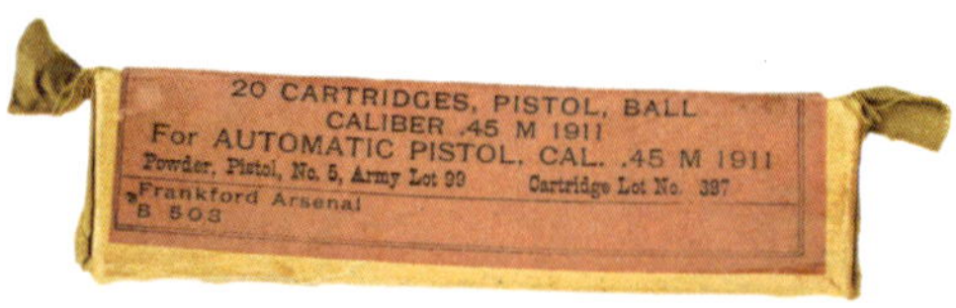

Jungle kit 38J5271 purchased under the emergency order. Unit demonstrates authorized components in their corresponding cutouts, excluding mosquito headnet. Headnet was either a commercial or QMC type and placed on top of components. Initial production examples used sponge rubber padding, transitioning to horsehair. Examples can be found with hand-stamped drawing number, or contract data, or an actual contract data tag sewn to the outer container. Also shown is the Collins #18 machete, .45 cal. cartridges and engineer prismatic compass (as prescibed by T.O.), Two Air Corps Emergency Rations 41D4293 and three match safes of quinine, tincture of iodine and matches.

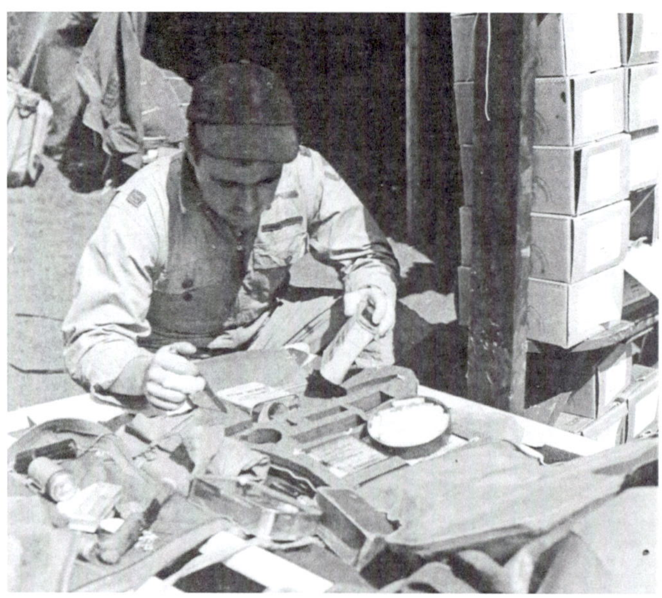

Emergency jungle kit 38J5271 was used in Asia-Pacific and remained in service for the duration of the war. Assistant Operations Officer of the 47th FS, Capt. Roy Obenshain, checks his emergency flying equipment, Iwo Jima, March 1945. He modified felt insert of a 38J5271 to accommodate two AN-W-5A drinking water cans. With the introduction of Type B-2 jungle kit and revised Technical Order 00-30-51, the 38J5271 was updated to comply to new components. Air units would modify emergency parachute kits to suit their preferences. In and around the kit is the E-3B aids kit, a B-4 first aid insert (under it is signaling mirror 40653 in box), 17L16 life preserver light, AN-C-101 match safe compass, fusee flare 43B3554, Type-5 fishing kit, and parachute ration. (National Archives)

In October 1937, Headquarters 1st Wing, GHQ Air Force at March Field, Riverside, California stated that the bombardment group commanders expressed interest in the development of a standard emergency safety kit for cross-country flights. Pilot's emergency kit 34G3416 and the jungle kit 35D1751 were found unsatisfactory for use in most crew positions aboard bombardment airplanes. The general specifications outlined that the container be of a lightweight material, air- and watertight, and buoyant. The container should also provide rings for towing and attachment. The proposed components included emergency rations, first aid packets, waterproof matches, a hunting knife, fishhooks and line, a flashlight, and a pocket compass. Water would be carried in canteens available to each crew position. The proposal was oriented toward flights over water to support the individual and supplement the life raft.

In response to this and other commands' requests for a similar unit, the Materiel Division expedited development incorporating revised components of the 34G3416 emergency kit to a new type of container. This new kit was detailed under Dwg. No. 39G2156 titled as Case-Airplane Emergency. This unit would include two of the new emergency rations 38D2666 to supplement one man for two days, intended for use in bombardment aircraft.

The 39G2156 underwent minor revisions further defining its operational use. Adopted as standard on 28 May 1940 was the Kit-Flyers Emergency Sustenance Type F-1, detailed under Specification No. 94-40314. It would be available as a complete assembly under stock number 8300-559275. This kit was comprised of the Case-Flyers Emergency Sustenance Type F-1 under Dwg. No. 40G7404, stock number 8300-081500 and the Container Assembly-Flyers Emergency Sustenance (aluminum tin) under Dwg. No. 40D7626, stock number 8300-126930. The Type F-1 was to be used in the event of forced landings in mountainous terrain, desert, or on water. Air Corps Circular Letter 65-116, addressing the Type F-1, states that it had been standardized for use within the continental limits of the United States and Alaska. It officially superseded pilot's emergency kit 34G3416. The established basis of issue was to be one kit per aircrew member of aircraft engaged in cross-country flights, distributed under the same procedure as detailed for the 34G3416 and 35D1751. Jungle kit 38J5271 was issued for flights over jungle terrain. The Type F-1 could be requisitioned as a complete unit, sealed and ready for issue, or by individual pieces and assembled by the requisitioning organization. Technical Order 00-30-49 was prepared for those commands assembling the units. Detailed for inclusion was: one case 40G7404, one container 40D7626, two emergency rations 38D2666, flashlight TL-122A, a 5" hunting knife, a pocket compass, a match safe with matches 35A5510, a fishing kit (15 yards of fishing line with two hooks), an ounce-and-a-half tube of tannic acid burn ointment, a Carlisle first aid packet, one box of iodine (six swabs), and two 2" by 6 yard gauze bandages. A revised version of the Type F-1 could also be requisitioned for cross-country flights that included a mosquito headnet under stock number 8300-559300.

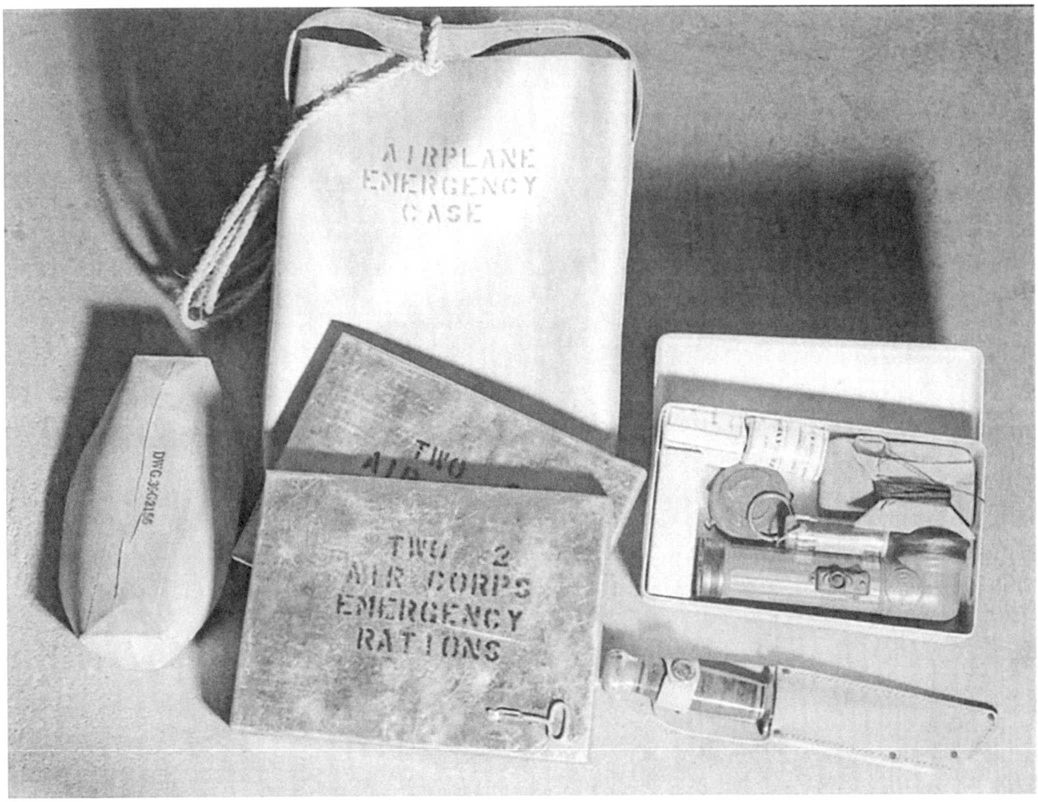

Emergency airplane kit 39G2156 included buoyant waterproof container and two emergency rations 38D2666; remaining components were stored in aluminum container (except knife). One kit was provided to each crew position on bombardment aircraft. In an emergency, a rope was provided, serving two functions: the kit could be tied to parachute harness, and when in the water it would either float with the individual or be lashed to life raft. (USAF)

Type F-1 kit with components stowed in aluminum container. Difference between this unit and 39G2156 is that hunting knife was stored in aluminum container and not separately in case. Type F-1 was placed on inactive status in 1942 and superseded by parachute emergency kits Types B-1 and B-2, then cross-country sustenance kits Types E-1 and E-2. Procurement data unknown. (National Archives)

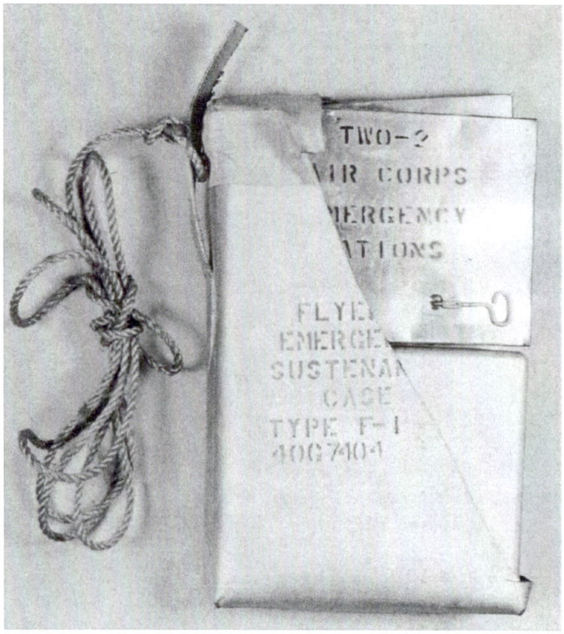

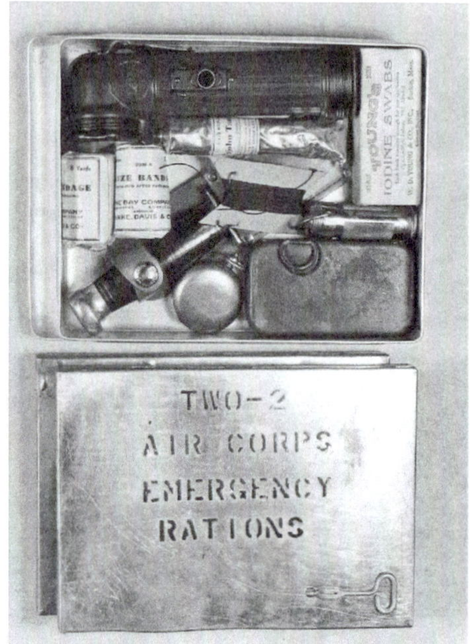

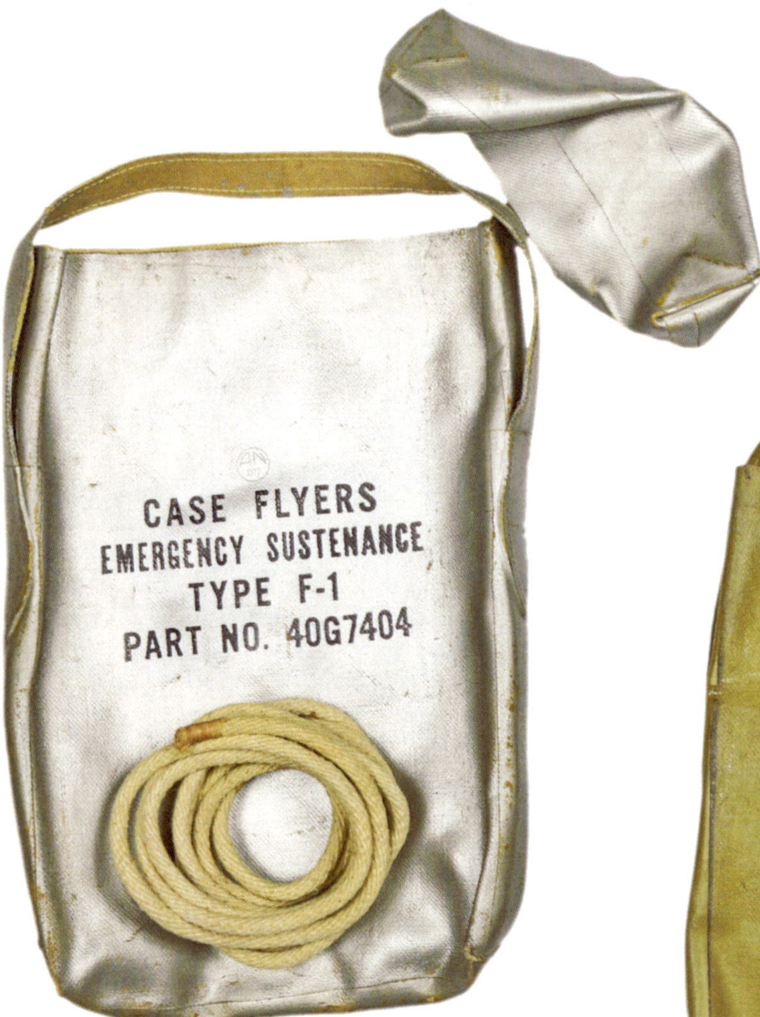

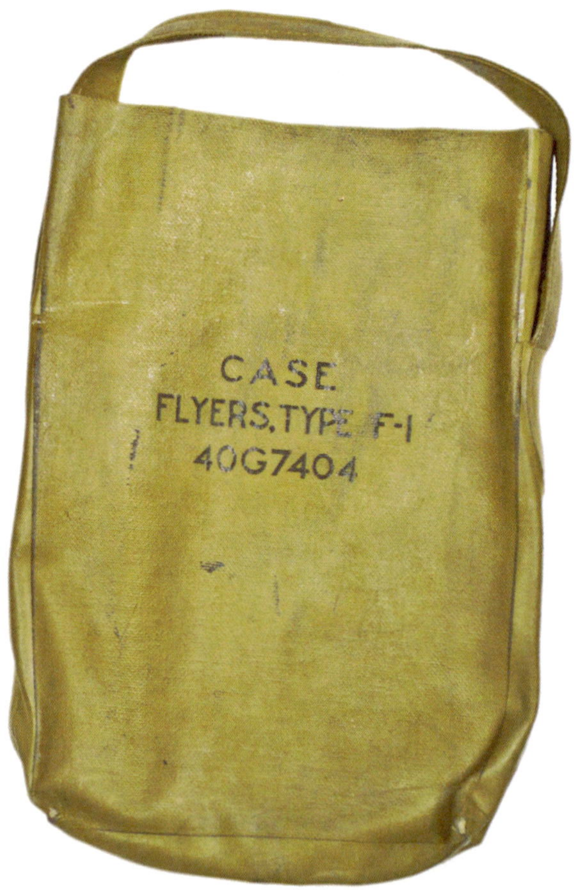

Type F-1 consisted of a waterproof doped canvas case 9" x 15" x 2" and aluminum container 9" x 7" x 2". When assembled it was sealed by tape 2" wide. Rope (9') was provided for attachment to the individual or lashing to life raft. Typical color was silver, but OD green variants can also be found.

Type F-1 aluminum container, Dwg. No. 40D7626, 9" x 7" x 2" with detail of the Army-Navy inspection stamp.

The development of seat-type parachute emergency kits was intended to satisfy the requirements on cross-country flights operating over isolated territories. They were quickly found to be nonoptimal, in part considering the types of airplanes in service. For jungle use, a book type back pad kit was developed, and, to accommodate space restrictions in bombardment airplanes, a more compact unit was developed. Upon issuance of Circulars 65-115 and 65-116, drafted in November 1939, there was a defined distinction as to which flying units were to be issued when operating over isolated territories. The new 38J5271 jungle kit was determined to be satisfactory, sufficiently incorporating into the new closed cockpit airplanes. Units in Panama and the Philippines were operating with the understanding of providing emergency equipment on an individual basis for forced jumps or landings, but felt improvements could be made. The Type F-1 was designed to overcome the deficiencies of the seat-type kit, designed around the direct parameters suggested by some bombardment groups. However, this unit was considered inadequate, prompting further suggestions for a more suitable sustenance kit to provide for multiple individuals. Units destined for Alaskan territory and operating over swampy terrain in the United States were requesting the issuance of the jungle type back pad kit but were restricted by authority as per Circular 65-116. It was the unanimous opinion by these units that this country was comparable to that of the jungle, with the 38J5271 desirable in both its contents and container.

In March 1940, the Chief of the Materiel Division evaluated the situation, concluding that an emergency kit attached to the parachute harness by a rope could prove hazardous, and discouraged the practice. In review, the back-type kit was a form-fitting unit having no discernible changes in the external dimensions over the standard back cushion. This configuration also added no obstructions to the proper functioning of the parachute. From these findings, special authority was granted to supply the 38J5271 for units preparing for embarkation to the Alaskan territory. Limited supplies were allocated since the bulk of supply was reserved for flights to Panama and shipment to the Philippines.

Work then began for the design of a new type of back pad kit for arctic use. The Materiel Division requested that Ladd Field recommend the most essential items required in Alaska and other cold climates. A back pad kit like the standard emergency jungle kit was developed for operating in the Alaskan territory as a supplemental emergency supply kit. Samples were assembled with items detailed by Ladd Field; it was submitted for field testing in January of 1941. At a conference held at the Materiel Division in June of 1941, the proposed kit was found adequate and meeting service requirements. Further service tests were waived to expedite a quantity procurement for use by flying personnel in Alaska during the 1941-1942 winter.

Adopted as standard on 7 August 1941 was the Kit-Alaskan Emergency Parachute Back Pad, Type B-1 under Specification No. 94-40365, detailed under Assembly Dwg. No. 42J2465. The kit was designed to provide navigational equipment, protective clothing, food, cooking utensils, and other supplies for emergency use. The basis of issue was one kit per flyer in Alaska or other cold climates, and it was only to be issued to organizations stationed in arctic regions. The Type B-1 could be requisitioned completed under stock number 8300-558600 or for the container only under stock number 8300-126725. For local assembly, Technical Order 00-30-86 was released on 27 November 1941. As per specification and technical order, Type B-1 was to include a pocket compass, one pair of gloves Type D-2, two match safes 35A5510 (one each for a quarter-ounce vial of iodine and matches), two emergency rations 41D4293 (improved over type 38D2666), a two-ounce bottle of insect repellent (Sta-Way commercial brand), one ounce of camphor gum paste, 12 bouillon cubes (Steero brand), a mosquito headnet with face frame, one hunting knife, a fishing kit (with 10 yards of fishing line, three fish flies, and six hooks), a cooking container 42B2059, and 20 .45 cal. cartridges.

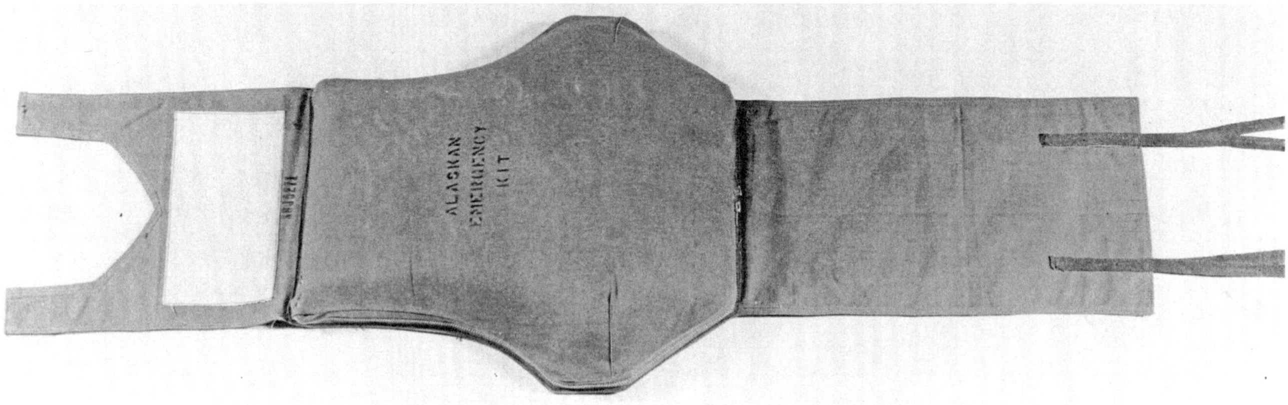

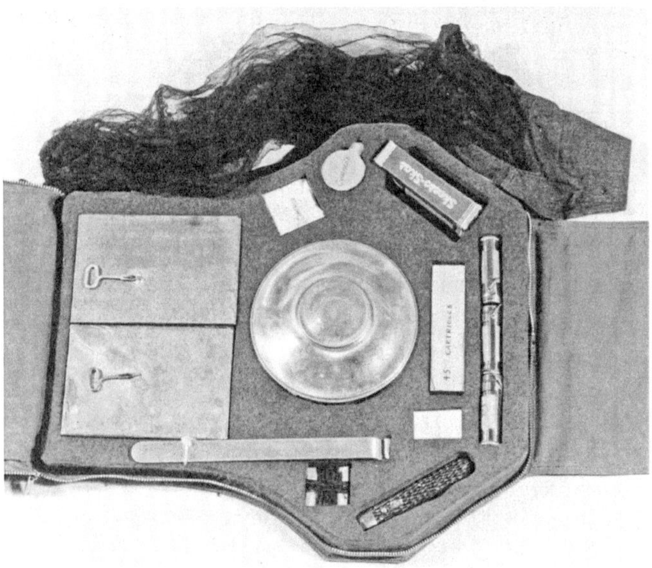

(Above) Outer container is identical to Jungle type 38J5271; only felt insert differs. (USAF)

(Left) Experimental Alaskan parachute emergency kit. Unit is configuration sent to Ladd Field for field testing in January 1941. (USAF)

(Right) Experimental Alaskan reconfiguration after standardization, integrating features from proposed jungle emergency kit and some components from Type B-1 such as knife, bouillon cubes, frying pan, insect repellent, camphor, fishing line, compass, and match safes. (USAF)

Type B-1 kit, when closed, measures 17" x 18-1/2", kidney-shaped, externally identical to 38J5271. Includes felt insert with cutouts to accommodate components, and horsehair-filled back-pad; sponge rubber was eliminated for WPB conservation purposes. Quantity of 15,000 was procured under three purchase orders: 42-6468, 42-14537, and 43-9344 awarded to the Aircraft Appliance Corp.

(Right) Detailed by drawing 42J2465, the components were: 1) emergency rations, 2) fish flies and hooks, 3) bouillon cubes, 4) cooking pan handle, 5) cooking pan, 6) cartridges, 7) pocket compass, 8) camphor gum, 9) fishing line, 10) hunting knife, 11) insect repellent 12) match safes. Gloves and headnet were stored on top.

Frying pan developed for Type B-1 was referenced as Pan Assembly-Alaskan Emergency Kit, Cooking under Assembly Dwg. 42B2059 and consisted of three pieces: handle Dwg. No. 42B2060, pan Dwg. No. 42D2061, and lid/plate Dwg. No. 42D2062. Each was stamped with the corresponding drawing number.

Bracket on side of frying pan for attachment of handle.

(Left) Frying pan assembly measures 6" diameter x 1½" deep, with a 9¾" handle and was manufactured of blued steel.

(Below) The knife specified to be included was a switchblade manufactured by Shapleigh Hardware Co., referenced under part number Model No. A-10. The knife measures 5" when closed, and 8¼" with blade extended. The knife is provided with a guard and imitation stag plastic grips. In Dwg. 42J2465 this knife was referenced as a "hunting knife," which led to erroneous designations. "Hunting knife" was misinterpreted and applied to the 5" fixed-blade type. Supply catalog details the Type B-1 to include a 5" hunting knife but that is an error. After failures, switchblade was replaced by the standard 3-blade Scout knife, stock number 7900-437900 in revised Technical Order 00-30-86 of 14 September 1942.

This kKit includes emergency ration 42B10959, plastic match safe 42A6930, and fishing kit packaged in manila envelope. Tin container of 10 bouillon cubes produced by Oxo Limited of London, England. (National Archives)

The Type B-1 was requisitioned empty or complete, so components varied. Kit includes headnet 43G3868, Type D-2 mechanics gloves, fishing kit 41A9711, match safe 35A5510, emergency ration 41D4293, cooking pan, pocket compass, and insect repellent. (National Archives)

Remaining components for the Type B-1. 12 bouillon cubes were contained in a commercial tin as shown. 20 rounds of .45 cal. cartridges were from standard ordnance stock, and Type D-2 gloves from USAAF stock. The fishing line was spooled on a cardboard sheet and secured in place by friction. Packaging of hooks and fishing flies was not detailed. Commercial headnet with face frame and commercial 2 oz. bottle of Sta-Way insect repellent were replaced during production with headnet 43G3868 and another brand of repellent. Not shown is camphor gum, which was provided as a paste. Because of its alcohol content, it was used as a fire starter but also for medicinal purposes such as lip balm. A pocket compass manufactured by the Keuffel and Esser Co. was included, cited by specification.

(Above)) Match safe approved for inclusion in Type B-1 was similar to the Marbles #18 but was easier to open. It was manufactured by Shanklin Mfg. Co., a subsidiary of Park Sherman Co., known as the Everdry. It was also referenced under Part No. 35A5510.

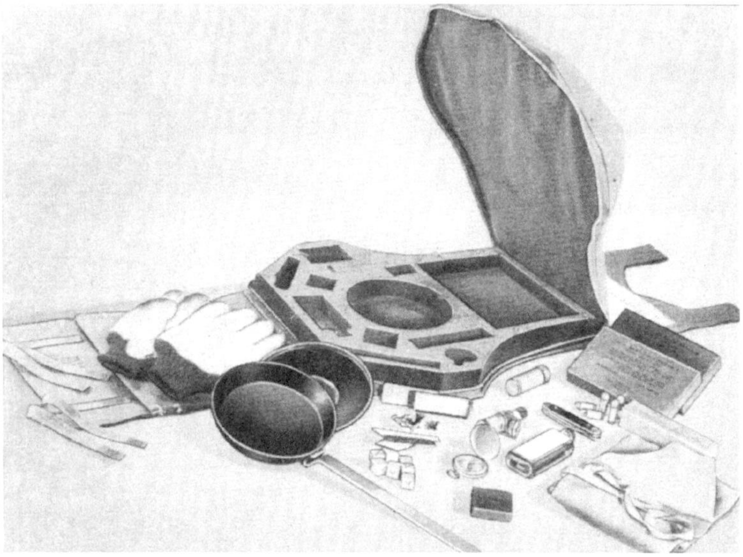

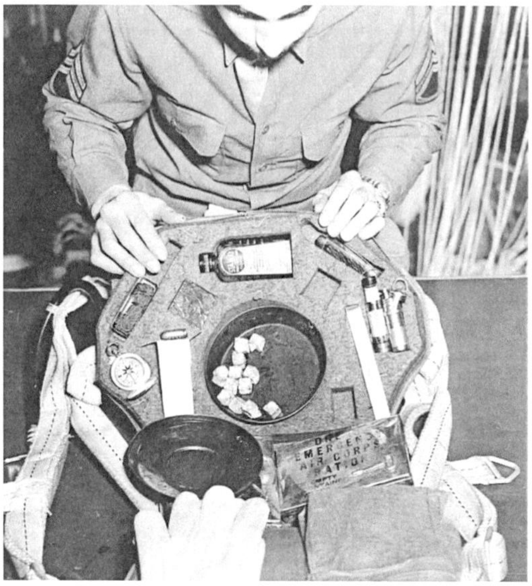

(Left) USAAF catalog Illustration of the B-1 showing its complete complement of survival items.

(Right) A USAAF parachute rigger examines the Type B-1, updated per Technical Order 00-30-86 with headnet and Scout knife but still including emergency ration 41D4293 and fishing tin 41A9711. With operational units consistently revising parachute kit components, the cooking container proved valuable for stowing additional items. (Library of Congress)

With the introduction of the Type B-1, revisions for the jungle kit were already underway, stemming from a long-running conversation. Back in 1937 and 1938, Albrook Field had requested some type of distress signal, preferably a fusee type. Part of the discussion concerned inclusion of a .22 cal. automatic pistol for hunting small game. Coinciding with the introduction of the 38J5271 jungle type back pad kit, the 9" machete was determined to be unsuitable for jungle use. This resulted in development of a new, improved machete. The list of recommendations grew through 1940, such as for the inclusion of an insect repellent and a water bottle. Six experimental kits were assembled by the Materiel Division and sent to Albrook Field for evaluation in December of 1941. The kit included additional first aid items, a rubber water bottle, a .22 automatic pistol, a Scout knife, a sharpening stone, insect repellent, and 15-minute fusee flare. Albrook Field submitted its field test report in January of 1942. It stated that the position of the machete and fusee flare made it impossible for the back pad to bend, and the rubber water bottle was not durable. Albrook Field commented that the .22 automatic pistol was a nice light gun but impracticable for use in the dense jungle. The heavier striking power of the .45 was more desirable. Although a pocket compass was provided by the Materiel Division, Albrook Field rebutted that the prismatic compass was much more accurate and suggested that it be retained.

On 4 March 1942, the Materiel Division finalized the design of the new jungle kit. The machete and fusee flare were repositioned, permitting bending of the back pad. A machete of a new design would be included, replacing the 9" type. The rubber water bottle would be replaced by a canvas water bag. The pocket compass was considered satisfactory as a directional compass (the prismatic compass, being a precision instrument, was in limited supply), and the .22 automatic pistol would be omitted but additional .45 cal. ammunition would be included. Immediate action was initiated to adopt the revised jungle emergency kit to replace the 38J5271. Adopted as standard on 9 March 1942 was the Kit-Jungle Emergency Parachute Back Pad, Type B-2 under Specification No. 94-40401, detailed under Assembly Dwg. 42J7251.

Due to demand for the Type B-2, procurement was expedited on 13 March 1942. Technical Order 00-30-51 was revised to reflect the new unit detailing its issuance and requisition the same day. What was not foreseen was the wartime conditions and requirements for such equipment, which changed the context in which this type of equipment was to be issued and handled. By the spring of 1942, Tables of Organization and Equipment for fighter and bombardment squadrons were being revised to reflect that new environment. Emergency parachute kits were now restricted toward operations outside the continental United States, and, with the rapid growth of the USAAF and vast expansions of air fields and air bases, it was no longer necessary to equip cross-country flights within the continent. Unlike the previous emergency parachute kits, for which in peacetime there was no urgency, the Type B-2 kit was expedited and placed on procurement before the availability of many of the proposed components. For much of 1942, contractors delivered the kits with as many of the components as possible; eventually, the kits would be assembled as prescribed by specification and technical order.

Like the Type B-1, the Type B-2 could be requisitioned complete or completely empty. Initially, both units were intended to be assembled by the contractor before shipment to central supply points. It wasn't until after the standardization of the Type B-4 that Type B-2 would be available empty as. A recommendation was made to cancel the existing contracts for both kits and place their components back into the supply system. The compromise was to instruct the contractors to deliver the remaining outstanding kits on order empty or partially assembled; they would then be available forrequisition and assembled at the discretion of the local commands.

In December of 1942, the time of the recommendation, 4,000 Type B-1 kits had been recorded as delivered, and 94,453 Type B-2 kits had been delivered. All their contracts would be completed by September of 1943. Placed on procurement, 349,495 Type B-2 kits, equally split under four purchase orders: 42-16361 to the Aircraft Appliance Corp., 42-16362 to Nathan Products, 42-16363 to Knight Leather Products, and 42-16364 to Cable Raincoat Co. (Purchase Order No. 42-10551 was issued to Cable Raincoat Co. but the details for this award are not known.) Once available, Type B-2 could be requisitioned complete under stock number 8300-559050 and the empty container could be requisitioned under stock number 8300-128100. As detailed per technical order and specification, Type B-2 was to contain three match safes 42A6930 (one each for matches, quinine, and a vial of iodine), emergency ration 42B10959, machete 42D7462, five-minute fusee flare 43B3554 (or 15-minute), a pocket compass, a headnet, a fishing kit (10 yards of fishing line, six hooks, and three flies), a two-ounce bottle of insect repellent, a three-bladed Scout knife, sharpening stone, a pair of gloves Type D-2, 40 .45 cal. cartridges in two boxes, a first aid kit (two Carlisle first aid packets, six band-aids, tourniquet, box or bottle of ammonia inhalants), box of iodine, and a tube of tannic acid ointment.

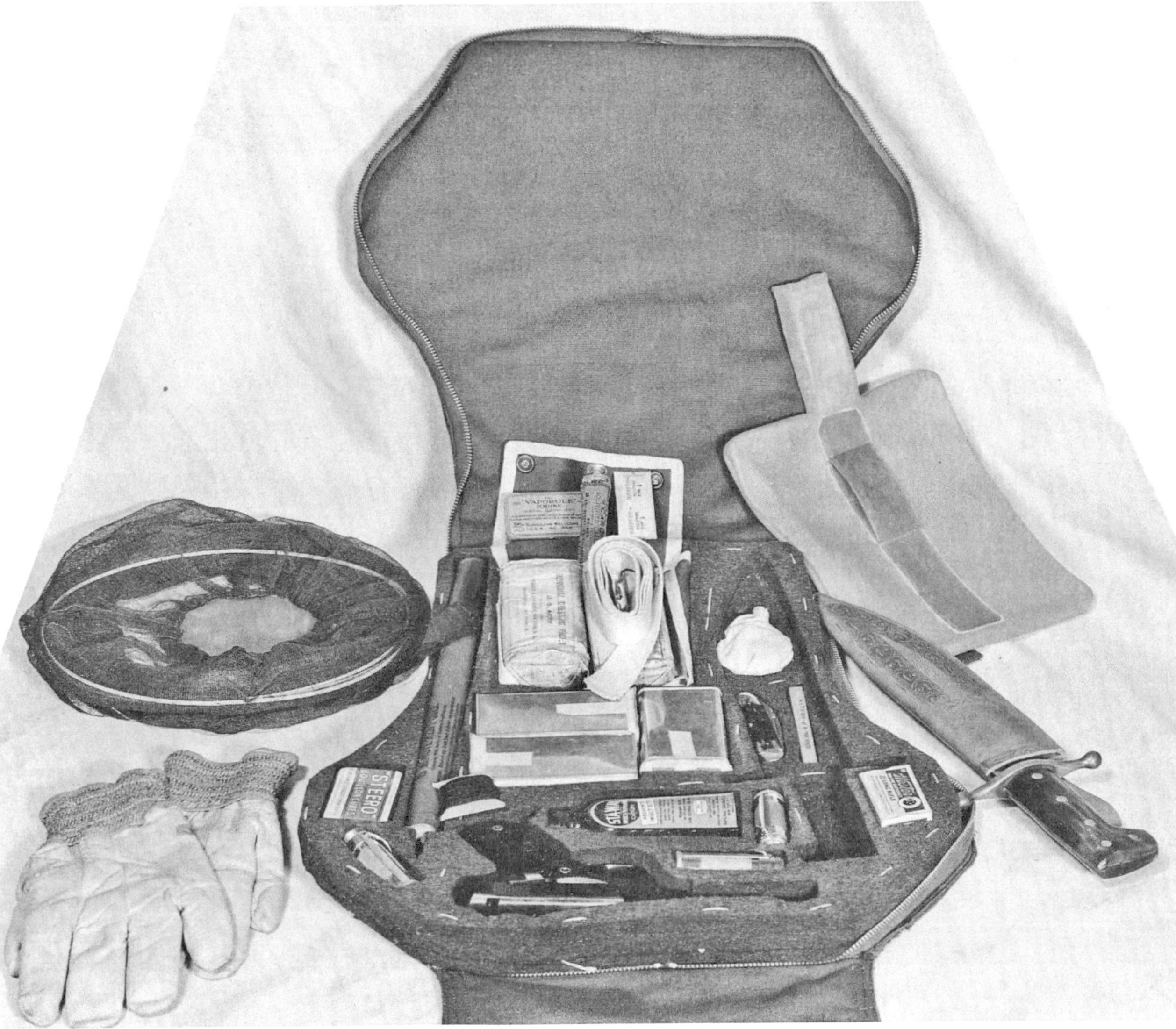

Experimental jungle back pad kit as submitted to Albrook Field in December 1941. The kit was a combination of Type B-1 and jungle kit 38J5271 with additions. Kit was later rearranged and integrated into standardized Type B-2. Machete and fusee flare were placed vertically on each side, which did not allow flexibility. These items were repositioned horizontally in Type B-2. (National Archives)

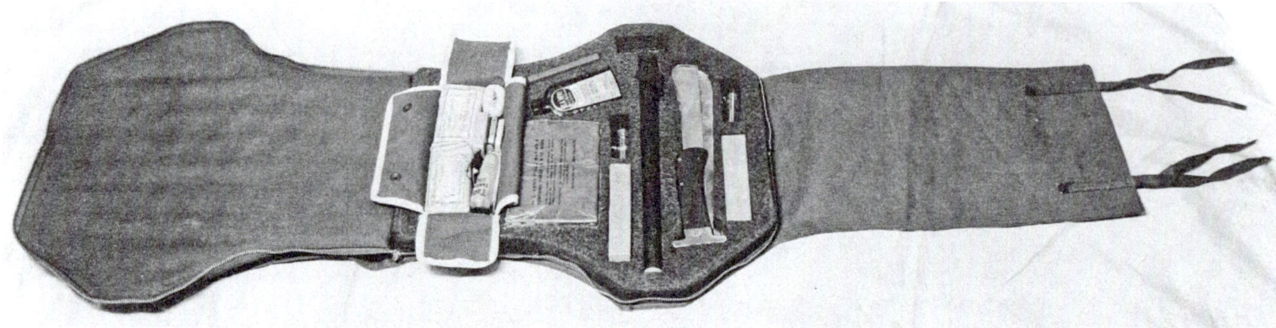

Externally, the Type B-2 was identical to its predecessor and Type B-1. All three had the felt insert with cutouts for components. Shown is the Type B-2 as standardized. Components varied during its service life. Included was new ration 42B10959, and machete 42D7462, which replaced 9" type. (USAF)

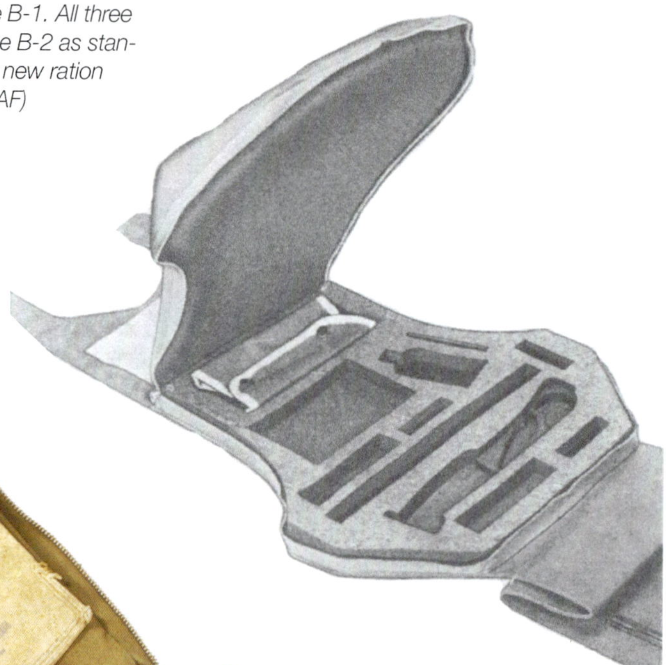

Type B-2 jungle emergency kit assembled with components as originally detailed upon standardization. Procurement of these kits was executed before many items became available, so items were added during production as they became avialable. Example: Kit was to include 10" folding machete 42D7462 but production of the machete was delayed, so contractors were directed to deliver the kits with the 9" type until 42D7462 became available.

Three new cellulose acetate match safes (Dwg. No. 42A6930) were added to new jungle kit. One contained sticks of waterproof matches, one contained iodine vial, third held quinine tablets. Glass vials held 0.25 oz. Shown are surviving examples of match safe and glass vial.

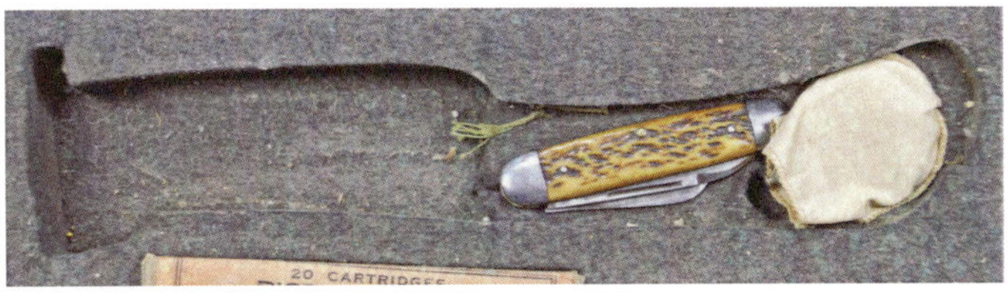

Secured in the cutout for the machete were the 3-bladed Scout knife and pocket compass, both standard AAC items.

(Right) The Type B-2 kit was to include new chocolate ration as detailed under Dwg. No. 42B10959. It comprised three 4 oz. D-Ration cakes packed in wax-coated chipboard.

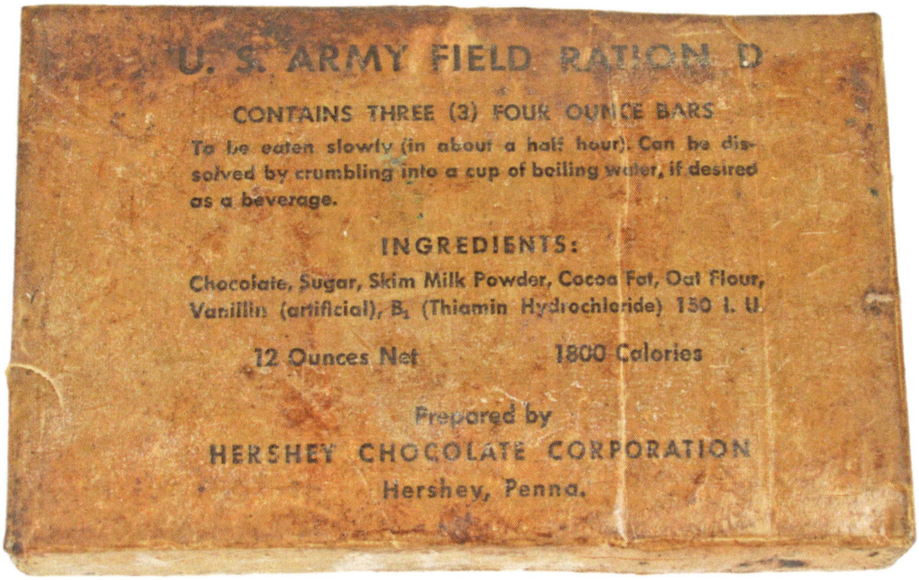

(Below) Prior to standardization of Type B-2, the Personal Equipment Laboratory worked on a revised fusee flare for parachute kits. The original was a 15 min. fusee, which created flexibility issues. Adopted was a 5 min. type half as long, referenced under Dwg. No. 43B3554. Slot in felt insert was cut to accommodate the 15 min. type since contractors were directed to supply it until the 5 min. fusee became available. Once in stock, Type B-2 would include one 43B3554 as standard.

(Right) The original Type B-2 included a 2 oz. bottle of Sta-Way insect repellent (pictured on page 75). This formula was toxic, so a new formula was procured from Davis Young Soap Co., Summer 1942. Sta-Way bottle retained commercial label and was CFE; the Davis Young Soap Co. bottles included P.O. number on label and were also GFE. USMD 2 oz. insect repellent could be requisitioned for replenishment.

The Type B-2 included 40 rounds (2-boxes) of .45 ball ammunition, pair of Type D-2 mechanics gloves, sharpening stone (4" x 3/8" x 3/8"), mosquito headnet and water bag. In initial deliveries, commercial headnet was provided, later updated with headnet 43G9868, supply permitting. While the Personal Equipment Laboratory searched for a suitable water container, a water bag was included as an interim. Eagle Brand water bag (1 gal.) was commonly observed. This was later replaced by vinylite water bladder 43B9963.

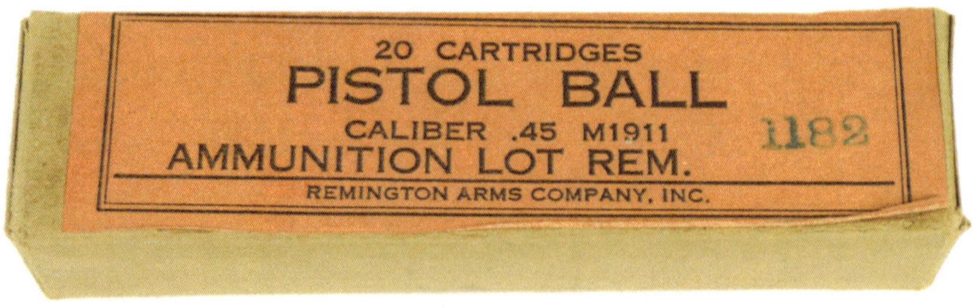

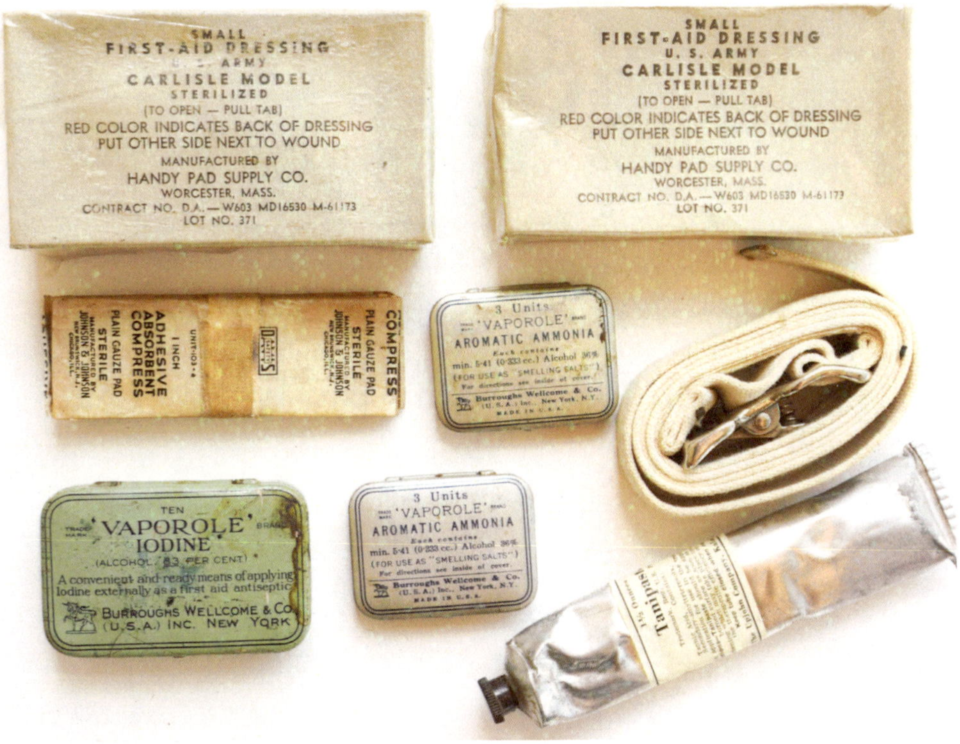

Type B-2 medical items were described generically. Dwg. 42J7251 lists two Carlisle first aid packets, six Band-Aids, tourniquet, box of ammonia inhalants, box of iodine, quinine tablets, and tube of tannic acid ointment. All but Carlisle bandages were provided by contractor and therefore a variety of items were included that met these descriptions. Documented statements mention tins of inhalants and iodine by Burroughs Wellcome & Co. under Vaporole brand, tin of ten iodine swabs and two tins of three inhalants. Tube of tannic acid was 1.5 oz. Iodine and quinine tablets were placed in glass vials then stowed in the acetate match safe. Another method was to place iodine swabs and quinine tablets in the match safes without the glass vials, with an instruction label. In 1943, for restocking purposes, Materiel Command circulated a list of approved medical items: tube of tannic acid ointment 91018, box of aromatic ammonia 91025, package of sulfadiazine 91204, box of iodine swabs 91122, six 1"adhesive bandages 92000, two Carlisle bandages 92060 and field tourniquet 93780. Quinine was excluded.

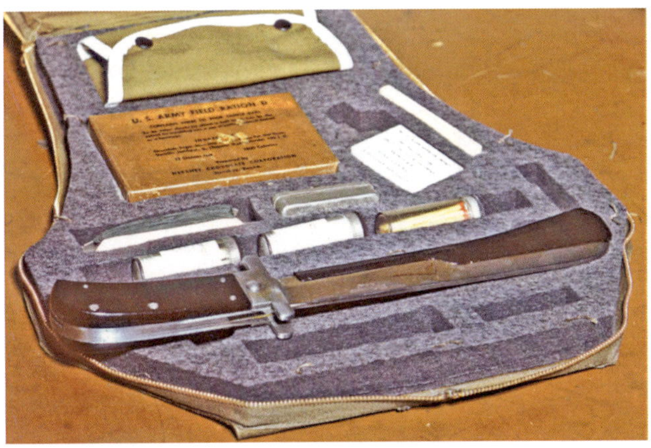

Wartime photo showing medical items stored in matchsafes with labels attached.

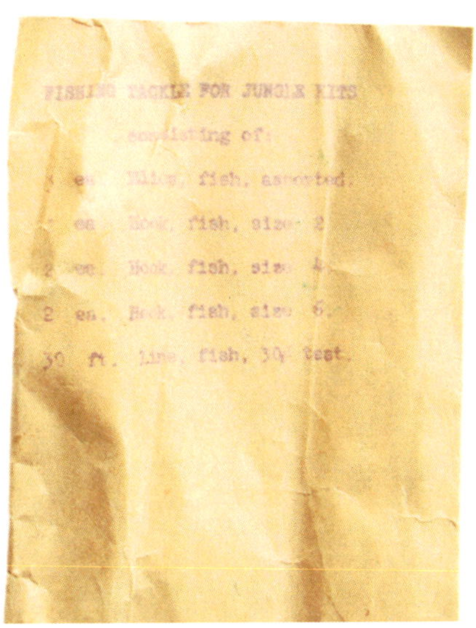

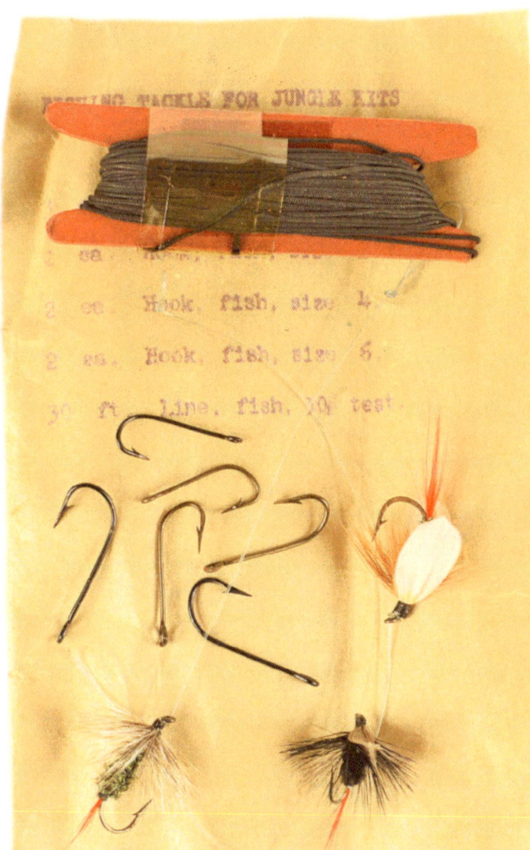

(Right and far right) Tin conservation pressure necessitated packaging the fishing kit in a manila envelope. The 41A9711 was used until supply was exhausted. Fishing kit with 10 yards of fishing line, three fish flies and six hooks.

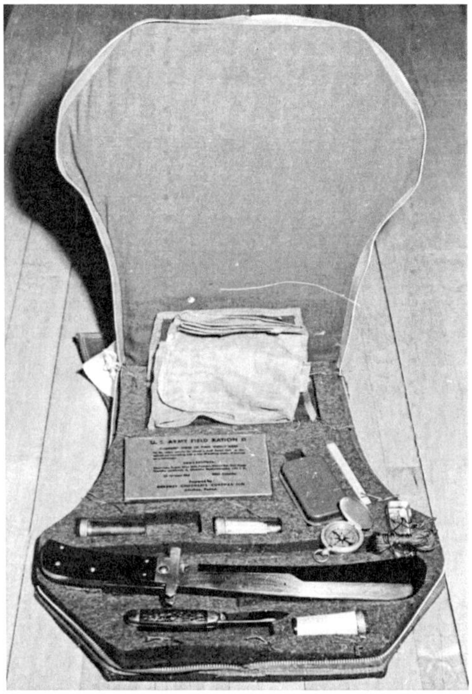

(Far left and left) These two wartime images exhibit deviations from the prescribed components. One Type B-2 includes the fishing line on a large spool. With kits shipped from the contractor partially complete or empty, a variety of articles can be observed. They retain the same items by generic description. These kits lack ammunition, added prior to issue. (National Archives)

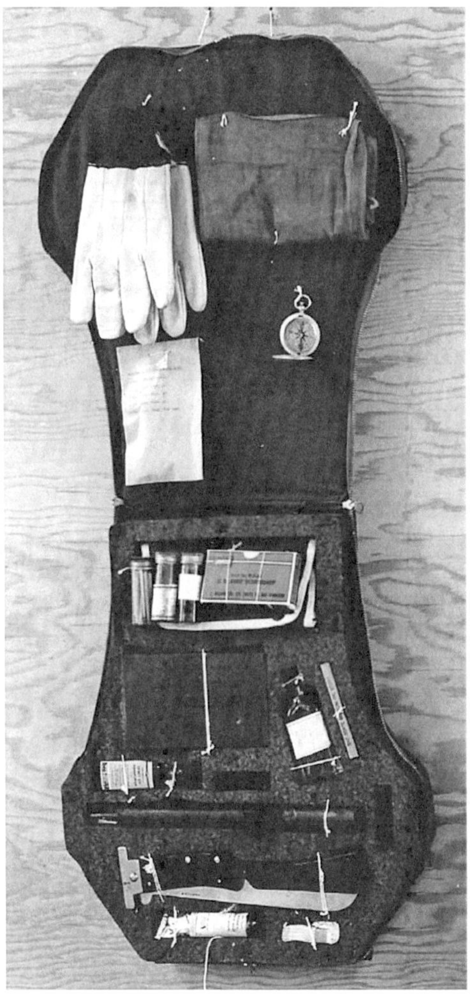

(Above) Wartime image with deviations from the prescribed components. The supply pocket had the three match safes and a boxed tourniquet by E.D. Bullard, glass bottle of aromatic ammonia in slot for ammuntion. (National Archives)

(Below and opposite page, top) Technical Order 00-30-51-1 dated 22 February 1944 directed all emergency parachute kits be updated. New items included signaling mirror 40653, 2 boxes of .45 cal. M15 shot, and survival manual Jungle Desert Arctic Ocean Emergencies. First aid frying pan insert 43B10436 was to be included in Types B-1 and B-2, replacing all previous medical items. The Type B-2 kit below contains those updates. Other emergency manuals had been utilized, and after July 1944 all emergency kits were to be updated with the AAF Manual 21W. In records where local commands discuss the Type B-2 kit, they mention addtional items such as leggings, candles, Atabrine, tracer ammunition, maps, and various Evasion & Escaoe items. (National Archives)

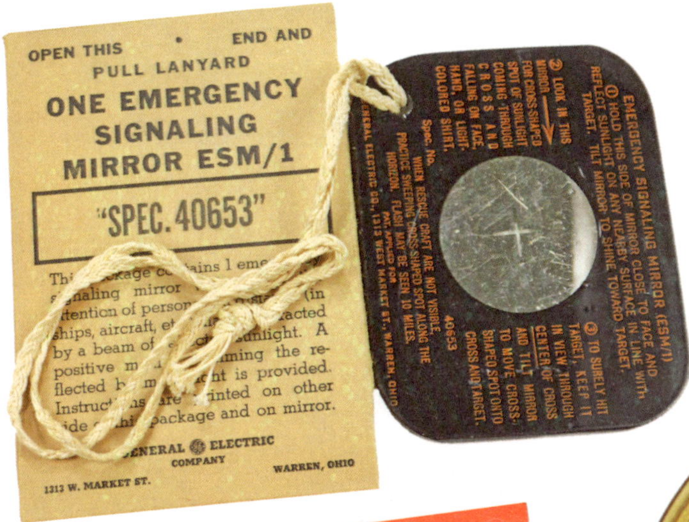

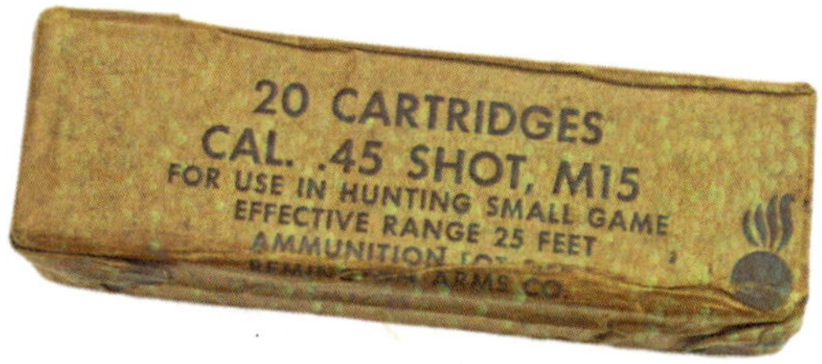

A training volunteer examines frying pan insert 43B10436 in Type B-2. Note inclusion of 15-minute fusee flare, c.1944. Parachute emergency kits were issued as standard allowance per T/O&E for all flying squadrons in the forward area. Each crew position operating outside the continental US was issued an individual emergency kit.

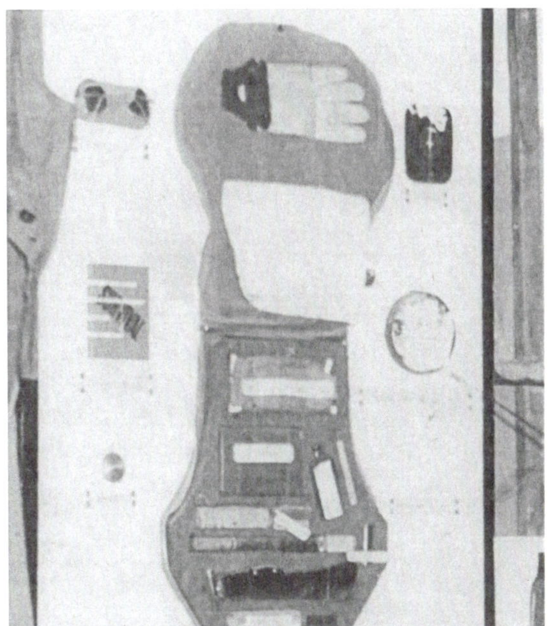

This Type B-2 hangs in a lecture room in New Guinea, c.1944. Signaling mirror 40653, 2 boxes of .45 cal. M15 shot, survival manual Jungle Desert Arctic Ocean Emergencies and first aid frying pan insert 43B10436 are present. Added, a pair of googles with tinted lenses. (National Archives)

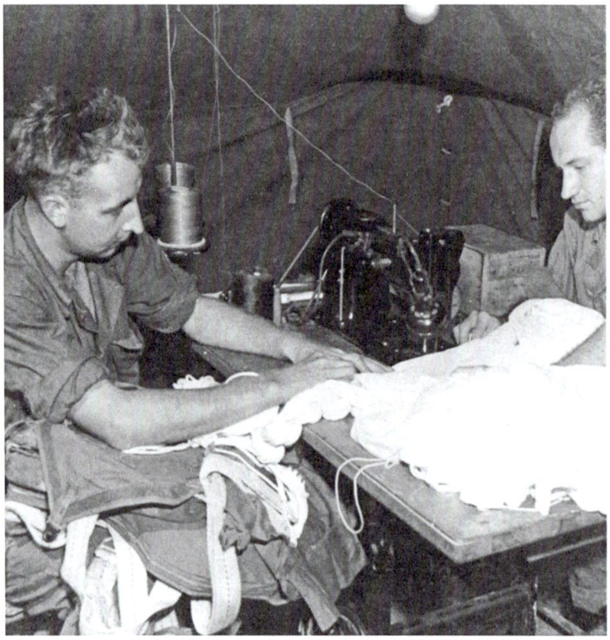

(Above) Parachute rigger of the 163rd Liaison Squadron, 7th AAF, mending a parachute on Okinawa, c.1945. Attached, is a parachute emergency kit.

(Right) P-47 pilot Lt. Lloyd Millet stocking cockpit with his safety equipment, including parachute, pararaft, and emergency back pad kit. (National Archives)

(Below right) Parachute rigger Sgt. Howard Jacobs fitting parachute harness with back pad kit. Individual being fitted is Sgt. James Whitmore of the 6th Photo Group, SWPA, c.1944. (National Archives)

(Below) Parachutes with back pad kits issued to pilots operating in Panama. (National Archives)

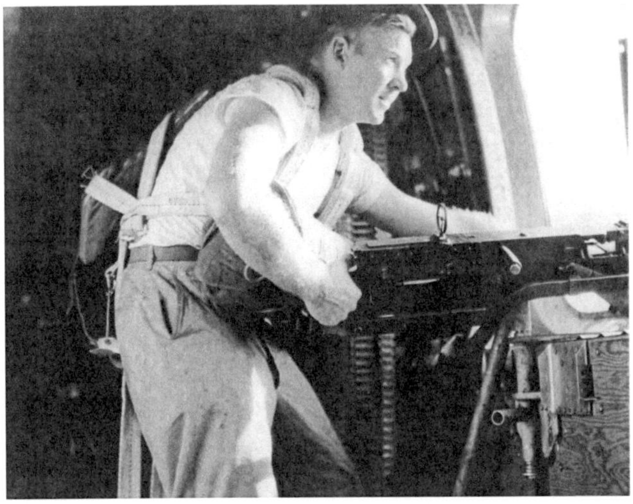

(Above) B-24 waist gunner of 13th AF, S/Sgt. Charles Held watched the skies over the Philippines, c.1944. Attached to QAC parachute harness was an emergency back pad kit, identifiable by the perimeter zipper. (National Archives)

(Right) B-29 aircrew member of the 11th Combat Camera Unit utilizing a QAC parachute harness with emergency back pad kit, Guam, c.1944. (National Archives)

(Below) Two Vultee L-5 pilots of 163rd Liaison Squadron, 7th AAF reviewing map of flight path over Okinawa, c.1945. Attached to standard seat type parachutes are back pad kits. (National Archives)

(Right) 2nd Lt. William Findley of the 9th Photographic Reconnaissance Squadron in cockpit of Lockheed L-5, India, c.1944. Standard flying complement included emergency back pad kit. (National Archives)

(Below) USAAF Lt. Coyd Yost helped train pilots of the Chinese Air Force in India, June 1944. Attached to parachute harness was an emergency back pad kit. (National Archives)

(Below) P-47 pilot handing flight equipment to ground crew member after landing on Saipan, June 1944. Flying complement was a standard seat type parachute with a pararaft and back pad kit. (National Archives)

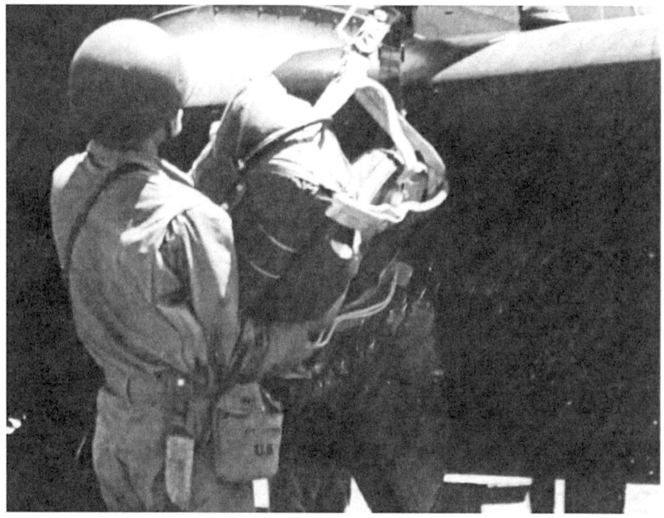

All types of parachute emergency kits were shipped to all combat theaters regardless of climate zone. Since external features are identical on all back-type kits, it is impossible to determine the exact model when reviewing wartime images. Because these kits were issued to all combat squadrons, many were sent to the ETO/MTO per authorized allotment. Parachute emergency kits at Burtonwood USAAF Base Depot, England, c.1944. This depot was the primary hub and distribution point for supplying USAAF 8th and 9th Air Forces.

(Above) 379th FS commander Col. Joseph Laughlin wears a back pad kit with his flying safety complement, France, c.1945. Use of parachute emergency kits in ETO was not as common as in Asia-Pacific. However, operations from North Africa and staging from operating bases in Italy were conducted over Soviet-held territory and the Balkans. These regions had vast uninhabited areas where emergency parachute kits could prove useful. Elements of the 8th AF used parachute kits as well. Many evasion narartives from france mention carrying "jungle" kits attached to parachute. (National Archives)

(Right) 94th FS pilot wears a parachute, pararaft and back pad kit in Italy. (National Archives)

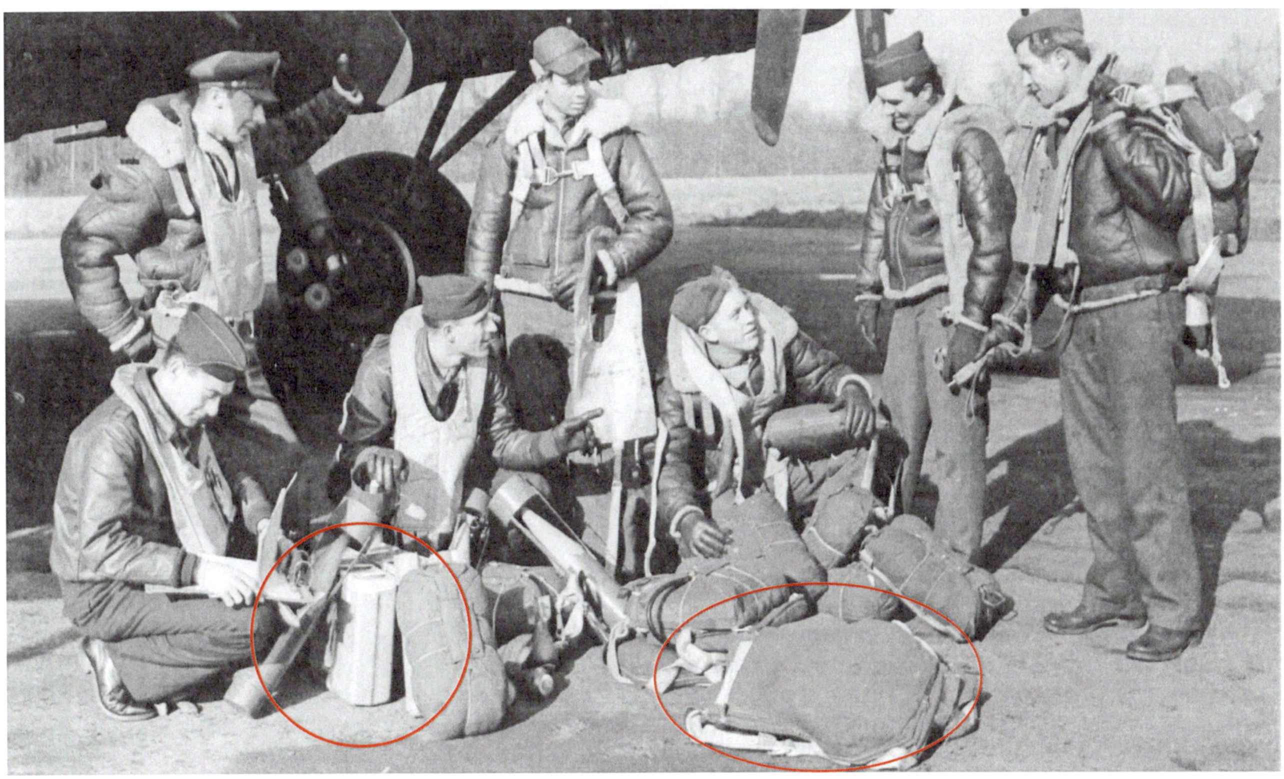

B-17 crew members check their equipment before taking off for a weather reconnaissance mission, England, February 1944. Back pad kit attached to a QAC parachute harness. RAF MK-IV ration boxes supplemented life raft equipment. (National Archives)

Standard parachute emergency back pads were designed to integrate onto the QAC and seat type parachutes but were not adaptable to the back type. A jungle kit was developed for personnel wearing the Type B-7 or other back type parachutes. Specifications were prepared and released 4 July 1942 for a unit designated as Kit, Emergency,

Type B-3 (Jungle-Parachute Seat Pad), under Specification No. 94-40443, detailed under Assembly Dwg. 42H15730. This kit used all the same components as the Type B-2 but was assembled in a seat kit measuring 13" x 15". It was designed to secure to the parachute harness by tie tapes provided at each of the four corners through the loops formed at the lower end of each riser.

The Type B-3 was recommended for standardization with all appropriate data prepared for inclusion in technical orders and other distributed materials. However, authority for procurement was withheld pending the outcome of discussion of a new type of universal parachute kit. Ultimately, the Type B-3 was never procured with the adoption of the Type B-4 emergency parachute kit. Data was provided on the Type B-3 in such literature as the Class-13 supply catalog, though none were available.

Image provided in the Class-13 supply catalog and the drawing for the felt insert that accompanies Specification 40443.

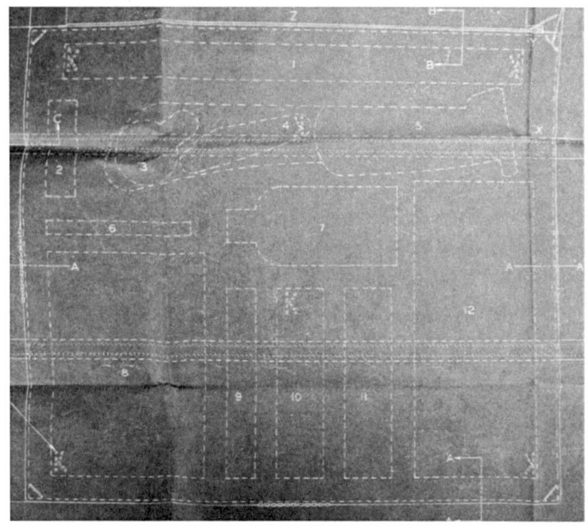

From the time of procurement of the Type B-2, Wright Field noted that there were frequent requests from commands for jungle emergency kits, which were not authorized except in the tropics. It was believed that some of the items contained in the kit were crucial in other areas. It was then recommended that one consolidated emergency kit be provided and available to all organizations regardless of their operating region. Before development progressed, it was deemed advisable to consult with scientists with field experience in either tropical or arctic regions, for the purpose of discussing the possibility of improvements or substitutions. Conferences were scheduled April 14-18 of 1942 in Baltimore, Maryland and New York City. A wide variety of individuals with experience was invited to these conferences from prestigious institutions, including Johns Hopkins University's School of Hygiene and Public Health, The American Museum of Natural History, The Rockefeller Institute for Medical Research, The Texas Company, and The Standard Oil Company. Individuals from all these entities had extensive field exploration experience and various backgrounds of geology, field study of mammals, anthropology, and medical knowledge.

The representatives from the Materiel Command presented the two standardized parachute kits, Types B-1 and B-2, for their evaluation. On 27 April, a report was compiled of the consolidated recommendations from these conferences. It concluded that similar articles were needed in both types of kits and that it was feasible to assemble a universal kit suitable for all regions. A universal kit was also preferred given the probability of sudden shifting of personnel from cold to warm climates or vice versa. From this date, all articles contained in the kits were either retained, revised or omitted. It was in these preliminary stages that a Type E-3 kit was considered to further consolidate all the necessary equipment for a true universal kit.

Most of the recommendations were applied to the design of a new universal parachute kit. Standard items from Types B-1 and B-2 were retained, and improved and new items were added. Adopted as standard on 3 October 1942 was the Kit, Parachute Emergency, Type B-4, under Specification No. 94-40463, detailed under Assembly Dwg. 43J10434, assigned stock number 8300-559075. The Type B-4 was designed to be universal to all climate zones and attachable to all parachute types. It officially superseded Types B-1, B-2, and B-3, putting them into Limited Standard status. Procurement of the B-3 was pending the results of the development of the Type B-4. With its adoption, the B-3 was no longer required, resulting in no procurements.

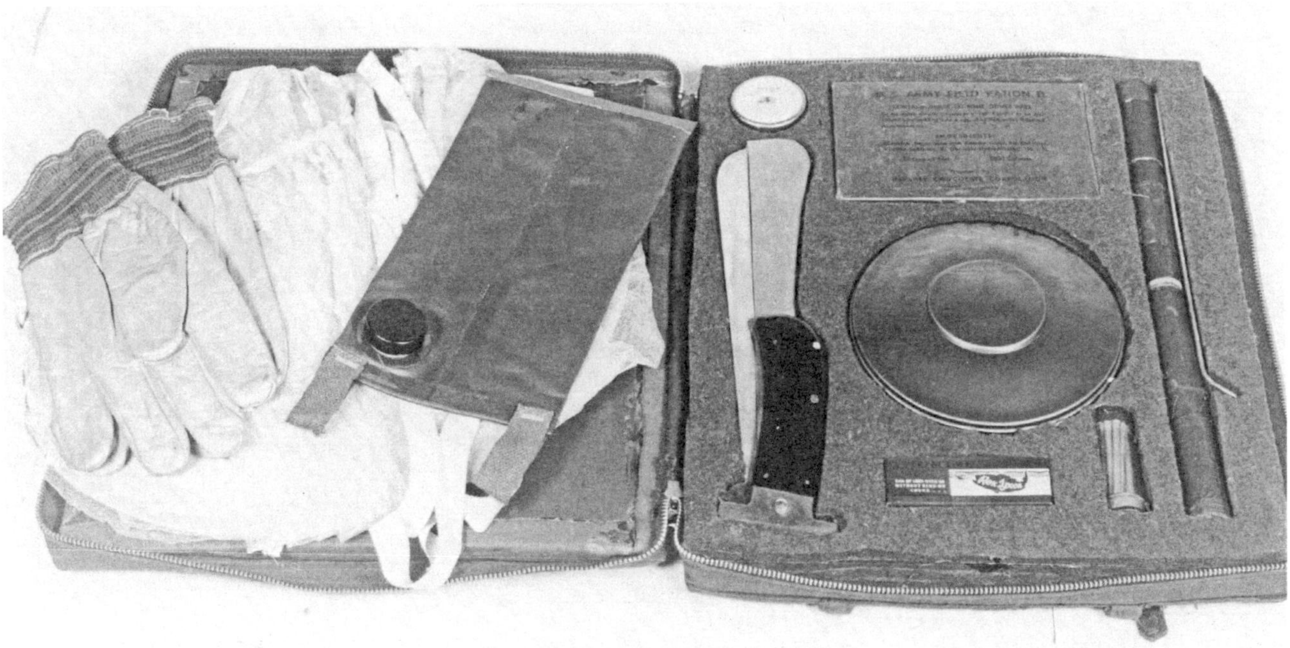

Preliminary Type B-4 parachute kit upon adoption. Not all components were available at assembly; it was a concept. Container for Type B-3 kit was carried over to the Type B-4 but modified for universal attachment to all parachutes. Type B-3 container had tie tapes at the four corners, whereas Type B-4 container included alternate securing methods. When assembled, Type B-4 measured 13" x 15" x 2-1/2" and weighed 11 lbs. (USAF)

Specifications for the Type B-4 received final approval on 10 November 1942 when it was released for bidding by potential contractors. The proposed procurement was for 75,000 kits. At this time, significant quantities of the Types B-1 and B-2 were outstanding. These quantities, along with those already delivered and in stock, were projected to meet service requirements of this type of equipment until 1944. It then became the opinion of the Procurement Division that procurement of the Type B-4 should not be authorized until existing stocks of Types B-1 and B-2 were exhausted. Alternately, a conversation was taking place on cancelling the contracts for the outstanding orders of the Types B-1 and B-2. Contractors replied that costs had already been incurred in obtaining components and that the bulk of those outstanding kits had already been fabricated. The contracts were allowed to run until completion. In February of 1943, those outstanding kits remaining was reported to be 53,000 Type B-2s and 5,000 Type B-1s.

As procurement of Type B-4 was being deliberated, work was still being conducted to refine and revise it. Technical Order 00-30-145 dated 17 February 1943, detailing the issuance of the new kit, was released for service-wide distribution. With this technical order in circulation, comments and inquiries began to stream into Wright Field on when this new kit was going to become available, as it was urgently needed. Those who knew that the Type B-4 was not yet on procurement argued that since the number of Type B-1 and B-2 kits in stock were considerable, flying personnel would continue to be issued inadequate parachute emergency kits for an indefinite time period. The Type B-4 procurement was strongly urged by commands in the combat zone. Pressured, the Procurement Division initiated a limited purchase of 25,000 kits to satisfy those immediate needs, under Purchase Order 43-32225 dated 4 June 1943, to Companion Bags Inc. The concept behind the limited purchase was that the Type B-2 was to be used and issued over preference to any other kit until supplies were exhausted, with the exception of those who used the back-type parachute; then Type B-4 would be issued. The Type B-2 was then shipped to all forward areas. There was reluctance for further procurement because there was concern the Type B-4 would be obsolete before it was even in use. Final action on any further procurements was pending information relative to the development of an emergency vest.

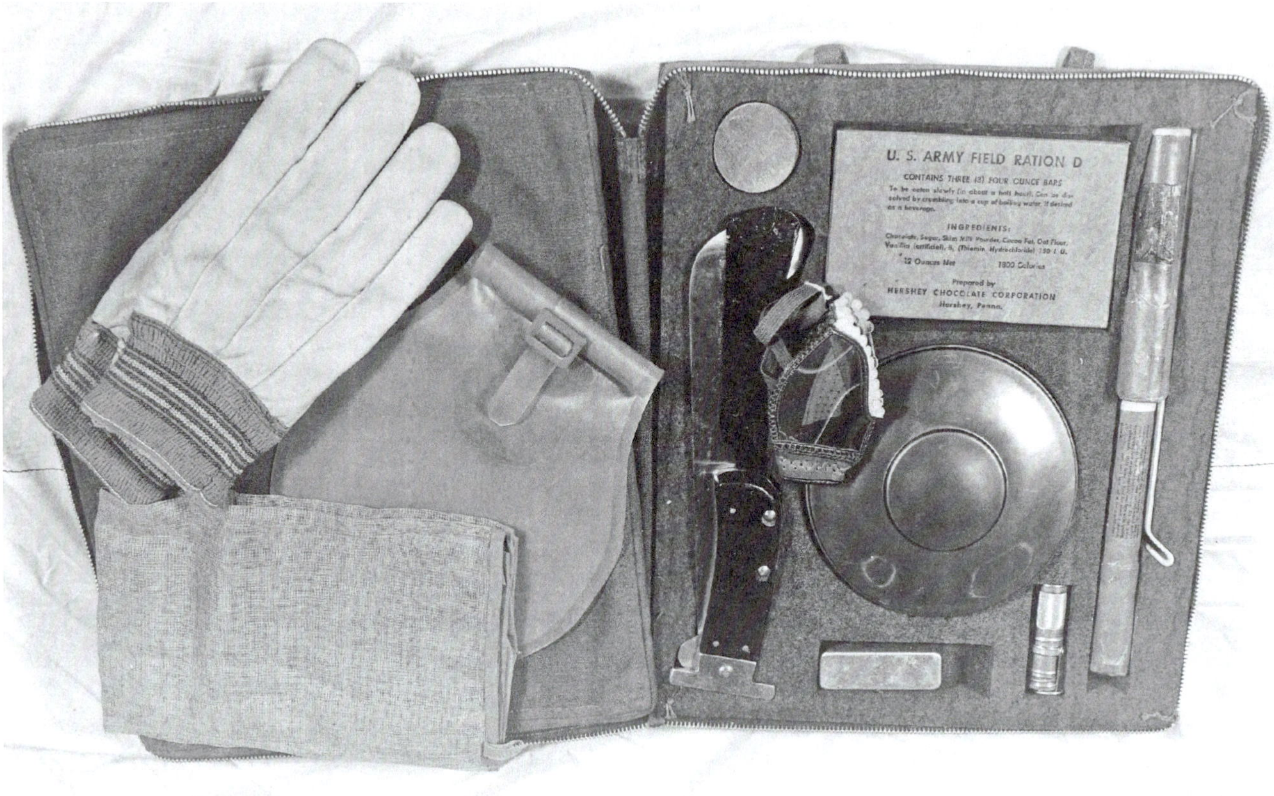

When Technical Order 00-30-145 was drafted, components for Type B-4 were pocket compass, match safe 42A6930, water bladder 43B9963, 2 5-min. fusee flares 43B3554, pair of gloves Type D-2, pair of goggles Spec. 3142, fishing kit 43B6987, first aid kit 43B10436, three-bladed Scout knife, machete 42D7462, mosquito headnet 43G9868, cooking assembly 42B2059, Field Ration D 42B10959, can of fire starter 43A12859, and signal panel 43A7607. Image distributed with the technical order with bulk of items listed except the signal panel. It includes older types of fishing kit and match safe. (USAF)

Included in the design was a signal panel following the details outlined in Specification 26756 for sails used in life rafts (see VOL-2). The panel included in the Type B-4 was blue on one side and an orange-yellow on the other, referenced in Dwg. No. 43A7607, with the unusual dimensions of 2' 11" x 8'. A series of messages was devised and adopted as standard for US-AAF sails. Shown here, signal panel for Type B-4 kit, eventually replaced by quilted poncho.

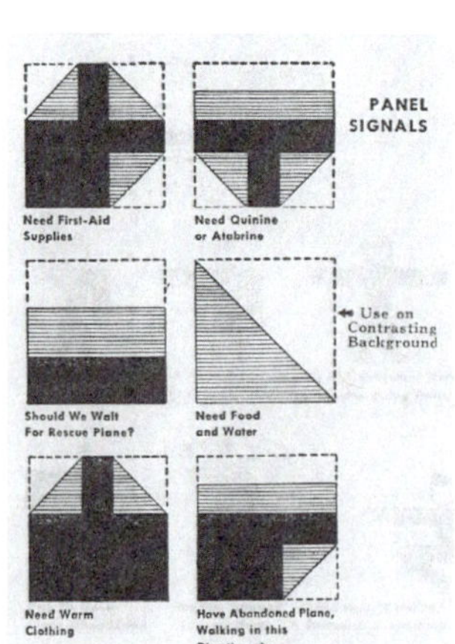

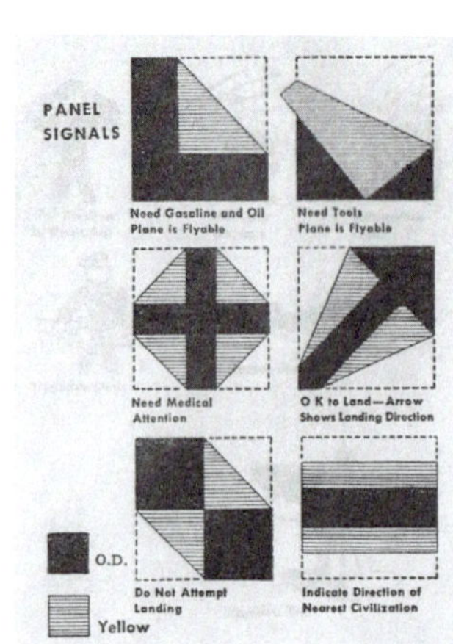

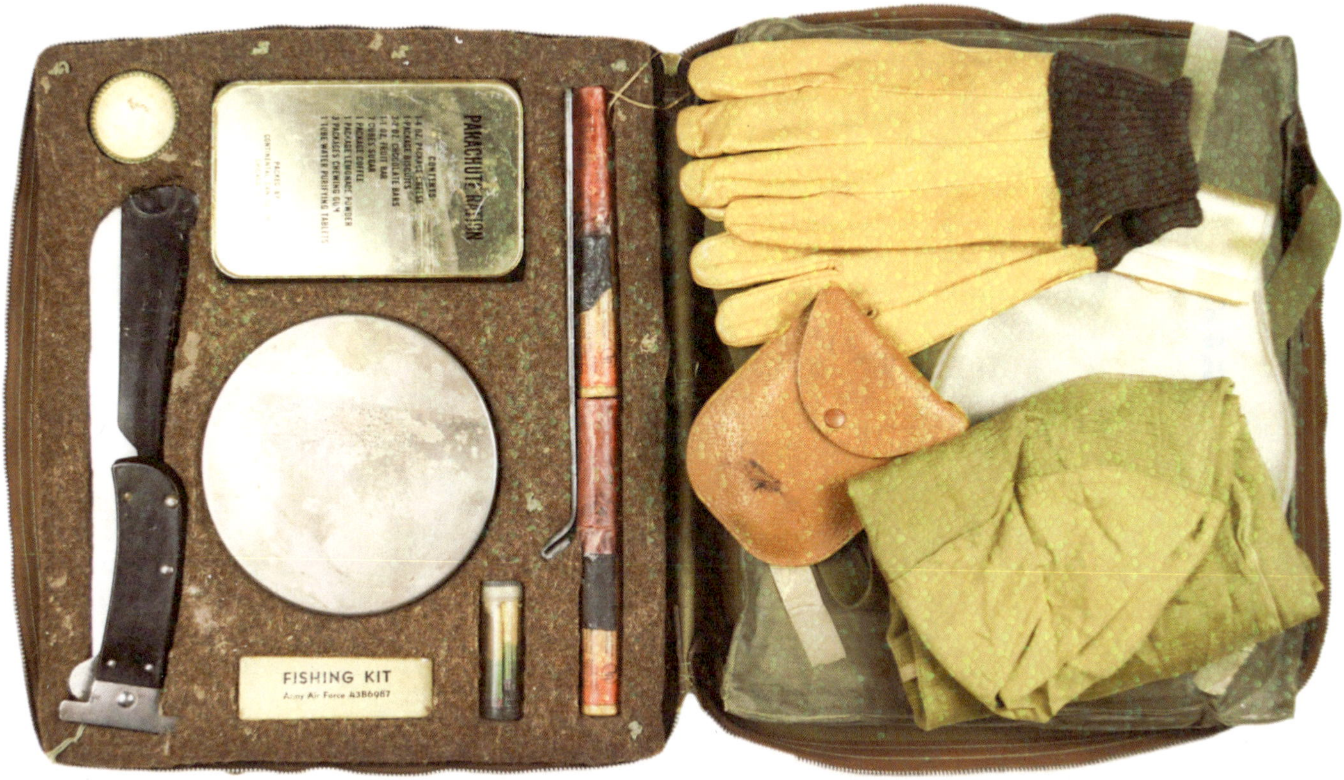

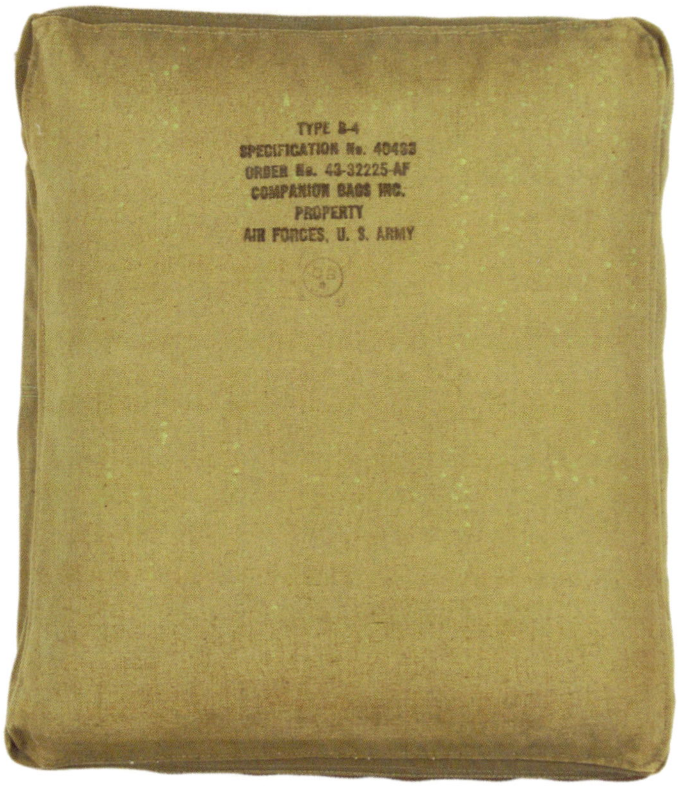

Type B-4 emergency parachute kit. Issued with the kit was a .45 cal. pistol, 3 clips of ammunition and a canteen. Cutouts under the folding machete for the Scout knife and pocket compass.

The Type B-4 had a cushioned top flap filled with horsehair like Types B-1 and B-2. While its procurement was being deliberated, the Personal Equipment Laboratory began investigating a quilted poncho presented by Thomas Quilt Factories. Immediate interest was expressed in it since it made the kit more comfortable and added buoyancy. The B-4's signaling panel was replaced by a quilted poncho in August 1943, olive drab on one side and yellow on the other, referenced under Dwg. No. 44D148. This poncho could be used as a life preserver, raincoat, sleeping bag, or shelter. The horsehair was omitted to allow space for it.

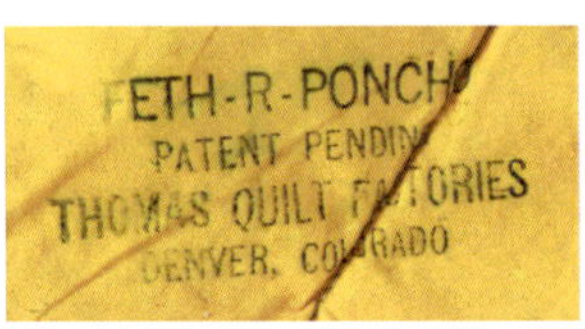

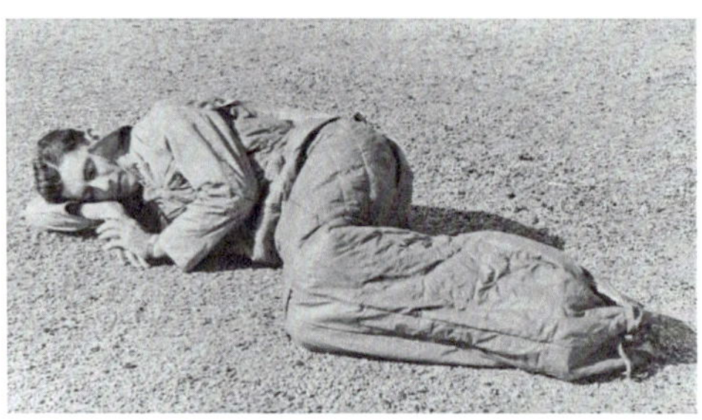

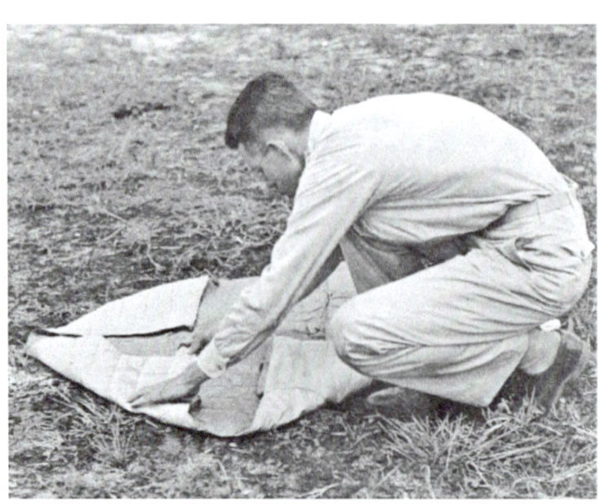

Quilted Poncho, Dwg. 44D148 measuring 42" x 65". It had a slot in the center, like a typical poncho, sets of tie tapes at the four corners, and single tapes along the perimeter for alternate securing methods. It was enclosed in a vinyl coated container 12"x 15" x 2-1/2", sealed and provided with a shoulder strap. With the quilt removed the container could be used for carrying items or gathering and storing water. Though the signal panel was officially superseded, some quantities were delivered.

To increase comfort, cooking assembly 42B2059 was revised to a flat plate; the original was domed, which became uncomfortable when seated on kit. All three pieces retained the same drawing numbers.

(Right) The most extensive critique during the conferences was of medical items. A suitable first aid unit was subsequently developed for the B-4 kit, referenced under Dwg. No. 43B10436, stored in cooking assembly.

(Below) Standard USAAF stock. water bladder 43B9963

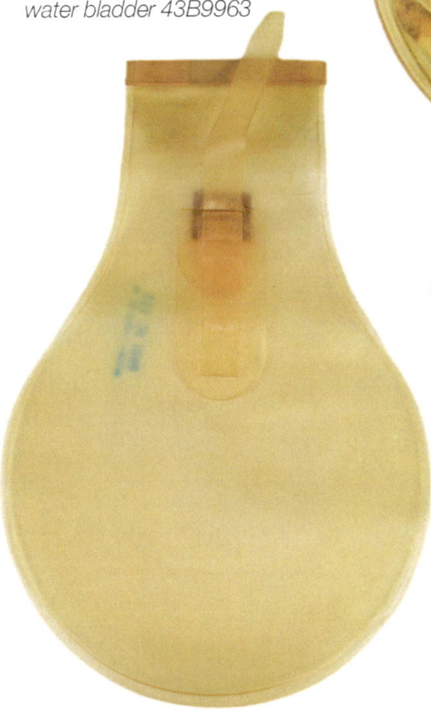

(Left) Solidified hydrocarbon, referenced under Dwg. No. 43A12859, replaced camphor gum as fire starter. Two types of tins were used, both 2"diameter x 0.75" D. Paraffin-dipped waterproof matches (Below) were included and contained in the plastic match safe (Dwg. 42A6930).

(Left) Standard USAAF stock included for body protection: mosquito headnet 43G9868 and gloves Type D-2.

(Right) Field Ration D was found inadequate. Developed with Type B-4 specifically in mind was the Parachute Ration, a revised K-Ration designed to integrate into all parachute emergency kits.

(Above) Conferences of experts recommended snow goggles, useful for both arctic and tropical locations. They consisted of a leather frame and green polarized cellulose lenses, referenced under Specification No. 3142, in leatherette case.

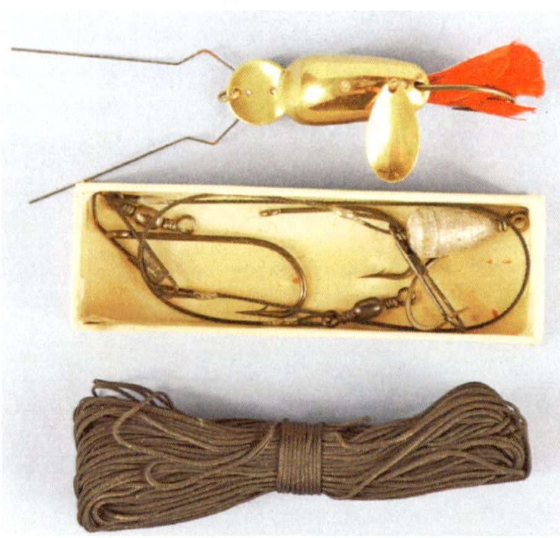

(Left) Fishing kit assembly, revised for Type B-4 under Dwg. No. 43B6987 was line, hooks and spoon jig in a manila box.

The remaining items included in the Type B-4 were either part of the original standard kits or of standard USAAF stock. Additional items include three-bladed Scout knife, two fusee flares 43B3554, machete 42D7462, and pocket compass.

(Right) The addition of the quilted poncho delayed delivery of the Type B-4; a limited quantity was available by the close of 1943. By the time the Type B-4 was available, work had progressed on an emergency vest whose standardization was certain. Type B-4 was obsolete before it became available. Technical Order 00-30-51-1 directed that all parachute emergency kits in service be brought up to the standards of the emergency vest. Type B-4 kit updated with the items detailed by that technical order, i.e. shot ammunition, emergencies manual and signal mirror. (USAF)

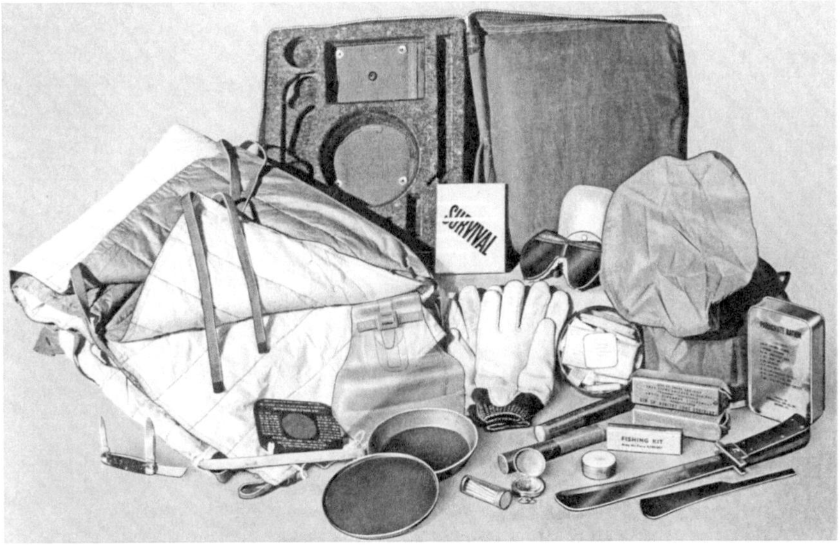

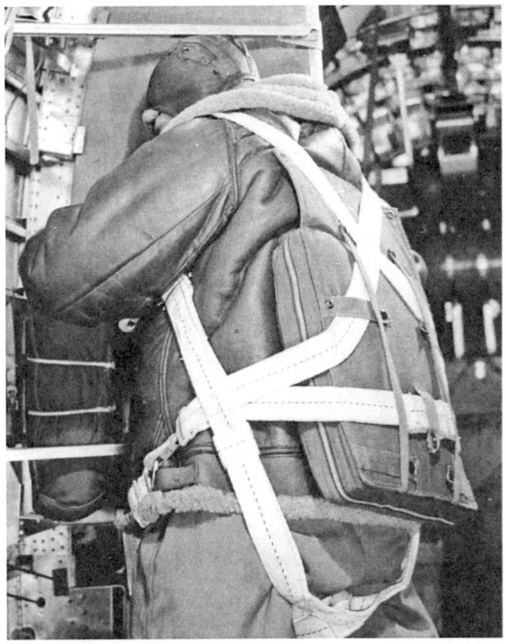

(Left) Proper attachment of the Type B-4 to either the QAC or seat type parachutes as a back cushion. Attachments to the back-type are the same as those prescribed for the B-3, through the loops formed at the lower end of each riser. Method could be applied to the QAC. (USAF)

(Below) B-24 aircrew members of 864th BS en route to the Carolina Islands, October 1944. Card table is B-4 parachute kit, extending its designation as universal. (National Archives)

(Above) P-51 pilot packs gear to his airplane; under right arm are Type C-2 pararaft and Type B-4 emergency kit. Iwo Jima, July 1945. (National Archives)

(Right) B-29 navigator Lt. Irving Bluefield checks equipment before a mission, sitting on two Type B-4 emergency kits, Saipan, c.1944. (National Archives)

With formal specifications drafted in January 1944 for an emergency sustenance vest, standardization was as all but certain to supersede the Type B-4. No further procurements of the B-4 would be made; however, it was felt that something in line with the Type B-4 could still prove beneficial. The Engineering Division began to redesign the kit to complement the new sustenance vest. The process included reverse-engineering the Type B-4, eliminating items redundant to those included in the vest. An initial idea was to provide supplemental items such as extra rations and first aid. This was quickly dismissed, and Materiel Command settled on essentials that provided flying personnel protection and shelter against the elements.

A quilted poncho that could be used as a sleeping bag, a hammock, a litter, a tent, a flotation device, and a signal panel was designed. To complement it, a quilted hood, pair of wool socks, and machete were added. When this assembly was field-tested in Florida in late May of 1944 it was found completely unsatisfactory, especially in wet weather. Still, it was found practical to continue development. At the time of the tests, the kit was designated as Type A-5; this was revised to the designation of Type B-5 with the machete omitted upon the procurement of 3,500 Type B-5 kits in July of 1944 for service testing. The contract was awarded to Thomas Quilt Factories, the designers of the quilted poncho, which they referenced as the FETH-R-PONCHO Type-2. Upon initial delivery, Type B-5 was adopted as Limited Standard on 13 September 1944 as the Kit, Parachute Emergency, Type B-5, referenced under Specification No. 94-40837, detailed under Assembly Dwg. 44D18823, assigned stock number 8300-559085.

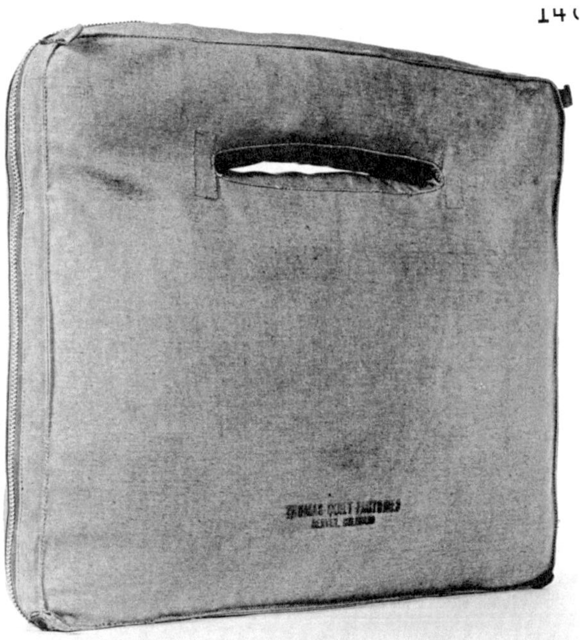

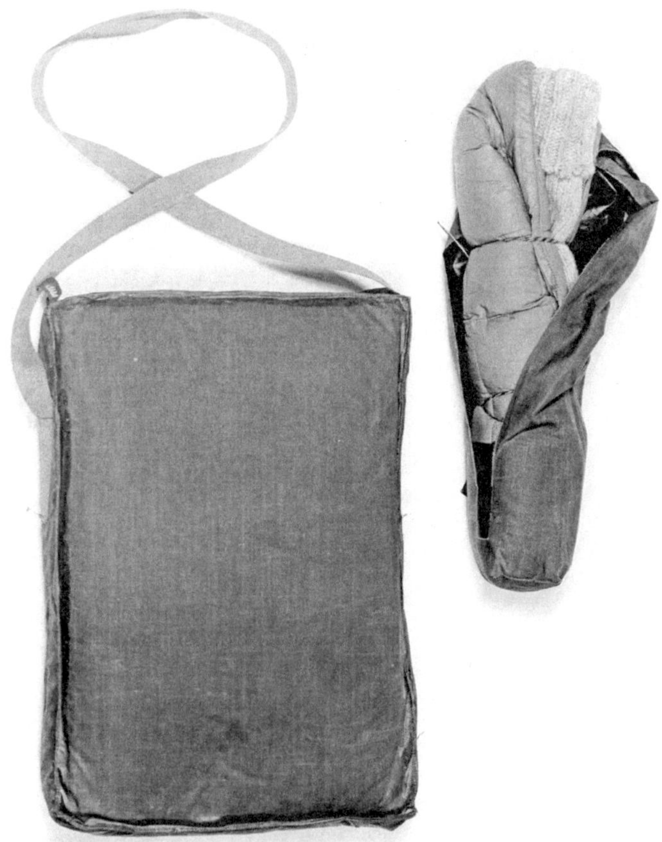

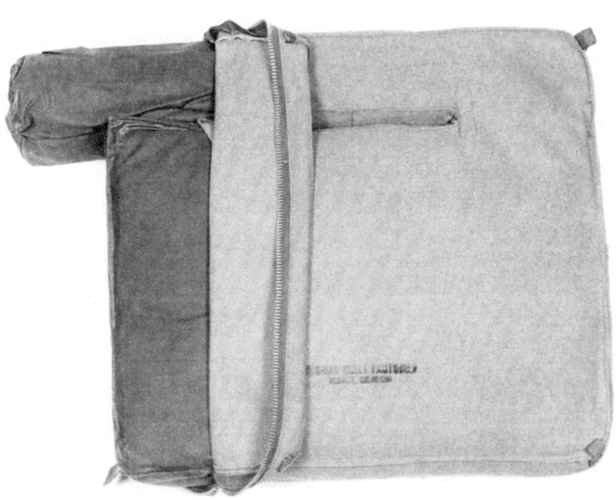

1944 experimental example of the Type B-5. Outer container was heavy duck fabric with same features as the standard seat cushion, including tabs at the four corners, but with a zipper along one edge. Kit included a quilted poncho, pair of wool socks Size 12, and quilted quick-attachable hood. Socks and hood were sealed in one waterproof container with the poncho in another, like Type B-4. Type B-5 was designed to be used with Type C-1 sustenance vest as a supplemental unit. (USAF)

Type B-5 seat cushion measuring 13" x 16"x 4". Standard parachute seat cushion was essentially repurposed, and it shares its characteristics. Slot made for a convenient two-compartment container; smaller slot for hood and socks, with another for poncho. Zippered closure on one end provided for removal of contents. The poncho was packed in a sealed waterproof container 10 -1/2" x 15" x 4" with carrying strap, which could be used for alternate purposes once poncho was removed. Included with the kit was a 12-page booklet with illustrations and instructions on various uses.

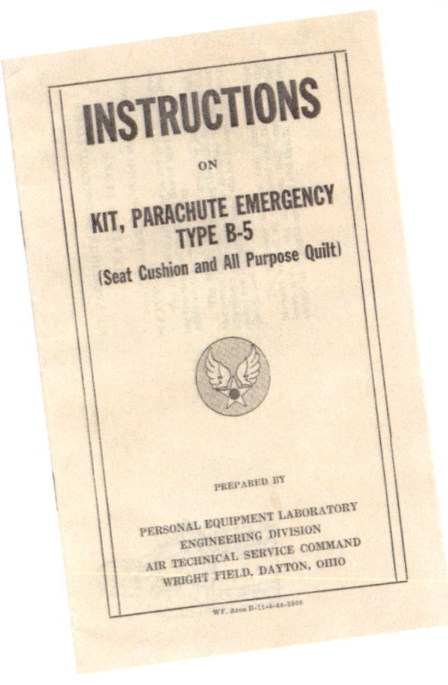

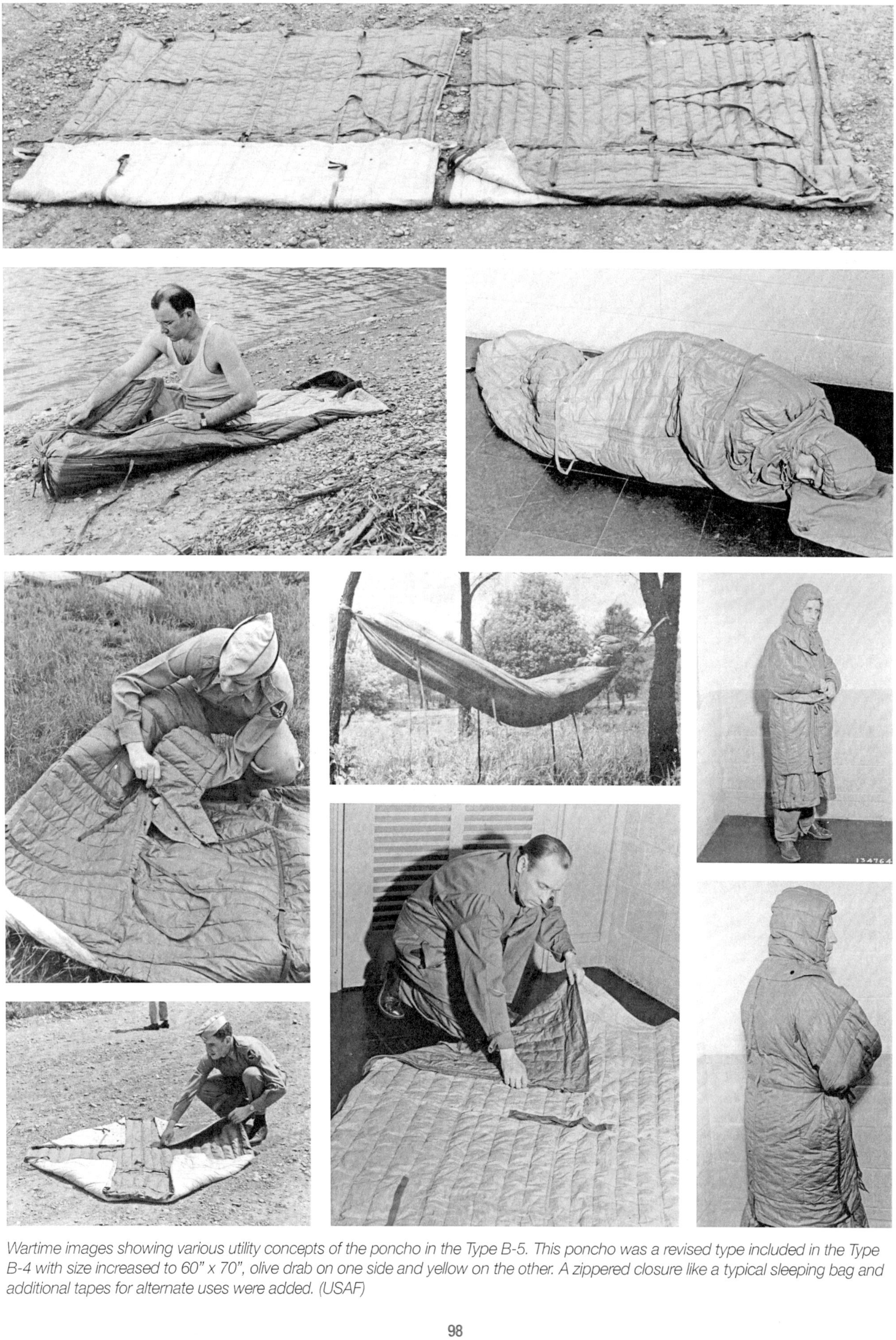

Wartime images showing various utility concepts of the poncho in the Type B-5. This poncho was a revised type included in the Type B-4 with size increased to 60" x 70", olive drab on one side and yellow on the other. A zippered closure like a typical sleeping bag and additional tapes for alternate uses were added. (USAF)

Requests continuously came into the Personal Equipment Laboratory for an emergency kit that would neither elevate a man in his seat nor push him forward. These were objectionable features of the present standardized kits, in addition to complaints about their bulk and weight. When seat kits were used in conjunction with the pararaft, there was insufficient headroom in certain types of aircraft and getting out of the airplane. There was also a danger in fowling the new Type B-8 back-type parachute, which was reported to be used by most pilots. It was found that the Type B-4 could be worn with the Type B-7 but could not be safely worn with the Type B-8 back-type parachute systems. A satisfactory solution to the problem was a proposed vest type kit. Development was initiated by George E. Peterson of the Personal Equipment Laboratory 10 April 1943. He worked in coordination with the Aircraft Belt and Trim Co. of Los Angeles, California, which engineered and fabricated a proposed vest. On 8 June 1943, after samples had been fabricated, a letter detailing the features of the vest were sent to O/C&R for consideration. In the following month, July, a directive was issued to develop a superseding line of sustenance kits. O/C&R requested that the General-Purpose Kit be a container or pack of a type to be worn with the seat, back or QAC parachutes, and by flyers in all types of aircraft. In its view, the proposed vest type kit met those requirements. The kit was assembled to provide each flyer with equipment and supplies for emergency use in the event of forced landings or abandonment of the aircraft in any climate region. The vest was designed so that only one size need be made; the adjustable features would accommodate all types of flying clothing fitting any person up to Size 48. It was to be worn over outer clothing and under the parachute harness.

The vest type kit was demonstrated by Peterson at the Army Air Force's School of Applied Tactics in Orlando, Florida on 15 July 1943. Twelve mockup vests were left for tests in combat tactics and in various types of aircraft. While the Personal Equipment Laboratory waited for the report on the outcome of the tests, the lab began development of materials and components. On 1 October 1943, the AAF Equipment Board released its report detailing findings on the proposed kit. It summarized that the experimental vest is the only kit, standard or experimental, which could be satisfactorily worn with the Type B-8 parachute. Other benefits noted were: no danger of losing the kit if forced down in water (because essential items were part of the flying clothing and not attached to the parachute harness in any way), it was less cumbersome for quick emergency exits especially for pilots with limited cockpit space, it was more comfortable and practical to carry and all items were waterproof or in acetate boxes.

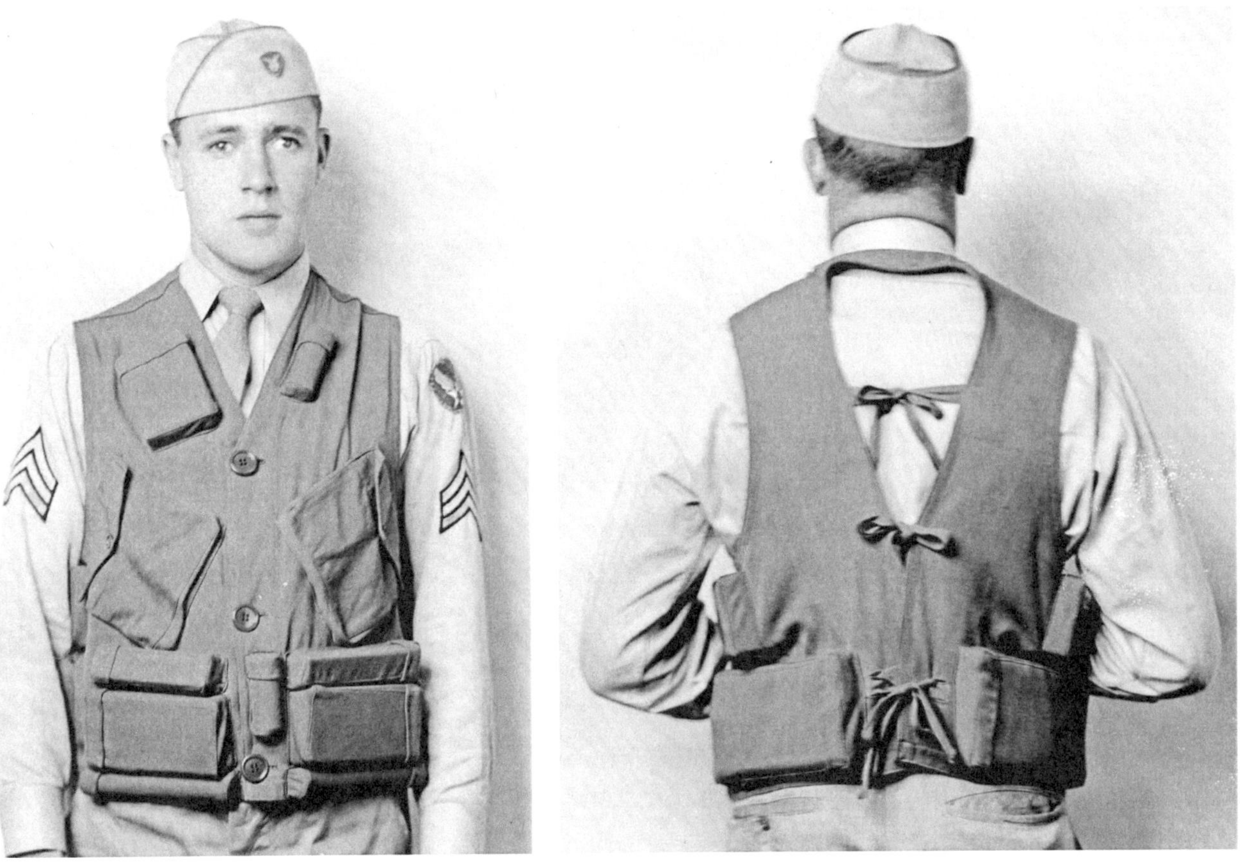

Vest type kit developed by Peterson and Aircraft Belt and Trim Co. over summer flying clothing; the rear view exhibits the universal adjustment method. (USAF)

Vest was to be worn under the parachute harness. Pocket arrangement had to assure comfort under the harness and could not interfere with its adjustment and operation. This volunteer demonstrated those attributes of the vest. He used AN-R-2A pararaft as part of assembly. (USAF)

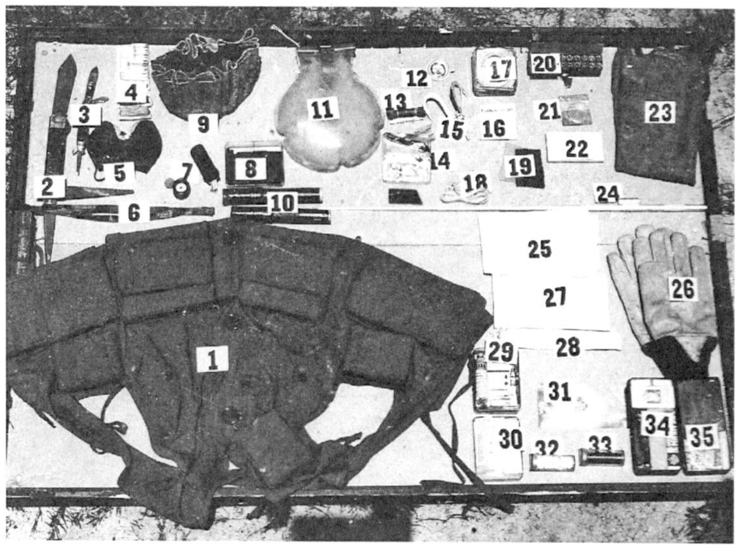

(Right and below) Pocket configuration as originally patterned with proposed contents. Photographs were taken during field tests conducted at AAFSAT and were attached to the Equipment Board report. Pockets 5 & 11 were for rations, pockets 6 & 12 were for accessory items such as sunhat, cooking container, tissue, and gloves. Pockets 7 & 13 were for the 5 min. fusee flares, pocket 8 was for the first aid kit, pockets 3 & 10 were for the large knife and shot ammunition, pockets 1 & 2 were for match safes containing supplemental medical items, and pockets 4 & 9 were for fishing kit and accessory fire starter. (below) Worn over winter flying clothing. (USAF)

A volunteer demonstrates proper application of the vest, front and back views. As per T/O&E, flying personnel had the .45 cal. pistol with ball ammunition and an M-1910 canteen. Items 2 & 3: Experimental padded holster with pouch containing supplemental water bag. It was thought that padded holster would improve comfort while flying and was designed to remain closed during parachute descent. It placed pistol in a position that would not interfere with combat crew duties. To further reduce bulk, collapsible water bladder replaced standard M-1910 canteen and cover. (USAF)

(Left) Vest worn under Type B-4 life preserver and Type H-1 bailout bottle. This individual displays the proposed standard complement with the addition of the Type B-8 parachute and pararaft attached. (USAF)

(Below) Assembly concept. Two types were tested. Image shows a back pack type (with pistol and water bladder removed), but the same configuration applied to the type worn on the thighs. (USAF)

Quickly following the report, preliminary specifications and drawings were drafted covering the construction details by the end of the month. The proposed kit was then designated as the Vest Type Emergency Kit. On 13 December 1943, action was initiated for a service test order of 3,000 vest type kits. To assure the vest underwent proper evaluation, samples were to be sent to various combat theaters: 1,000 to the 5th Air Force in the SWPA, 500 to India for the 10th Air Force, 500 to China for the 14th Air Force, 500 to Alaska for the 11th Air Force, and 500 to the Air Transport Command.

A questionnaire would be sent to obtain detailed information on the equipment. From these surveys, recommendations for further improvements and recommendations could be obtained to work toward a standard item of AAF equipment.

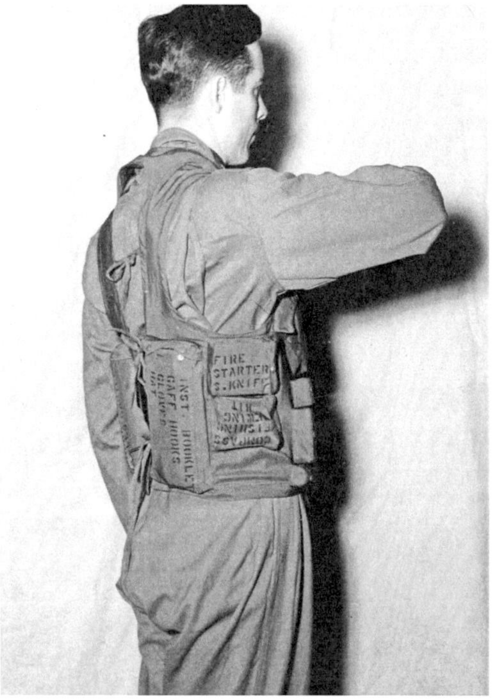

(Left) For production of 3,000 service test samples, modifications were made to the pocket arrangement. Pockets for sunhat, cooking container, tissue, gaff, goggles, gloves, toilet tissue, and mosquito headnet were moved to vertical pockets positioned on the back. Other pockets were placed in a more orderly fashion. It was decided that a shoulder holster be integrated securing the pistol under the left arm in a position that would not interfere with crew duties and would remain in position during parachute opening. Images illustrate first working prototype of Vest Type Emergency Kit. (USAF)

(Right) On 11 January 1944, O/C&R approved the design, detailed under Specification No. 94-3026, now titled Vest, Emergency Sustenance, Type C-1. By then, many components had been revised and updated, which had to be accomplished before O/C&R would accept it for standardization and future service requirements. Still, Materiel Command was negotiating with potential contractors and approving others for all aspects of the vest. By Spring 1944, 3,000 vests on the service test order from December 1943, in the modified form, were reaching destinations for evaluation. But before recommendations were received from overseas stations, local evaluations found the vests bulky and uncomfortable. Requests were made to remove the two vertical pockets on back and move those items into two breast pockets inside. A service test order of the new design was initiated 1 March 1944 for 3,000 vests executed under Purchase Order 44-5039E. These vests were shipped as follows: 500 to the 5th AF, 600 to the 6th AF, 250 to the 8th AF, 500 to the 10th & 14th AF, 200 to the 11th AF, 250 to the 12th AF, 200 to the 13th AF, 250 to the 15th AF, and 100 to the Cold Weather Testing Unit in Colorado. (USAF)

Upon issuance of the AAFSAT report, the Personal Equipment Laboratory refined many of the components in experimental form for the new vest kit. Displayed are items proposed for inclusion in Type C-1: a 2-part allotment of items from the Type-K ration, .45 cal. shot cartridges, large knife, spit & gaff assembly, sewing kit, fishing kit, match safe compass, pair of gloves, instruction booklet, mosquito headnet, pair of goggles, fusee flares, signaling mirror, fire starting kit, a small knife, whistle, burning lens, cooking container, water bladder, and a first aid kit that included two separate match safes with iodine and quinine. Pockets were placed on the vest for the two match safes. (USAF)

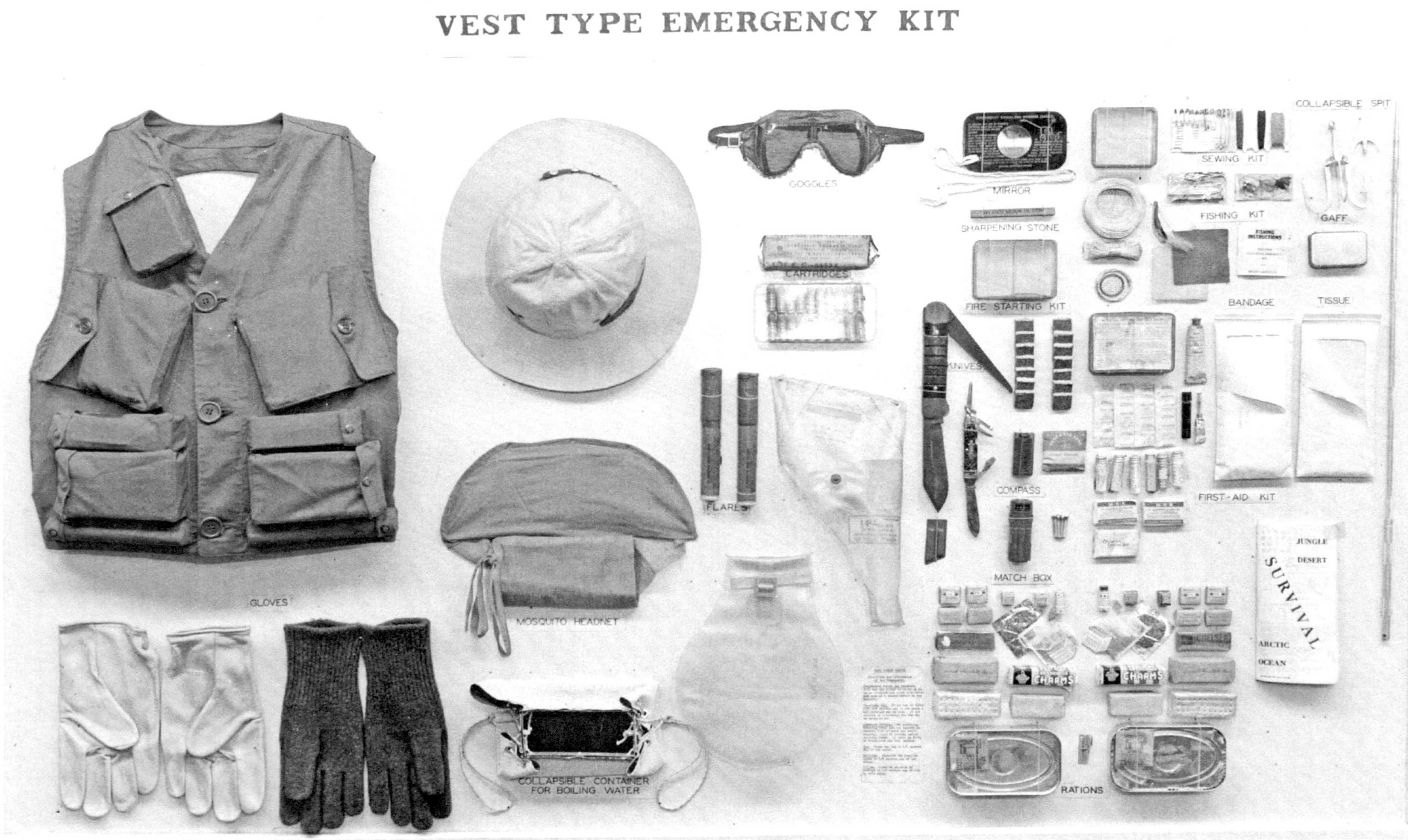

Adopted as standard on 3 May 1944 was the Kit, Emergency Sustenance, Type C-1, referenced under Specification No. 94-3026, Assembly Dwg. No. 44G9387, stock number 8300-939600. Unit superseded Types B-1, B-2, and B-4. This display represents the components approved for the vest upon standardization, some of which remained unchanged while others were revised during its production lifecycle. Three technical orders were later drafted and approved for circulation: Technical Order 13-5-46 dated 5 September 1944, Technical Order 00-30-196 dated 9 March 1945, and Technical Order 13-1-24 dated 13 March 1945. (USAF)

In June 1944 it was requested that Type C-1 vest procurement be expedited. Contracts were negotiated immediately and awarded in July for a quantity of 53,600 vests. This order was split between the Breslee MFG. Corp. and Cappel Macdonald and Co. Breslee and Cappel had a working relationship. (Prior to the service test order in March, Cappel had negotiated a contract the following month under a fixed price contract, 33038-AC-038.) The manufacturing and assembly were conducted by Breslee under Purchase Order 44-5039E.

The Type C-1 was standardized while it was still undergoing service and field tests. Due to the urgent requirement for this item, it was felt beneficial to place them in service and amend any deficiencies as they may arise. For the purposes of discussion and description of the Type C-1, the iterations that follow will be referenced as first, second and third productions.

(Left and below) First production of Type C-1 procured under service test order 44-5039E and quantity procurement under Purchase Orders 45-411, and 45-412. The empty vest was referenced under Dwg. No. 44H8830. The same pattern vest was used on the two July 1944 purchase orders. Also illustrated is data stamped on the March 1944 P.O. and one of the labels from the July 1944 contracts. Delivery of the C-1 under Purchase Orders 45-411 and 45-412 was completed in December 1944. Chart for the first quantity production of the Type C-1 sustenance vest.

Contractor	Award Date	Contract/ P.O.	Quantity
Breslee MFG. Corp.	July 1944	45-411	26,800
Cappel Macdonald and Co.	July 1944	45-412	26,800

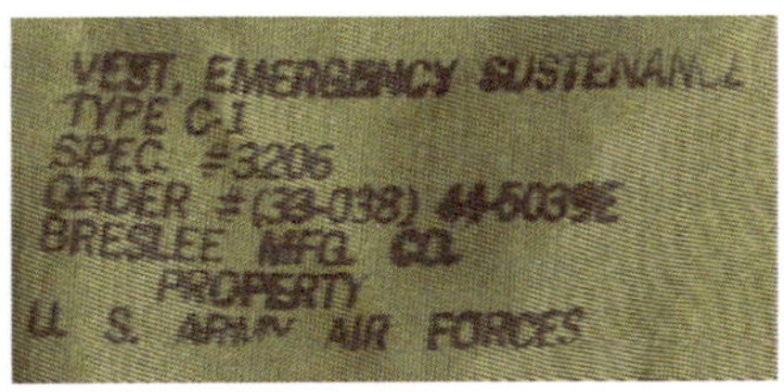

(Below) During negotiations for expedited procurement, the first scheduled field tests of Type C-1 were concluded. AAF Proving Ground submitted its report on 9 June 1944. On its recommendation, Specification 3206 was revised as Amendment-1, dated 17 June 1944. AAF had tested vests from the first 3,000 service test order, which included much of the older generation components and were not complete. The vest received satisfactory reports and reported deficiencies were thought to be remedied by new items not yet available but in production. The most significant recommendation was to add insect repellent, added under Amendment-1. A pocket for a 3 oz. can of insect repellent was added to the vest under the pocket for the fishing-sewing kit. This addition was applied to the second production after redesign. The second production of vests was negotiated for a procurement of 155,686. Below, Type C-1 that includes the insect repellent. AAF insignia would be applied per Technical Order 13-1-12 dated 1 February 1944, which also detailed that all flying equipment would be marked for the purpose of identification, outside and inside. (Regulation to this effect had been established in October 1943.) This production included new contracted components, differing from those in previous purchases. Also included are examples of labels on second production vests. Chart detailing procurements for second production run of the Type C-1 as detailed in Amendment-1. All these contracts were negotiated at the same time; award date depends on when the potential contractors had all appropriate documentation approved.

Contractor	Award Date	Contract/P.O.	Quantity
Aircraft Appliance Corp.	August 1944	33038-AC-4198	20,000
Aircraft Appliance Corp.	October 1944	33038-AC-5415	26,000
Lite MFG. Co.	October 1944	45-2316	10,000
Sears & Roebuck Co.	November 1944	33038-AC-5411	40,000
Breslee MFG. Corp.	November 1944	33038-AC-5413	39,686
Cappel MacDonald and Co.	November 1944	33038-AC-5414	20,000

The first significant change to the Type C-1 derives from the final report submitted by the AAF Proving Ground on 9 September 1944. This report would be reviewed multiple times up to the spring of 1945. It was then the most complete and comprehensive assessment of the vest and its complement of accessories. The recommendations would extend further thought on the redesign of the vest down to minor details of each component and additions.

Following the award of contracts for the second major production, those quantities delivered and on procurement would not meet future requirements through the year 1945. To meet those new requirements, an additional 112,153 vests would be essential, necessitating a third production. Negotiations commenced in November of 1944.

For these contracts, changes to the physical features of the vest would be made. The cloth of the vest was changed from a cotton-backed rayon to nylon. Nylon weighs less per square yard, has increased porosity, and a higher tear strength. Overseas commands reported that, when wet, components were difficult to remove from the pockets. Pockets were therefore slightly enlarged to accommodate shrinkage.

The most distinguishable change was to the breast pockets, Nos. 2 & 6. On the first and second production vests they were slanted and secured by a single button. In the third production, the design was revised to a vertical position that included two ring snaps. There was no reported discussion on the reason for the change.

These changes were officially addressed in Amendment 2 of Specification 3206, dated 25 January 1945. Manufacturing and assembly of the second production vests spanned these changes; vests under those contracts would be manufactured under Amendments 1 and 2. Records indicate that the assembly of the third production patterned vests commenced around March-April 1945. The assembly of those third production vests also spanned the deletion of the asbestos cooking vessel; the inner pocket of these vests can be found with and without the stenciling for this component or blanked out.

(Opposite page and below) Third and final production of C-1 sustenance vest with examples of labels from those contracts. Chart details contracts for third production run of the C-1.

Contractor	Award Date	Contract/P.O.	Quantity
Aircraft Appliance Corp.	December 1944	33038-AC-6679	8,153
Clark WLM Inc.	December 1944	33038-AC-6680	9,000
Lite MFG. Co.	December 1944	45-5128	16,000
Breslee MFG. Corp.	January 1945	33038-AC-6639	35,000
Sears & Roebuck Co.	January 1945	33038-AC-6640	44,000

Profiles of the third production C-1.

Each pocket on the Type C-1 would be numbered and include a brief description of its contents. The numbers on the pockets were consistent but the labeling varied. However, most of the variations are merely contractor variations. In the original assembly, the instruction book (21W) was contained in pocket No.15, with the cooking vessel in pocket No.14. Upon deletion of the cooking vessel, the instruction book was transferred to pocket No.14. Another notable change was to pocket No.12, which contained the fire starter. Some manufacturers listed the contents as burning lens. Once this was deleted, some vest manufacturers listed razor on the pocket. There were multiple suppliers of the empty vests; known subcontractors included Breslee MFG. Corp., the Ben Greenholtz Co. and Belber Trunk & Bag. The Sears & Roebuck Co. subcontracted assembly of the vests to Belber.

Type C-1 vest worn under the QAC and seat type parachutes per Technical Order 13-5-46, which detailed attachment methods of the Types B-2, B-4, and C-1 parachute emergency kits. It was imperative the vest be worn over all flying clothing and under all parachutes. Images demonstrate its use with winter and summer flying clothing. Chest pockets 2 & 6 were left unassigned to accommodate the fastening of the chest strap of the parachute harness. Back pockets 10 & 11 originally contained components transferred to two internal chest pockets 14 & 15. The back pockets were retained but left unassigned for utility purposes. (USAF)

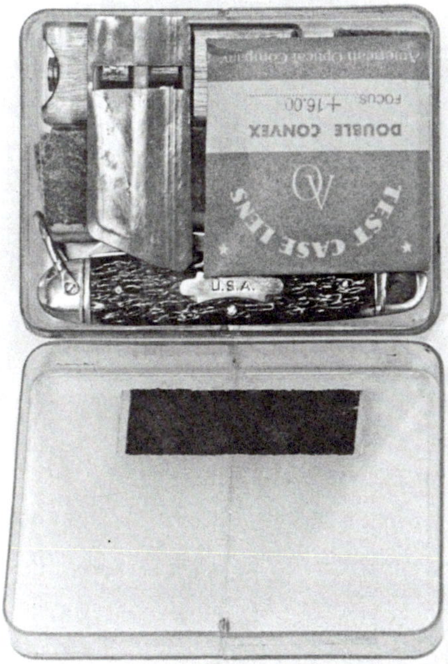

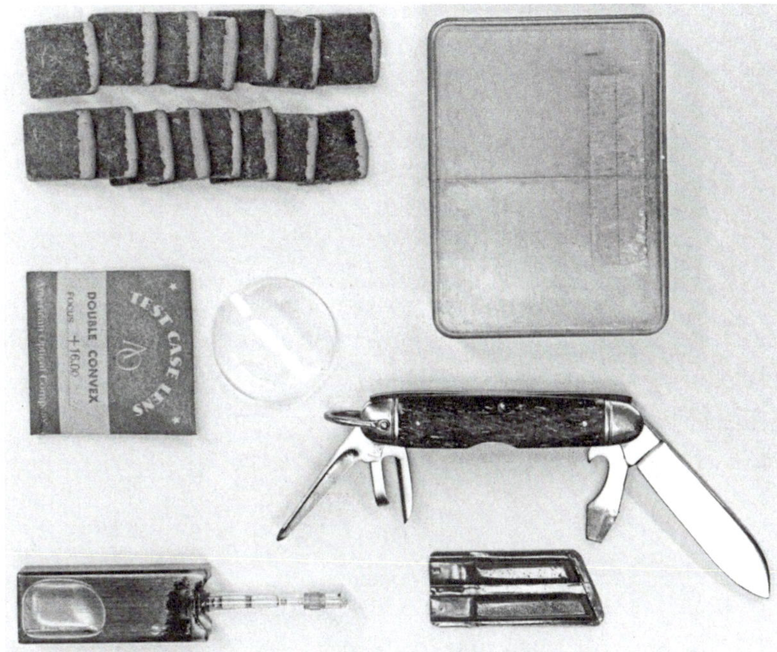

An issue with previous standardized kits was that components were not secured sufficiently, often coming free and bunching at the bottom of the back pads. The items were intended to be secured by a friction fit into the slots in the felt insert. Type C-1 overcame that issue with segregated secured pockets, which also allowed access to those items without having to open the entire kit. Integrated into the vest was the Kit Assembly-Fire Starting referenced under Dwg. No. 44A8835. The box was made of cellulose acetate plastic, and was waterproof when sealed, measuring 4" x 3"x 1". Also provided was a striking surface attached to the lid. Included was a pocket-knife, 14 tabs of fire starters, whistle Dwg. No. 44A8837, double convex burning lens, and oil container. (USAF)

The pocketknife, often mentioned in records as Boy Scout knife, was detailed by specification, drawing and technical order as the US Army Engineer type under Specification No. 17-170. Camillus Cutlery Co. was the principle contractor until late 1944, when a critical shortage of pocketknives was reported and Camillus could not promise additional quantity contracts. Contractors were directed to obtain them from other sources, which also could not promise deliveries, as they were consumed with US Navy contracts for the 41-J-5. Pocketknives were allocated from Navy contracts for the supply of Type C-1 vest contracts when available. The two contractors approved were the Pal Blade Co., and the Imperial Knife Co. The knives supplied were detailed under Navy Department Specification 41-J-5. Top-bottom: Camillus Cutlery Co. 17-170, Pal Blade Co. 41-J-5, and Imperial Knife Co. 41-J-5. The knife at the very bottom was made by Kingston, a subsidiary of Imperial. It was a commercial Scout style knife, stamped Camp Knife. This pocketknife was removed from a sealed kit for referencing purposes. With the critical shortage, these companies would ship a variety to maintain projected delivery schedules.

Items included in the fire-starting kit. In September 1944 delivery of a significant number of vests was delayed, pending receipt of overdue burning lenses. Because the C-1 was urgently required by staging areas and overseas theaters, the AAF Board decided to delete the lens in favor of a more useful item. Contractors were notified in November 1944 to switch to a plastic safety razor. Following this change, a large order of lenses was received by contractors and continued to be included until the razor became available. The change was officially addressed in Amendment-2 of Specification 3206. Plastic razors were provided by the Dillon-Beck Mfg. Co. in a variety of colors. The convex lenses were provided by C.L. Smith and Associates.

Fully assembled fire-starting kits 44A8835. L to R, pre- and first-production examples (44-5039E, 45-411, 45-412), second production example (remaining in production until change to plastic razor), and a late production assembly with razor.

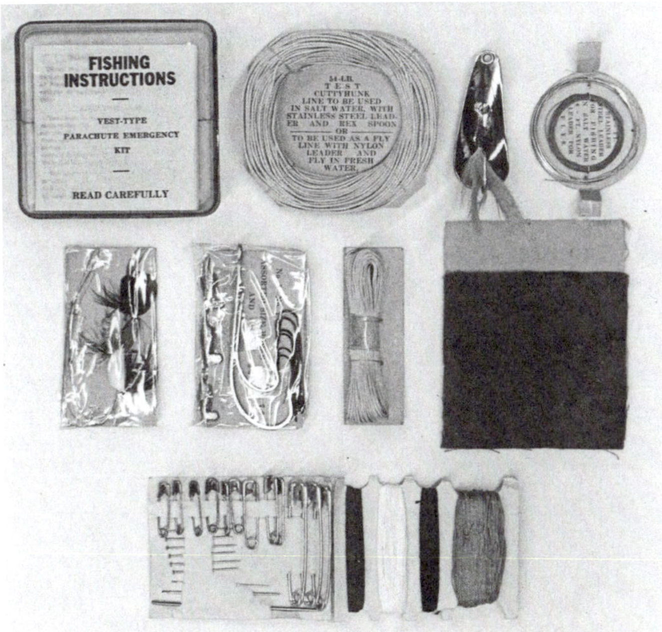

Another unit included was a compact combination fishing-sewing kit, referenced under Assembly Dwg. No. 44B9156, which used a waterproof cellulose container 3" x 3" x 1". Included were three small flies, spoon lure, two cloth patches (yellow and red), 75' of 54 lb. test line, 75' of 18 lb. test line, 6' of stainless steel leader, 6' of nylon leader, eight assorted fish hooks, three sinkers, instruction booklet, 12 safety pins, three needles, six straight pins and 4 types of thread. The contents remained unchanged throughout production of the Type C-1. The original contractor was the S. Buchsbaum Co., and in October 1944 Pachner & Koller Inc. was approved for the assembly.

(Below) Production variants of fishing kit 44B9156, left to right: (44-5039E, 45-411, 45-412), second production and late production types. Instruction sheet in the first variant was printed by the Materiel Command (far left). (USAF)

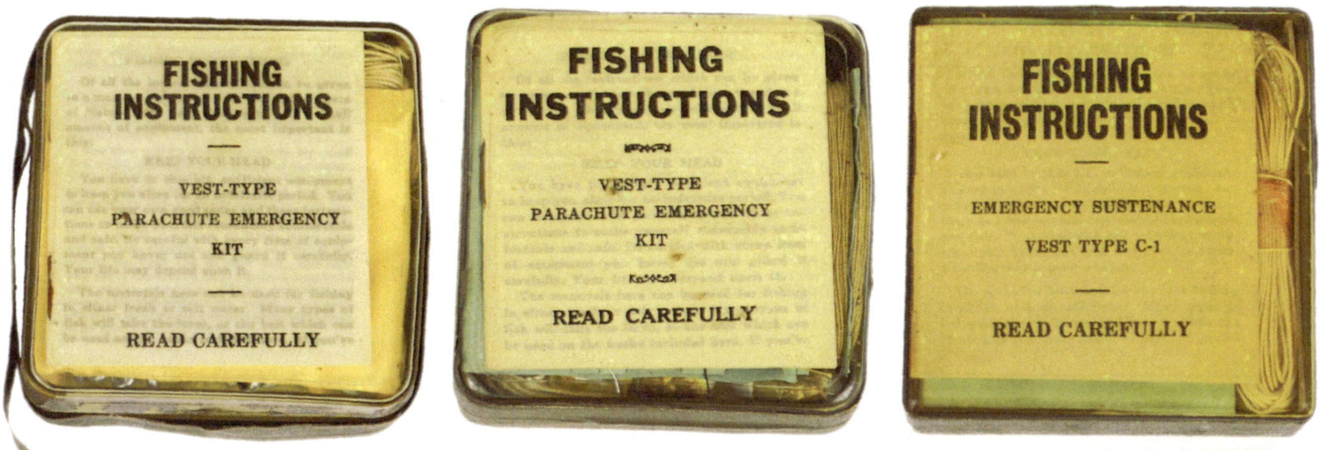

(Below) The first aid complement provided in Type C-1 was a two-part supplement. The bulk of the items was contained in a sealed cellulose container 4" x 3" x 1", referenced under Assembly Dwg. No. 44D9362. This unit included an instruction sheet, two 2" bandages, bar of soap, four adhesive bandages, morphine syrette, vial of 30 halazone tablets, vial of eight salt tablets, vial of 12 Atabrine tablets, vial of eight sulfadiazine tablets, vial of six Benzedrine sulfate tablets and vial of iodine. These were assembled by Burke Drug Supply. Unit on left supplied with 44-5039E, 45-411 and 45-412 production vests.

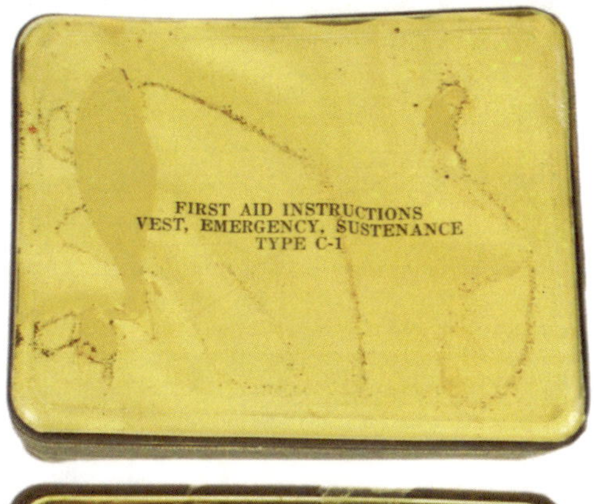

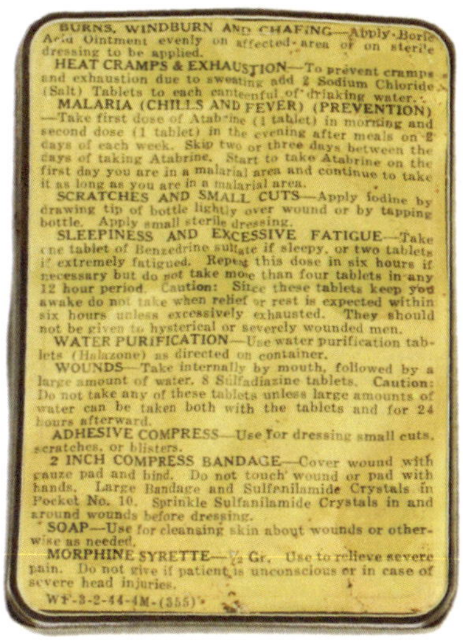

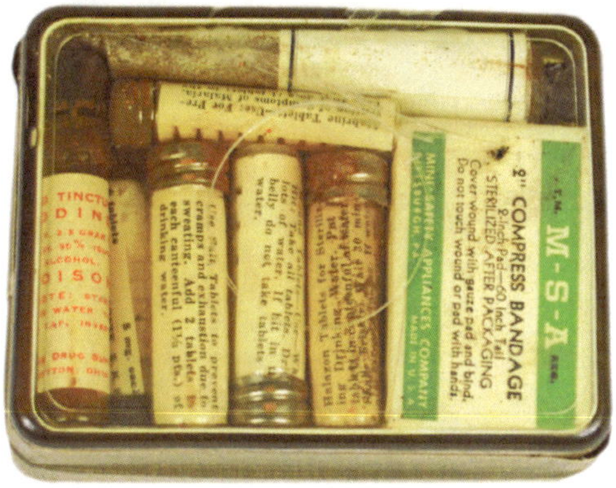

(Left) Supplementing the celluloid first aid container, a sealed manila package 3-3/8" x 7-7/8" x 3/8" contained 3" x 6 yd. and 10 yd. gauze bandages with an envelope of sulfanilamide.

Sunhat 44G5062, green and yellow side out.

Sergeant demonstrating equipment issued to B-29 aircrews at Roswell, New Mexico, May 1945. The M-1942 headnet is part of the Type C-1 being demonstrated.

(Above) For protection against the elements, the sunhat Dwg. No. 44G5062 and mosquito headnet 43G3868 were included with the Type C-1, as well as a pair of goggles and a pair of gloves. Following service test report submitted by AAF Proving Ground at Elgin Field, Florida in September 1944, it was recommended that a mosquito headnet with a face frame or something to keep the netting away from the face be added. Integrated into the assembly of the vests was the QMC M-1942 headnet. Production of the C-1 would continue to include the olive drab cotton mosquito headnet, with the later integration of the green nylon type and M-1942. (USAF)

The three types of headnets issued with the C-1 Vest:ODcotton, the green nylon type, and the M-1942.

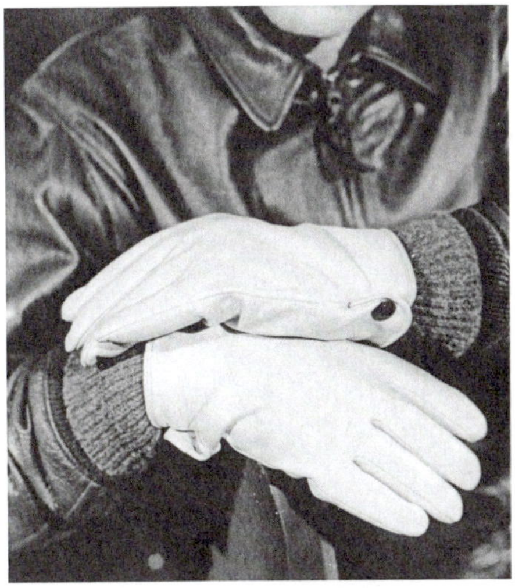

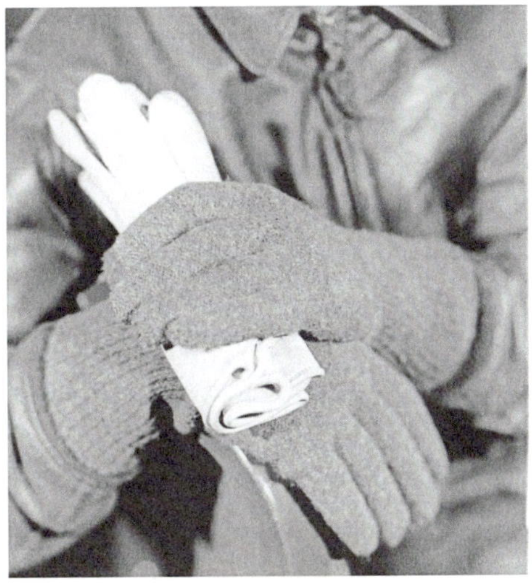

(Left) Design concept of the Type C-1 included Type D-2 mechanic gloves, a transfer from their standard use in emergency parachute kits. When the C-1 was standardized, a more useful glove was adopted, the Gloves-Mechanic Winter, Type D-3, adopted as standard on 30 March 1943. These were comprised of a leather shell with woolen inserts. Wright Field images illustrate their design.

(Below) The Type D-3 gloves underwent a series of amendments to increase comfort and facilitate finger movement. On 14 August 1944 Amendment-A was made effective, revising the securing method with a pull tape, which allowed the glove to be secured more effectively around the wrist, providing protection without the insert. The Type D-3A gloves would supersede the use of the Type D-3 in the Type C-1 sustenance vests, supplied as they became available. Several manufacturer variations of the Type D-3A gloves included with the vest. (USAF)

(Above) Two variations of D-3 gloves.

(Right) Wool inserts for the D-3 gloves.

(Left) In Summer 1944, it was reported that due to extreme shortage of the D-2 and D-3 gloves, Materiel Command authorized the substitution of Type A-10 flying gloves. These Gloves, Flying Winter, were leather with a wool wristlet, adopted as standard on 20 July 1938. A-10 gloves were provided in the first prodcution vests. (Right) On a display board for the components of a Type C-1 is a pair of Type A-10 gloves, part of a visual aid for personal equipment lectures provided by the 2nd Medical Establishment, FEAF, on New Guinea, c.1944.

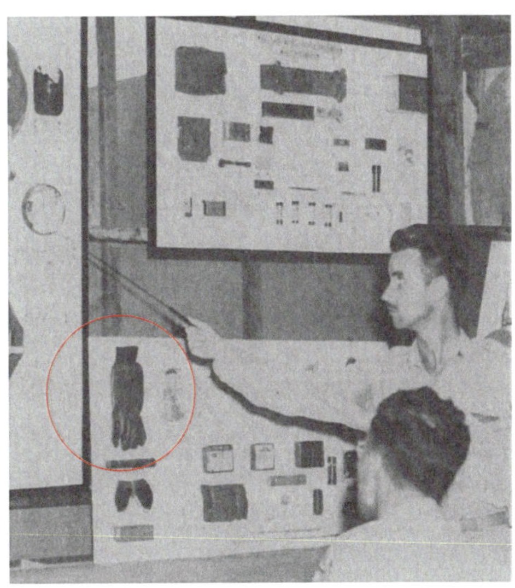

(Right) Insect repellent provided in the Type C-1, referenced under Dwg. No. 45B9008. Because Sta-way insect repellent had proven to be highly toxic and insect repellent provides only a brief reprieve, it was not included in the Type B-4 parachute kit. Following field tests conducted at Eglin Field in late May 1944, Eglin recommended inclusion of an insect repellent, and a 3 oz. can was added under Amendment-1 of Specification 3206. A formula produced by The Davis-Young Soap Co. was approved

(Below) The goggles in the Type C-1 were as specified for the Type B-4 parachute kit. Contractors were authorized deviations to include the QMC M-1943 type, later to be designated as the Type H-1, under USAAF Specification No. 3227. Various packaging methods were used.

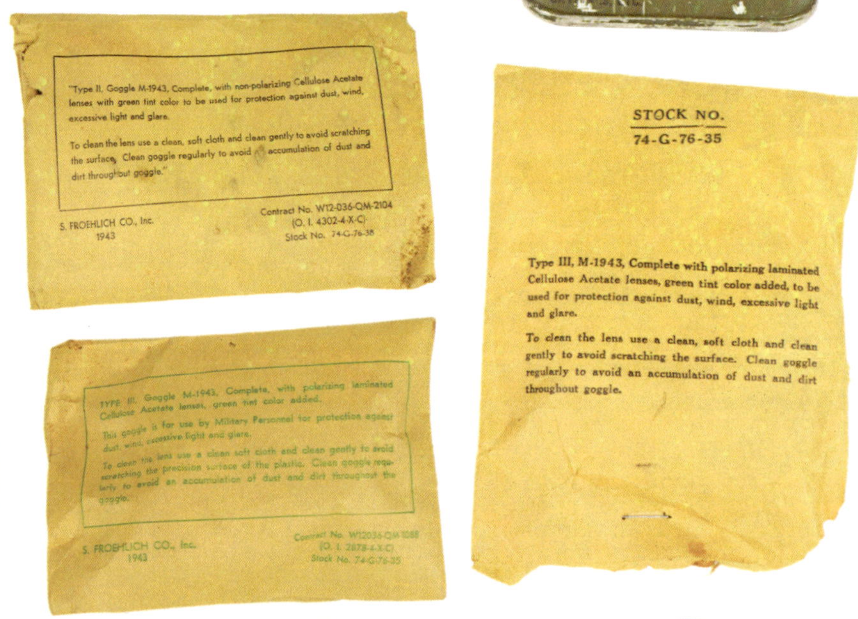

(Below) For fire starting and navigation, a match safe compass was integrated into the vest. The type included in the concept design was the BQD 82. (left) An improved version was adopted upon standardization of the Type C-1, manufactured by Livingston Plastics Corporation. In January 1945, Dupage Plastics Co. (right) was approved for manufacture under the newly approved USAAF Specification No. 94-27610, referenced under Dwg. No. 45A9033. Its version included a flint striker. Both types were manufactured in what was referenced as Olive Drab No.108.

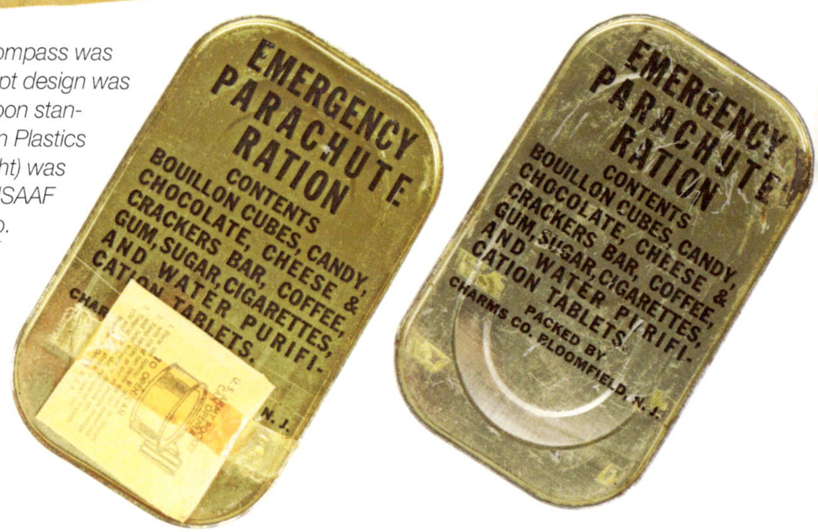

(Above) A new ration was developed that included standard QMC components of the Type-K ration. Concept design had ration in acetate plastic containers. It was immediately viewed as unsatisfactory since it didn't fully provide adequate protection of the perishable items in long-term storage. The QMC Subsistence Laboratory revised the K-Ration and Parachute Ration to a new type for integration in the vest kit. This new ration was adopted as the Emergency Parachute Ration, CQD 302. Two rations and two can openers were included as standard for the Type C-1 in a packet measuring 3-1/2" x 5-1/2" x 1-1/4".

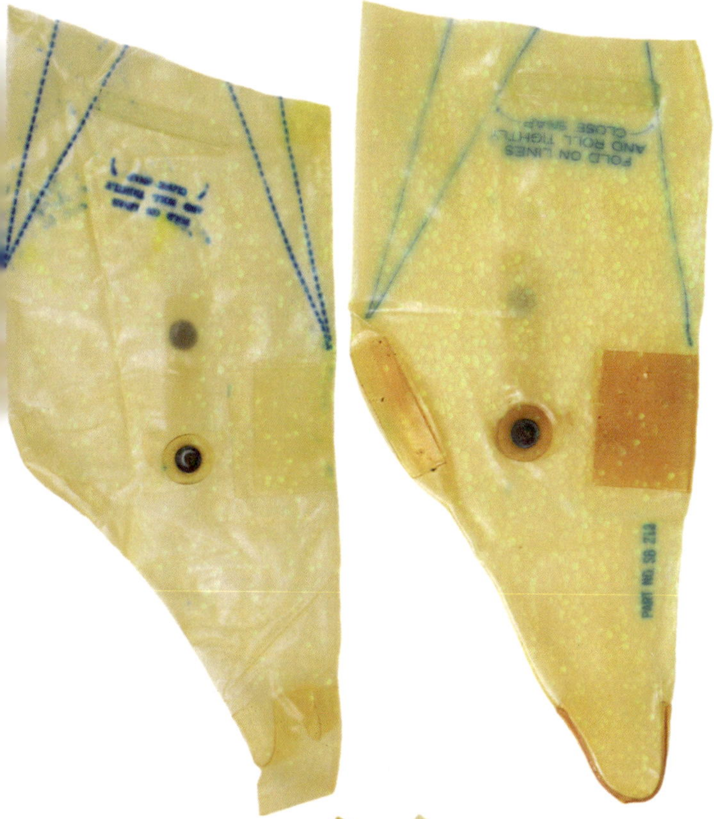

For protection against corrosion of the pistol, a vinylite cover was provided. National Carbon Co. was the sole provider until 23 February 1945, when S. Buchsbaum Co. was approved as a supplier. L-R, National Carbon Co. Part No. 339C, and S. Buchsbaum Co. Part No. SB212.

(Below) Toilet tissue (50 sheets) was provided in a sealed manila package 3-3/8" x 7-7/8" x 3/8", the same as the gauze bandage pack. These two items were stored together in the vest. The packets were detailed to be legibly marked but examples can be found without labeling. The ability to keep clean was vital to staying healthy and maintaining morale. Soap was included in the first aid kit and was one reason the plastic safety razor was chosen.

(Below) Rations were provided in the vest for sustenance, but a fishing kit was also provided, along with a spit & gaff assembly, referenced under Assembly Dwg. No. 45B9011. This item was developed based on reports that fishing kits were often found useless and survivors wished they had a means to spear sea life. Gaff assembly consists of a tin, an instruction pamphlet, and adapters for the spit.

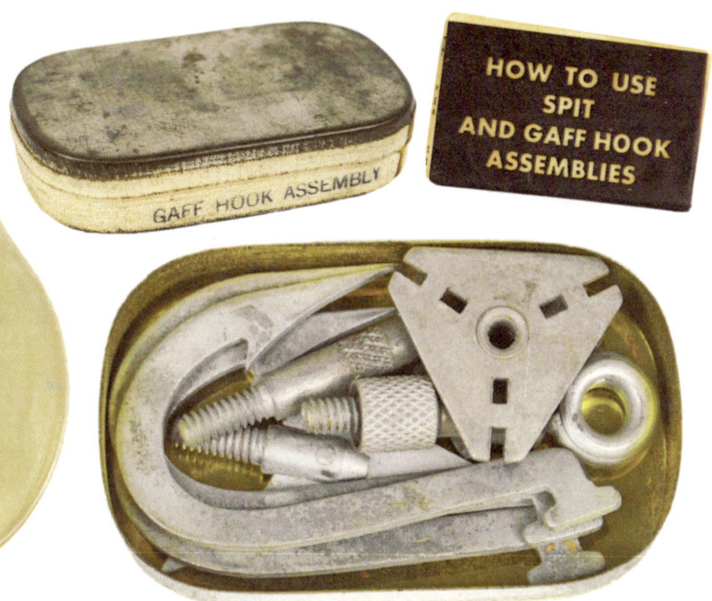

For water storage and purification, a 1 qt. water bladder was provided, referenced under Dwg. No. 43B9963. This bladder had been used in previous parachute kits. National Carbon Co. was the sole provider, referenced under Part No. 233. In 1945, S. Buchsbaum Co. was approved for supply under Part No, SB210. L-R: National Carbon Co. and S. Buchsbaum Co.

(Above and right) Spit assembly consisted of five interlocking cylindrical shafts. Variants of the packaged spit assembly.

(Below) Food preparation was found essential but the frying pan assembly needed to be revised for vest integration. A design of a collapsible cooking vessel presented by U.S. Rubber Co. proved to be the most satisfactory. Adopted as a standard accessory was a collapsible neoprene-lined asbestos vessel for boiling water and cooking food, referenced under Dwg. No. 44D9332. The sole contractor was U.S. Rubber Co. However, AAF Proving Ground found the cooking vessel subject to burning, quickly destroyed by a large fire enveloping over the sides. It recommended modification. Later discussions addressing vest redesign resulted in removing the cooking implement. This order was effective 7 May 1945. Examples of the 44D9332, measuring 12" x 10," with grommets along the edge. Images illustrate vessel suspended over a fire and damage that had occurred during testing.

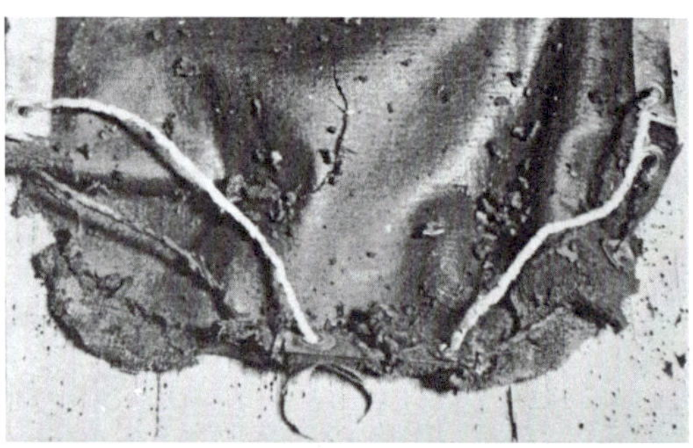

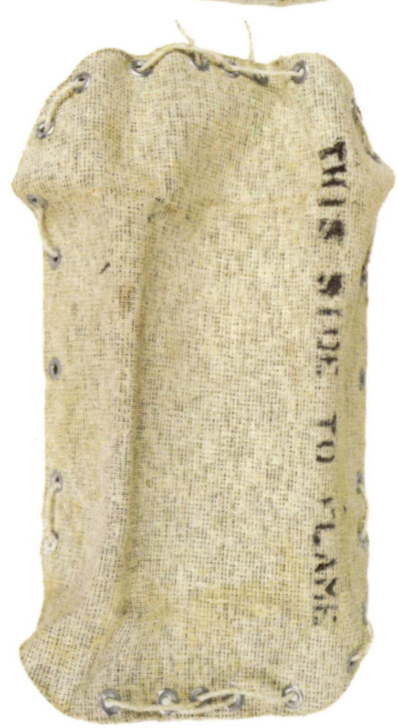

With the emphasis on food gathering and small game hunting, USAAF requested development of shot type cartridges for the .45 cal. automatic pistol. Prior to standardization of Type C-1, the Army Ordnance Department had developed a 7--1/2 chilled shot cartridge designated as M12 and later revised to M15. A 20 round box was provided in the vest. For service test procurements and first quantity procurements (45-411, 45-412), the M12 was provided; when supply permitted, M15 was provided. L-R: Boxed M12 and M15 shot cartridges

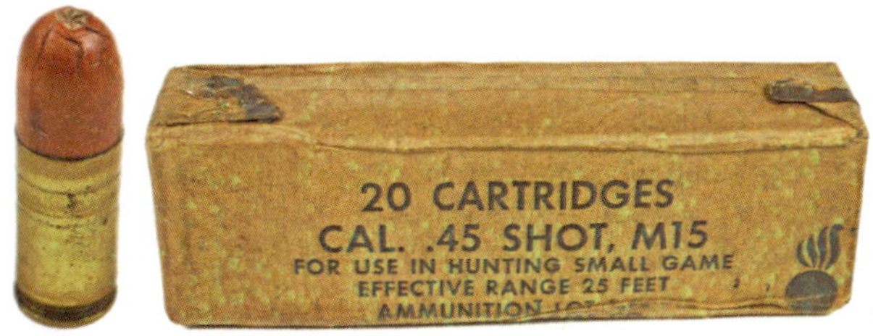

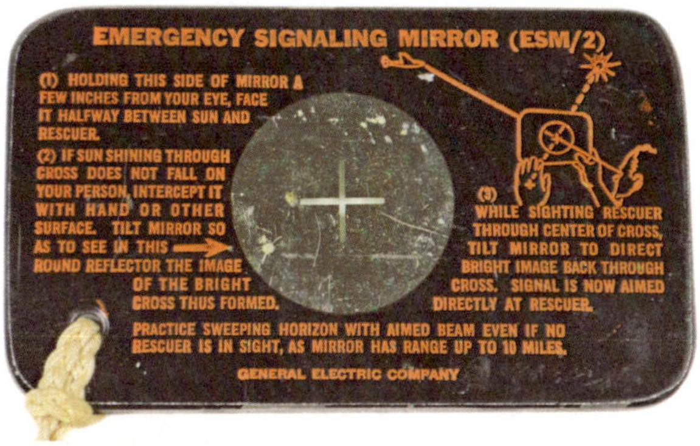

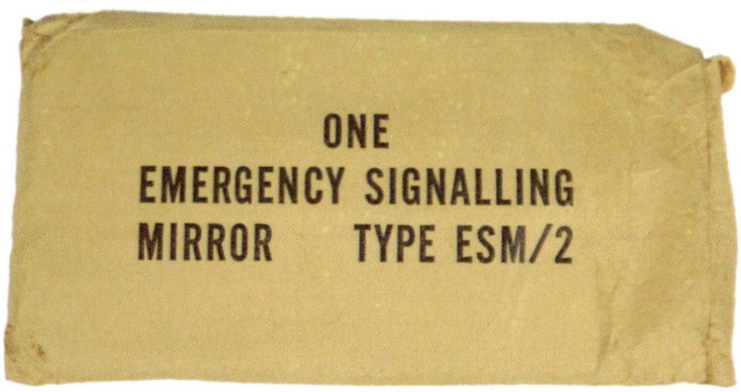

A signaling mirror was added. Initially chromium plated, it was replaced by tempered glass type referenced under Specification No. 40653 of the Size-15 (ESM/2).

For sharpening the knife shgown below, an India medium stone was provided that measured 4" x 3/8" x 3/8".

For distress signaling, the standard complement of the 5-minute fusee flare for parachute emergency kits was included, referenced under Dwg. No. 43B3554. Two were included as standard.

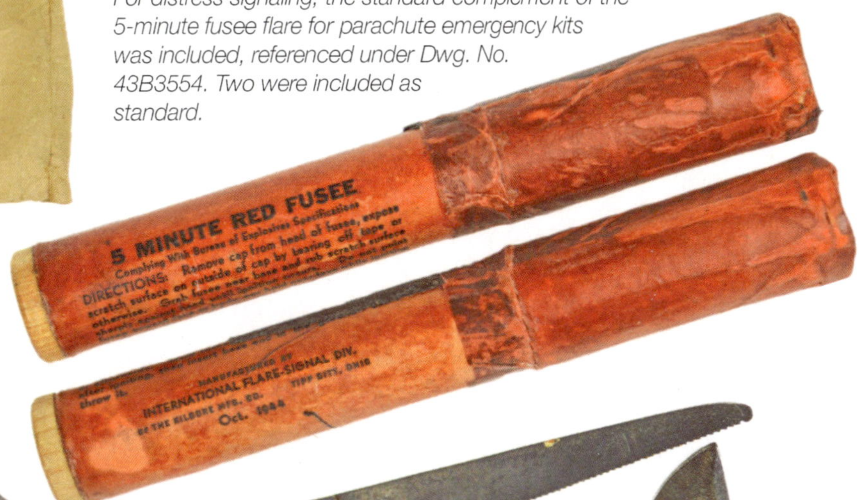

Developed specially for the vest kit was a large folding hunting knife. The main blade was 5" long and made of tempered steel to withstand hammering for wood splitting. A 5" saw for cutting wood and metal was part of the assembly. The knife was referenced under Dwg. 44D5159 but later revised to Dwg. 45D9060. Colonial Knife Co. was sole contractor until January 1945, when United Machine Tool Co. was approved for manufacturing. Shown are examples by Colonial Knife Co., and United Machine Tool Co.

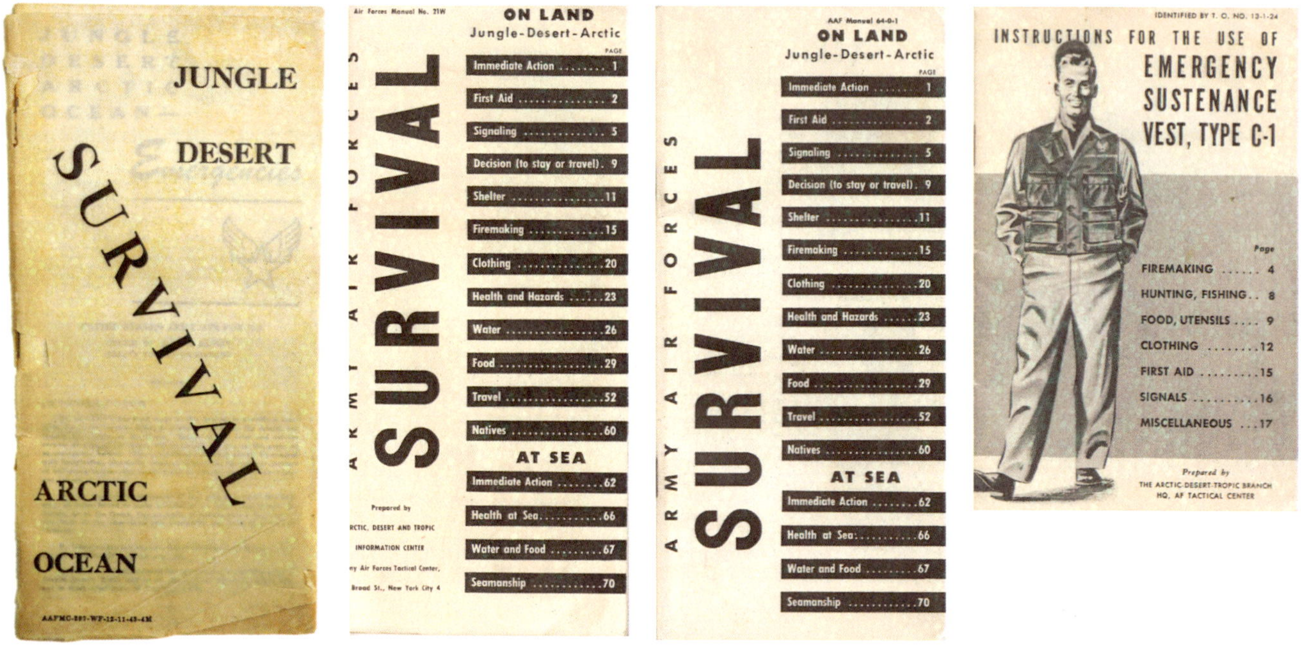

When Type B-4 kit was procured, survival literature had been prepared and was in circulation for service distribution. The emergencies booklets were placed in all emergency kits. When specifications for Type C-1 were approved, a new survival manual Jungle, Desert, Arctic, Ocean Emergencies was approved for inclusion, published by the Office of Flying Safety. Materiel Command developed a condensed version of the manual, which was approved for mass publication, becoming the official AAF manual for emergency use, as the AAF Manual 21W. After July 1944, it was directed to be included in all emergency equipment, service test, and quantity procurements of the C-1 used older emergencies manuals until the Manual 21W became available. L-R: Materiel Command-printed manual, AAF Manual 21W, and AAF Manual 64-0-1 (late production vests would include this revised version). Following reports that individuals were not clear on use and care of components, an instruction booklet was prepared, designated as Technical Order 13-1-24. Upon availability, these booklets were provided with the vest accompanying manual 21W.

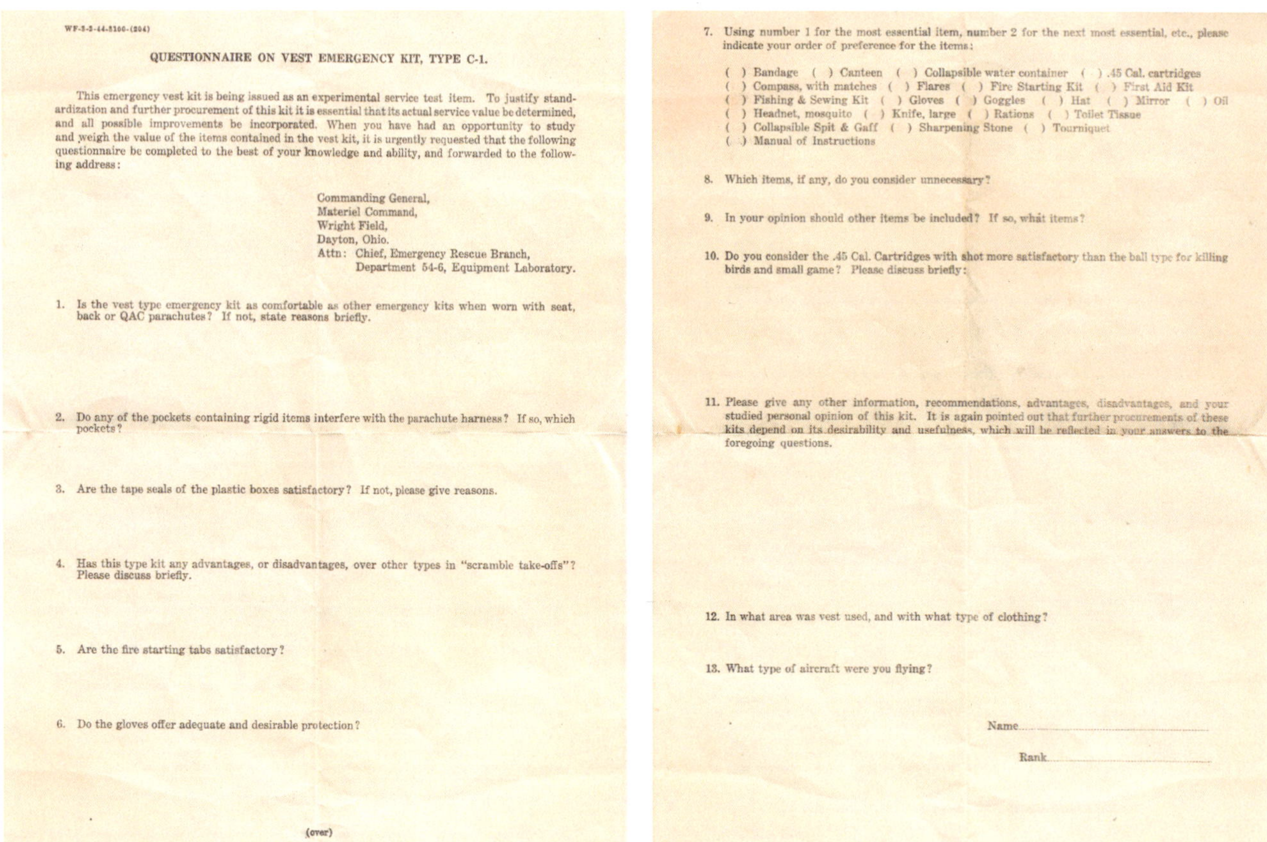
WF-3-3-44-3100-(204)

QUESTIONNAIRE ON VEST EMERGENCY KIT, TYPE C-1.

This emergency vest kit is being issued as an experimental service test item. To justify standardization and further procurement of this kit it is essential that its actual service value be determined, and all possible improvements be incorporated. When you have had an opportunity to study and weigh the value of the items contained in the vest kit, it is urgently requested that the following questionnaire be completed to the best of your knowledge and ability, and forwarded to the following address:

Commanding General,
Materiel Command,
Wright Field,
Dayton, Ohio.
Attn: Chief, Emergency Rescue Branch,
Department 54-6, Equipment Laboratory.

1. Is the vest type emergency kit as comfortable as other emergency kits when worn with seat, back or QAC parachutes? If not, state reasons briefly.

2. Do any of the pockets containing rigid items interfere with the parachute harness? If so, which pockets?

3. Are the tape seals of the plastic boxes satisfactory? If not, please give reasons.

4. Has this type kit any advantages, or disadvantages, over other types in "scramble take-offs"? Please discuss briefly.

5. Are the fire starting tabs satisfactory?

6. Do the gloves offer adequate and desirable protection?

(over)

7. Using number 1 for the most essential item, number 2 for the next most essential, etc., please indicate your order of preference for the items:

( ) Bandage ( ) Canteen ( ) Collapsible water container ( ) .45 Cal. cartridges
( ) Compass, with matches ( ) Flares ( ) Fire Starting Kit ( ) First Aid Kit
( ) Fishing & Sewing Kit ( ) Gloves ( ) Goggles ( ) Hat ( ) Mirror ( ) Oil
( ) Headnet, mosquito ( ) Knife, large ( ) Rations ( ) Toilet Tissue
( ) Collapsible Spit & Gaff ( ) Sharpening Stone ( ) Tourniquet
( ) Manual of Instructions

8. Which items, if any, do you consider unnecessary?

9. In your opinion should other items be included? If so, what items?

10. Do you consider the .45 Cal. Cartridges with shot more satisfactory than the ball type for killing birds and small game? Please discuss briefly:

11. Please give any other information, recommendations, advantages, disadvantages, and your studied personal opinion of this kit. It is again pointed out that further procurements of these kits depend on its desirability and usefulness, which will be reflected in your answers to the foregoing questions.

12. In what area was vest used, and with what type of clothing?

13. What type of aircraft were you flying?

Name..........

Rank..........

Two-sided questionnaire accompanied the vests under service test procurements. Individuals using vest were instructed to complete form to provide feedback on basic operational performance. It was sent to the Materiel Command for dissemination.

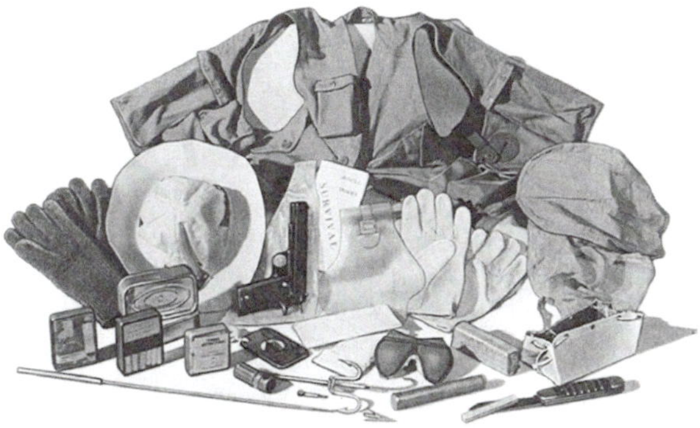

A reported issue with all emergency kits was missing items, either due to improper inspection or packing, or pilfering. During vest kit development, sealing the pockets was stressed. Lead seals were first considered, but the adopted method was securing the pockets with a thread and a speed-lock seal. E.J. Brooks Co. was the approved supplier of the speed-lock seals, which were adapted from commerce for attaching tags to garments. Once the thread was weaved through pocket and flap, the seal could be easily closed with pressure of thumb and first finger. These locks helped prevent theft and allowed spot inspection for compromise of components. Technical Order 00-30-196 states When a seal is found broken, it would be the duty of the inspector to report at once in writing to the Supply Officer, who is responsible for a subsequent investigation and who would inspect the kit, replace all missing articles and reseal it. Here, examples of speed-lock seals and secured pockets. Seal on first aid pocket on a Type C-1 worn by B-29 navigator, c.1945.

(Right) Personal and lifesaving equipment displayed for lecture, 500th BG, Saipan, July 1945. On the floor is Type C-2 pararaft, Type E-5 sustenance kit, and components of SCR-578. On the wall, second production Type C-1, recognized by the slanted chest and insect repellent pockets. Note envelope for goggles of the type provided for the Type E-1 Dark Adaptation, but "Dark Adaptation" has been blacked out.

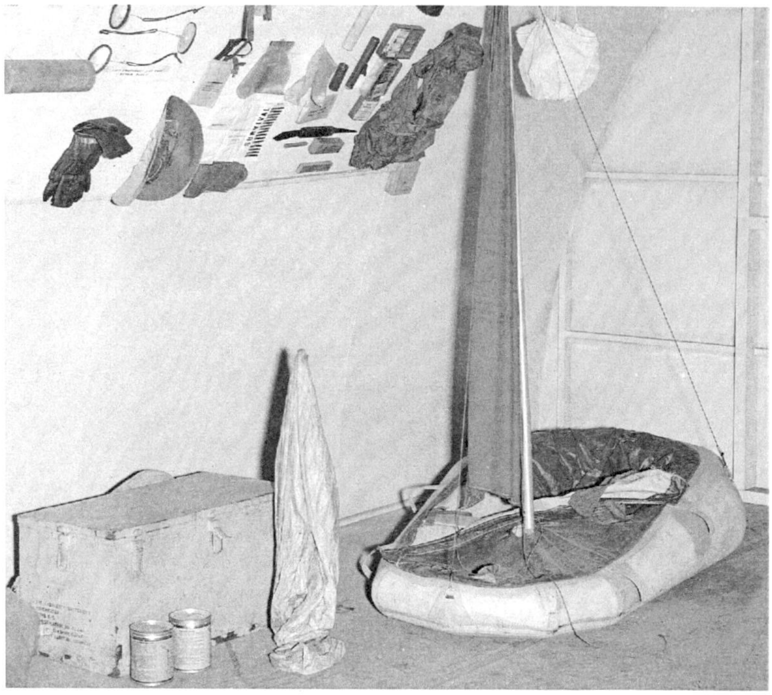

(Below) Volunteer exhibiting flying safety complement during lecture. Intended to supplement the Type C-1 were personal aid kits Types E-3 or E-17. Canteen was a supplemental aid. As of March 1946, Types E-17 and E-3C were still classed as Standard.

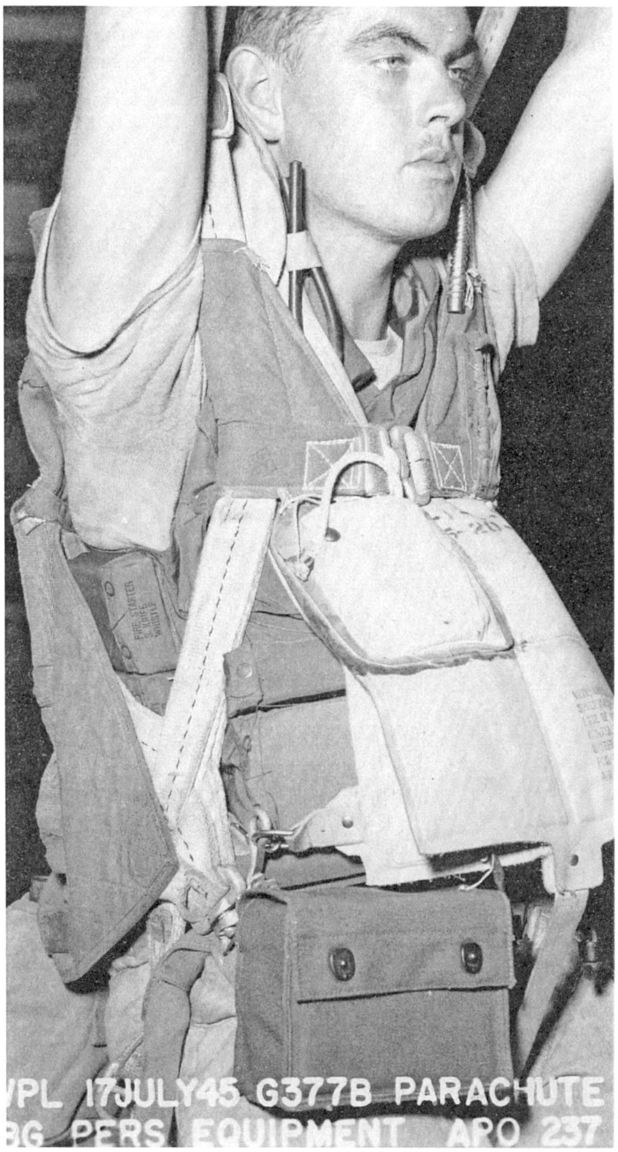

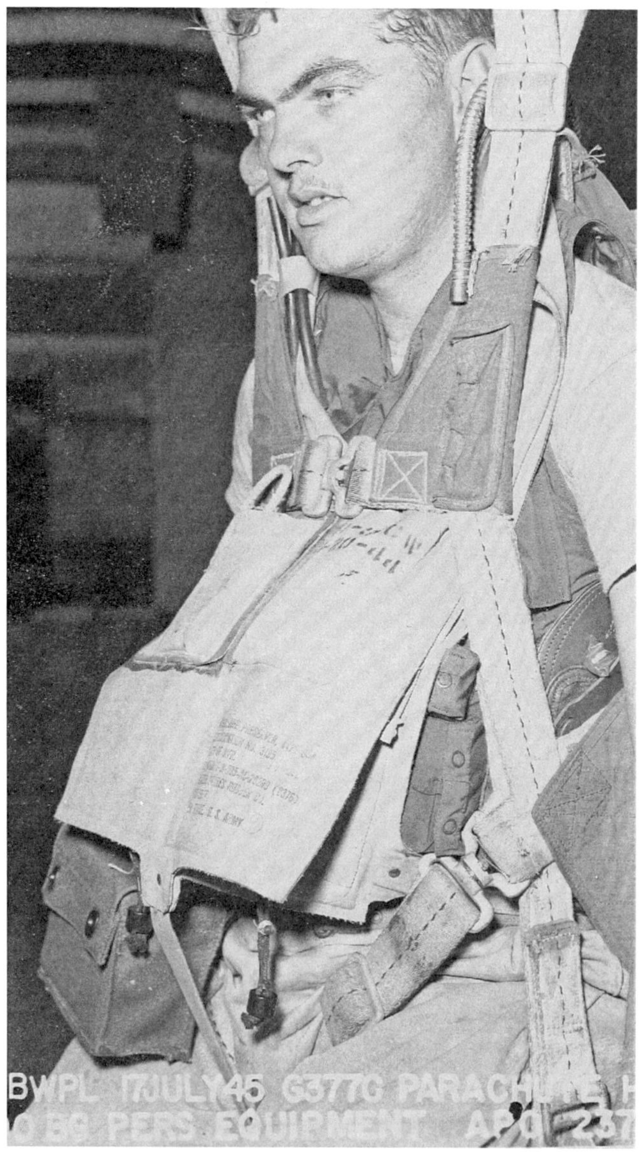

B-29 aircrew members pose in front of their airplane at Iwo Jima after being forced to crash from damage during air raids over Tokyo, March 1945. Six wear Type C–1. Reflection of inspection seals can be observed on some pockets; many can be seen on vest of the member at far right. (National Archives)

(Below) B-29 aircrew members of the 468th BG with C-1 sustenance vest, India, November 1944. Some seemed to prefer retention of M3 shoulder holster worn under vest. Without vest provision for two extra clips, standard allotment of pistol belt and magazine pouch were retained. (National Archives)

(Below) B-29 pilot of the 330th BG, Maj. Howard McClellan returns from mission over Japan, c.1945. C-1 vest under back type parachute.

The third production vest had components from the earlier procurements in a progression of integration as new items became available. Though not a great quality photograph, this display illustrates that aspect well. Some features were the pair of Type D-3 gloves and the survival manual provided by the Materiel Command.

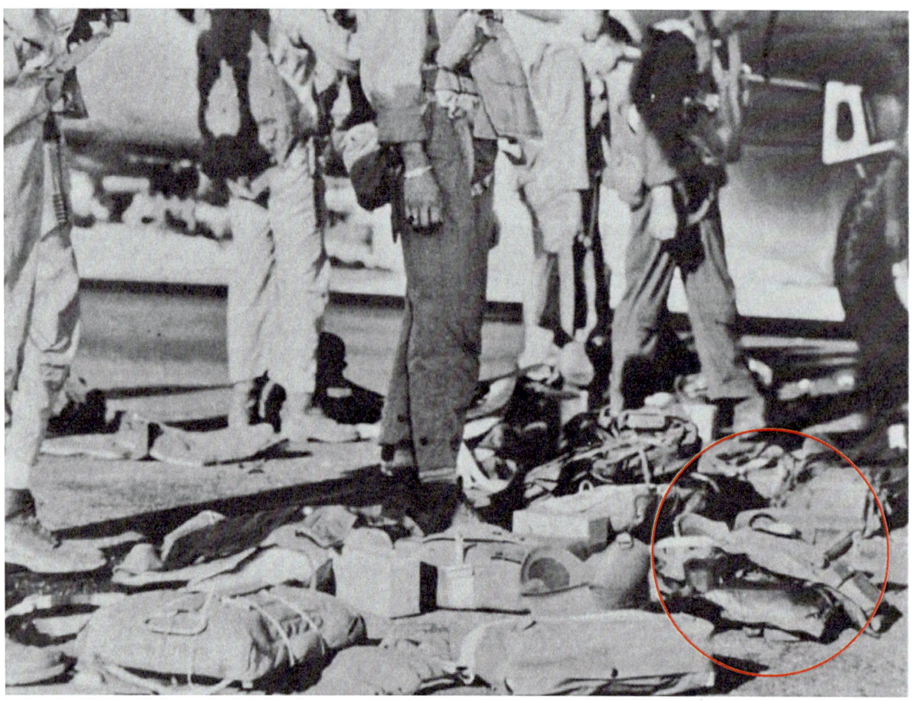

(Above) One of the principal points of Type C-1 issuance was at Staging Areas for overseas embarkation, where all new available items were allocated as per T/O&E. Shipments were also expedited to overseas stations. B-29 crews with flying equipment laid out for inspection prior to bombing mission over Tokyo from base in Guam, March 1945. Note modified Type C-2 pararaft cases and Type C-1 vests. (National Archives)

(Right) B-25 aircrew member going over last-minute details at base in India prior to a bombing run over Rangoon, Burma, c.1944. He wears a C-1 with features suggesting it is from service test procurement or first procurement. (National Archives)

Type C-1 sustenance vest was designed to overcome deficiencies of previous parachute emergency kits by providing all essential items a survivor may require as compactly as possible. It was intended to be universal for all world regions, over land or sea. Type B-5 was developed to supplement the vest by providing protection from elements. (above) Full flying safety complement. Assembly included Types B-8 parachute, C-1 vest, C-2 pararaft, and B-5 kit.

(Left) Simulated survival scenario.

(Top left) The Type B-5 assembly is used. (USAF)

On 20 February 1945, a conference explored the final report from AAF Proving Ground on the Type C-1, dated September 1944. Conference resulted in proposals to furnish vest with a basic complement and to have supplementary items requisitioned as required. This reduced weight and bulk. On 15 March 1945, AAF Board approved the research project and noted other problems with production of the Type C-1. Changes were to both eliminate and add new items. The new kit, designated Vest, Emergency Sustenance, Type C-2, retained from the standard complement the first aid kit, sharpening stone, one ration tin, signaling mirror, shot ammunition box, fire starter kit, fishing-sewing kit, water bladder, gauze bandage, match safe compass, and 21W manual. Items to be added were signals (2 of T-49 and 1 Mark 1, Mod 0 handheld smoke), water bladder, box of .45 cal. tracer, and hand-energized flashlight. The Proposed Type C-2 pockets were modified to accommodate the revisions referenced under Dwg. No. 45G18989.

To adapt to the conditions in any theater, supplemental items were available for requisition then placed in the vests at the discretion of the local command. At this time development began on a Basic Airplane Kit that would further supplement the Type C-2 vest. The large knife was eliminated in favor of a machete in the basic airplane kit and provided with a sheath for attachment to trouser or pistol belt.

By January 1946, all Type C-1 vest contracts were cancelled, so no further action was taken to standardize or procure the Type C-2. The basic airplane kit was later standardized as Type E-18 sustenance kit, with Type C-1 being retained as the standard. Optional supplemental items suggested for the Type C-2 were integrated into kits packaged in the Type A-13 sustenance containers.

(Below) Proposed Vest, Emergency Sustenance, Type C-2. Vest relatively remained unchanged except enlargment of the two pockets for distress signals, large enough to contain T-49 and Mark 1, Mod 0. One of each per pocket. Pocket for spit and large knife repurposed for storage of tracer ammunition. Articles on right are of the optional supplemental items for the arctic and jungle which were to be stored in attachable containers. (USAF)

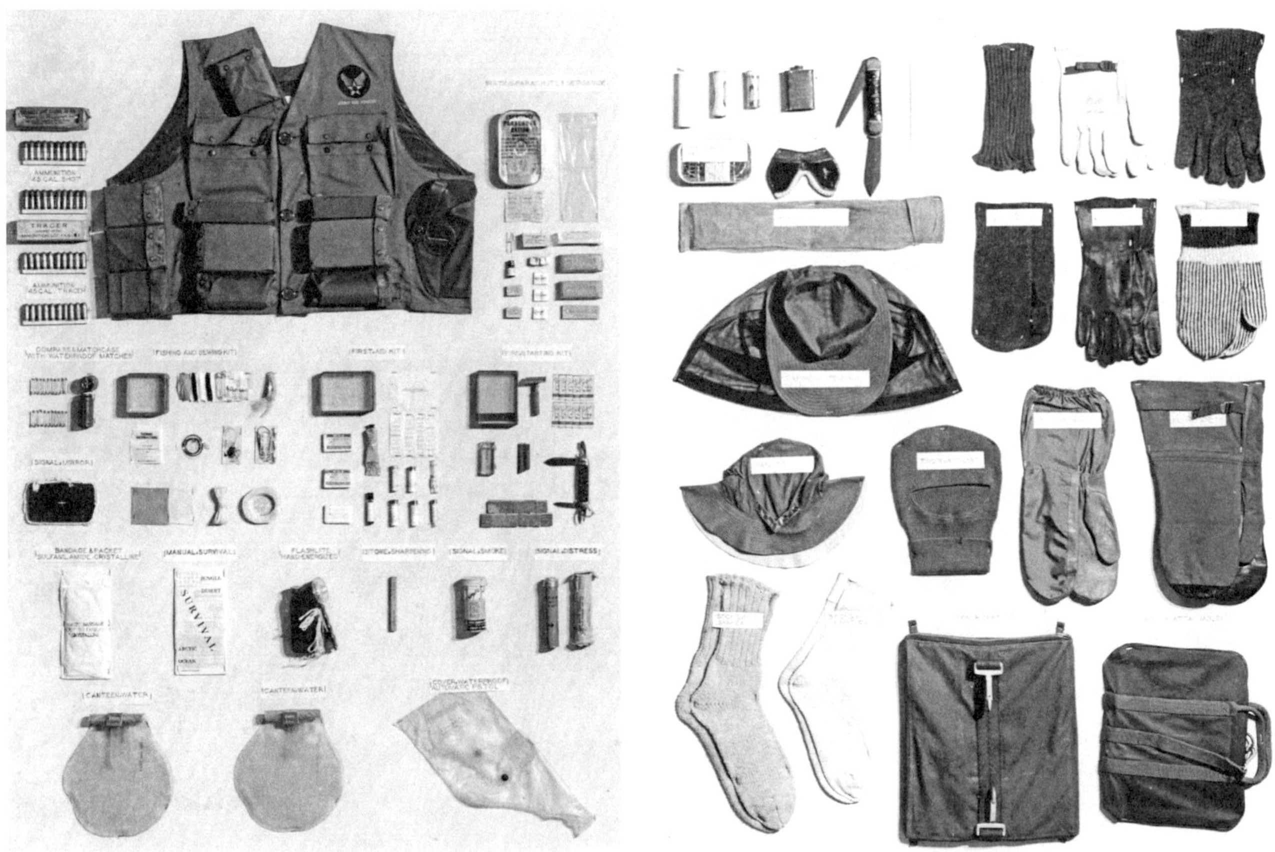

By late 1944, theater commands suggested some form of a bag to attach to parachute harness to carry the C-1 vest because many found the vest too hot for constant wear and were disinclinedto wear it at all. By having the vest available separately in a quick-attachable container, it would eliminate fitting the parachute harnesson top of the vest. When the vest was not worn, the already-fitted harness presented a safety issue. The Personal Equipment Laboratory then intiated a research project to develop the tentatively titled Pack, Quick Attachable.

It was concluded that the vest should always be worn to assure that essential survival items remain with the individual during an emergency, and a standard quick-attachable bag was abandoned. However, alternative to wearing the vest was illustrated in the revised Reference Manual for Personal Equipment Officers 55-0-1 of June 1945.

Drawing included in the manual was that of the standard musette bag as a quick-attachable container. The illustration included the material requirements, method of sewing and method of attachment.

The Air Technical Service Command makes note that they could not recommend that you should or should not carry the C-1 vest in the musette bag in the illustration, which was intended merely to provide a carrying case in the event your Air Force or Command authorizes the use of the C-1 vest in that fashion. They further state that it should be borne in mind that the reason for designing the emergency kit in the form of a vest was to insure that it would be with the person at all times. The decision was for organizations in the theaters as to whether it could safely serve its purpose if carried in the attachable bag.

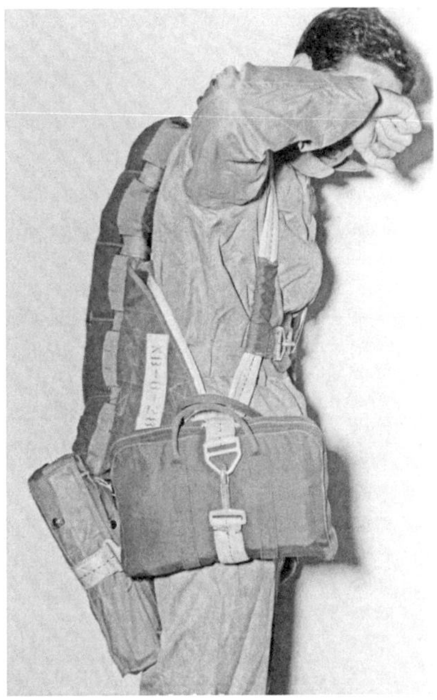

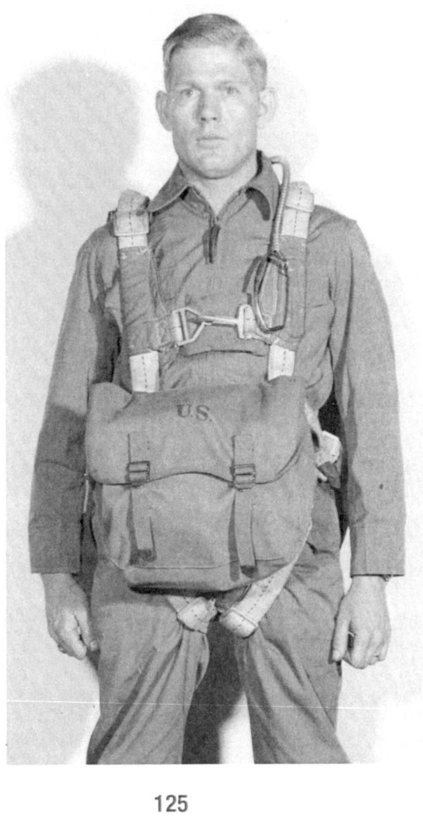

A variety of concepts tested by the Materiel Command from November 1944 to February 1945 using standard QMC musette bag and a specially fabricated zippered bag for carrying the Type C-1 sustenance vest. (USAF)

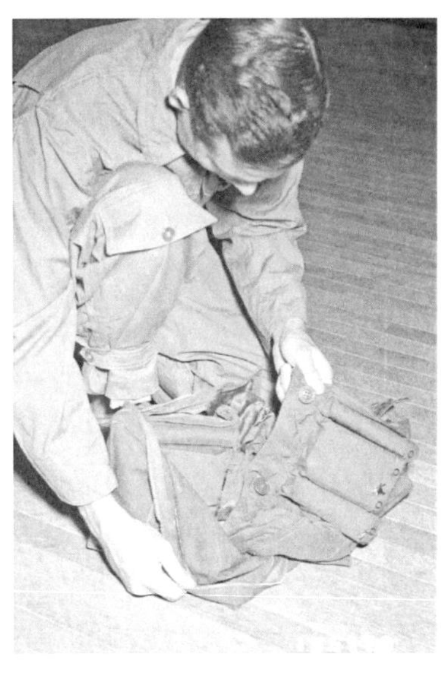

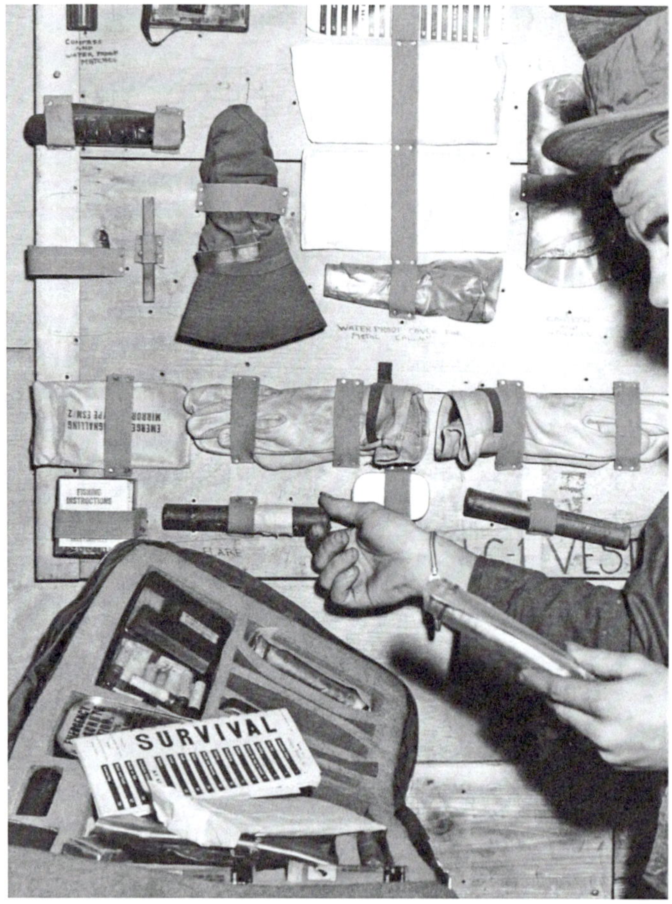

The Type C-1 remained the standard individual emergency kit in the post-war years to the Korean conflict, supplemented by the Type E-18 sustenance kit, and other jungle and arctic assemblies. On 14 July 1950, Specification 3206 was redesignated as Specification MIL-V-6833. One year later, 17 July 1951, the C-1 sustenance vest was declared obsolete and replaced by new line of arctic and jungle kits better suited for ejection seat systems. (Left) USAF member undergoing survival training in Alaska, wearing a 3rd production vest and sunhat with M-1942 headnet, D-3A gloves, and holding water bladder. (Right) Components of the Type C-1 on the display board in a lecture room, Korea, c1952. More Type C-1 components seen packed in CNU-1/P sustenance kit container or similar, c.1951. (National Archives)

INDIVIDUAL EMERGENCY KITS

Part II.
U.S. NAVY

In April 1937, a VJ utility squadron operating from Fleet Air Base Coco Solo, Panama Canal Zone, devised an emergency kit for use in conjunction with the standard seat type parachute, titled Emergency Jungle Seat Pack. This kit integrated a canvas case attached to the parachute pack and contained a .45 cal. 1911 automatic pistol with five clips, hunting knife, machete, sealed tin with three chocolate bars, hard crackers, two tins of concentrated beef, a first aid kit, fishing line with assorted hooks, cigarettes, matches, salt and a mosquito headnet. This was the BuAer's first documented attempt to provide protection and aid for individuals in a forced landing. The kit was influenced by USAAC jungle kits in use by adjacent squadrons operating in the same region. Its significance was as a personal kit.

However, this philosophy was not shared by the command of the BuAer when the pack was submitted for review and standardization was recommended. The decision not to pursue development of a personal emergency kit was largely influenced by the fact that aircraft conducted limited operations over land. Most flight paths were over water, and each aircraft was supplied with a life raft and life preserver for aircrew members. This met the requirements at the time. The suggested emergency kit would be ahead of its time, becoming relevant only later when Naval and Marine aviation units began operations in the SWPA in 1942.

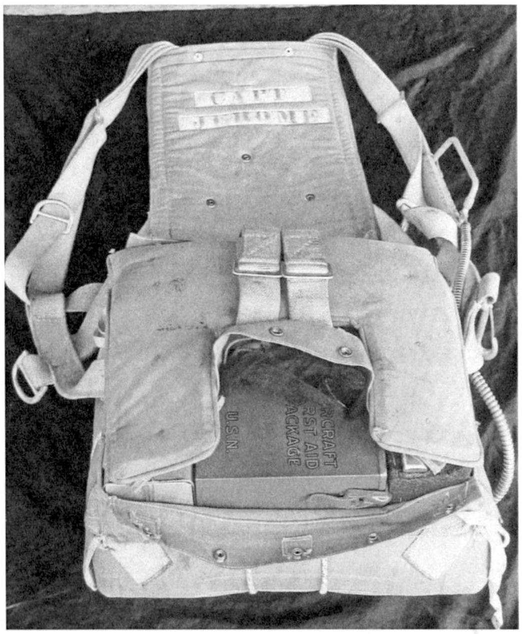

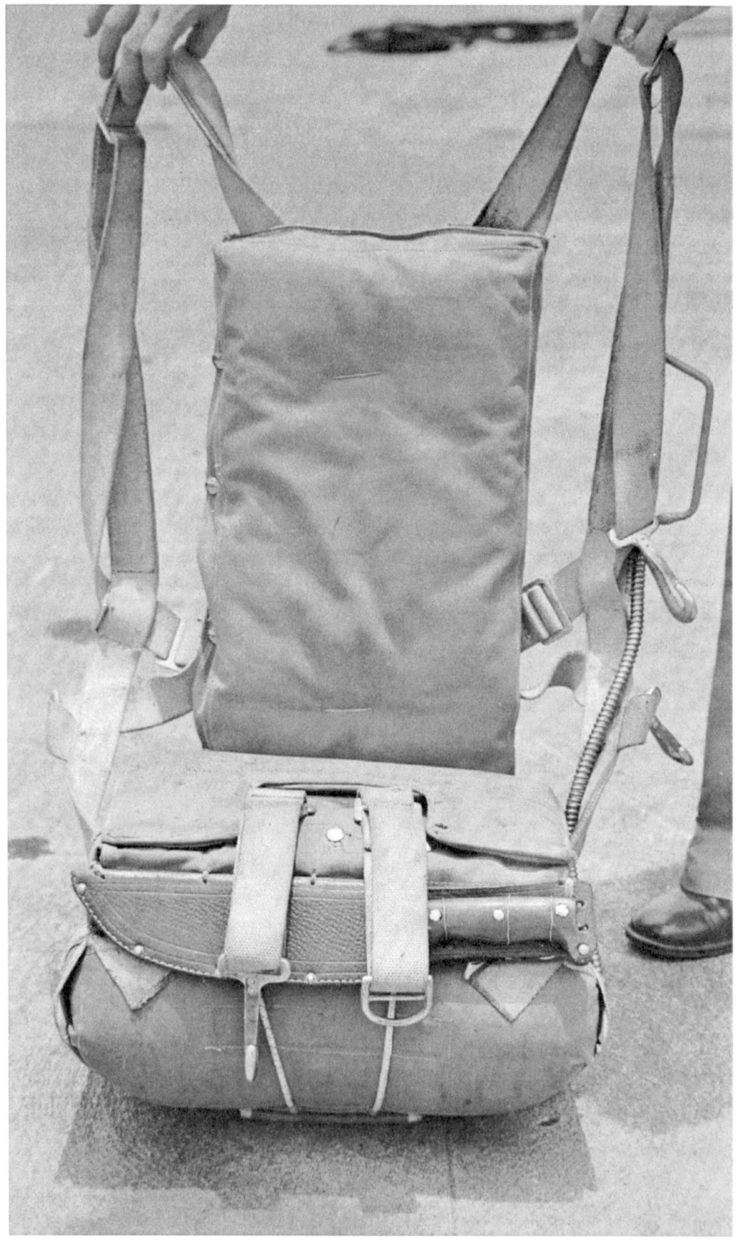

Emergency jungle seat pack developed at Coco Solo, C.Z., April 1937. Developed in a coordinated effort between the Flight Surgeon Lt. Comdr. D.V. Armstrong, Operations Officer Lt. J.P.W. Vest and Lt. C.W. Crawford. Designed to replace the standard kapok seat cushion on parachute packs. (National Archives).

Development of personal emergency kits was not taken seriously by the BuAer in the prewar years. Efforts were focused on over water equipment specifically oriented toward aircraft life rafts. Emergency life raft provisions were significantly improved by the BuAer in 1941, when it updated the M-3 series and introduced a new concept of a one-man parachute type. In April of that year, the emergency equipment container was introduced to the M-3 series. It contained a first aid kit, emergency rations, sea marker and smoke grenades. Subsequent revisions added water, a fishing kit and a sailcloth. This equipment container was a separate unit stored with the raft package and attached by a lanyard. All four raft sizes under Specification M-3 used the emergency equipment container; the only difference was the quantity of rations and water. Of particular interest is the M-3 MK-I one-man life raft, installed as standard equipment in the F4F Wildcat, which was the primary US Navy and Marine fighter from 1940-1943.

The pararaft was intended to replace the MK-I, allowing the raft unit to be secured to the user. To make a suitable raft pack to be used in conjunction with the parachute, the unit had to be streamlined, limiting space for emergency provisions. Up to this point, the BuAer was only considering personal emergency equipment, but, because of expanded combat engagements in the SWPA, a new service requirement determined it necessary that all pilots and aircrew be provided with a personal complement. Storing life rafts and emergency provisions in a location separate from the individual proved highly problematic, especially in the event of a bailout in which none of these items could be retained, leaving only the life preserver.

An individual only has a few seconds to escape an aircraft in a ditching or a bailout. There is no time to gathering emergency equipment. By having all the necessary items attached to the user, chances of survival and sustaining life would increase. As part of the initial emphasis of the pararaft, the MK-I emergency equipment container, with its complement of provisions, would be retained for use and storage aboard the aircraft.

In June 1942, to further assist service requirements, the NAF began design of a personal emergency kit to be worn with the parachute harness. The idea was to take the provisions from the equipment container and place them in a backpack that would remain with the individual. In July, 500 test samples intended to replace the standard parachute back cushion were fabricated under Naval Aircraft Factory Dwg. No. 69233, Revision-2, designated as the Back Pad Equipment Kit. In August 1942, limited service test units were available to the fleet departing the continental US for the combat zone. Kits were also sent to the fleet operating in the forward area via expedited shipment.

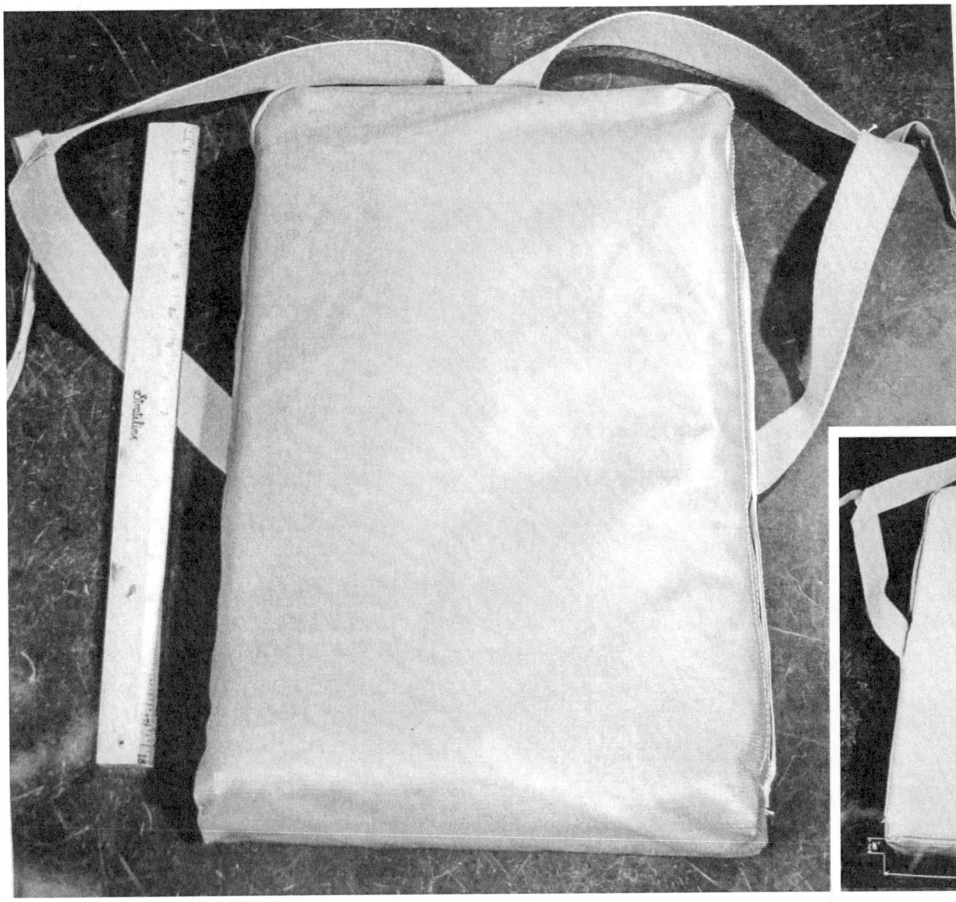

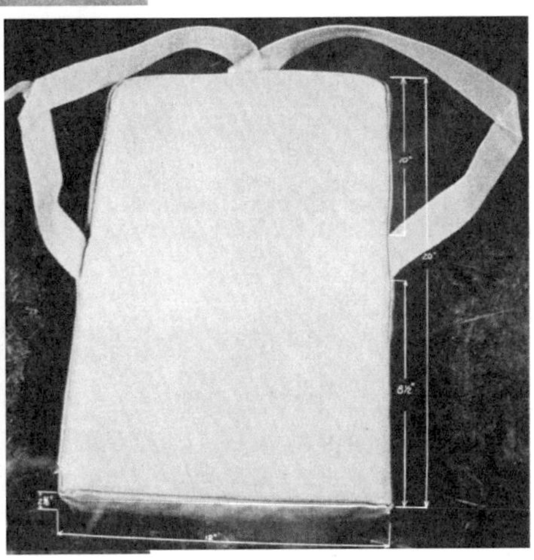

Front and back views of the NAF-69233 back pad emergency kit. Dimensions: 20" x 12" x 2-3/8". An 18" zipper is located only on one side to reduce water seepage into the kit, and shoulder straps were adjusted by tie tapes across the chest. (National Archives).

(Right) Right side view of the NAF-69233 illustrating the 18" zipper closure. (National Archives)

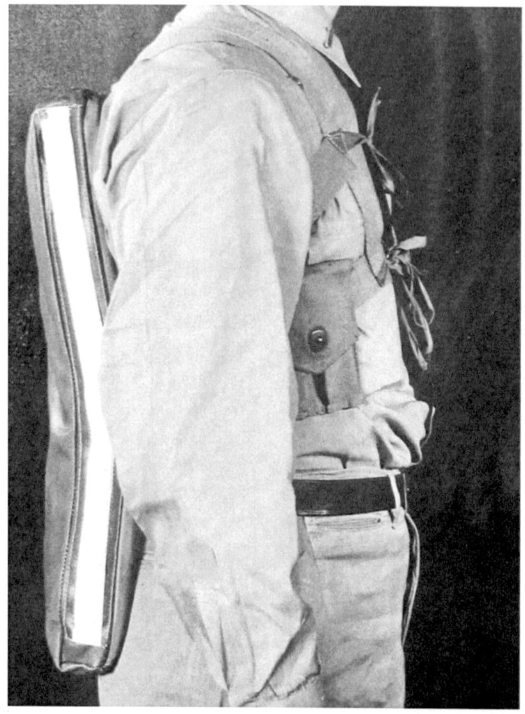

(Below) NAF-69233 kit broken down to show its construction and components. It included two red MK-I Navy Lights, two U.S. Army Field Rations D, fishing kit, first aid kit, specially fabricated canteen, match safe compass, whistle, signal mirror, Boy Scout knife, 50' of fishing line, 24" wire leader, mosquito headnet, can of preservative oil, the booklet Castaways Baedeker and an OD poncho fabricated under NavAer Specification M-549. Kit was fabricated with waterproof khaki fabric, sponge rubber back pad and felt inserts dividing three compartments. (National Archives)

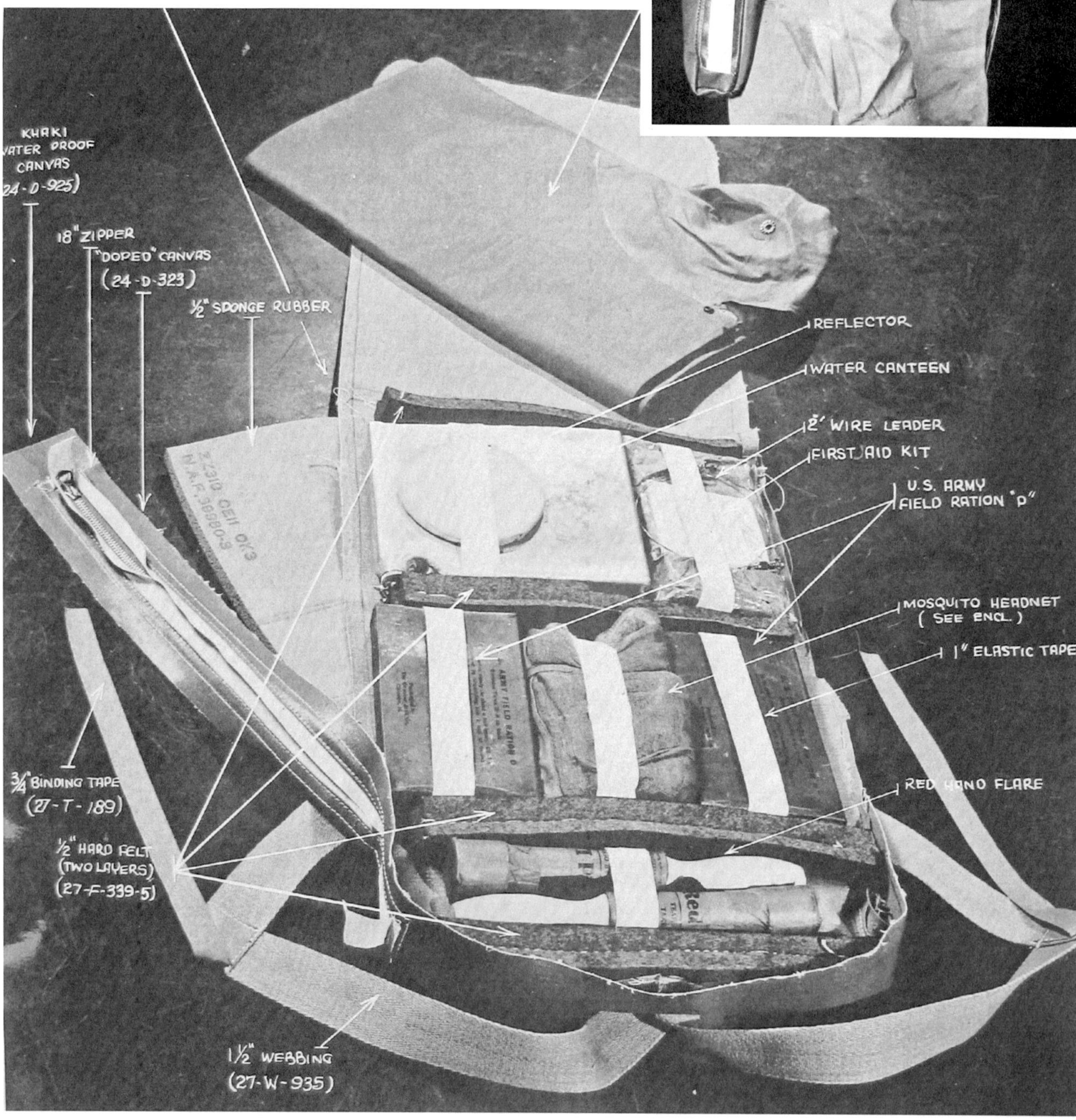

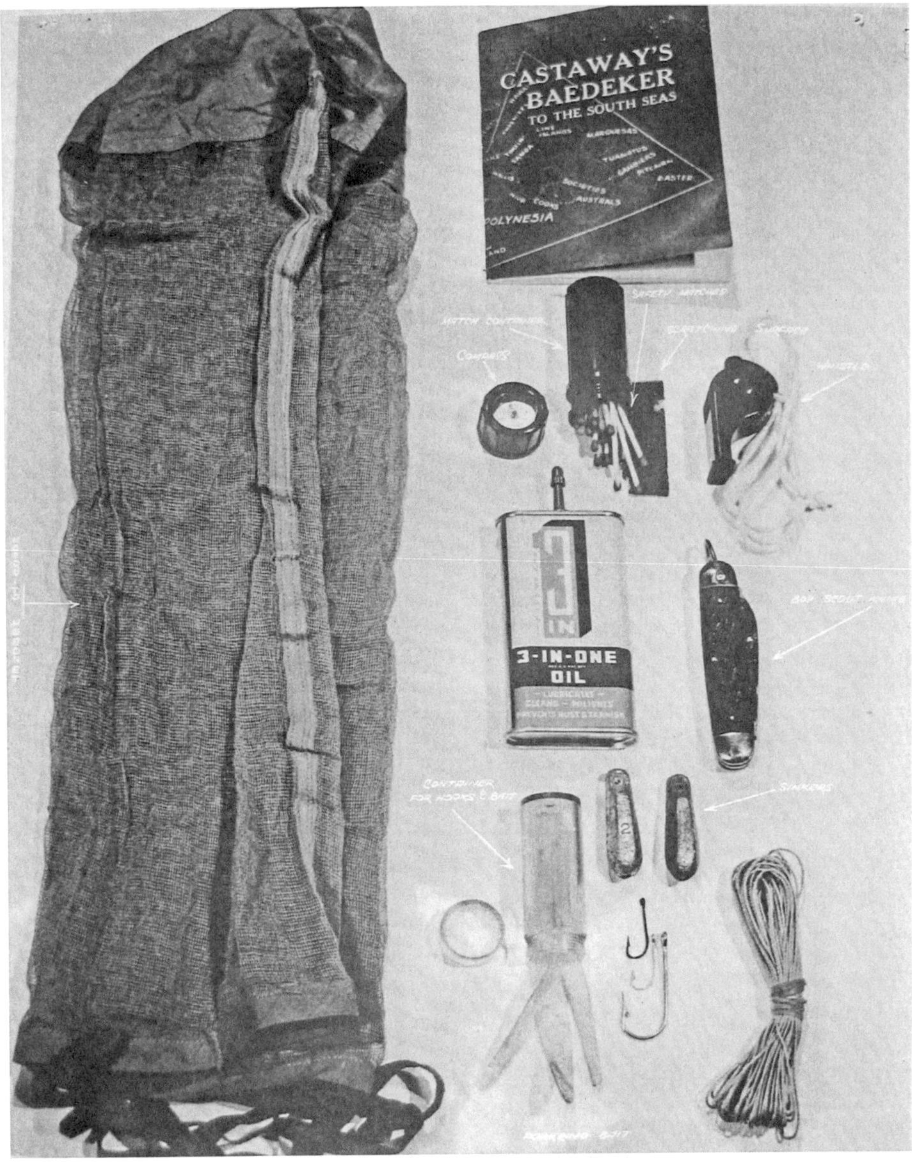

Small items were bundled with the headnet, USAAF type referenced under Dwg. No. 43D3868. These items include: the match safe compass, whistle, oil, Boy Scout knife, a fishing kit, Castaway's Baedeker and waterproof match safe referenced under USAAF Dwg. No. 42A6930, used for storage of fishing hooks, sinkers and bait. The Scout knife predates specification M-575 and would follow requirements detailed in NavAer Specification M-3. This unit is the revised version. Originally it contained a pocket compass, which was replaced by the match safe compass M-554. The Castaways Baedeker was also a later addition. (National Archives)

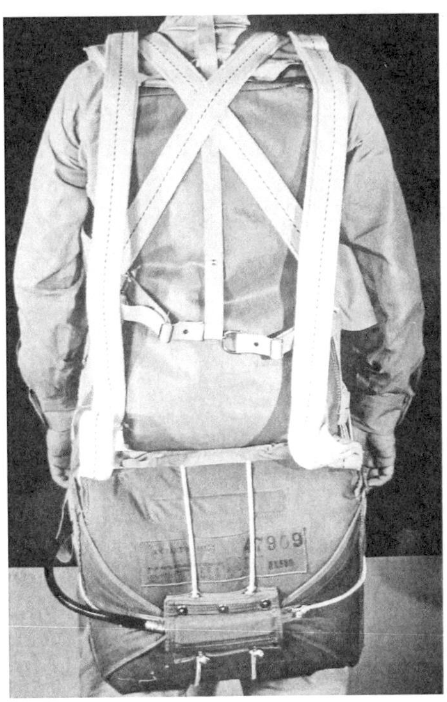

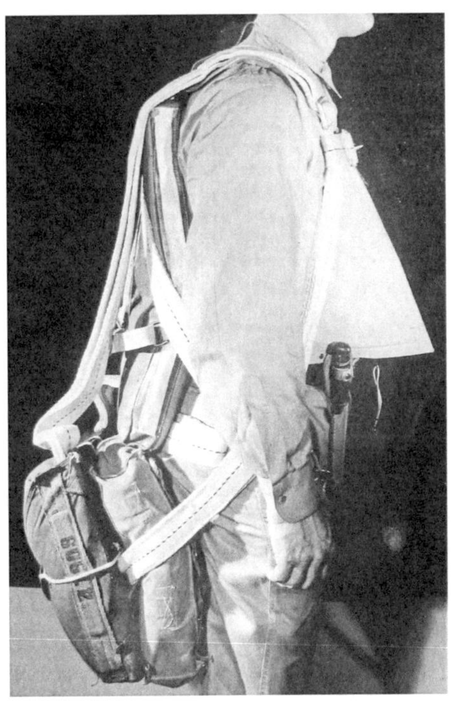

(Left and far left) The NAF-69233 back pad kit is demonstrated with the standard seat type parachute, pararaft M-524 and the 23P6 life preserver. The standard back cushion would be removed from the parachute harness when the back pad kit was used. (National Archives).

Test sampling of the NAF-69233 back pad kit was divided among Naval carrier and Marine land-based units. There were consistent requests for more. By February 1943, 1,021 kits were reported to be in service. Due to demand, the BuAer then outsourced the manufacturing, negotiating a contract with the Sears & Roebuck Co. on 18 December 1942 for 5,000 units under a new designation Kit, Jungle, Emergency for Parachutists. The formal contract was awarded 15 January 1943 under contract number 288-10340. The manufacturing was conducted by Belber Trunk & Bag, a sub-contractor for Sears. The 69233 was available from the ASO under stock number R74-K-35.

Manufacturing commenced prior to the formal award of contract with a reported delivery of 250, 23 January 1943. Sears makes note during the negotiation that a partial delivery would be made within 60 days, with a completion of contract in 120 days. This contract was terminated with the development of a new superseding type parachute back pad kit. Exact date of termination unknown. Contracts for this new kit were awarded in February of 1943. Those quantities of the 69233 delivered were directed to be stocked under the limitation symbol, indicating obsolete material, and issued until exhausted. The 69233 was to be removed from service when it could be replaced by the new type, designated as the Original Back Pad Kit. Records indicate some quantities were delivered by Sears of the negotiated 5,000 before its termination, but it is not clear how many. The ASO reported, as of 12 July 1943, 1,500 were on hand. By July, significant deliveries of the Original kit were reaching supply points. It would had been issued in lieu of the 69233. Production of the 69233 is estimated between 1500 to 2500 but because of the rapid transition to new types, only a limited amount of those produced would actually see service.

For the base design concept, the rectangular back cushion on a QAC parachute harness was used. Its ½" sponge rubber pad was integrated into a waterproof khaki canvas providing the back pad, then formally replacing the standard back cushion worn between the user and parachute harness. The folded poncho stowed in the kit provided additional comfort. Internally, 1" felt spacers were at each end and between the three rows of equipment, the components would be secured by elastic bands. These were included in order to assure a mean thickness of 2-3/8". The sponge rubber pad would rest against the user and the poncho adding additional cushion resting against the seat pan. The shoulder straps were prescribed to be of a 1-1/2" webbing. For a universal fit, buckles were eliminated with adjustment being accomplished by tie tapes across the chest. For waterproofing, the 18" zipper was to be covered with a light film of oil and be provided with a fly over the zipper made of a doped airplane fabric. The zipper assembly was to be completed with a strip of adhesive tape, sealed until use.

The three most unique items included in the kit were the poncho, canteen and first aid kit. The aluminum canteen had a capacity of 24 fluid ounces measuring 7-1/2"x 6"x 1", manufactured by the NAF referenced under Assembly Part No. 69233-3. In further development of emergency back pad kits the flat AN-W-5 water cans would supersede this design.

The poncho was adopted as standard 6 July 1942 under NavAer Specification M-549, referenced under stock number 74-P-230. Its construction consisted of an olive drab two ply cotton fabric treated with a vinyl resin for water repellent properties measuring 90"x 66". Five ball-and-socket snap fasteners were attached to each long-sided edge. The neck opening was described as having an elliptical shape 23" in width and a standing collar. The collar was tall enough to protect the back of the head and the side of the neck. The neck closure could be adjusted by a tab secured under the chin in three different positions. The neck closure was the same as provided on those ponchos procured by the USMC, and later, those camouflaged types by the BuOrd for the US Navy. The final significant feature of the poncho M-549 was that it was to be ink stamped with the name of the contractor, date of contract and property marked U.S.N. ¼" in height. Its production was limited, in further evaluation for a more suitable design it was recommended that one side be finished with the color yellow. Following a more defined list of prescribed uses for the poncho, its design would be significantly amended and superseded by the poncho P-61 that was included in the adopted back pad kit M-592.

At the request of the BuAer, the BuM&S developed and adopted a new first aid kit designating it Jungle Kit, First Aid, stock number S2-1057. It included a 36" tourniquet, a 4" x 4" compression bandage, a glass vial containing 24 quinine tablets and two morphine syrettes, one tube of tannic acid jelly, a package of iodine and one battle dressing. One S2-1057 was included in this kit. This unit would also evolve with the overall development process of the back pad kit, the medical components changed. This list was the final prescribed components. It would have a short life span, being replaced by the S2-1055 and the 57K1, these became the new standard in all emergency equipment.

The final approved list of components upon the negotiated contract with Sears included two red hand flares (MKI Lights), two US Army field rations Type-D, first aid kit S2-1057, fishing kit, a canteen water flask, poncho M-549, match safe compass M-554, a whistle, reflector NAF-214299, magnifying lens, Scout knife, 25' of cotton line, mosquito headnet 43D3868, a 3oz. can of oil and the booklet Castaway Baedeker.

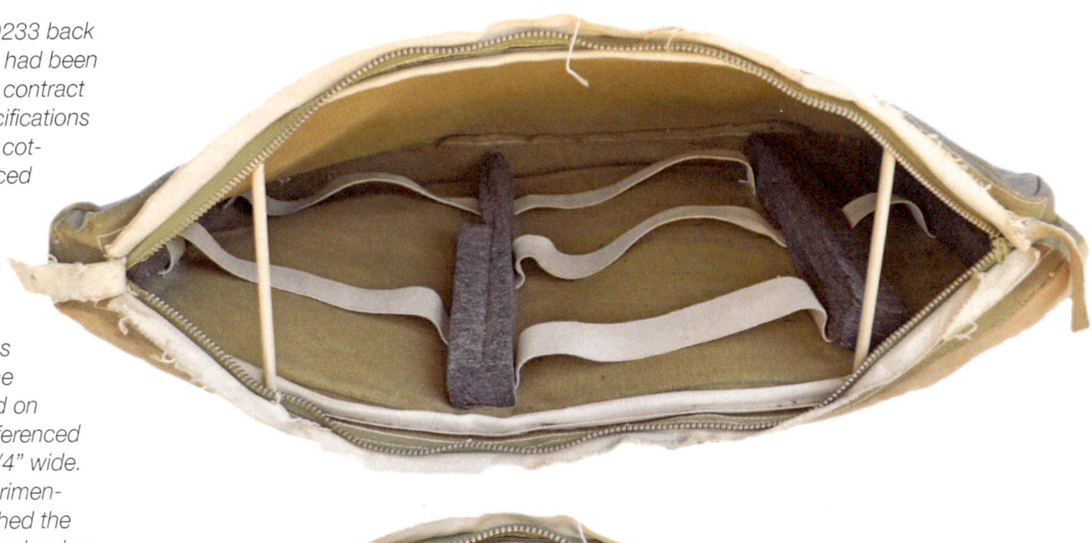

This example of a 69233 back pad kit is believed to had been manufactured under contract with Sears. The specifications they used detailed a cotton webbing referenced under Specification AN-JJ-W-151. It covers a variety of types of webbing related to parachutes, seatbelts and restraint use. The shoulder straps used on this example was referenced as the Type VIII, 1-3/4" wide. This kit was an experimental item that established the foundation for further development. It has been painted with a blue/grey aircraft color, factory material was khaki. Components could only be accessed from one side. Elastic bands secured the components. Felt inserts separated the three rows of accessories.

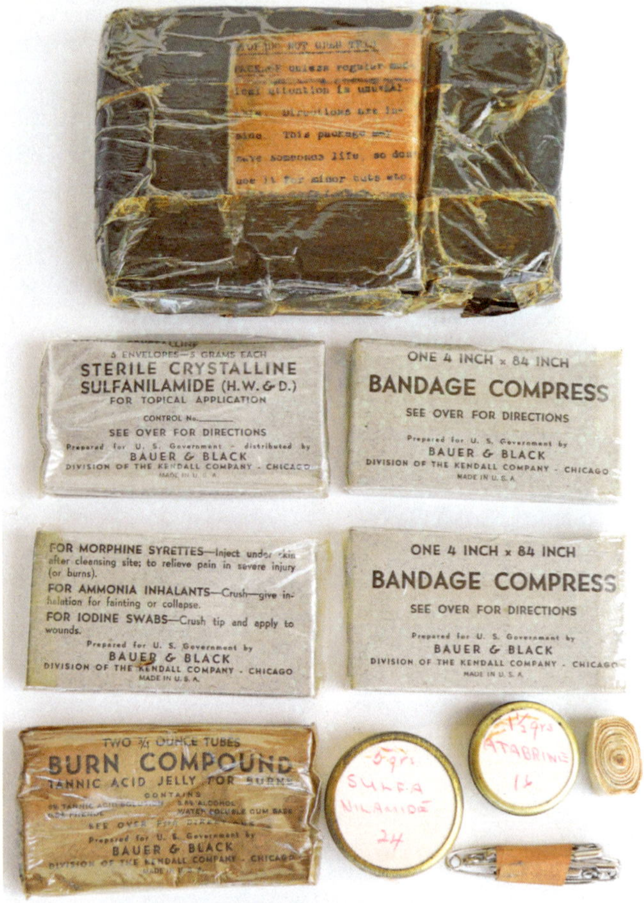

(Left) First aid component. Updated to include packages of the standard 5-unit S2-1055. Additions include two tins each containing tablets of Atabrine and sulfanilamide, bundle of safety pins and a roll of adhesive tape. The unit is sealed in asphalt lined paper then wrapped with cellulose tape for waterproofing.

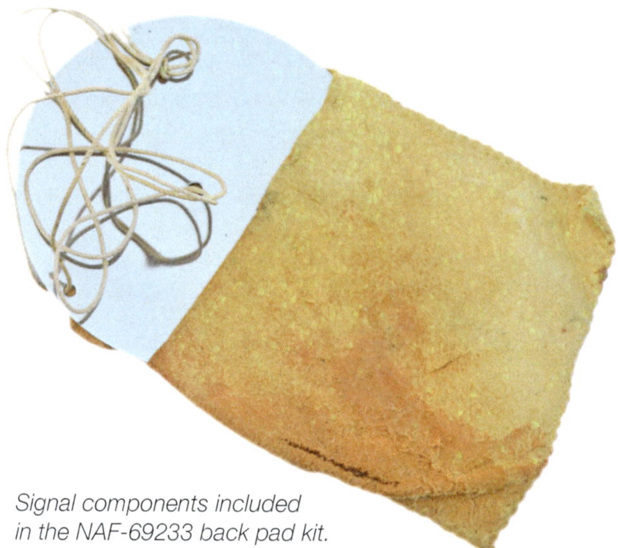

Signal components included in the NAF-69233 back pad kit. One chromium signaling mirror (above) NAF-214299 and two (2) red Mark-I lights (below left). At the time of development, the Mark-III or MK-IV hand projector had only recently been adopted and in production.

(Below) Short of a suitable type, NAF fabricated a canteen, or perhaps more accurately described as a water flask, for inclusion. At the time of development, the existing options were the M-1910 canteen or the cylindrical water cans AN-W-5A, neither were optimal for the strict dimensions in thickness. The complete assembly was machined and fabricated from aluminum.

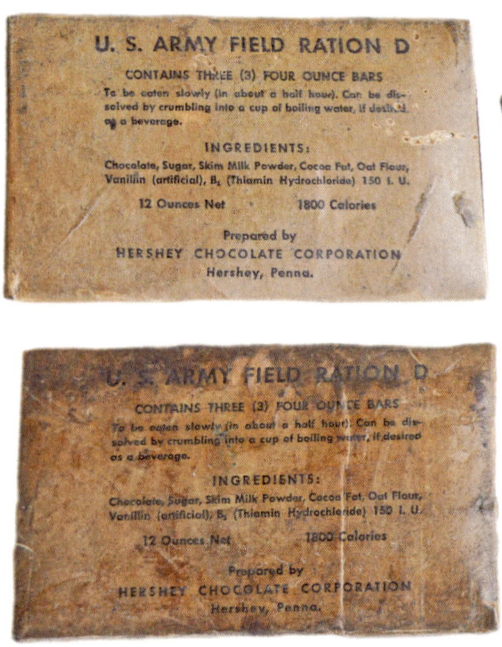

(Left) Additional ration component included two US Army Field Rations D, USAAF Dwg. No. 42B10959.

In January 1943, the Aviation Training Division of the BuAer produced a training film titled Forced Down at Sea. This short film strip discussed basic emergency procedures on operation of life-saving equipment provided in aircraft, standard equipment at the time of production. They highlight the types of life rafts and their accessories to include the one-man parachute type. The film goes on to highlight the NAF-69233 back pad kit, illustrating the proper wear and demonstrates its components. The camera pans over the layout providing short descriptions of each item included in the kit.

The drawing specifies a 3oz. can of oil, but in the in film they were utilizing a 1oz. can. Another variant included the use of a metal Marbles match safe in lieu of the USAAF acetate type. The fishing kit component in the film included two wire leaders, one with a jig or lure attached. The type of lure present was that referenced as a Paw-Paw type. The other notable observation with its components was that of the jackknife. The assembly of this back pad kit predates the standard BuAer jackknife M-575, all specifications require a Boy Scout type. The jackknife in the film was an official Boy Scout knife with round shield and BSA crest, a type associated with the manufacturer Remington.

Accessory items for the NAF-69233 emergency back pad kit. All would be bundled in headnet and stored between the D-Rations. (National Archives)

Pilots of VMF-213 gather around the operations tent on Henderson Field, Guadalcanal in June 1943. Many were wearing NAF-69233 emergency back pad kits. Two pilots on each end were wearing the kits viewing the left- and right-hand profiles. (National Archives)

Pilots of VMF-213 gather around the operations tent on Henderson Field, Guadalcanal in June 1943. Unidentified pilot, far right, wears a NAF-69233 emergency back pad kit. NAF-69333 kits were also noted to had been used by VF-11, which were stationed on Guadalcanal from May to July 1943. (National Archives)

As combat operations in the SWPA began to ramp up, an increasing number of pilots and aircrew were operating over vast regions of jungle-laden islands and ocean, creating a new set of needs for emergency equipment. It became immediately apparent that individuals should be provided with an emergency kit in case of a parachute jump or crash over land, particularly in uninhabited areas, establishing the emphasis of "survival." To partially address this issue, it was reported by Commander Air Force, Pacific Fleet that at least one squadron was being equipped with USAAF jungle kits, and additional squadrons were provided the NAF-69233. The pararaft was not a prime factor in the initial design of an emergency kit, as it was more oriented toward jungle survival, reflected in the new nomenclature. Both kits would be used but would not fully complement each other. In November 1942, ComAirPac submitted a design for a jungle emergency kit combining the best attributes of the NAF-69233 and USAAF jungle kit, conducive to space and weight conservation. One overseas squadron was equipped with this unit for service testing by the end of the year.

Modified USAAF 38J5271 jungle back pad kit used by undisclosed Naval aviation unit in 1942. Rations and machete were organic to the kit; remaining provisions added by local command. The additional components include cotton, gauze, fishing line with hooks, bicarbonate of soda, adhesive, aspirin, iodine, compass and quinine capsules. From various accounts and interviews, aviation squadrons reported to use USAAF back-pad kits include VMF-124, VF-23, and VF-10. (National Archives)

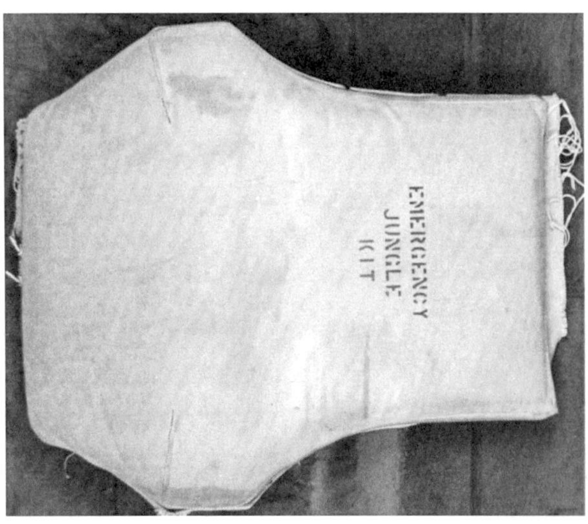

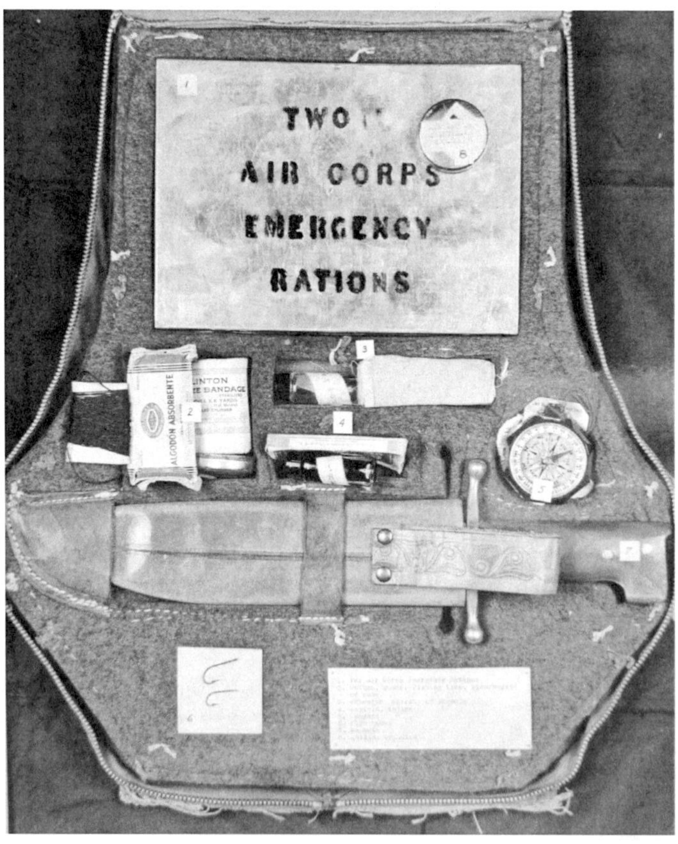

The assembly submitted by ComAirPac copied the internal design features of the USAAF jungle kit but incorporated the standard US Navy kidney-shaped parachute back cushion. A sponge rubber insert was provided with cutouts for the components. It provided optional features: shoulder straps for securing to the user and buttoned tabs for securing to the parachute harness. The kit transferred the bulk of the components contained in the NAF-69233 to the new container.

ComAirPac emergency kit submitted to the BuAer for evaluation. This kit was converted from a standard parachute back cushion by opening the side and bottom seams, then trimmed with a zippered closure. A sponge rubber insert was included to help with buoyancy factors and to segregate the components. This design, with the components spread out, allowed the kit to be more flexible, increasing comfort. It includes the pocket compass originally included in the NAF-69233. Small articles were no longer bundled with the headnet; fishing kit was transferred to two USAAF acetate 42A6930 match safes between the Type-D rations. (National Archives)

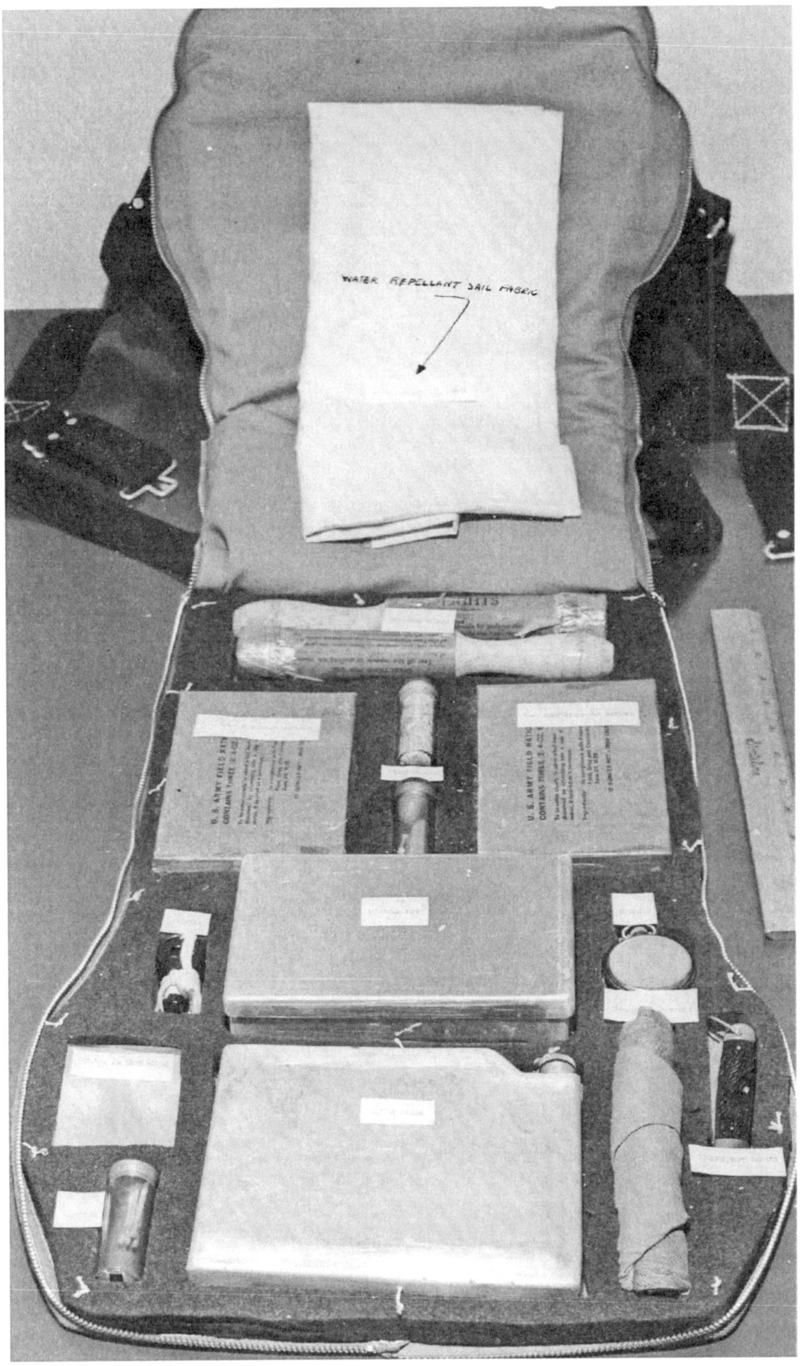

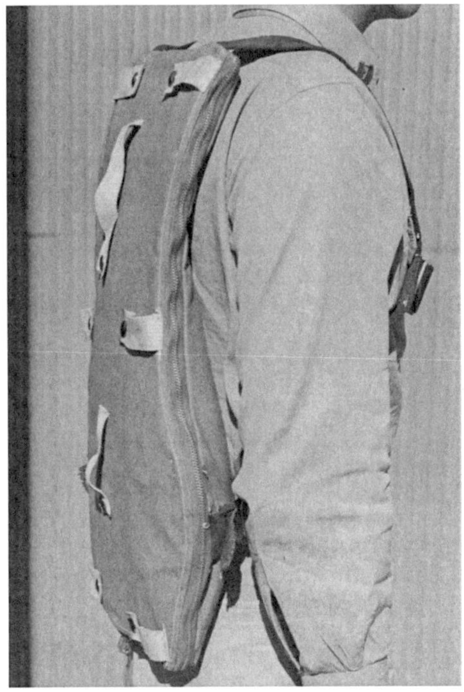

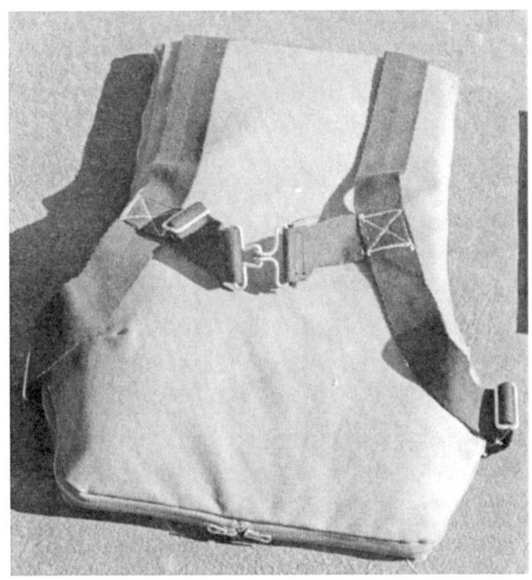

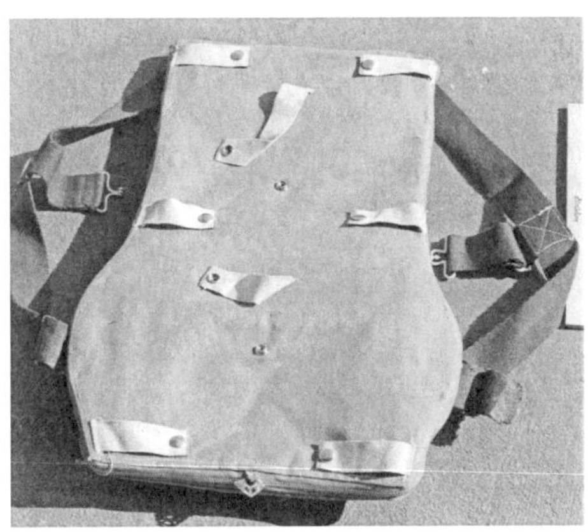

By the close of 1942, ComAirPac determined that pararafts and emergency equipment kits should be supplied for all personnel and all types of aircraft, making it a high priority issue. Influencing factors from combat engagements during 1942 included bailouts, rapid sinking of an aircraft after ditching preventing recovery of the life raft and emergency equipment container, and life rafts stored in the fuselage damaged from combat. These prompted the necessity for equipment to sustain life both at sea and in the jungle and for it to be issued to the individual.

With the basic parameters of a suitable emergency kit established, the BuAer coordinated designs for the next generation. To expedite the process, the Engineering Division built on the concept presented by ComAirPac. The unit was upgraded to be fabricated from a waterproof vinyl-coated fabric and to include components organic to standard stock. The class of components remained the same but updated to the newly available and standardized items. The kit would include a fishing kit, sailcloth, a mosquito headnet, a match safe compass, rations, water, a flashlight, a scout knife, one Type C battery, a chromium-plated mirror, 25' of line, a package of salt tablets, adhesive tape, safety pins, MK-IV hand projector with six red signals and an instruction booklet. This back pad kit was designed as a companion to the pararaft, and at the time of its design, Specification AN-R-2 was amended to include the medical provisions. By using standard stock items, procurement could be immediate to accommodate service demands. Contract 288-10870 was awarded for the manufacture of 20,000 kits to the New England Bedding Co. on 6 February 1943 and the quantity was established to equal the amount of pararafts in service and on order. This new kit was designated as the Original Back Pad Kit, Non-Standard (Kidney Shaped). In April 1943, the Original back pad kit became available for distribution under ASO stock number R83-K-520100. This kit officially replaced the emergency equipment container retained in VF class aircraft 7 July 1943.

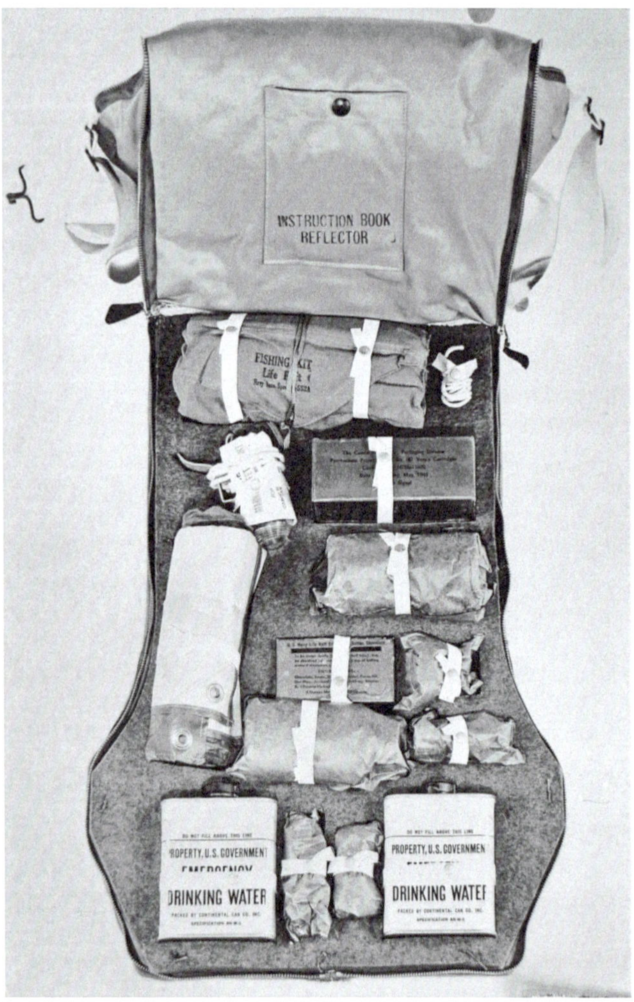

Components for the Original back pad kit were separated by cutouts in a 1" felt insert and secured by cotton straps with glove snaps. Kit measures 23" x 12" x 4" and weighs 12 lbs. In the color image, one can of water cans has been replaced by desalting kit M-613, which would have been an update conducted late in its service life. The waterproof asphalt wrappers have been opened for display purposes. (National Archives)

The Original back-pad kit as procured under contract 288-10870. It has characteristics of both the standard kidney-shaped parachute back pad and USAAF parachute kits. As detailed in the fleet circular of July 1943, the MK-1 emergency equipment container was officially deleted and replaced by this unit as supplies became available. It was, however, only relevant for the F4F Wildcat, the new fleet fighter; the F6F Hellcat would omit the life raft compartment and used the pararaft.

ITEM	STOCK NO.	SPEC. No.
Original Back Pad Kit (Complete)	R83-K-520100	None
Emergency Drinking Water (2)	R51-W-135	AN-W-5A (Flat)
Life Raft Rations	R56-R-6125	M-539
Fishing Kit	R37-K-300	M-552A/AN-K-2
Compass & Match Container	R37-C-2500	M-554/AN-C-101
Headnet & Fish Seine	R27-N-261	M-565
25' of 75# Test Line	21-R-150	T-R-571
Pyrotechnic Kit	R83-K-710309	None
Jackknife	R41-K-365	M-3/M-575
Whistle	R42-W-24000	None
Signaling Mirror	R83-R-20500/C.F.E.	M-3
Salt Tablets	C.F.E	C.F.E
Adhesive Tape	C.F.E	C.F.E
Safety Pins	C.F.E	C.F.E
Sail	R83-C-76950	M-551
Signaling Light	C.F.E	M-567
Spare Battery & Bulb	C.F.E	C.F.E
Instruction Booklet	C.F.E	C.F.E

(Left) Official list of accessories for the Original back pad kit with their corresponding stock and specification numbers. Initial procurement occurred prior to the standardization of some items, which is reflected by the dual specification numbers on some articles. C.F.E. signifies Contractor Furnished Equipment.

(Below left) Modifications of back pad kits were common as evidenced by this unit on the flight deck of the USS Lexington of Air Group 16, c.1944. Added was the booklet Castaway's Baedeker, 57K1 first aid kit, box of 50 rounds .38 cal. ammunition and two rolls of Lifesavers candies. (National Archives)

(Below) Original back pad kit on a display board in the operations tent of VMF-214, c.1943.

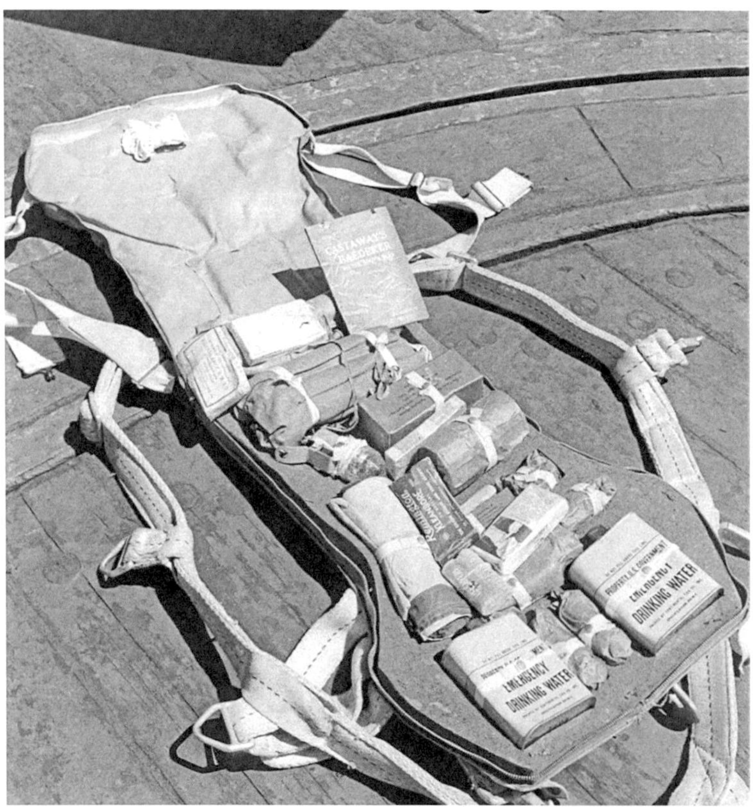

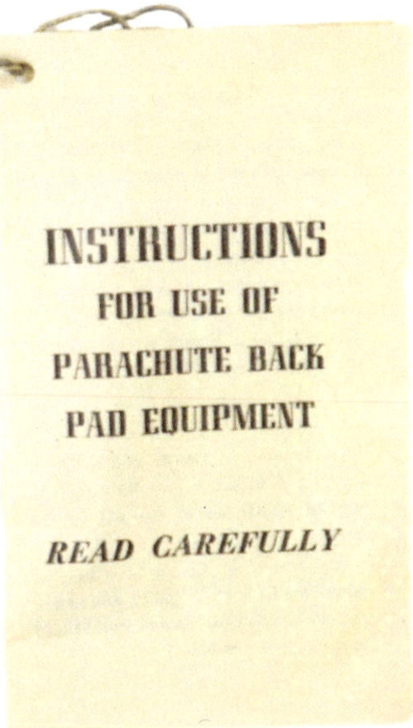

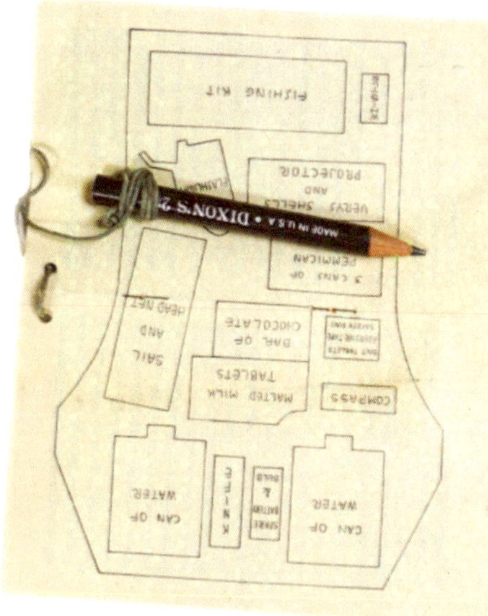

The components for the Original back pad kit were organic to standard supply. The round chromium mirror supplied to the Original kit had a white cotton case, as compared to the chamois type manufactured by the NAF. The match safe compass included in some kits has the red M-554 with a lanyard secured by a strip of friction tape. Later assembly would include the AN-C-101, the modification of the M-554 with lanyard was a common practice. Only item unique to the kit was the 3" x 5" instruction booklet. It includes 23 pages describing individual items and their prescribed use, and includes basic land and water survival principles.

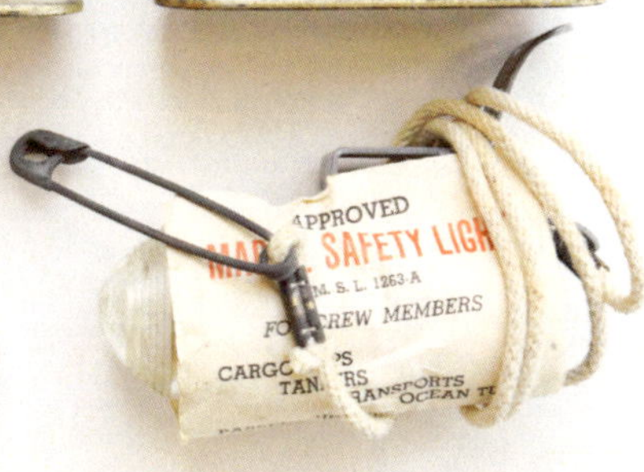

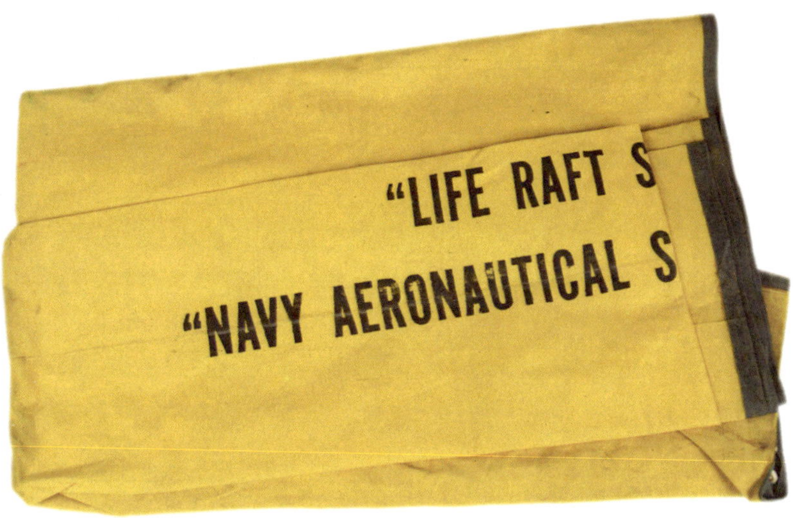

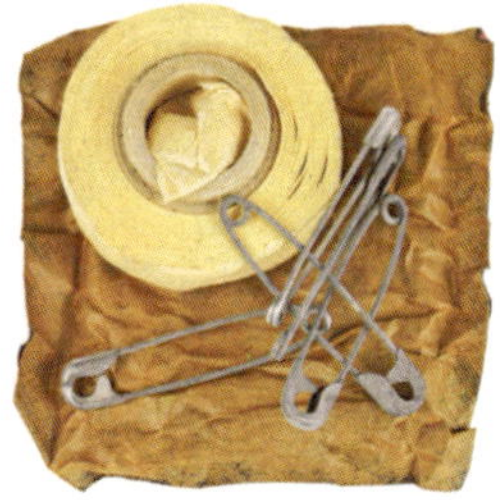

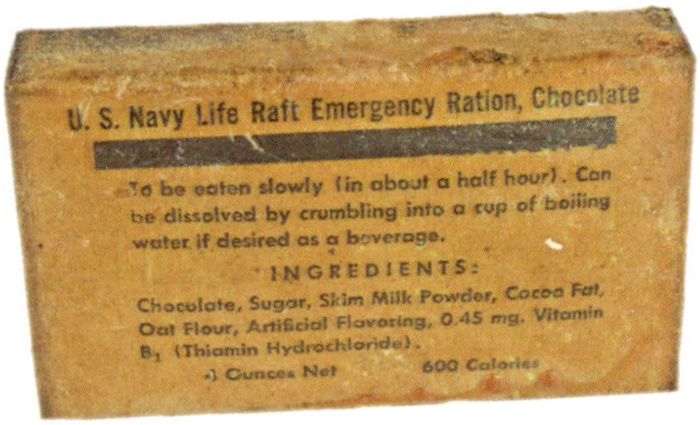

Additional components of the original back pad.

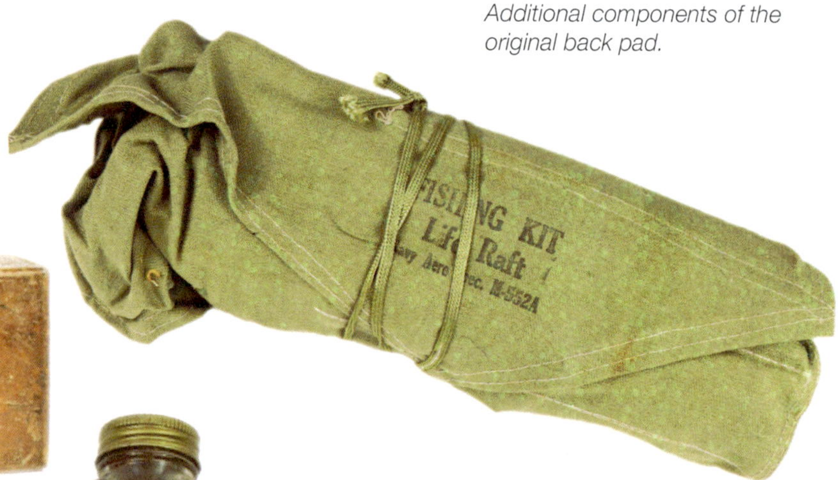

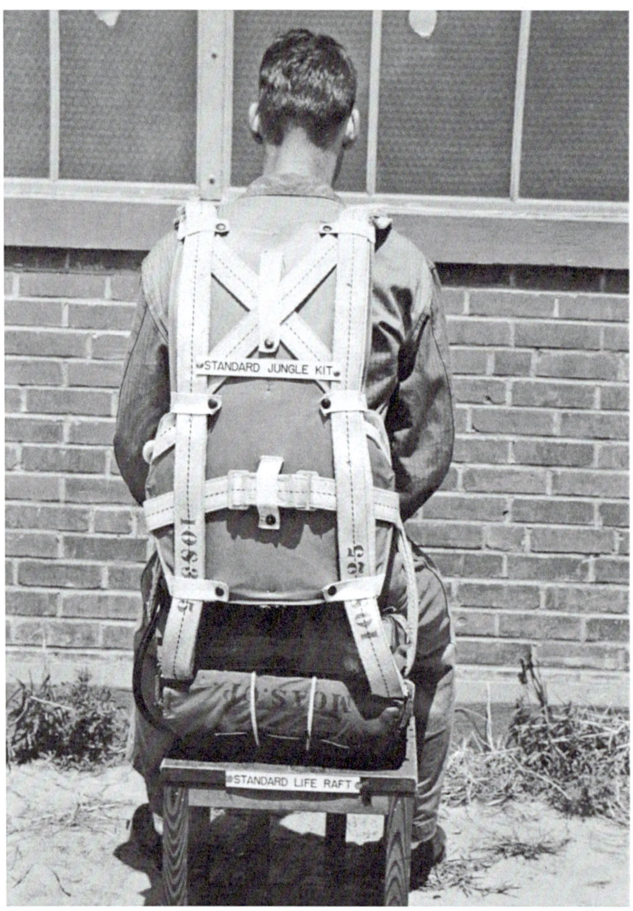

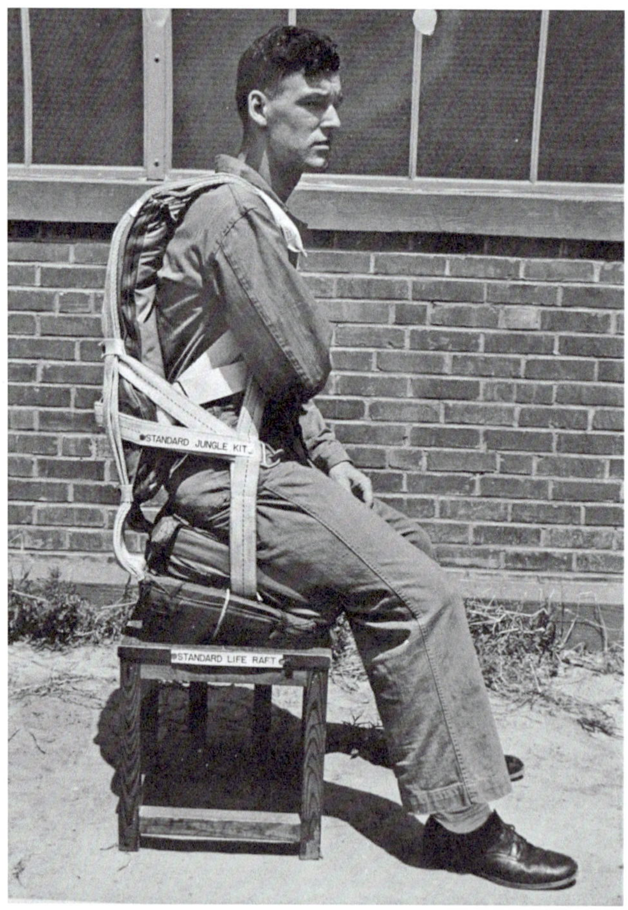

(Above) Configuration of the Original back pad kit, pararaft and standard seat type parachute, demonstrated by personnel at Marine Corps Air Station Cherry Point, August 1943. (National Archives)

(Left) SBD Dauntless pilot Lt. jg Henry Minot of VB-5 wears a typical standard complement of flying clothing and equipment of Naval aviators for the era, USS Yorktown, October 1943. The back pad kit was included as part of the complement provided for pilots, which in this case was the Original type. (National Archives)

(Below) Aircrew member of VT-5 aboard the USS Yorktown with Original back pad kit attached to his QAC parachute harness, October 1943. Back pad kits were only issued to aircrew if there was enough supply. (National Archives)

The issuance of emergency back pad kits was first delegated to VF squadrons, then to all other carrier-based aircraft, supply permitting. The allowances for Standard Aircraft Equipment was officially amended in November 1943 for the issue of back pad kits and pararafts to all pilots of carrier-based aircraft and VP squadrons. Tables of basic allowances establish that life rafts and emergency kits were issued to the aircraft and its assigned crew positions. Parachute allowances followed the same principle. Priority for the back pad kits and the pararaft were for airplanes of three persons or fewer. VP squadrons and others would receive this type of equipment when requested. Life preservers, hunting knives, and sidearms were issued as personal equipment and, once operational, it becomes a complete flying safety assembly.

The back pad kit had to meet a long list of requirements, but in development of aircraft emergency equipment, bulk reduction is the number one priority. The immediate and consistent complaint about the back pad kit was its bulk, placing the user too far forward and away from the armor plating. It also increased the chance of injury by hitting the gunsight upon a crash landing. To reduce the thickness, the New England Bedding Co. developed a modification kit by rearranging the components. The BuAer procured 10,000 modification kits on 28 September 1943, which included a ½" sponge rubber pad to replace the kapok padding in the top flap and a new ¼" felt pad with sewn tie tapes to replace the existing 1" felt insert. The components were rearranged and items removed. From the fishing kit, the net, fish knife and black line were removed. The remaining kit folded flat. The headnet was removed from the sail and secured separately. In turn, the sail was to be folded flat, and the pemmican cans were placed face up; this rearrangement significantly increased comfort.

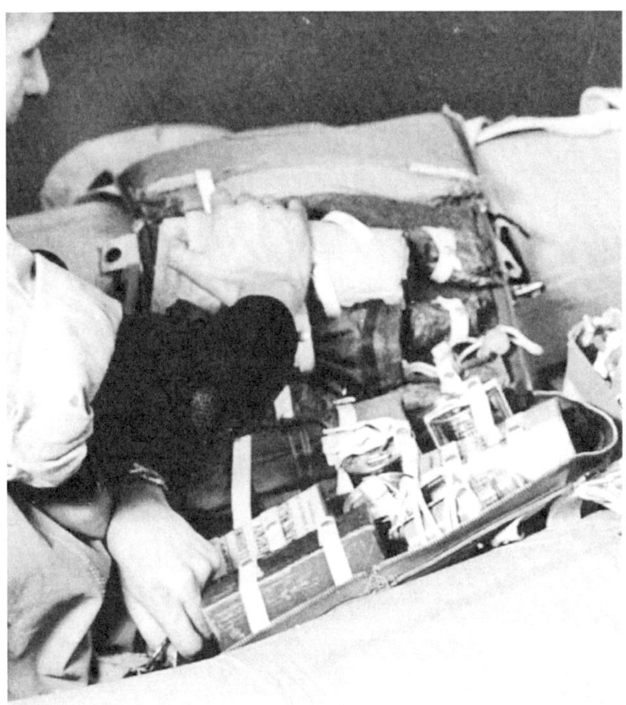

Demonstrating emergency equipment; an original kit with the modification insert. (National Archives)

(Above) TBF aircrew member Selbie Greenhalgh ARM2/c of VT-16 poses for a photo aboard the USS Lexington, June 1944. Greenhalgh and two fellow crewmembers were rescued by a floatplane from the USS Wichita after they had to ditch their Avenger off Anguar Island in the Palau Islands on 30 March 1944. Hanging behind Greenhalgh is a parachute harness with an Original back pad kit. (National Archives)

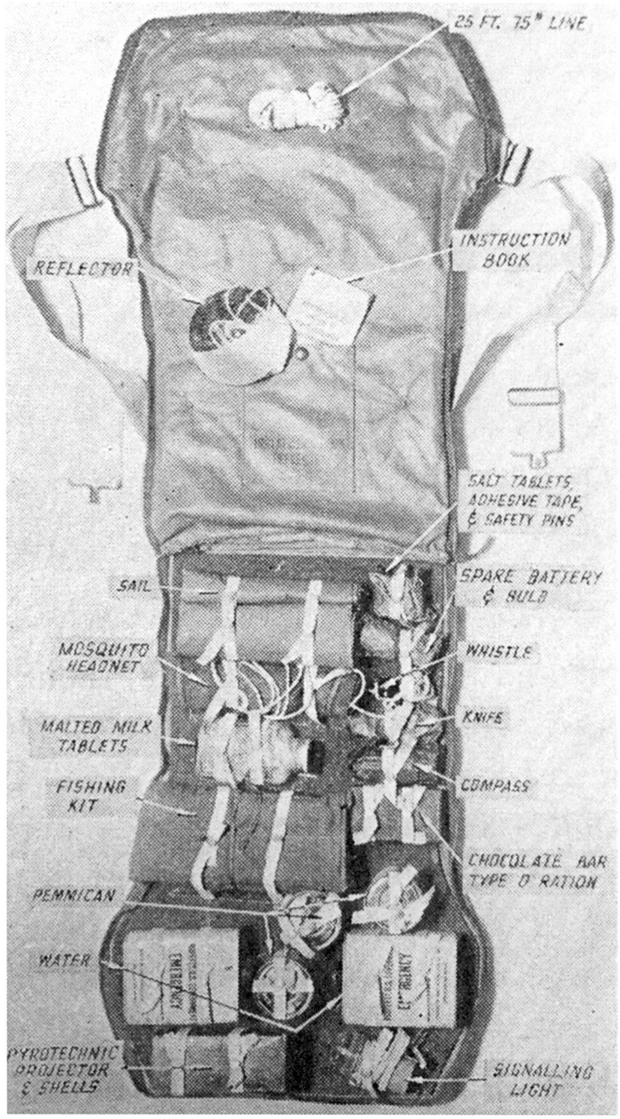

(Right) Original back pad kit with the modification insert featured in the Naval Aviation News Letter of November 1943. Only the padding was procured to modify those kits already in service. (National Archives)

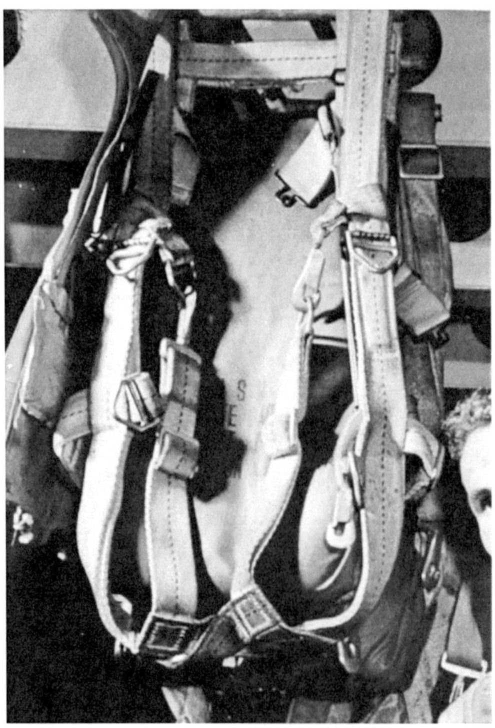

(Right) Since the Original back pad kit can be attached to the parachute harness, they are often seen hanging together. Two harnesses in the ready room of the USS Hornet, February 1945. (National Archives)

(Far right) VF-5 pilot Ens. Theodore Schofield is wearing an Original back pad kit aboard the USS Yorktown, October 1943. With the back-pad kit attached to the parachute harness, the chest strap is secured in front over the life preserver. This assures the individual can safely slip free of the parachute. (National Archives)

The Original back pad kit was a rushed measure to meet the growing service demands, new service requirements, and to issue the pararaft to all stations aboard combat aircraft. Now that the initial requirements could be met, the focus then turned to the development of a unit that could be standardized. A list of recommendations from the fleet was compiled and evaluated.

The container was the first item to be taken under consideration. It was important that the unit maximize space and provide all the necessary items to sustain life in the jungle and at sea. The container adopted would use waterproof fabric and shoulder straps, be divided into three compartments, and eliminate the provisions for attachment to the parachute harness. Items specially developed for this kit would include a combined poncho-raft cover and a 10" nonfolding machete. All other items included in the kit were available from standard stock or commercially. Like that for the Original kit, the container would include new standardized and updated articles such as the tablet rations M-539B, AN-C-101 match safe compass, M-575 jackknife, sunburn cream, M-580 signal mirror and the six-unit 57K1 first aid kit. The life preserver light M-567 was transferred to the list of personal equipment.

The finalized unit was fully adapted to be used on its own on land and complement the pararaft at sea. The first contract was awarded to the New England Bedding Co. on 19 August 1943 for 20,000 kits under contract number NOA-1409. An additional contract was awarded to the Sears & Roebuck Co., which subcontracted to Belber Trunk & Bag, on 4 September 1943 for 10,000 kits under contract NOA-1594. Deliveries were reaching central supply points in November 1943 and officially standardized the next month on 11 December 1943 as the Standard Back Pad Kit under NavAer Specification M-592, available under stock number R83-K-520100. Both the M-592 and the Original back pad kit were available under the same stock number and, when ordering, supply depots were directed to cite specification and/or nomenclature. Technical Note 1-43 for the inspection, packing and maintenance of the pararaft was revised to include the emergency back pad kit distributed as Technical Note 76-43, dated 12 November 1943, titled One-Man Parachute Raft and Back Pad Kits, Packing, Operation, Inspection and Maintenance Of. This note includes both the Original and M-592.

Two additional contracts were awarded for the M-592 in June 1944: one to each of the 1943 contractors. The New England Bedding Co., under contract 288-23689, would manufacture 15,000 more kits and Sears & Roebuck Co., under contract 288-23690, would manufacture 30,000 more kits through subcontractor Belber Trunk & Bag. Specification M-592 was amended on 1 May 1944, deleting the oil can and replacing it with a tube of Kant-Rust, applied to the June 1944 contracts. A subsequent change occurred on 14 September 1944, affecting all M-592 kits in service. Technical Order 115-44 was distributed detailing that the placement of the sharpening stone and 25' of line be switched. In the original configuration, the sharpening stone was between the two water cans, which it tended to puncture.

As per allowance, the back pad kit was issued to crew positions, first to the pilot, then to the aircrew. It was stated that the M-592 was to be first issued to pilots and the Original kits were distributed to aircrew. Eventually, the back pad kits filtered through the service, being utilized by all flying personnel in the combat zone or forward area. The year 1945 saw a decline in its use with the introduction of the combination Model-A and PK-1 pararaft kits and the streamlined method of carrying items in the flight suit.

1943 contract from the New England Bedding Co. M-592 back pad kit. Not shown are gloves and headnet, which would be stored with poncho. It also illustrates the change per Technical Order 115-44 placing the line between the water cans, with the sharpening stone repositioned at the opposite end. (National Archives)

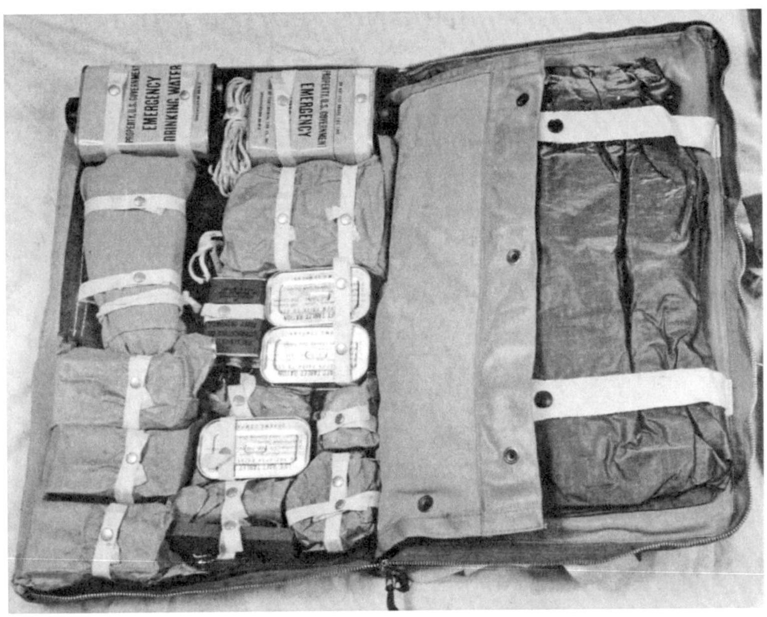

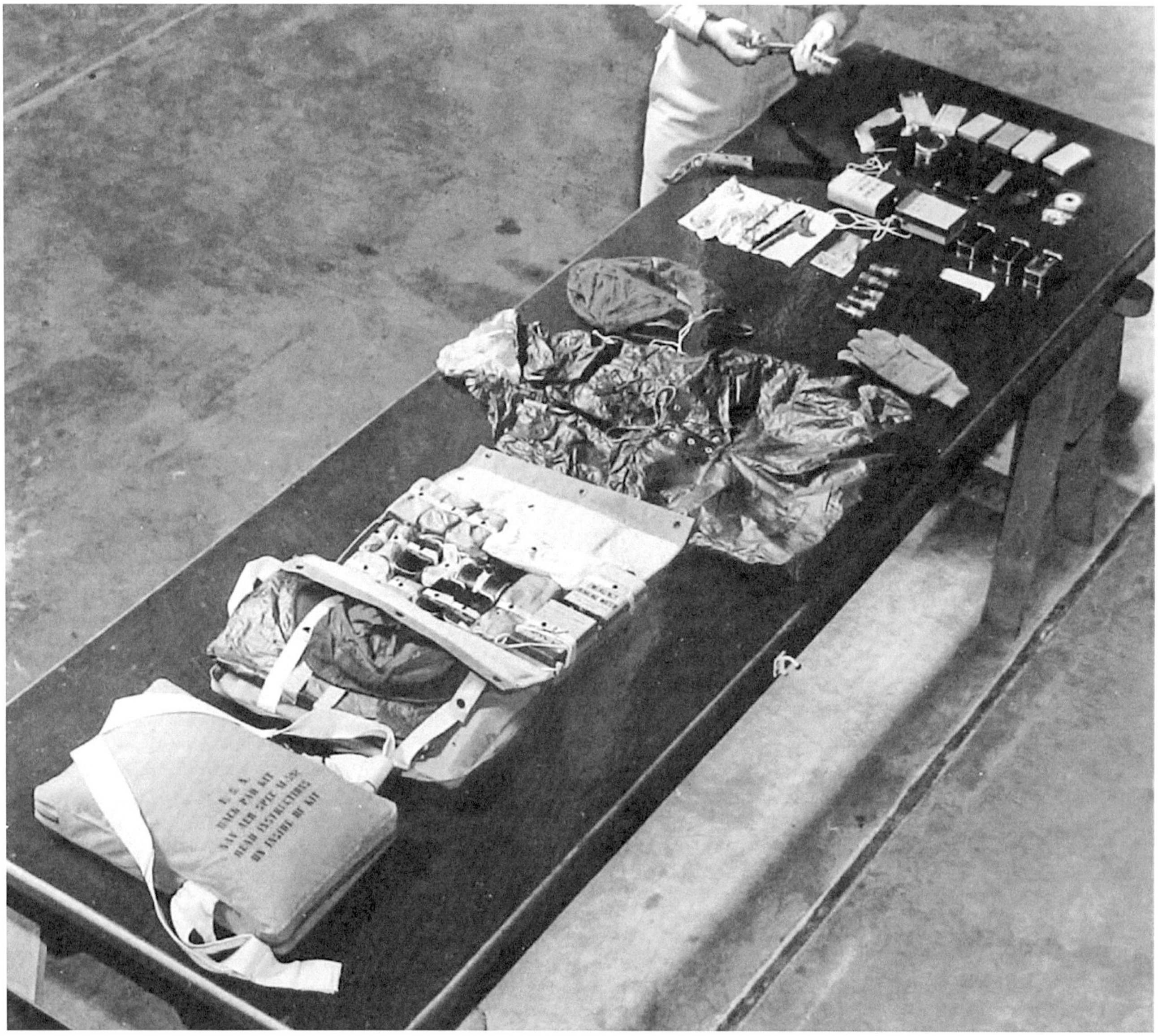

(Above) M-592 back pad kit in three separate views. Left to right: kit closed, kit open, individual components laid out, September 1944. (National Archives)

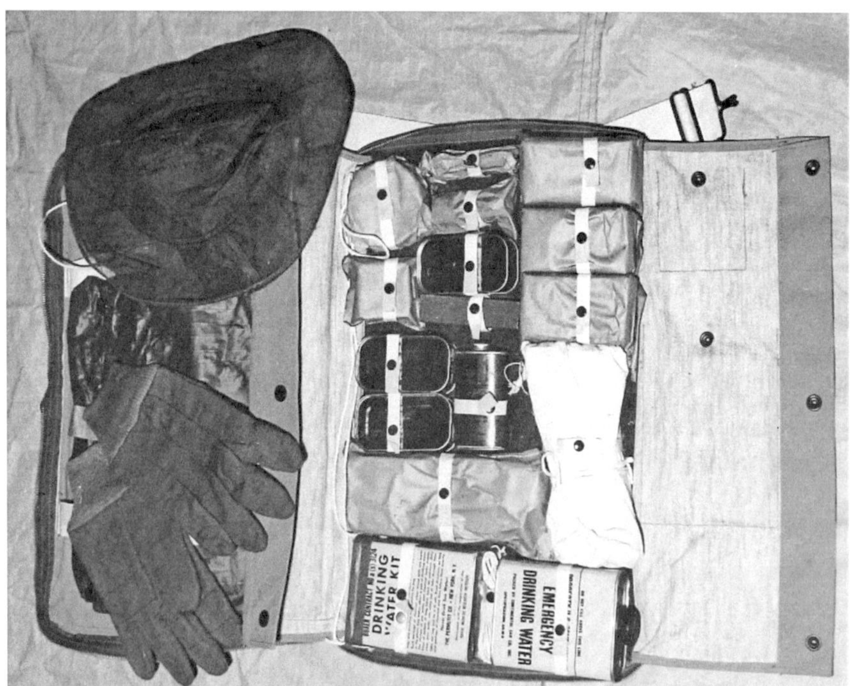

(Right) 1943 M-592 assembled by Belber Trunk & Bag. Kit had been updated to include desalting kit M-613. Official distribution detailed in Technical Note 100-44 dated 30 November 1944 for inclusion of the M-613 for all emergency kits, supplied at a ratio of one M-613 drinking water kit for every two AN-W-5 drinking water cans. (National Archives)

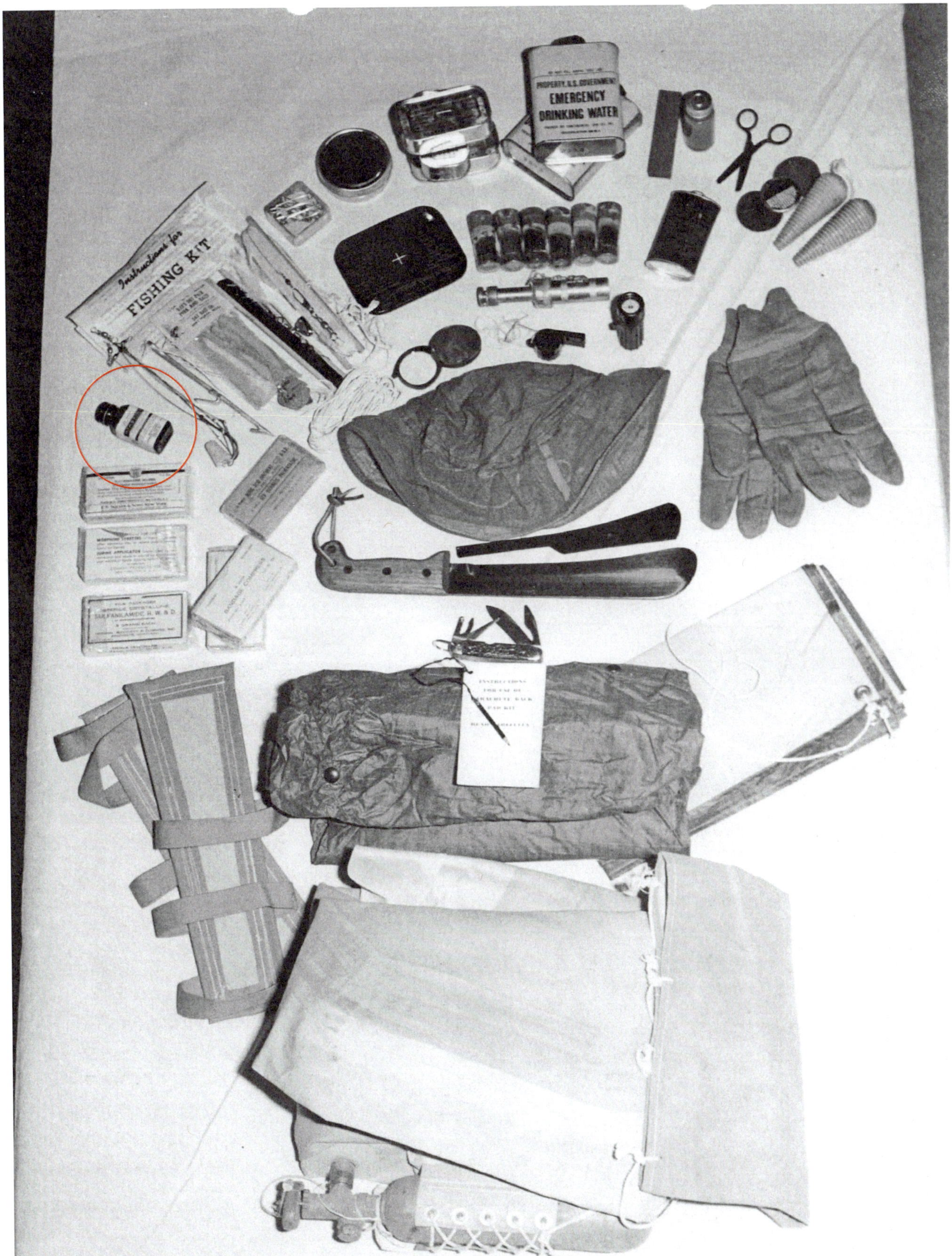

Nearly complete complement of the M-592 back pad kit and pararaft components for an individual. Not pictured are the water and sea marker contained with the pararaft. Also included is the addition of a bottle of Atabrine. Flight surgeons often added medicinal aids such as bottles of Atabrine, quinine, and/or salt to supplement equipment kits, viewed as vital medical aids for tropical use. (National Archives)

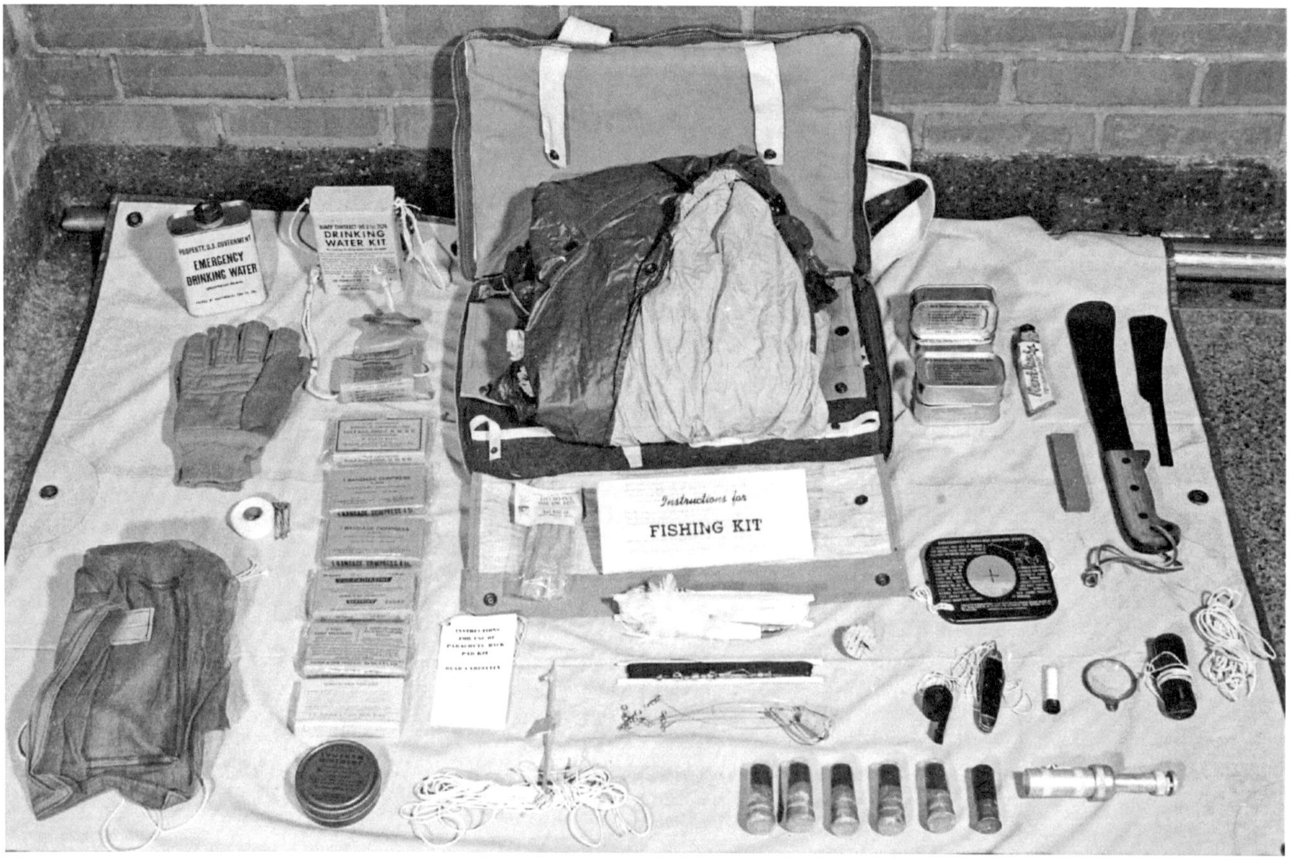

(Above) 1944 Sears & Roebuck Co. contract M-592 back pad kit at NAS Norfolk, c.1945. This unit updated as per Technical Note 100-44 for the M-613. (NNAM)

(Right) M-592 is demonstrated with the standard seat type parachute at NAS Patuxent River, January 1944. (National Archives)

(Below) Lt. W.C. Kelly of VF-6 dons an M-592 back pad kit in conjunction with a QAS parachute harness, January 1945. (National Archives)

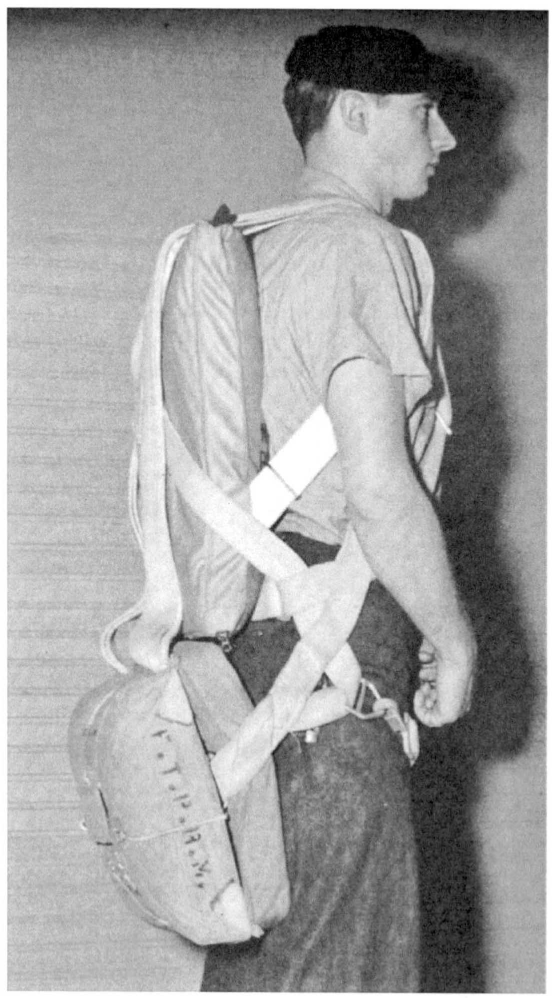

(Left) Air Group 80 pilots aboard the USS Ticonderoga receiving a briefing. Hanging on the wall of the ready room is an M-592 back pad kit, November 1944. (National Archives)

(Below) VC-20 pilot Lt. jg. E.A. Philips is wearing a Belber Trunk & Bag M-592 back pad kit aboard the USS Lexington, January 1945. Modifications to standard equipment were a typical occurrence among aviation units, often toward personal preference. Added to the kit worn by Philips are parachute harness retaining straps, a feature on parachute back pads and the Original kit. (National Archives)

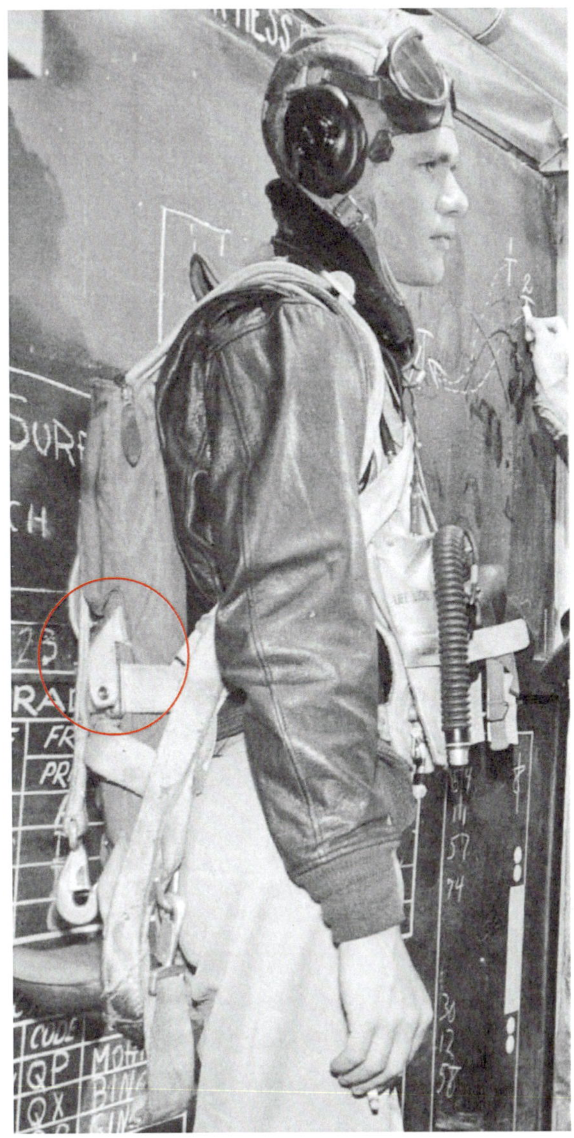

August 1943 contract for the M-592 awarded to the New England Bedding Co., contract No. NOA-1409. Features specific to this contract were a vinyl-coated fabric for the container and utilization of a rectangular oil can. Components in the M-592: poncho, gloves, and headnet secured in place by two straps under the front cover. Two internal flaps contained pockets: one for the machete, one for the mag lens, and one for the signal mirror and logbook. In the last compartment was a ¼" felt pad with straps securing the remaining accessories in conspicuous locations.

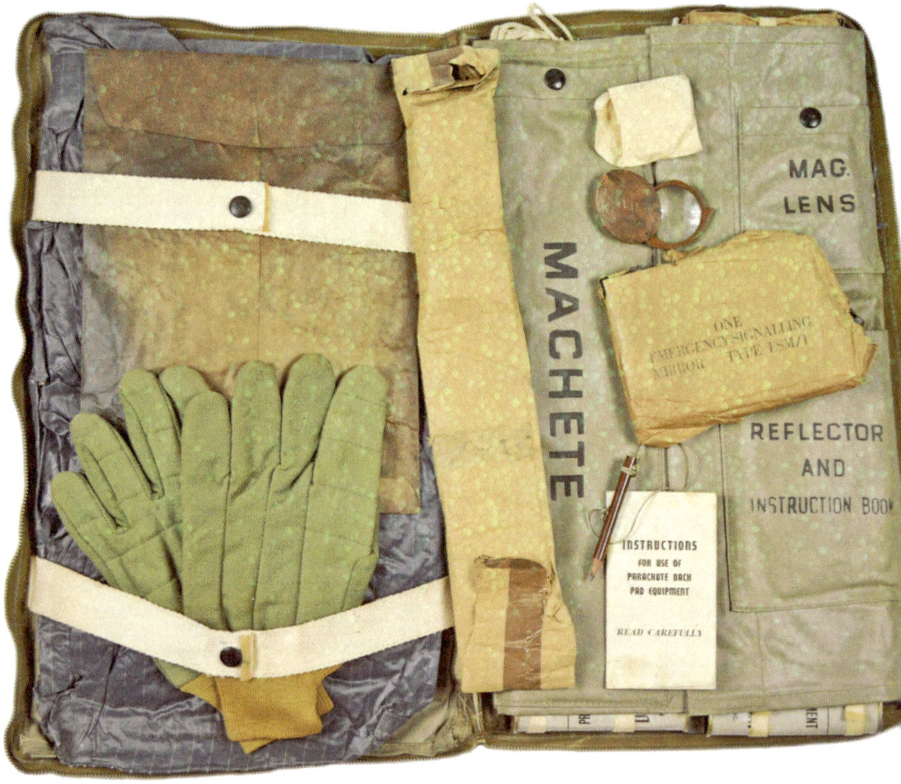

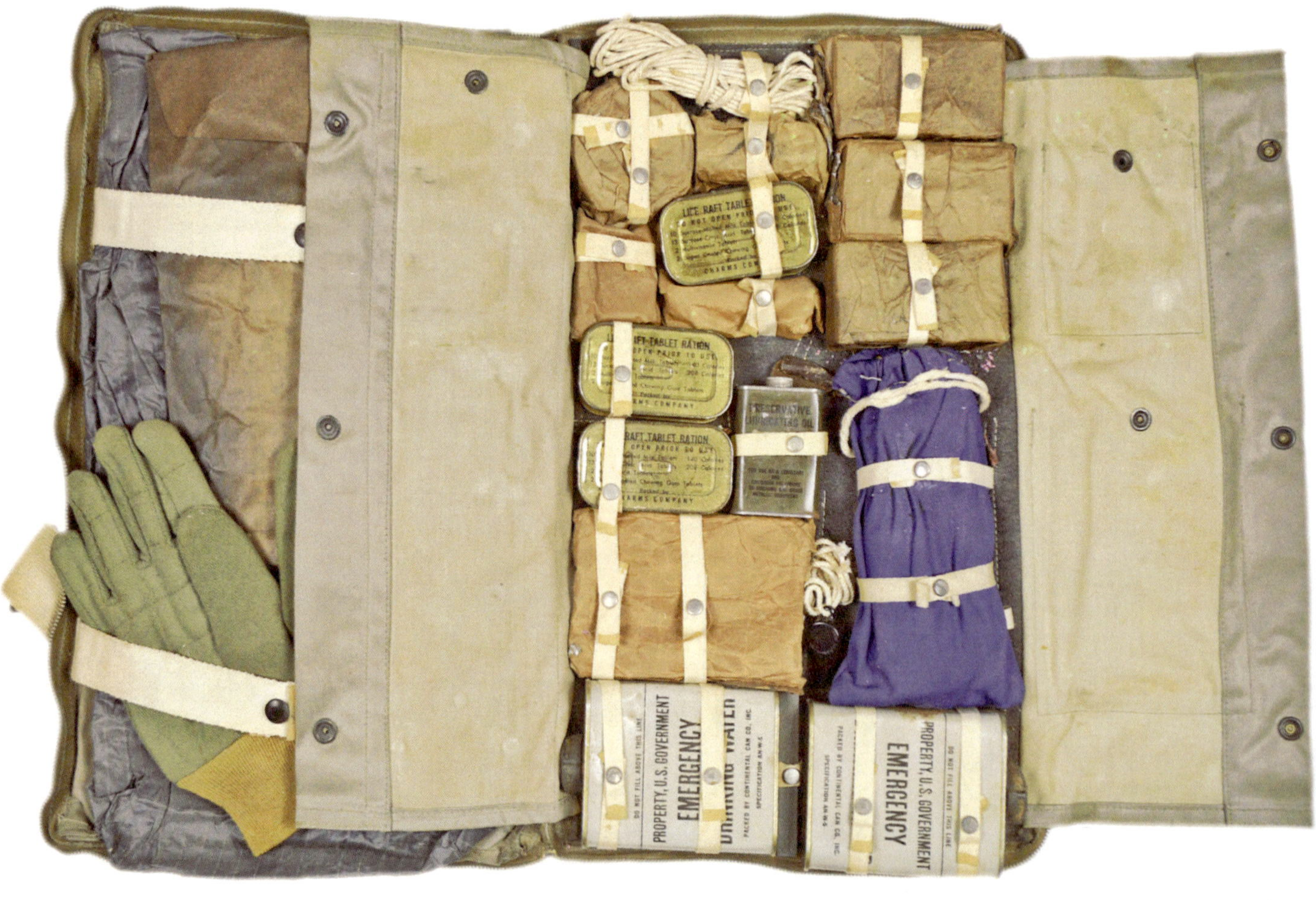

September 1943 contract awarded to Sears & Roebuck (Belber Trunk & Bag), contract number NOA-1594. This contractor used a neoprene rubber-treated fabric with a dull finish and an oval oil can. Note lack of labels on the three pockets containing the machete, mag lens, instruction book, and mirror.

June 1944 contract M-592 of the New England Bedding Co., contract number 288-23689. Per Amendment 1 of specification M-592, kit includes tube of Kant-Rust. Container is identical to that in 1943 contract except the 25' of line placed toward the middle of the kit adjacent to the Kant-Rust. For Illustration purposes, a drinking water kit M-613 is displayed with the unit. Two cans of water would be delivered with the kit from the contractor.

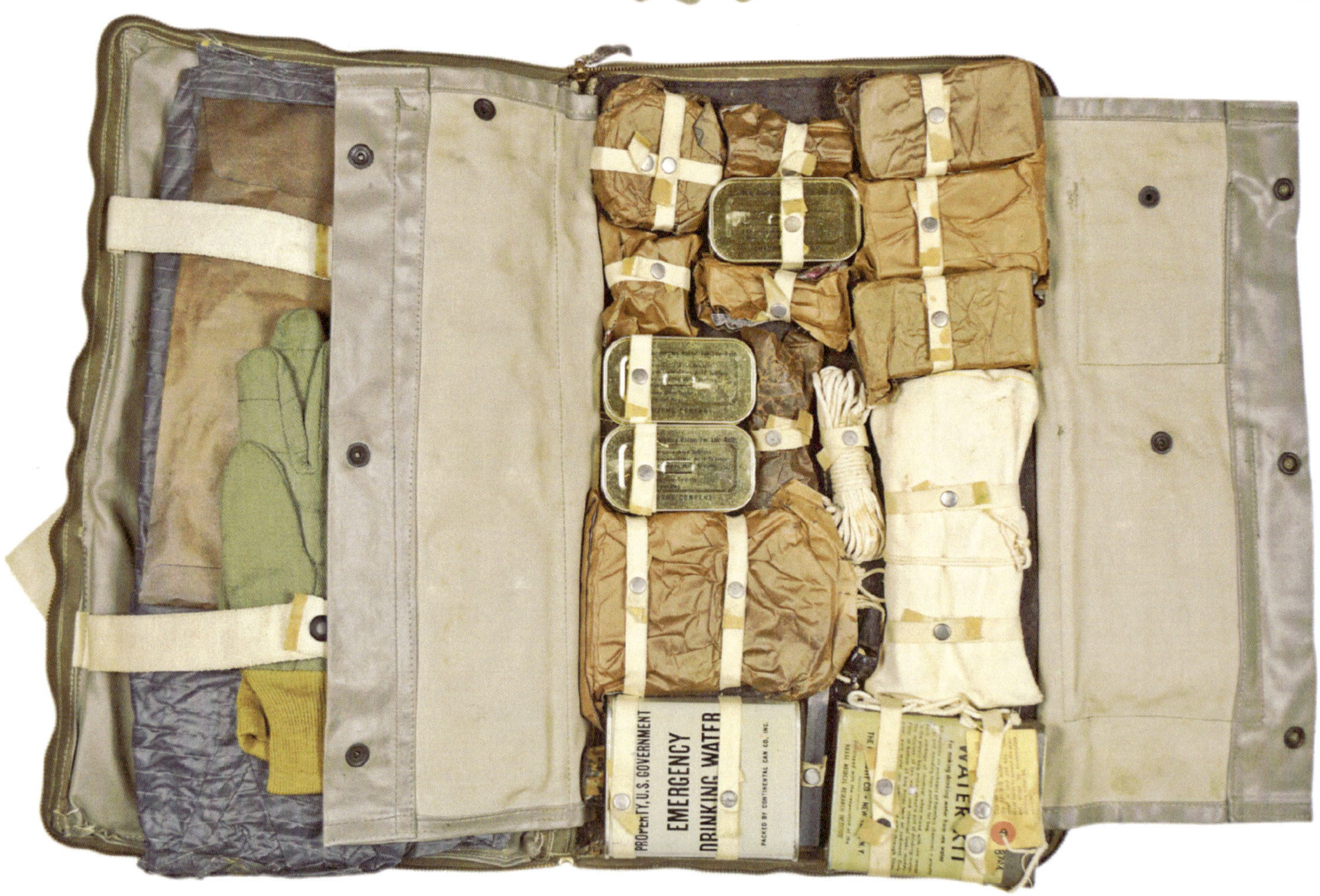

June 1944 contract M-592 of the Sears & Roebuck Co., contract 288-23690. Per Amendment 1 of specification M-592, kit includes the tube of Kant-Rust. The property mark USN was enlarged on the face and labels were added on the three inner pockets for the machete, burning glass, signal mirror, and log book.

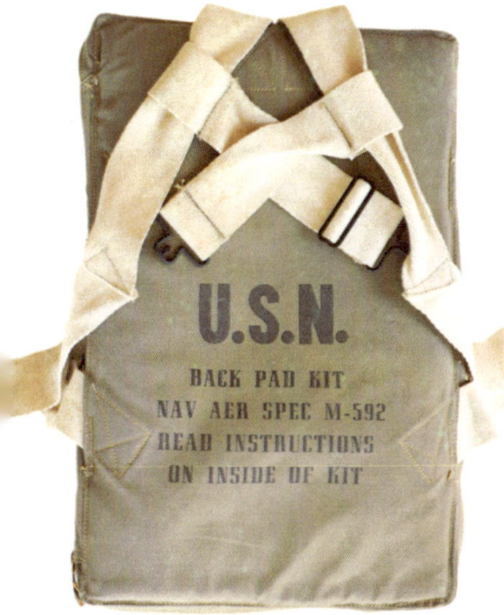

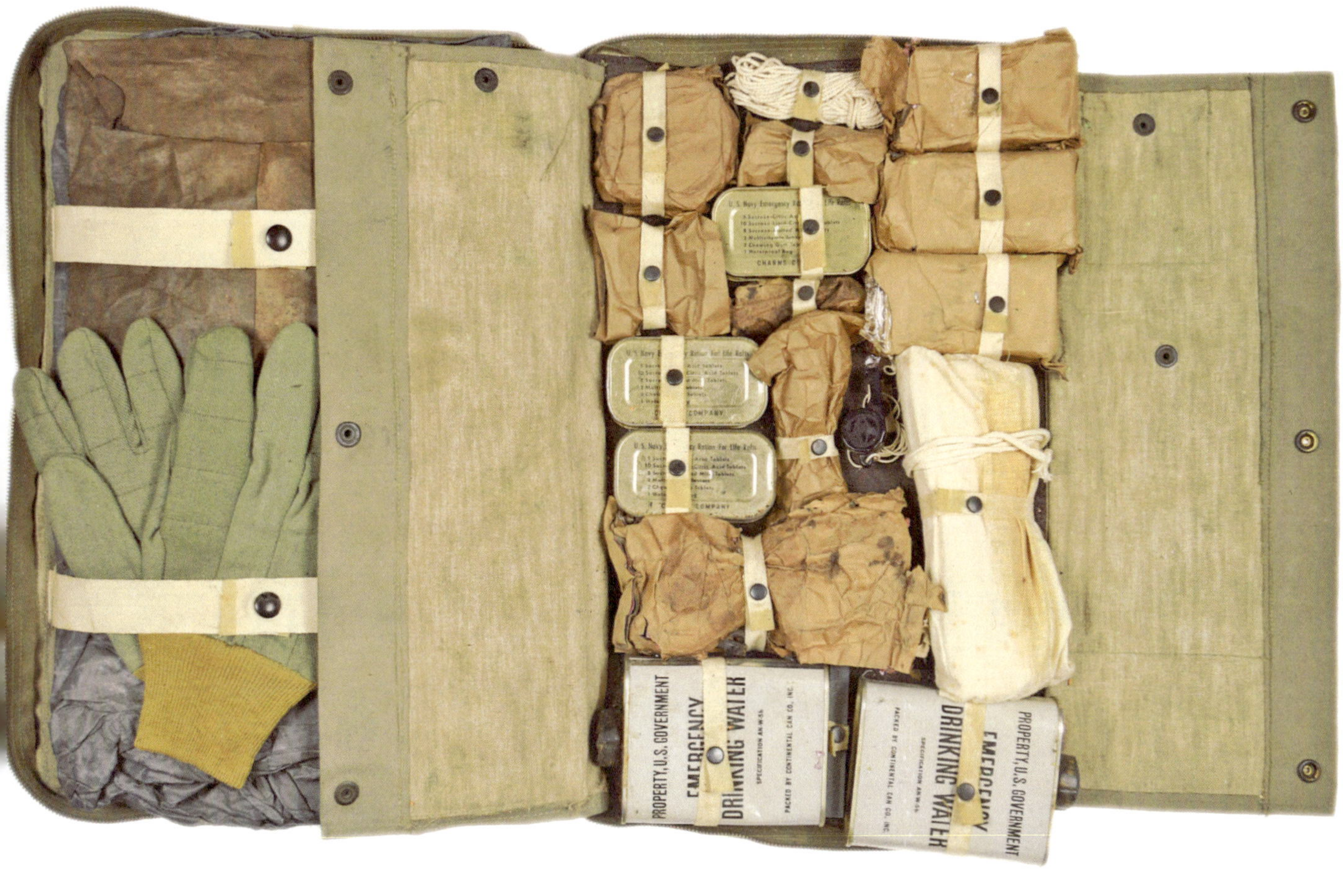

(Right) VC-21 fighter pilots Lt. jg. C.R. Bradford and Ens. W.M. Penny aboard USS Marcus Island, January 1945. Both have M-592 back pad kits; another hangs on the wall. (National Archives)

(Below) Comparative view of the original back pad and the M-592 showing differences in shape and size. (National Archives)

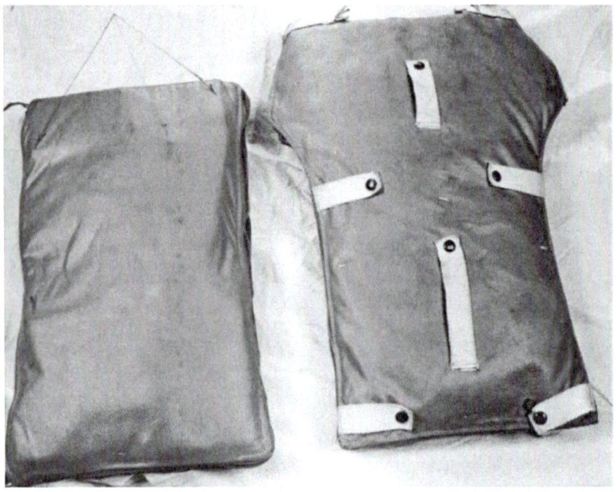

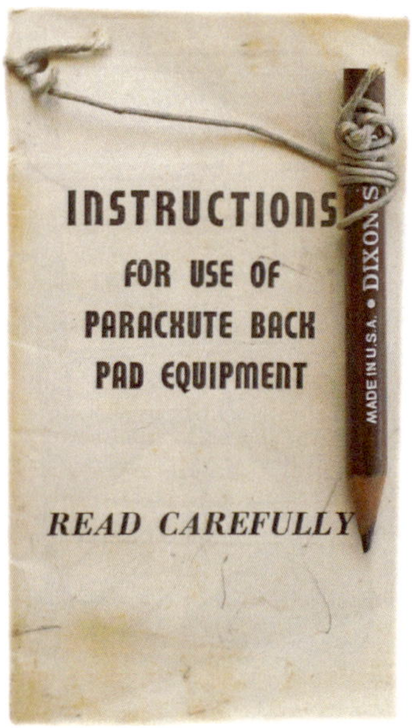

(Left) Instruction booklet provided in the New England Bedding Co. M-592 contracted kits. Same booklet was used for both contracts, not amended to include the tube of Kant-Rust. Booklet from Original kit was used, amending the drawing and descriptive information for the new items.

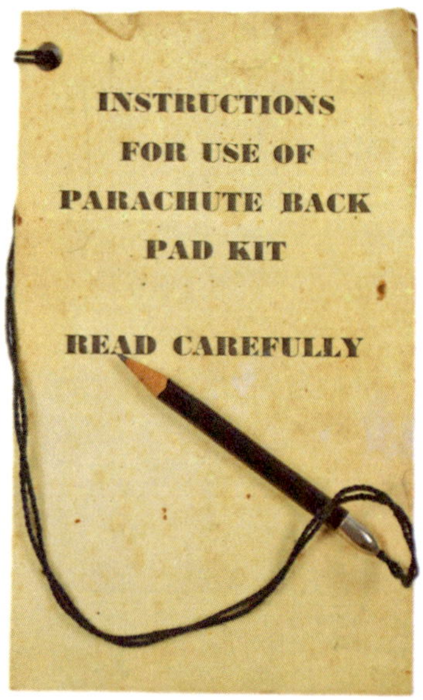

(Left) Instruction booklet provided by Belber Trunk & Bag (Sears & Roebuck Co.) on 1943 M-592 contract. Several pages of the booklet provided for the kits were left blank for a recording log; all booklets include an attached pencil.

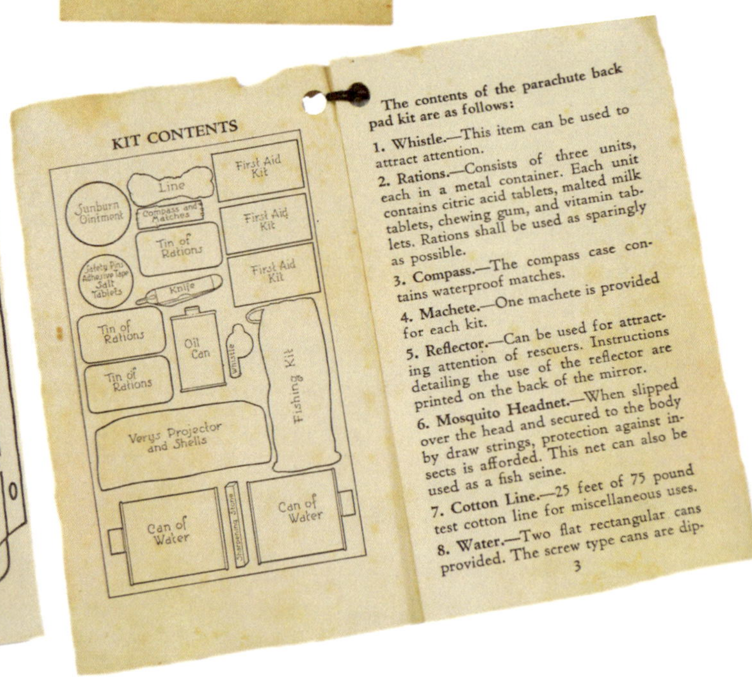

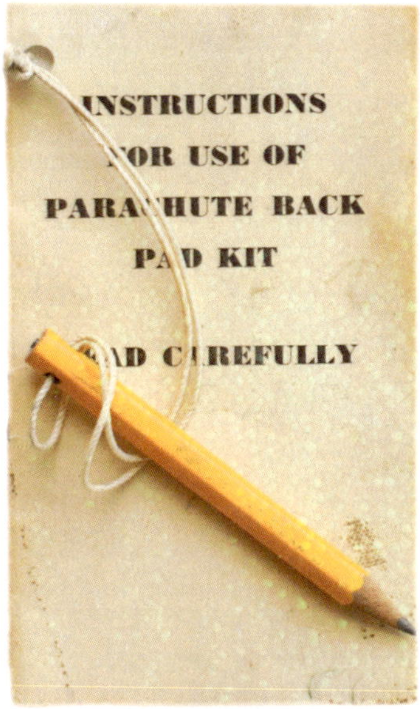

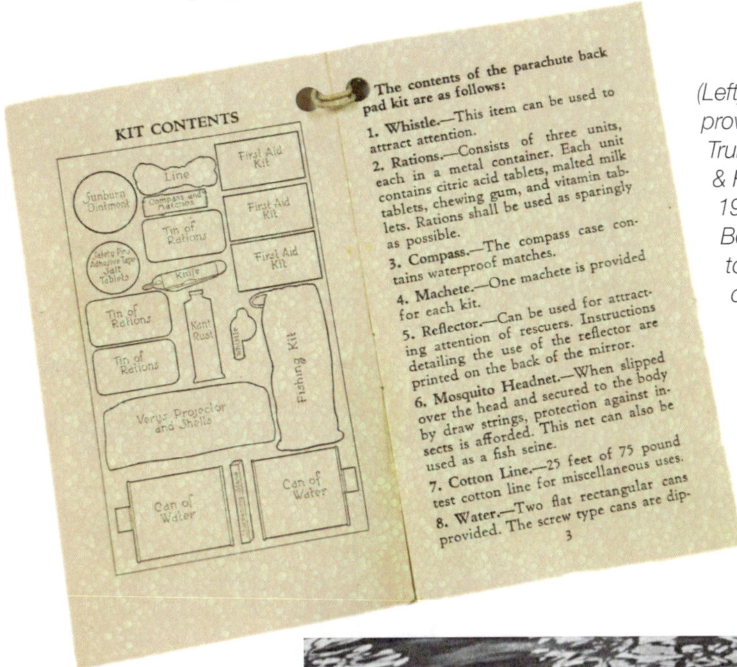

The contents of the parachute back pad kit are as follows:

1. Whistle.—This item can be used to attract attention.
2. Rations.—Consists of three units, each in a metal container. Each unit contains citric acid tablets, malted milk tablets, chewing gum, and vitamin tablets. Rations shall be used as sparingly as possible.
3. Compass.—The compass case contains waterproof matches.
4. Machete.—One machete is provided for each kit.
5. Reflector.—Can be used for attracting attention of rescuers. Instructions detailing the use of the reflector are printed on the back of the mirror.
6. Mosquito Headnet.—When slipped over the head and secured to the body by draw strings, protection against insects is afforded. This net can also be used as a fish seine.
7. Cotton Line.—25 feet of 75 pound test cotton line for miscellaneous uses.
8. Water.—Two flat rectangular cans provided. The screw type cans are dip-

3

(Left) Instruction booklet provided by Belber Trunk & Bag (Sears & Roebuck Co.) on 1944 M-592 contract. Booklet was updated to reflect the inclusion of the Kant-Rust.

Specially designed NavAer Specification P-61 Poncho, Raft Cover included in the M-592 back pad kit. Poncho was made from a nylon ripstop fabric, sea-blue on one side and orange/yellow on the other. The poncho's oblong shape covers the AN-R-2 pararaft with elastic edges at the front and back, forming a snug fit. Access slots were provided on each side for the extension of the arms with an additional slot out the rear for the sea anchor. When used on land, snap fasteners were provided at the front and rear edges for securing. The poncho was stored under the front compartment forming the back cushion.

Standard components for the M-592 back pad kit for all contracts included a pair of cotton work gloves, V-44 machete, match safe compass AN-C-101, sunburn ointment, jackknife M-575, a sharpening stone, mosquito headnet M-565, signal mirror M-580 and a pyrotechnic kit with a drawstring bag.

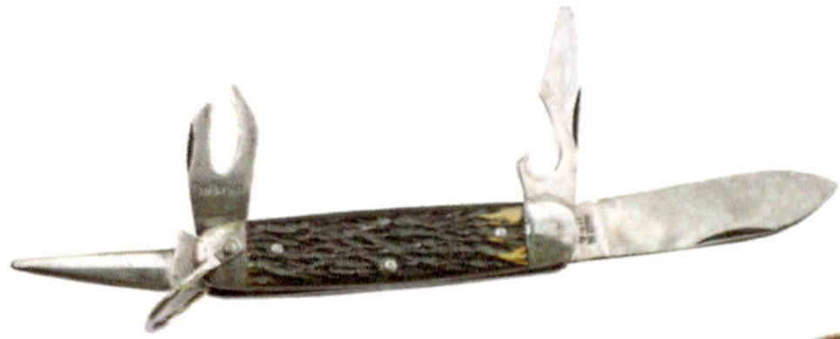

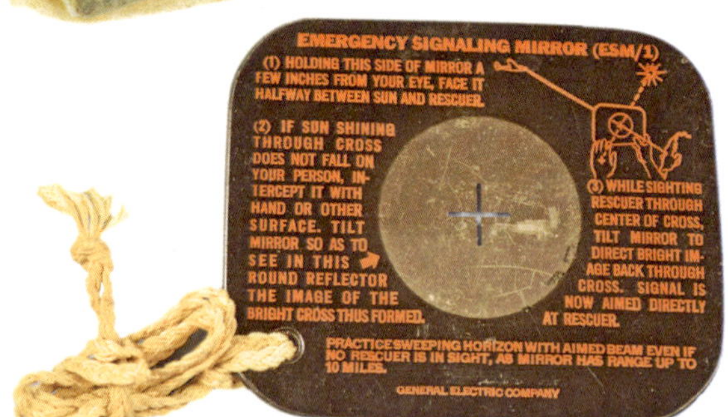

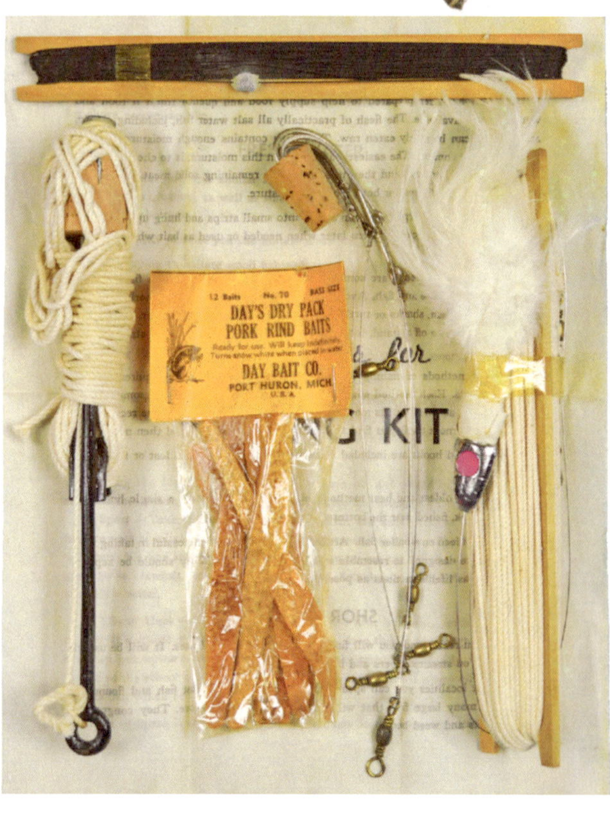

(Left) Fishing kit provided in the M-592 is in accordance with AN-K-2 but only provided with 100' of 15 lb. nylon line, 100' of 24 lb. shoreline with feather jig, five assorted hooks with wire, 12 pork rinds and instruction sheet provided in a cotton drawstring bag. The bags provided by Belber Trunk & Bag are lightweight white cotton. New England Bedding Co. provided an assortment of heavy duck cloth colored bags, to include white, red, blue, tan, and sea green. Same applies for the drawstring bag for the projector kit – white only from Belber, and a variety of colors from New England.

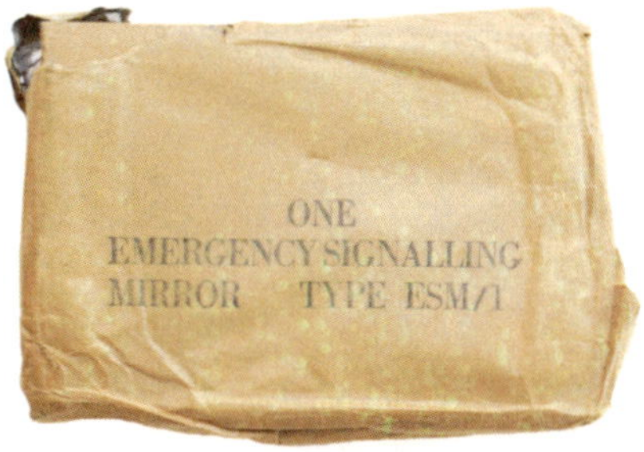

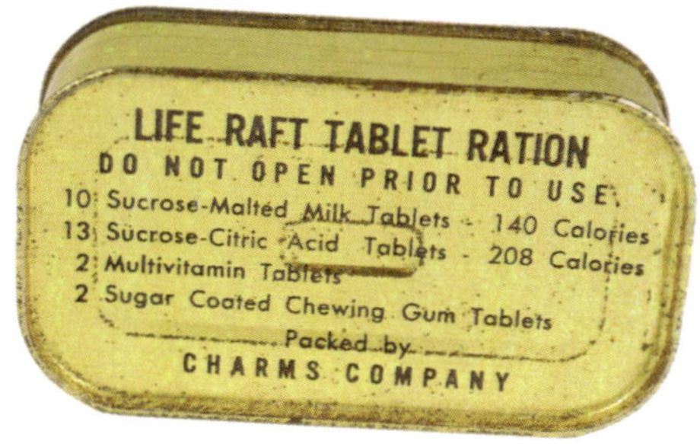

1943 contracts for both the New England Bedding Co. and Belber Trunk & Bag utilized the flat AN-W-5A water can, expedited procurement M-539B tablet ration, and a Bakelite frame magnifying lens. Additionally, the waterproof wrapper for the M-580 would be provided with an ink stamped nomenclature for these early kits.

(Right) New England Bedding Co. and Belber Trunk & Bag amended the assemblies of the M-592 for the 1944 contracts. Included were the AN-W-5B water can, standardized M539B tablet ration, and a simplified magnifying lens with cellulose acetate rim. Packaging for the mirrors in these later editions would eliminate the stamped nomenclature.

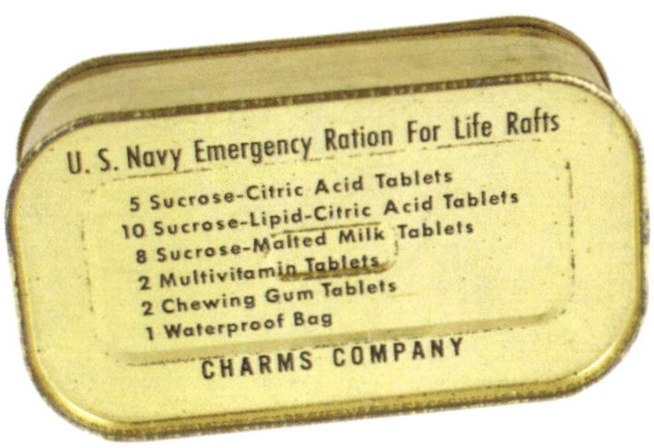

(Below) Both contractors utilized a 3 oz. can of AXS-777 oil. The New England Bedding Co. provided a rectangular type (middle), and Belber Trunk & Bag provided an oval type (right) on 1943 contracts. Both contractors included a tube of Kant-Rust in the 1944 contracts (left).

(Above) Both contractors also had signature articles they provided in the kits, further distinguishing them. (below) Whistle and cotton line provided in back pad kit assemblies by the New England Bedding Co. (above) Whistle and cotton line provided in the M-592 by Belber Trunk & bag.

The six-unit first aid kit provided for the M-592 conforms to BuMed Specification 57K1 (Interim). This unit has multiple variations; one combination here. Two boxes would be packed together and waterproofed with an asphalt-coated paper wrapper creating three separate units.

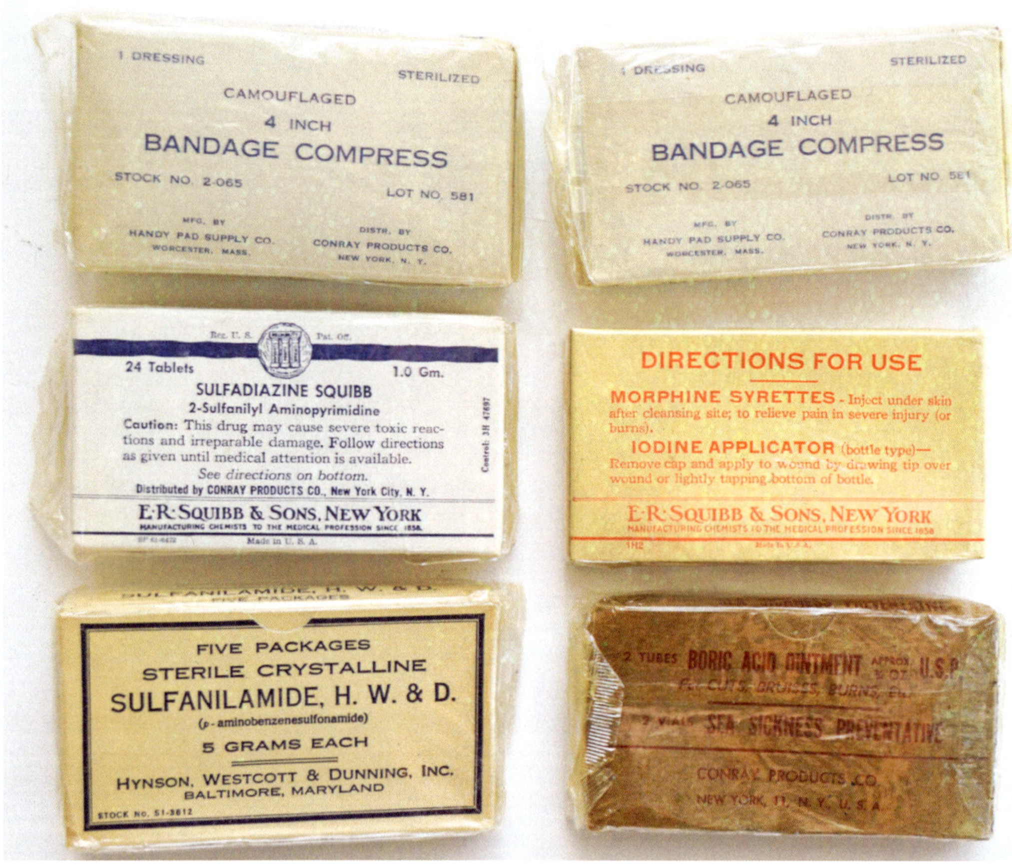

Each M-592 was provided with 5 yards of ½" adhesive tape, eight salt tablets, and six 2"safety pins in a waterproof wrapper.

1943 contract Belber Trunk & Bag kits utilized a 2" x 2" x ¾" clear cellulose acetate container (right). The New England Bedding Co. and the 1944 Belber Trunk & Bag kits packaged the tape, pins, and salt in a waterproof wrapper only, eliminating the use of the cellulose container. Two variations are shown (below), the difference being the salt tablets in a cellulose vial vs. a paper wrapper wedged into the tape roll.

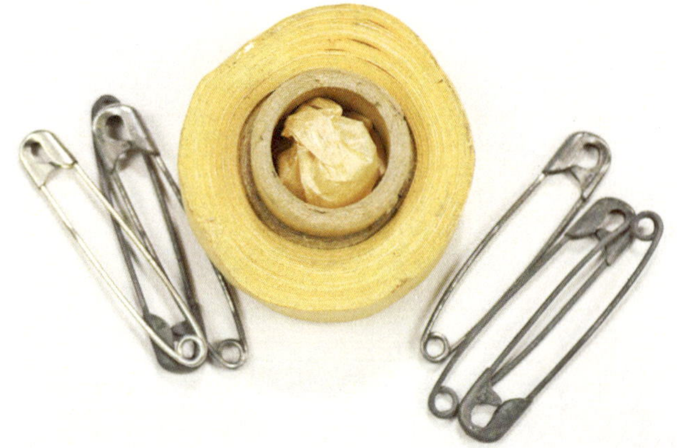

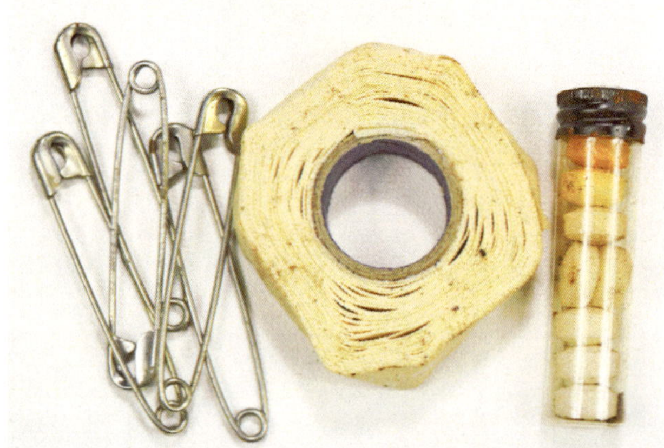

(Above) The design emphasis for the M-592 was on ocean survival and tropical use. It was not uncommon to add supplemental aids. Two typical items were the addition of salt to combat sodium loss through sweat, and Atabrine for malarial resistance.

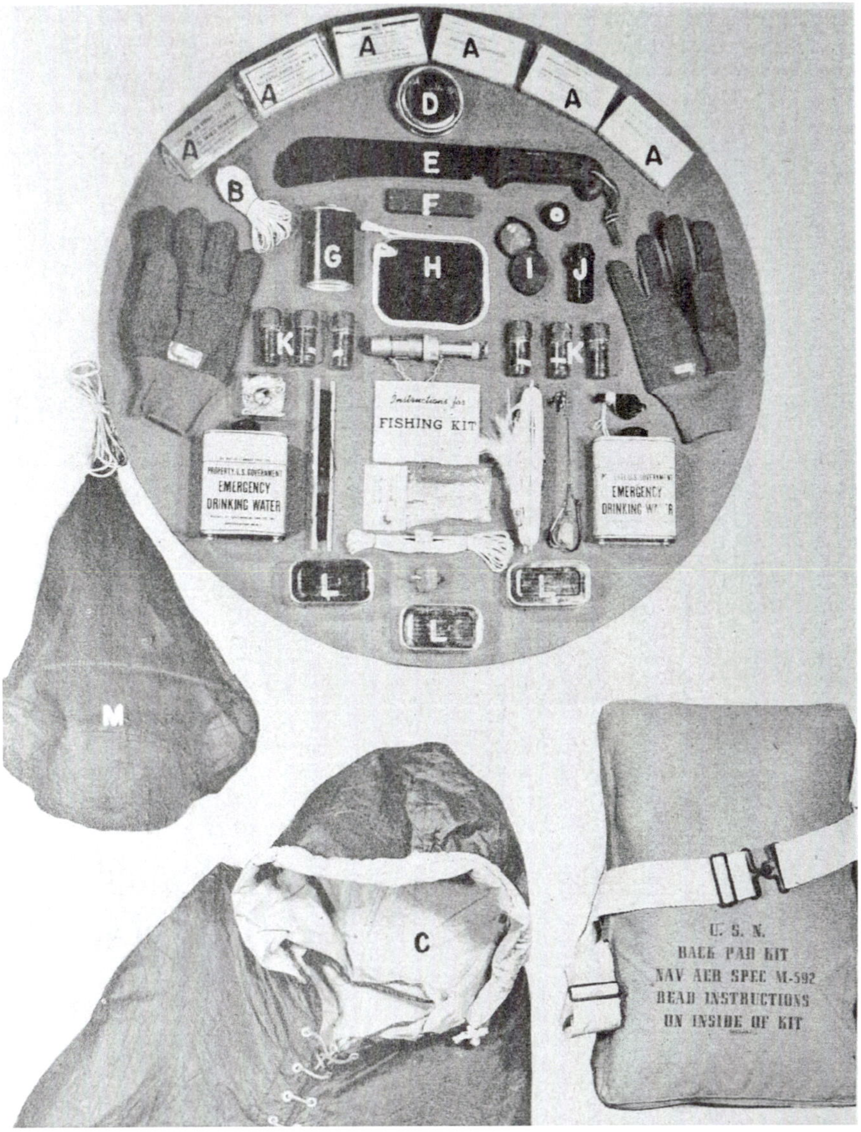

(Left) Additional look at display of components for the M-592 back pad it. The M-592 is from the 1943 contract Belber Trunk & Bag with distinguishable external markings and oval oil can. In some images kits are using the QMC M-1944 mosquito headnet, observed hanging in the image designated by the letter M. The BuOrd catalog for landing forces stocks the M-1944 headnet, an item in naval supply. Specifications for the M-592 remain consistent for the inclusion of the headnet M-565.

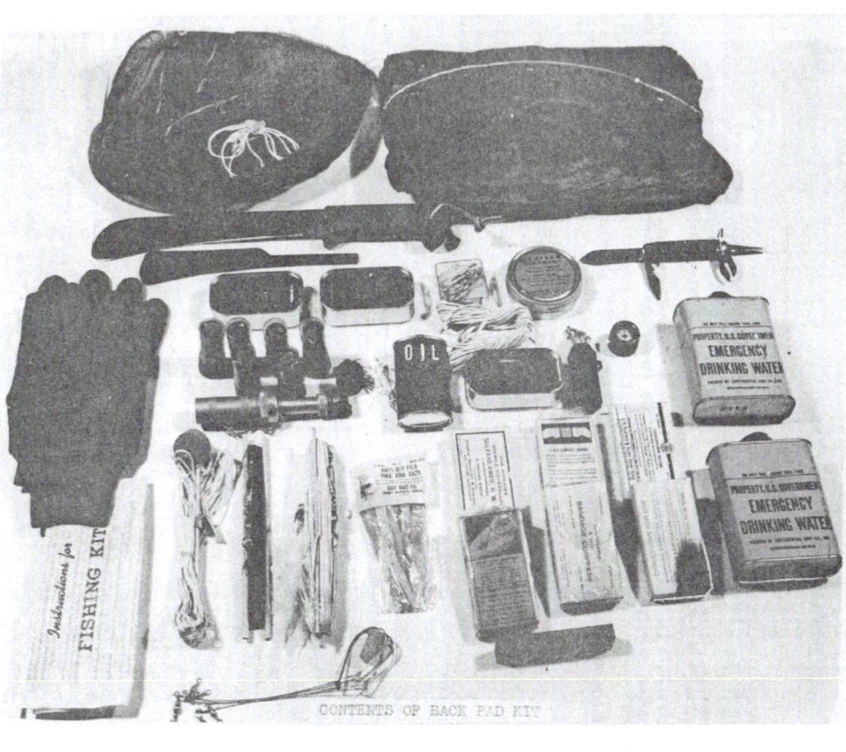

(Left) Another view of components for the M-592 back pad it.

ITEM	STOCK No.	SPEC. No.
Standard Back Pad Kit (Complete)	R83-K-520100	M-592
Emergency Drinking Water (2)	R51-W-135	AN-W-5B
Life Raft Tablet Rations (3)	R56-R-6300	M-539B
Fishing Kit	R37-K-300	M-592*
Compass & Match Container	R37-C-2500	AN-C-101
Headnet	R27-N-261	M-565
25' of 75# Test Line	21-R-150	T-R-571
Pyrotechnic Kit	R83-K-710309	M-592
First Aid Kit	R57-K-8525	57-K-0366 (BuMed)
Jackknife	R41-K-365	M-575
Whistle	R42-W-24000	M-592
Non-Folding Machete	R74-K-200	BuAer Dwg. 8662
Signaling Mirror	R83--M-525525	M-580
Adhesive Tape	C.F.E.	C.F.E.
Salt Tablets	C.F.E.	C.F.E.
Sharpening Stone	C.F.E.	SS-S-736
Poncho Cover	R83-P-650050	NAF Spec. P-61
Cotton Gloves	37-G-2868	73-G-3
Sunburn Ointment	R57-O-680	P-40469 (BuMed)
Magnifying Lens	C.F.E.	C.F.E.
Corrosive Preventative	C.F.E.	AXS-777/ Kant Rust

The official list of accessories for the M-592 back pad kit with their corresponding stock and specification numbers. C.F.E. signifies Contractor Furnished Equipment.

Some design characteristics of the New England Bedding Co. M-592 as compared to the the Belber Trunk & Bag M-592 kit. Bellber straps were sewn parallel to each other, perpendicular to the top edge of the pack (top left), while on the New England Bedding Co version, the straps were sewn at a 90° angle to each other (bottom left).

The leather zipper pull thongs provided on Belber packs were sewn in place (top right), while New England Bedding Co.units had the leather thong knotted onto the metal zipper pull (bottom right).

The Original and M-592 back pad kits were developed by the Engineering Division within the Bureau of Aeronautics designed to accommodate service requests. The intent was to have kits useful both on land and sea, complement the pararaft, and provide essentials to an individual in almost any scenario. This was a goal over-reach that would lead to operational deficiencies.

With the deliveries of the Original type, fleet commands almost immediately found these kits to have bulk issues forcing the user into a forward leaning position, which created considerable discomfort during long flights. The kits were also found to place the user 3-1/2" further from the armor plating reducing angular coverage of protection. Furthermore, the components in the Original kit were found to be too lumpy, addressed through the procurement of modification kits, with no attempts to formalize a revision for the M-592. By the time of the M-592's introduction, the BuAer was already working on the concept of a combination seat pack inspired by a design introduced by MCAS Cherry Point, titled the Mark-I Parachute Raft and Back Pad Kit Combination. Various other service commands took it upon themselves to also correct these deficiencies, submitting designs and recommendations to the BuAer. Based off trial and error and revised service requirements, a new emergency seat pack would be introduced, later known as the Model-A and PK-1 pararaft kits. The demand for emergency kits continued to grow steadily and to meet that, additional quantities of the M-592 were procured in 1944 as a stopgap measure until a suitable design of a combination seat pack could be finalized.

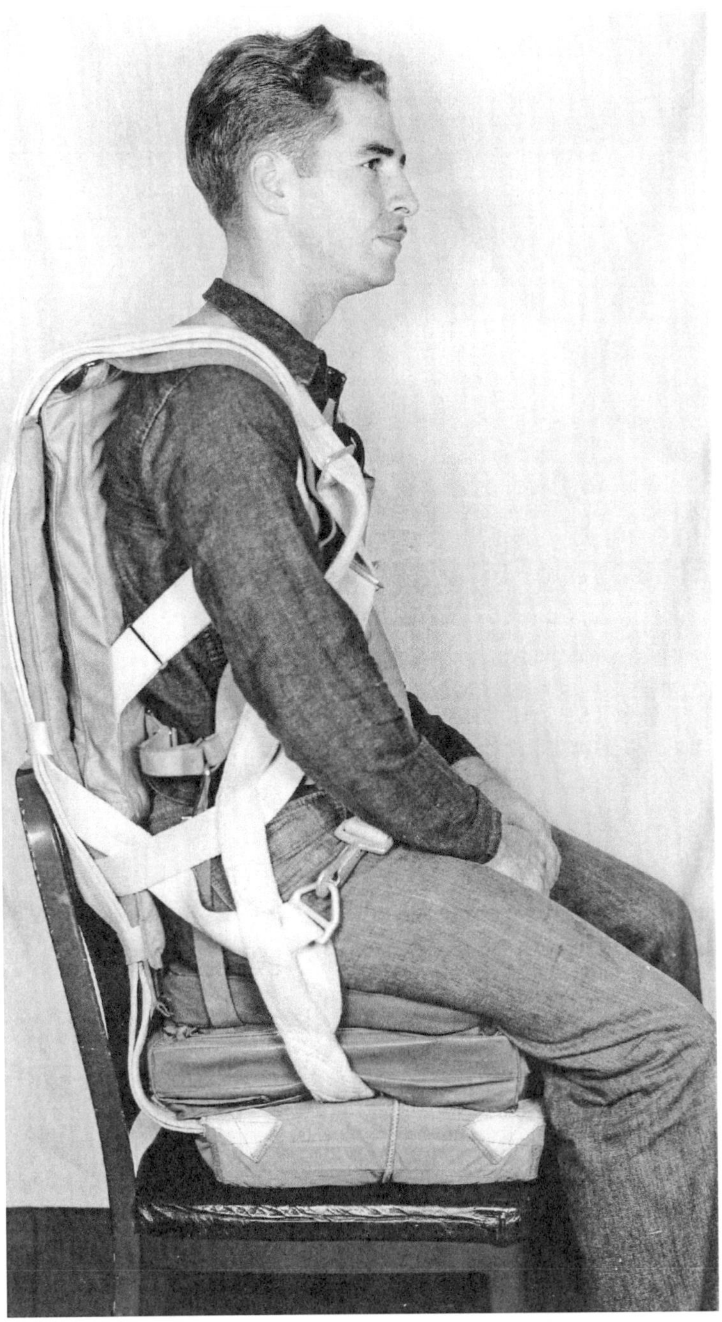

(Left) Standard assembly consisting of the parachute harness, pararaft, seat cushion, and M-592.

(Below) User demonstrates how the back pad kit places him into a forward position creating a gap between the pilot and armor plating. The issues encountered would depend on the airman's crew position and size. Smaller pilots and aircrew reported fewer problems, but the opposite was expressed by larger individuals. The most effective solution was the elimination of the poncho, which greatly reduced the thickness of the M-592. Other items deemed unnecessary were also removed. (National Archives)

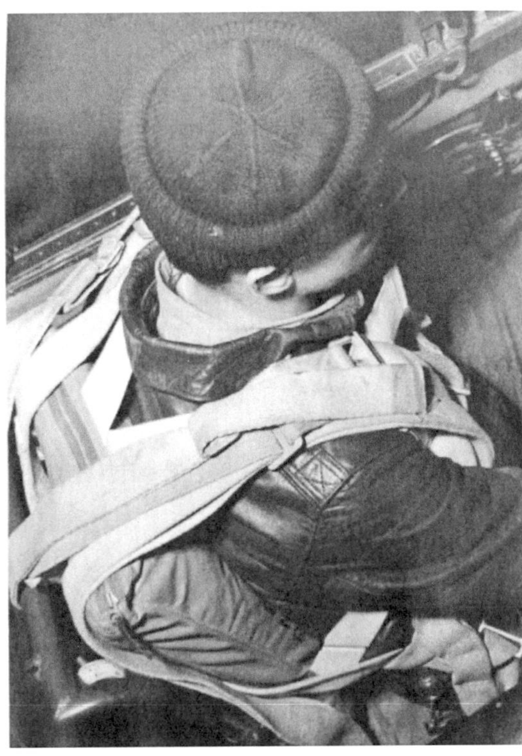

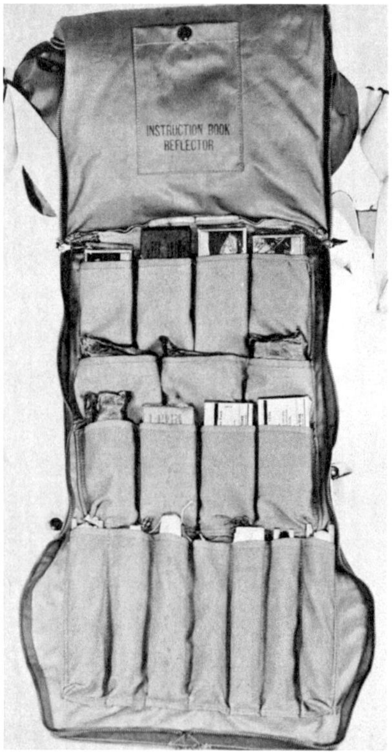

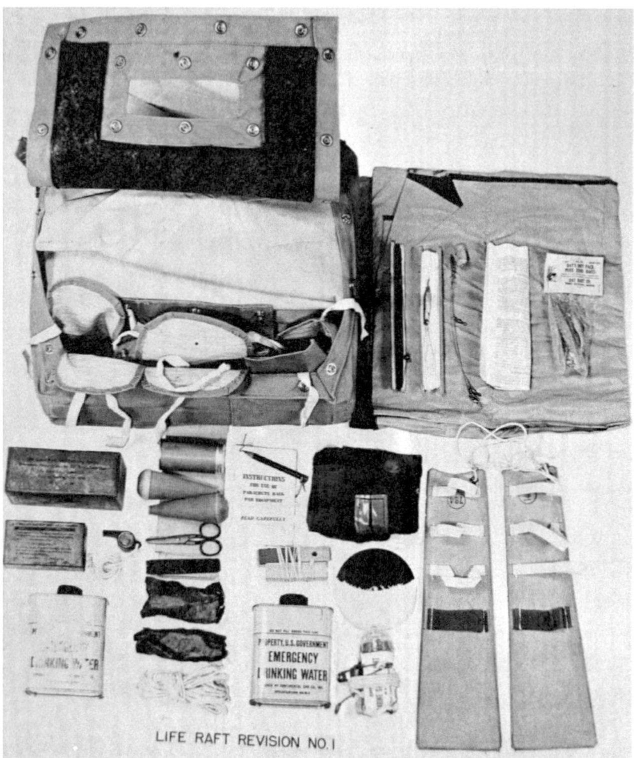

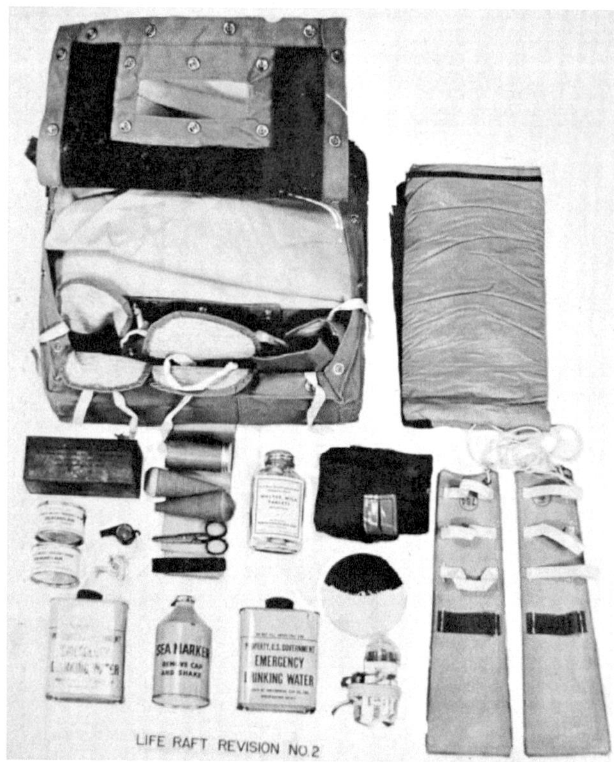

Several concepts under the designation of Mark-I Parachute Raft and Back Pad Kit Combination were devised at MCAS Cherry Point in August of 1943. The revision here used all the standard components from both the pararaft case and the Original back pad kit. Vertical pockets were installed in the back pad for each individual component lying as flat as possible but allowing some flexibility. All other components were stored in standard pockets provided in pararaft case. This configuration slightly reduces the backpad's thickness, with minimal increase to pararaft. A felt pad was sewn under the top flap of the pararaft case to help with comfort and eliminate the use of the standard parachute seat cushion. An additional revision removed several articles, to include the first aid kit and pemmican rations. With these items eliminated, it was possible to pack all other items into pararaft case. This was the most streamlined method and, as seen far above, a standard parachute back pad could be used placing the user's back to the original position in a cockpit; there was no increase in the thickness of the pararaft. This modification proved optimal for all types of aircraft and crew positions. Service test reports detailing the findings were submitted to the BuAer in October of 1943. (National Archives)

Following service test reports, the BuAer pursued the idea of a combination seat pack. In coordination with MCAS Cherry Point, an additional modified kit was developed retaining the title Mark-I Modified Parachute Raft and Back Pad Kit Combination. Though the combined seat kit was preferred, it eliminated vital articles of necessity. In this new modified unit, designers attempted to include all items but in a new configuration by placing the raft, sail, mosquito headnet and repair kit in the back pad. The pararaft case was modified to retain all other items, essentially reversing the designs. This unit was service tested in early 1944 and found better than the standard combination but still met with the same concerns over bulk and discomfort.

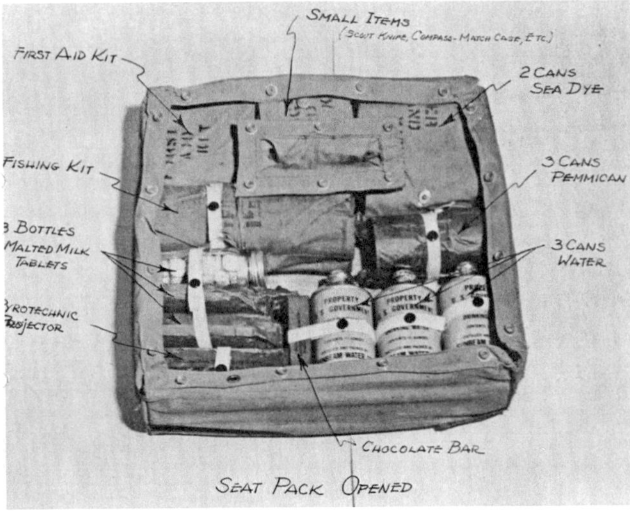

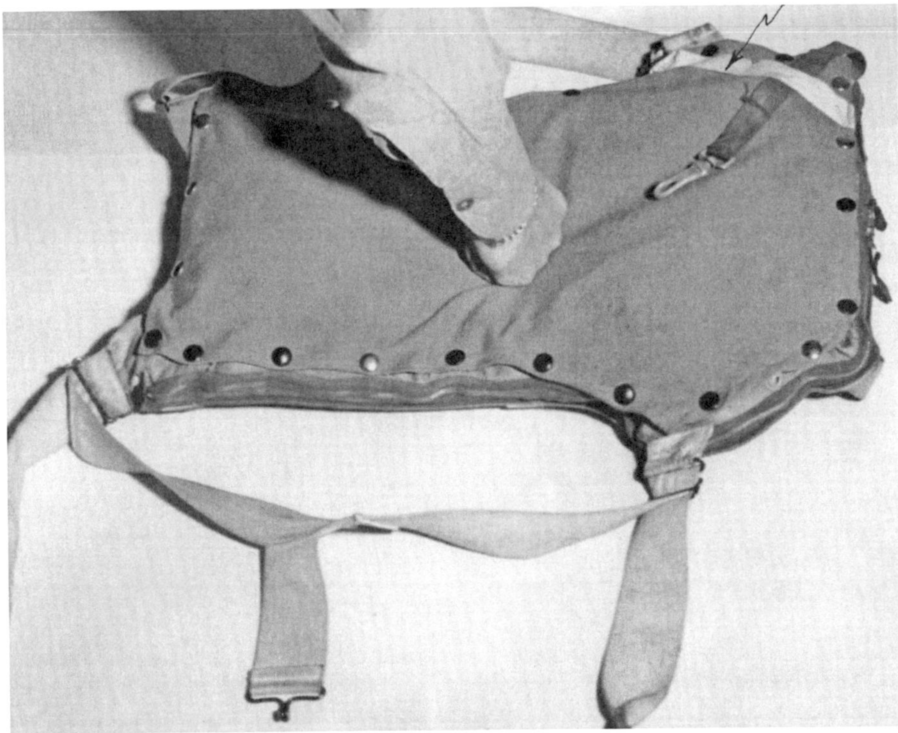

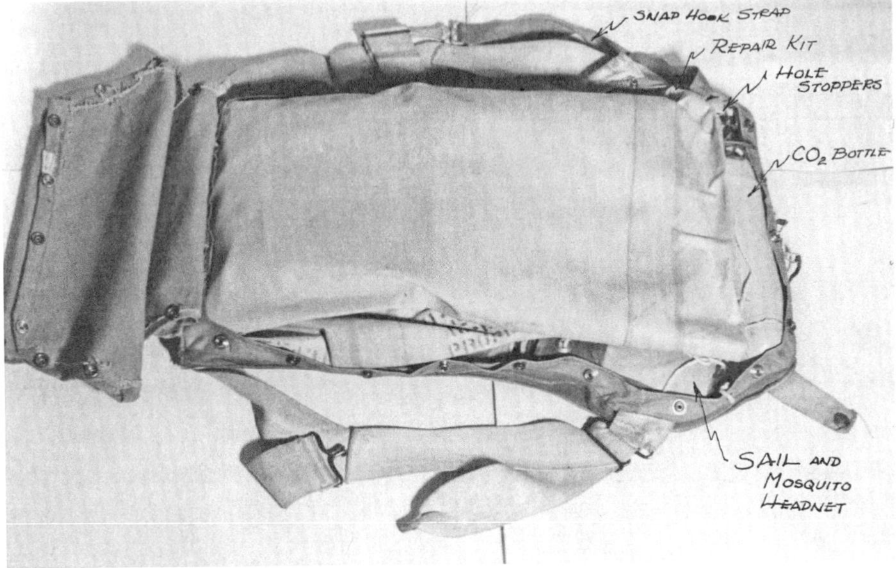

The concept of a combination seat pack was laid out, but the variety of components did not allow for a compact streamlined configuration. As new items were introduced such as the M-627 fishing kit, M-539B tablet ration and the individual camouflaged first aid kit, the combination seat pack would soon become a reality. 1/3 of the seat pack pictured is consumed by the old type fishing kit, rations, and first aid kit.

(Right and below) The original back pad kit was found to be more comfortable than the M-592, so it was a common practice to modify these units. Modification conducted by Air Group 51 in July 1944, adapting the Original kit to house the components from the M-592 with the poncho raft cover as the back pad stored in the top flap. Lt. T.H. Bridges of VF-51 demonstrating the kit on the deck of the USS San Jacinto. (National Archives)

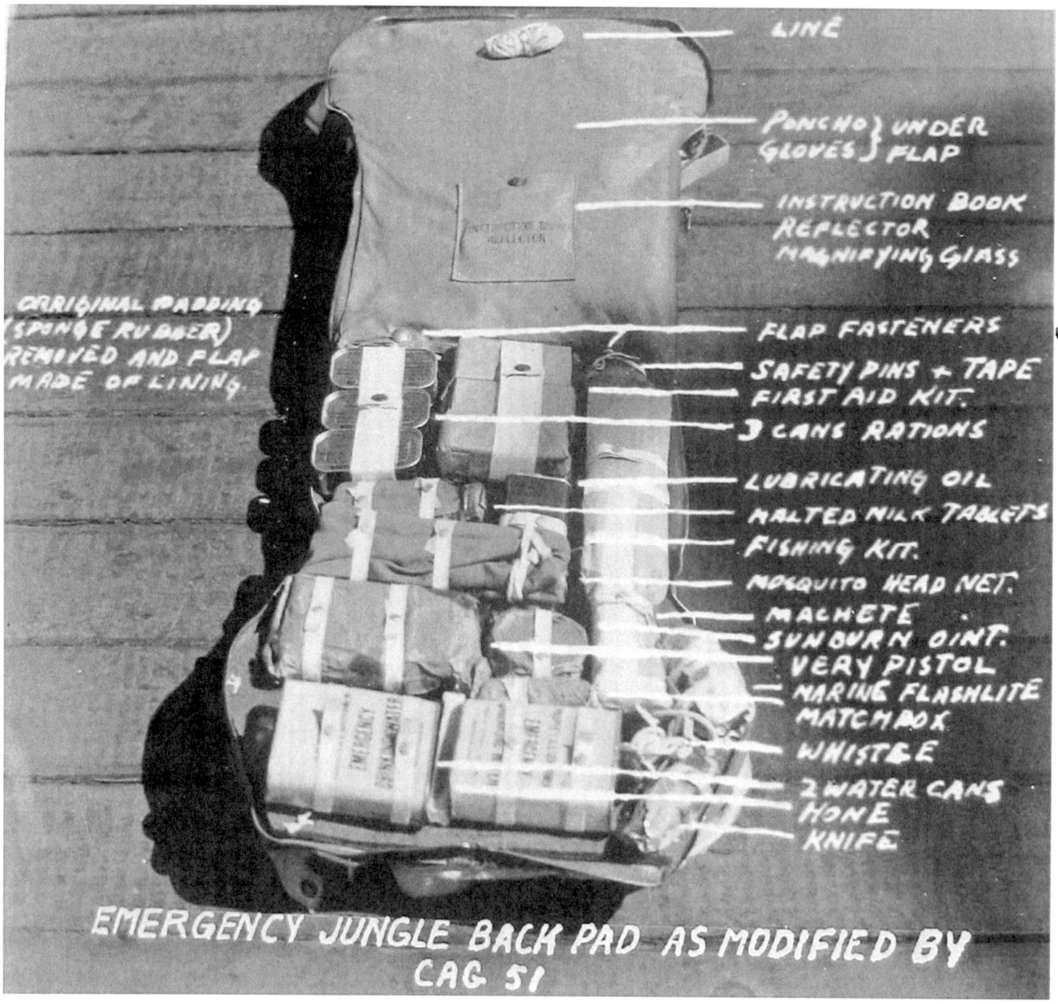

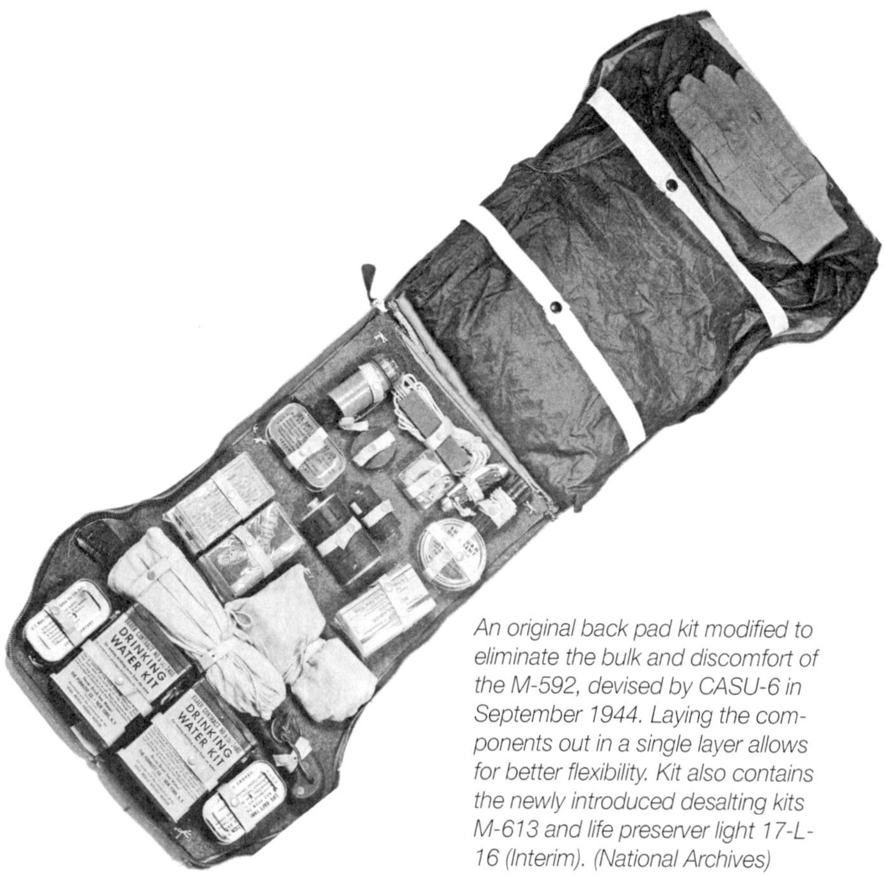

An original back pad kit modified to eliminate the bulk and discomfort of the M-592, devised by CASU-6 in September 1944. Laying the components out in a single layer allows for better flexibility. Kit also contains the newly introduced desalting kits M-613 and life preserver light 17-L-16 (Interim). (National Archives)

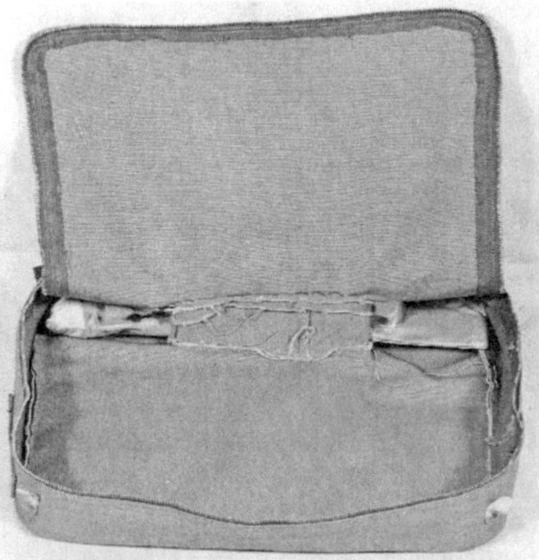

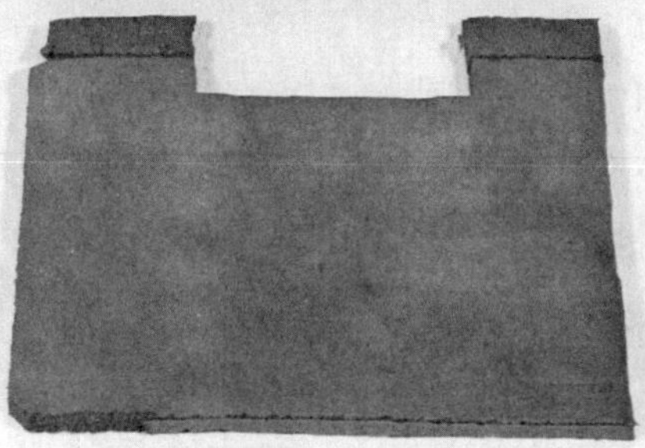

While some operational units were modifying the back pad kits, others were attempting to eliminate them. In March 1944, CASU-4 adapted the standard parachute seat cushion to house the components from the M-592. By only eliminating the poncho raft cover there was no discernible increase in the thickness of the seat cushion. However, this configuration was reported to be problematic for tall pilots and aircrew. With modifications like these to the parachute system, adjustments were required to the aircraft seats and safety belts to accommodate the extra bulk. (National archives)

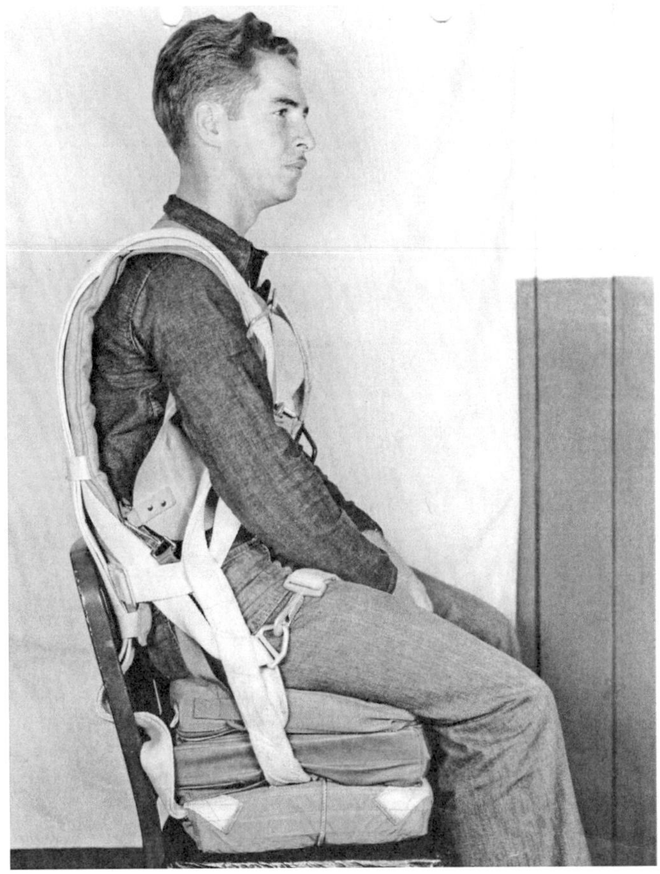

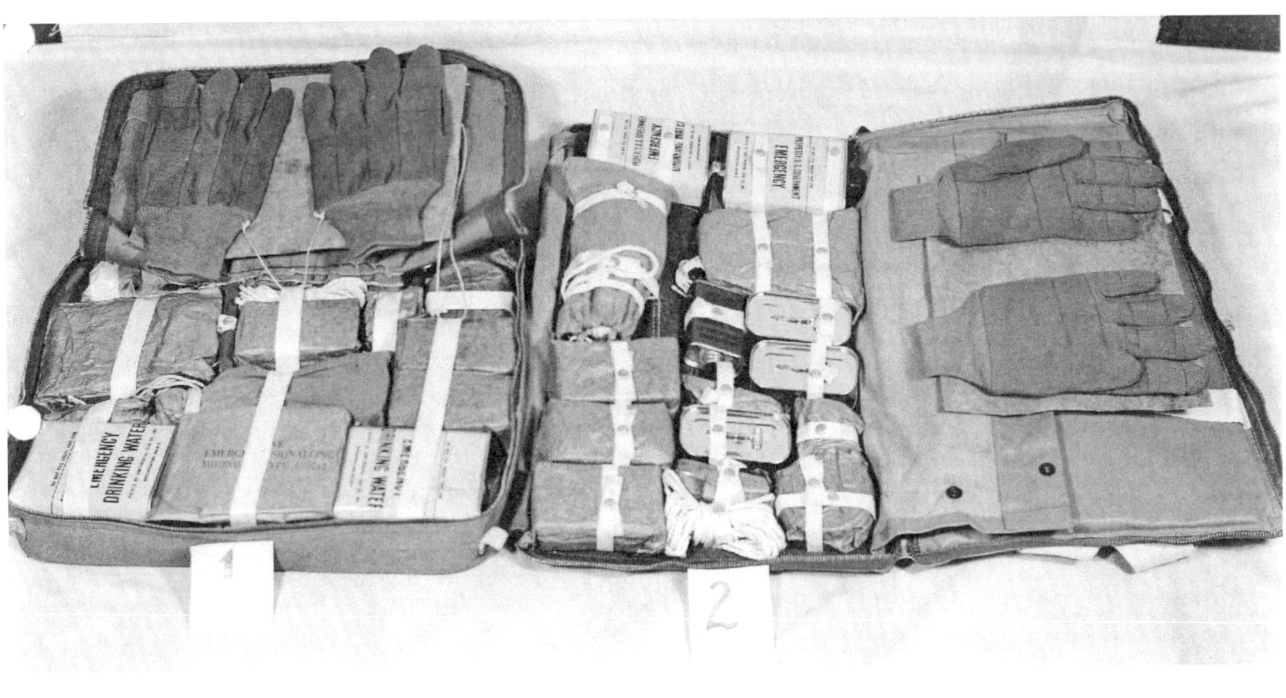

Additional concept of a seat kit developed by VF-39 to overcome the objections of back pad kit, June 1944. For these combination seat kits the poncho raft cover was eliminated. The raft cover was replaced with the sail cloth M-551 or AN-6521-1. The V-44 machete was also removed. Cigarette pack, useful for bartering with natives but also a morale booster, was added. Note inclusion of round chromium mirror; all the other items were standard to the M-592. These seat kits are not unlike the unit developed in 1937 in Panama. (National Archives)

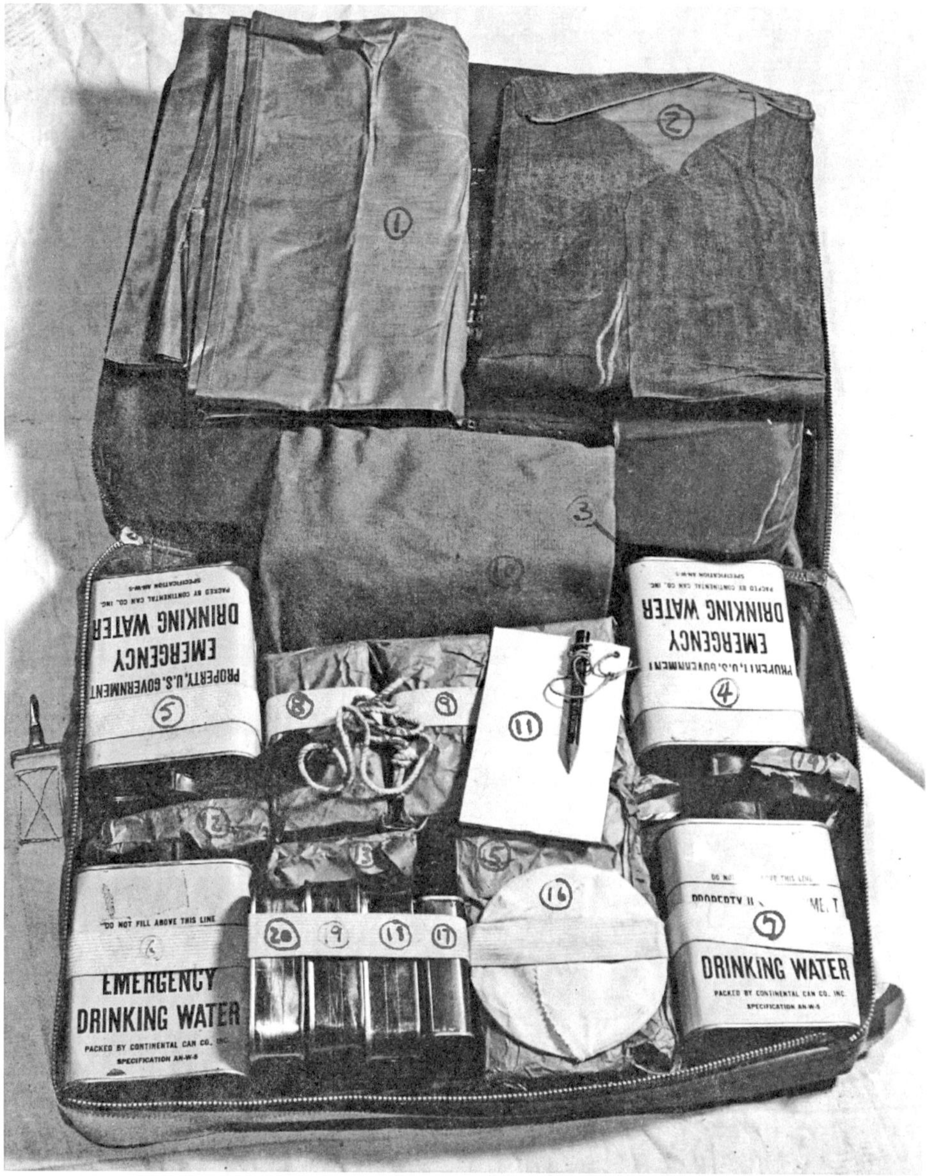

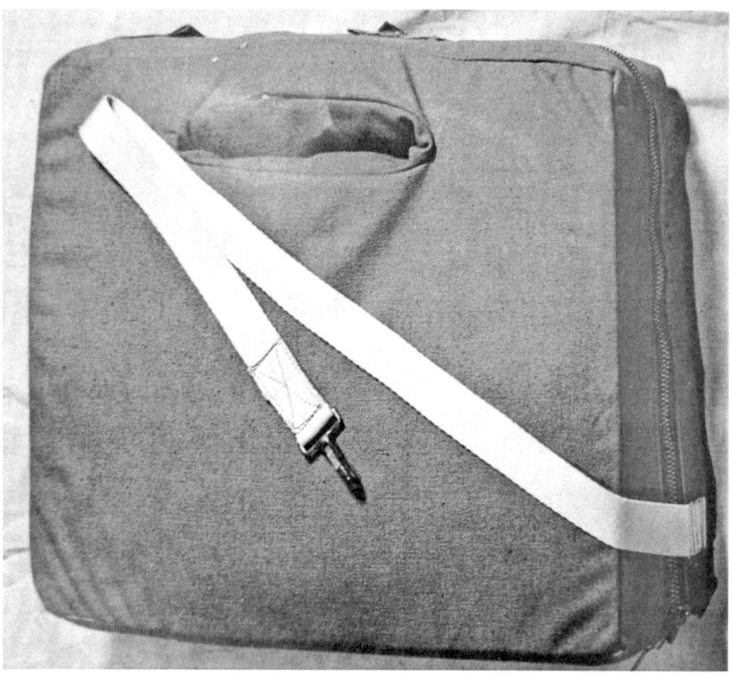

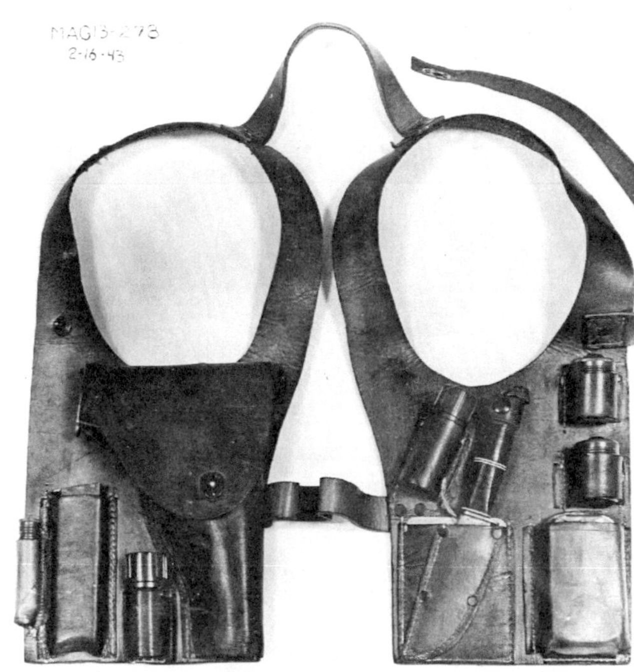

Lt. Col. Thomas C. Green, executive officer of MAG-13, demonstrates a personal emergency kit/shoulder holster developed to help meet demands necessary to sustain life in the jungle or at sea, February 1943. The well-fabricated leather shoulder holster contains fishing kit, basic medical items, vitamins, and a signaling mirror. (National Archives)

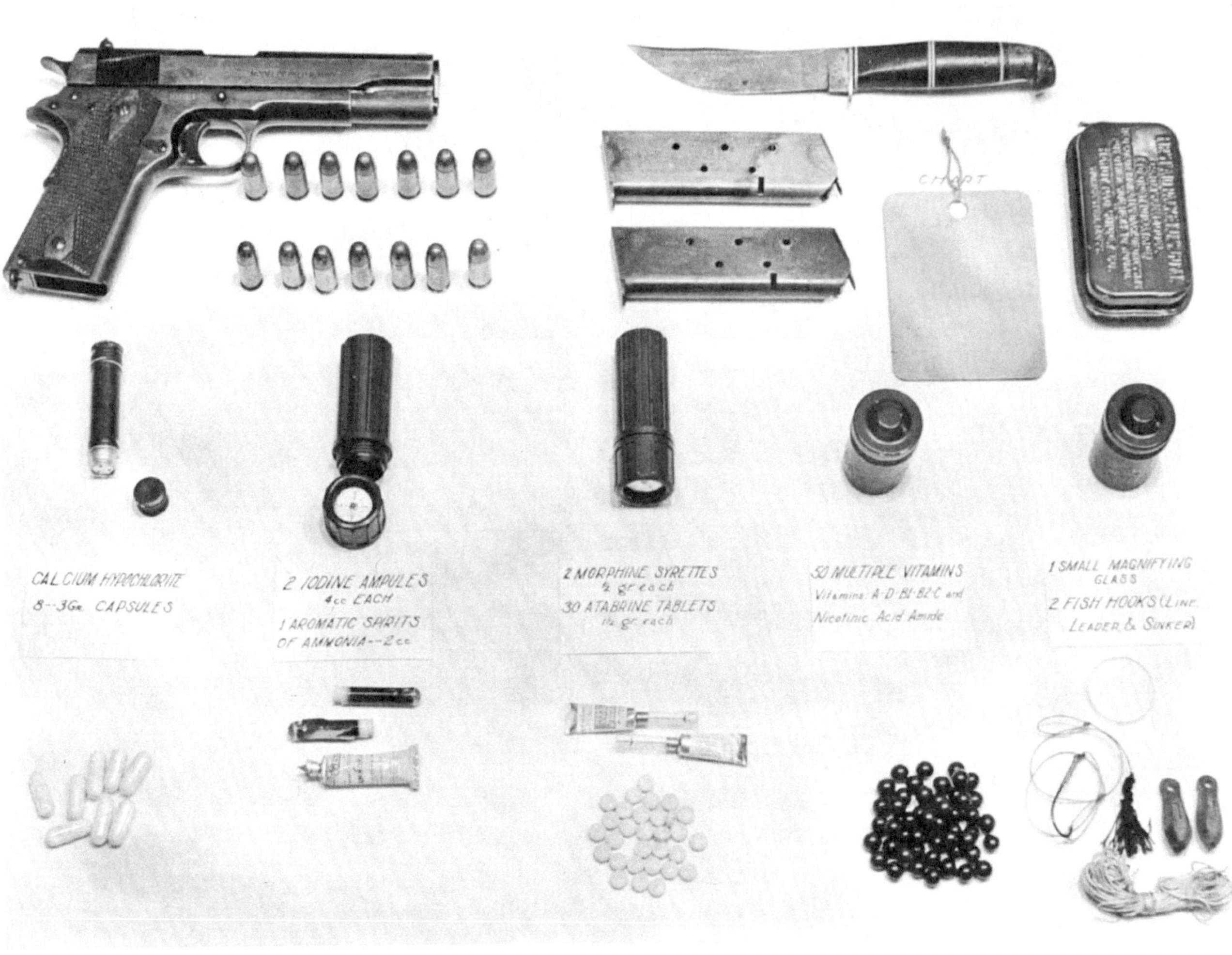

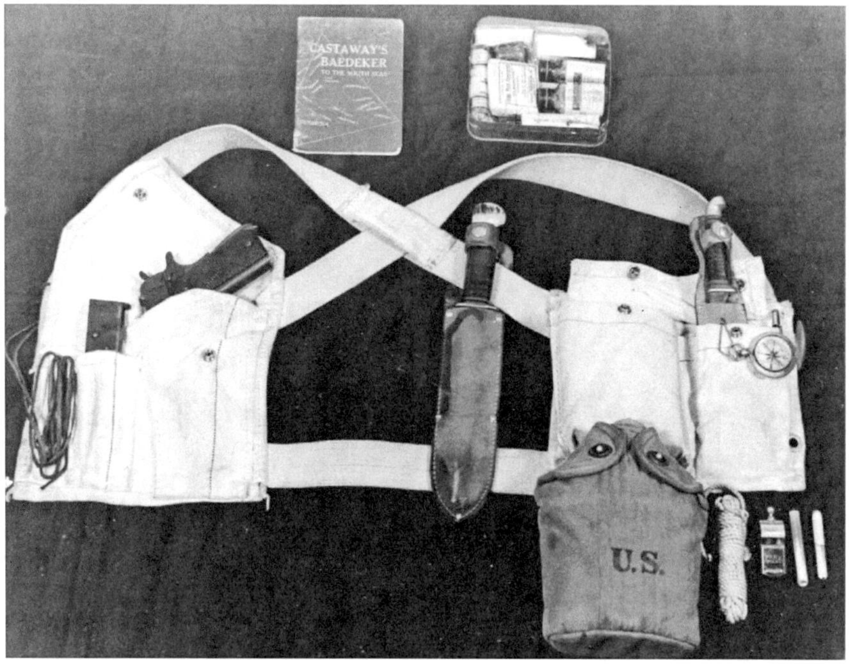

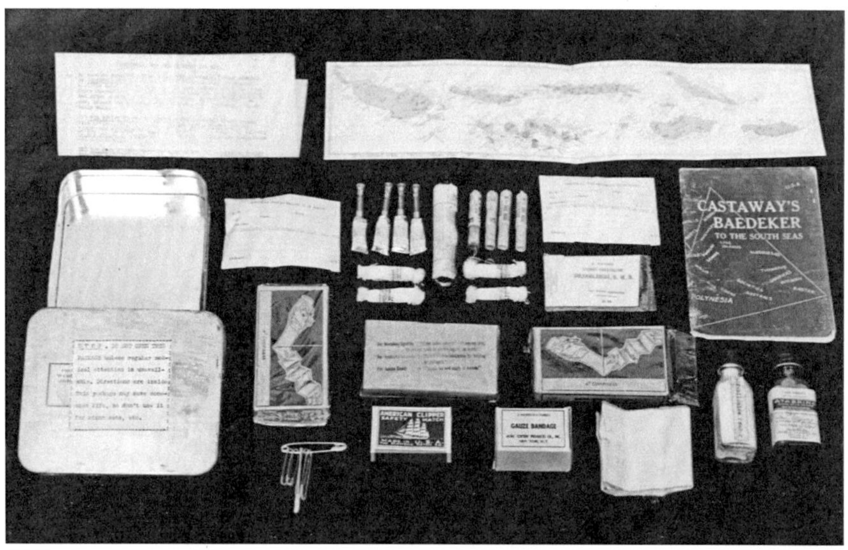

VF-28 took more dramatic steps to eliminate problems encountered with the back pad kit by developing a personal shoulder harness. Pilot Lt. John Pfeiler demonstrates his personal harness containing those items deemed necessary in a survival situation, August 1943. Unit is compact, comfortable, and secured to the user. The US Navy would adopt this concept in the 1950s with the SV-1 survival vest. (National Archives)

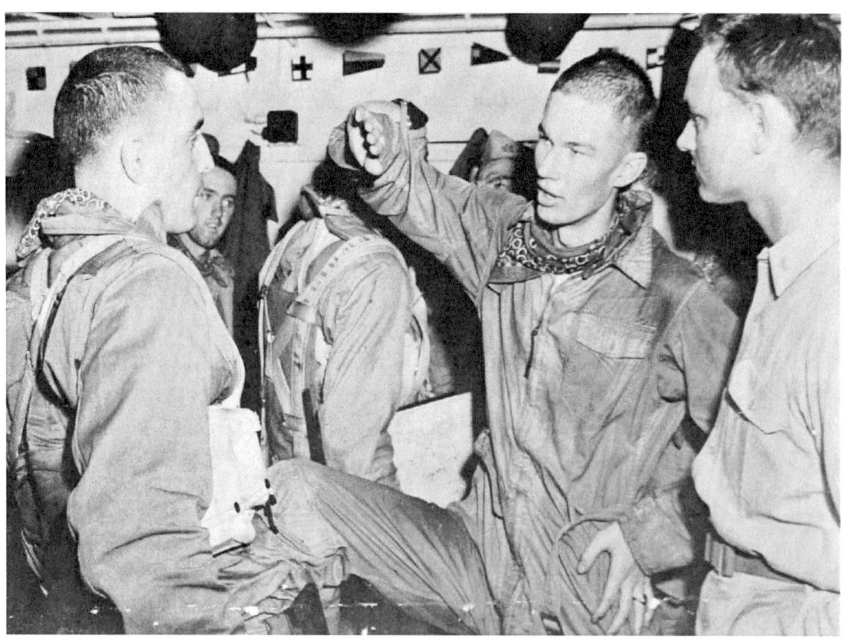

(Right) VF-13 pilot Lt. jg. J.H. Bowen is donning an emergency kit shoulder harness like that of VF-28, October 1944. Components included in these personal emergency kits would vary from unit to unit, but documentation indicates they contain similar items. Typically, they would include first aid items, rations, fishing tackle, barter items, matches and/or candles, compass, headnet, hunting knife and sidearm. The concept of implementing such kits would have originated within the Pre-Flight survival training curriculum where it was stressed that such items should be carried on the person. (Paul Faulkner Collection)

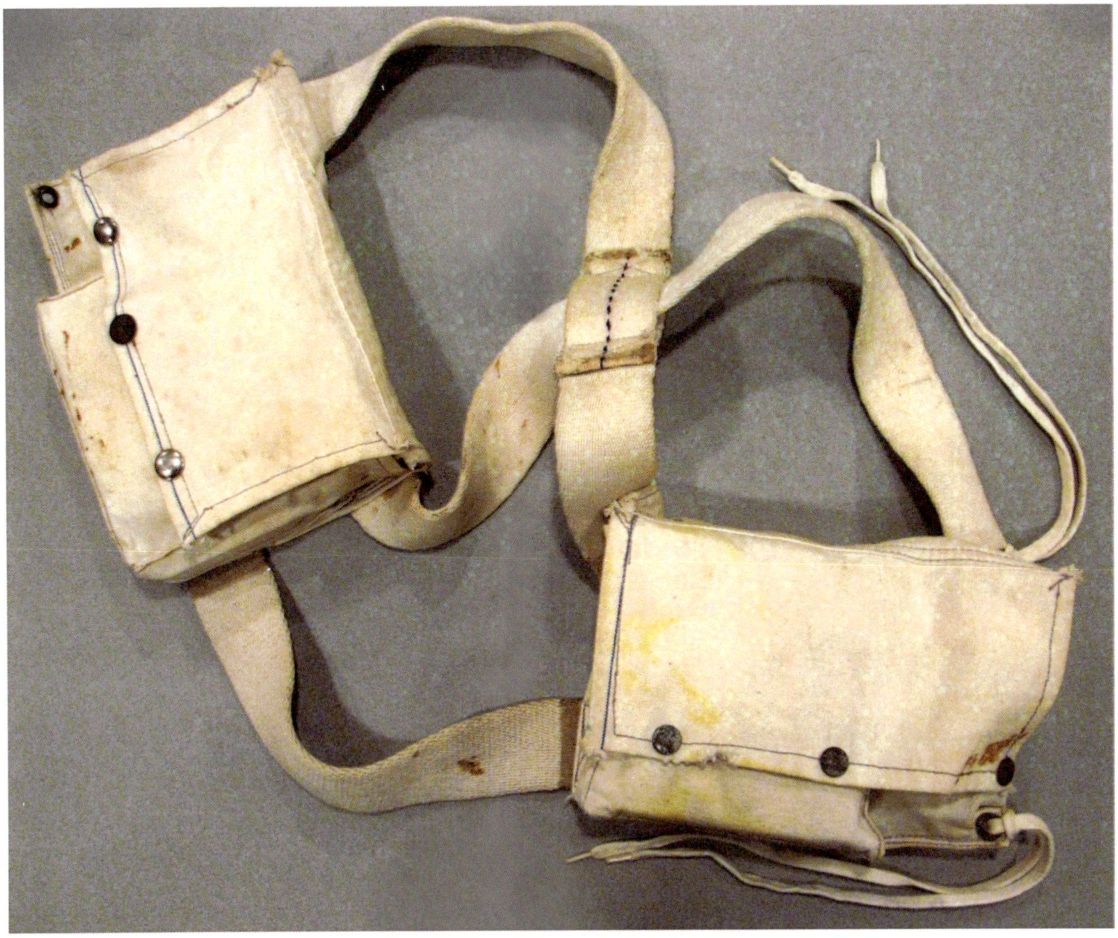

Existing example of the emergency kit shoulder harness belonging to Lt. jg. Francis Simrill of VF-13. (Paul Faulkner Collection)

(Above) VF-17 pilot Lt. jg. W.C. Lanreth is using an emergency kit shoulder harness, February 1944. An identical pocket was located on the opposite side and harness was worn as a vest. (National Archives)

(Below) Air Group 13 pilot LT. jg. Ralph Zender wearing a compact emergency kit on pistol belt, October 1944. A streamlined version in comparison to Bowen's.(National Archives)

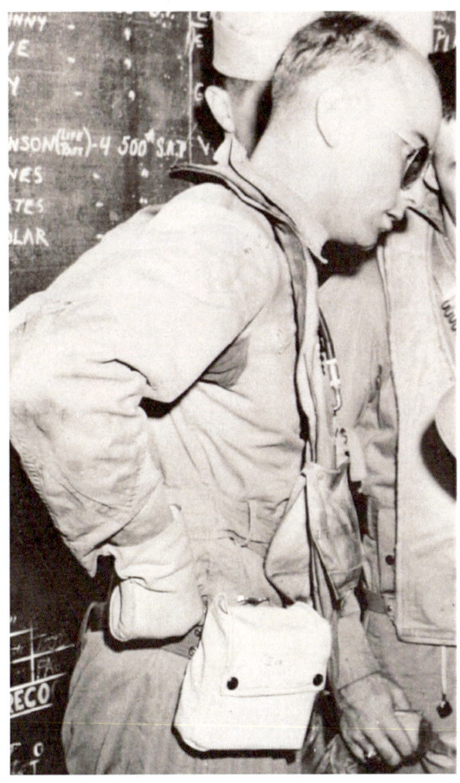

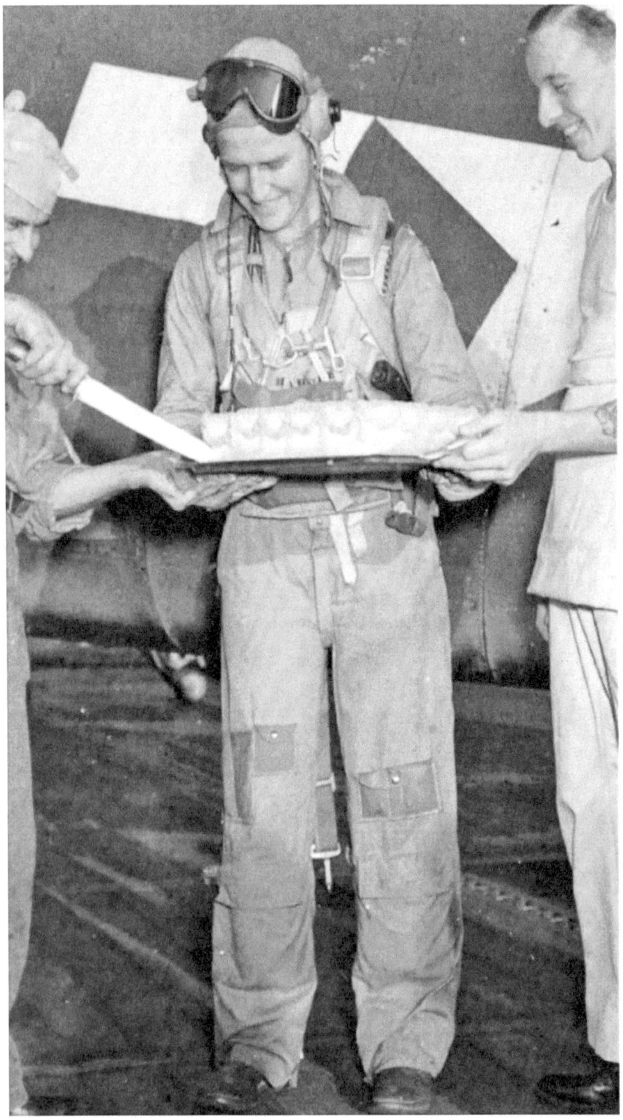

Hellcat pilot Ens. E.J. Hauser of VF-3 accepts a celebratory piece of cake after having the honor of making the 17,000th landing on the USS Yorktown. His flight suit is adorned with a series of small pockets for emergency items with pouch secured to his life preserver (obscured by the cake), February 1945. (National Archives)

TBF/TBM pilot Lt. George Cannon uniquely has applied the Mag Lens and Instruction sheet pocket off an M-592 on the left thigh of his flight suit; another pocket has been applied to the right thigh. Pockets applied to the thighs were convenient for storing the flask kit and evasion and escape articles such as cloth maps, money, and blood chits. He appears to have the fishing kit tied to the pistol belt on the right hip. Applications such as these were typically related to personal preference, leaving a variety of applications for observation. (Courtesy Gregory Pons)

In summary, personal emergency items were found to be a necessity and were addressed by the BuAer research and development facilities and units operating in forward areas. Use of them was then integrated into flight school survival training curricula. Early attempts at providing necessities were small pouches as aids, quickly remedied with the introduction of larger standard kits.

As bulk became a concern, service activities began to streamline personal emergency kits into compact single units in the form of various packs, pouches, vests, and harnesses. By the close of 1944 emphasis was given to carrying these items as personal equipment. Flight suits were amended for additional pockets to accommodate provisions. Service units took further steps to add pockets as exhibited here.

FIRST AID KITS

Development of first aid provisions for aircraft can be segregated into two categories: peacetime and wartime. The bulk occurs during war, of necessity. Initially, medical supplements were designed almost solely in a laboratory setting. By the close of WWII, however, first aid units were designed and implemented strictly on service recommendations.

At the start of WWII, first aid kits were developed based on existing requirements, with revisions adapting to the latest need. They were also adapted to serve the geographical regions of the arctic, jungle, desert and ocean. Furthermore, emergency kits had to be created to serve the layperson with simple applications and instructions to administer basic aid on themselves or others.

Most units for aeronautical use were designed to sustain wounded individuals until they could reach an aid station for proper treatment by trained medical personnel. Emphasis at first was placed on cuts, burns, and abrasions; later on bullet and fragment wounds.

There was a significant difference between the air arms of the US Navy and the US Army in the approach toward emergency rescue equipment. However, there are many similarities between the two services regarding infrastructure. Each had a unit or air group Flight Surgeon who worked in close coordination with the Bureau of Aeronautics Aviation Equipment Survival Officer or the AAF Personal Equipment Officer, to aslo include the Air Combat Intelligence Officer. The Flight Surgeon gave instructions to crews on such subjects as personal hygiene, sanitation, first aid, survival, and night vision. They also played a vital role in matters dealing with combat fatigue and regular examinations of general health. Flight Surgeons in both services were responsible for requisition and inspection of medical supplies for the station to which they were assigned.

The matter of supply varied between the two services. Naval first aid items, specifically for aviation purposes, typically have dual stock class numbers: one from the Bureau of Medicine & Surgery (BuM&S) and one from the Aviation Supply Office (ASO). The articles with an ASO stock class were maintained through standard BuAer supply channels located at Naval Air Stations or Naval Air Bases but alternately could be requisitioned from the BuM&S from its supply points.

Many Navy and later Marine air groups operated from carriers, with the medical supplies maintained through the ship's medical department, creating a coalition between the group Flight Surgeon and the ship's medical staff. Shore-based units had a similar arrangement and often had additional support from local general depots and aid stations.

The USAAF maintained a single US Army Medical Department stock class number for medical items. Medical supply in the USAAF was most often maintained by a local Aviation Medical Supply Platoon, which, in turn, was supplied from the medical section of General Depots or Branch Medical Depots from the theater Services of Supply. Additionally, medical items were provided, packed, or installed by the contractors for aircraft, life rafts, sustenance kits and parachute kits. For procurement from contractors, the US Navy BuM&S and the US Army Surgeon General's Office negotiated and awarded most all contracts.

Two primary institutions were involved in the development process: the Naval Medical Research Institute in Bethesda, Maryland and the USAAF Aero Medical Laboratory at Wright Field in Dayton, Ohio. All aeronautical items were typically first evaluated by these two institutions before they were submitted for approval to be standardized. The BuAer and the AAF Materiel Command worked in close coordination with their respective organizations in the development process: generating the need, coordinating design and testing, then, after manufacture, facilitating the distribution. Once an item was in service, it was evaluated forever in a perpetuating cycle, not unlike any development process.

The origin of first aid supplies for aircraft dates to World War I when, in October of 1918, a crashed German CZ aircraft was examined in Commercy, France. The German aircraft had a medical kit containing medication and bandage material suitable for an individual to administer on himself. Discovery of this kit prompted the US Navy and the US Army to investigate and experiment with the idea of providing aircraft with basic medical provisions as a standard. As World War I ended, the question of the aircraft's role in the future of warfare was arguable because at that time it had limited capabilities. Therefore, providing first aid items to aircrew was left to local commands and no serious effort was made during the 1920s to standardize an aeronautical first aid kit. By the dawn of the 1930s aircraft were flying farther, faster and higher, presenting new medical challenges regarding overall flying safety. The initial practice was to install a first aid kit only on aircraft conducting cross-country flights, but this was left to the discretion of the local command.

From 1932 to 1934 the USAAC Materiel Division, later known as the Materiel Command, investigated the requirements of a first aid kit for use with the Air Corps. In December of 1934 it released a report outlining conclusions and recommendations. The Command determined that a first aid kit should be permanently installed in all aircraft at the rate of one kit per two-passenger airplane, two kits per 3-12 passenger airplane, and four kits per passenger airplane for 12 or more. The report stated that most crashes occur far from medical aid where it could be days before survivors are reached. Alternately, they found that there was no difference in being one mile or a thousand miles from help when wounds are severe, and care was critical within the first 30 minutes of the crash. The following year, 1935, the Materiel Division standardized the requirements and guidelines of an aeronautic first aid kit for aircraft. Materiel Division leaders visited and corresponded with several companies for fabrication and supply. The kit exhibiting the best attributes was the Burroughs Wellcome & Co. No.731 Tabloid First Aid Avion Kit.

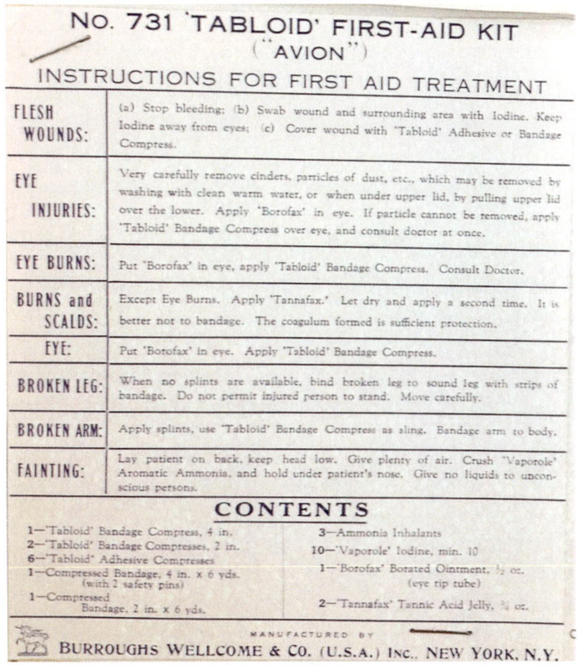

NO. 731 'TABLOID' FIRST-AID KIT
("AVION")

INSTRUCTIONS FOR FIRST AID TREATMENT

FLESH WOUNDS:	(a) Stop bleeding; (b) Swab wound and surrounding area with Iodine. Keep Iodine away from eyes; (c) Cover wound with 'Tabloid' Adhesive or Bandage Compress.
EYE INJURIES:	Very carefully remove cinders, particles of dust, etc., which may be removed by washing with clean warm water, or when under upper lid, by pulling upper lid over the lower. Apply 'Borofax' in eye. If particle cannot be removed, apply 'Tabloid' Bandage Compress over eye, and consult doctor at once.
EYE BURNS:	Put 'Borofax' in eye, apply 'Tabloid' Bandage Compress. Consult Doctor.
BURNS and SCALDS:	Except Eye Burns. Apply 'Tannafax.' Let dry and apply a second time. It is better not to bandage. The coagulum formed is sufficient protection.
EYE:	Put 'Borofax' in eye. Apply 'Tabloid' Bandage Compress.
BROKEN LEG:	When no splints are available, bind broken leg to sound leg with strips of bandage. Do not permit injured person to stand. Move carefully.
BROKEN ARM:	Apply splints, use 'Tabloid' Bandage Compress as sling. Bandage arm to body.
FAINTING:	Lay patient on back, keep head low. Give plenty of air. Crush 'Vaporole' Aromatic Ammonia, and hold under patient's nose. Give no liquids to unconscious persons.

CONTENTS

1—'Tabloid' Bandage Compress, 4 in.	3—Ammonia Inhalants
2—'Tabloid' Bandage Compresses, 2 in.	10—'Vaporole' Iodine, min. 10
6—'Tabloid' Adhesive Compresses	1—'Borofax' Borated Ointment, ½ oz. (eye tip tube)
1—Compressed Bandage, 4 in. x 6 yds. (with 2 safety pins)	2—'Tannafax' Tannic Acid Jelly, ¾ oz.
1—Compressed Bandage, 2 in. x 6 yds.	

MANUFACTURED BY BURROUGHS WELLCOME & CO. (U.S.A.) INC., NEW YORK, N.Y.

(Right) Instruction sheet pasted inside the lid of No.731 Avion kit.

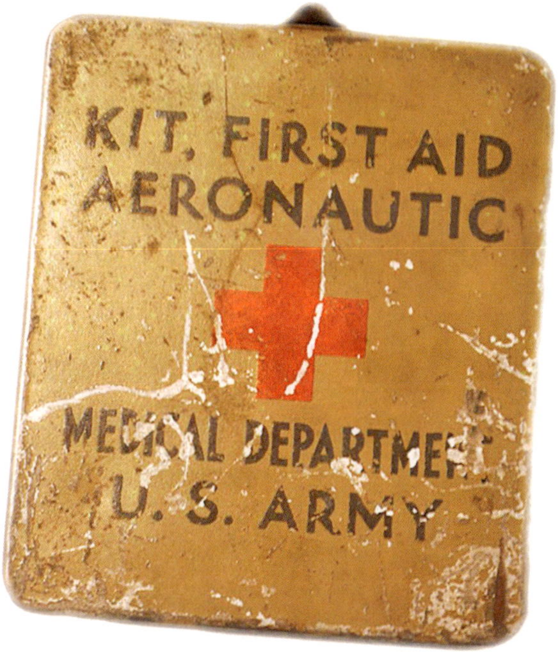

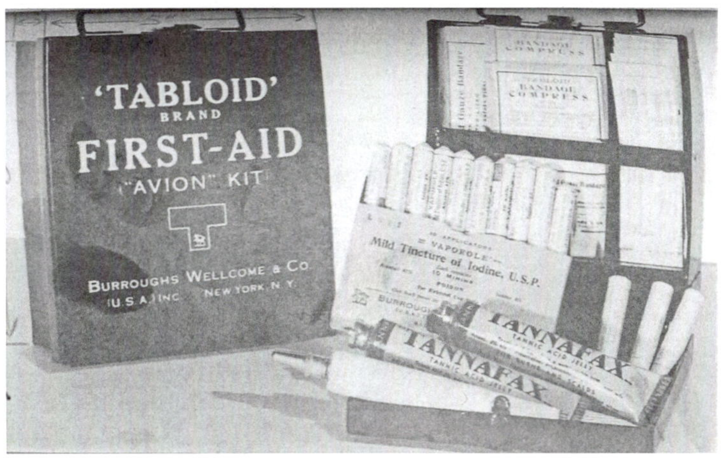

(Right) No. 731 Burroughs Wellcome & Co. First Aid Avion Kit. (National Archives)

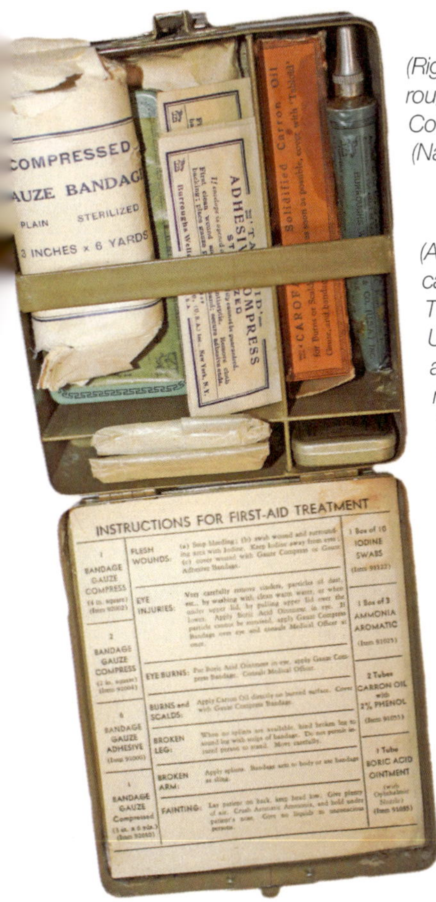

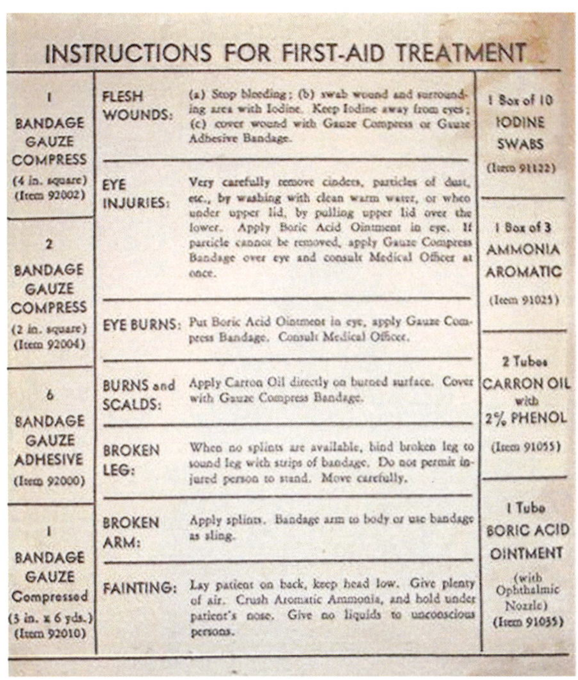

INSTRUCTIONS FOR FIRST-AID TREATMENT

1 BANDAGE GAUZE COMPRESS (4 in. square) (Item 92002)	FLESH WOUNDS:	(a) Stop bleeding; (b) swab wound and surrounding area with Iodine. Keep Iodine away from eyes; (c) cover wound with Gauze Compress or Gauze Adhesive Bandage.	1 Box of 10 IODINE SWABS (Item 91122)
2 BANDAGE GAUZE COMPRESS (2 in. square) (Item 92004)	EYE INJURIES:	Very carefully remove cinders, particles of dust, etc., by washing with clean warm water, or when under upper lid, by pulling upper lid over the lower. Apply Boric Acid Ointment in eye. If particle cannot be removed, apply Gauze Compress Bandage over eye and consult Medical Officer at once.	1 Box of 3 AMMONIA AROMATIC (Item 91025)
	EYE BURNS:	Put Boric Acid Ointment in eye, apply Gauze Compress Bandage. Consult Medical Officer.	2 Tubes CARRON OIL with 2% PHENOL (Item 91055)
6 BANDAGE GAUZE ADHESIVE (Item 92000)	BURNS and SCALDS:	Apply Carron Oil directly on burned surface. Cover with Gauze Compress Bandage.	
	BROKEN LEG:	When no splints are available, bind broken leg to sound leg with strips of bandage. Do not permit injured person to stand. Move carefully.	
1 BANDAGE GAUZE Compressed (3 in. x 6 yds.) (Item 92010)	BROKEN ARM:	Apply splints. Bandage arm to body or use bandage as sling.	1 Tube BORIC ACID OINTMENT (with Ophthalmic Nozzle) (Item 91035)
	FAINTING:	Lay patient on back, keep head low. Give plenty of air. Crush Aromatic Ammonia, and hold under patient's nose. Give no liquids to unconscious persons.	

(Above and left) 99518 aeronautical first aid kit for the USAAC. The No.731 unit supplied to the USAAC was a commercial first aid kit modified to include stock medical supplies with an instruction sheet pasted on inside of front. It had printed government nomenclature, property markings, and camouflaged paint scheme. It was assigned USMD stock number 99518 and stored in a tray secured to the fuselage. In 1935 the standardized kit measured 5" x 6" x 2". USMD government label, revised to include USMD stock numbers for the contents with instructions for first aid treatment remaining the same. (Kriss Usherwood Collection)

At that time, commercial units were simply adapted for military use in a peacetime environment. However, the Materiel Division noted in its 1934 report that in war, requirements would increase.

In the genealogical process of the aeronautic first aid kit, adoptions and revisions generally coincided between the Army and the Navy. In 1935 the BuM&S was working on the adoption of an equivalent kit, which it titled Aircraft, First Aid Package, US Navy. However, it did adopt a unit in 1923 titled Kit, First Aid under Specification 57-K-1A. The original date of the drawing was 10 May 1923.

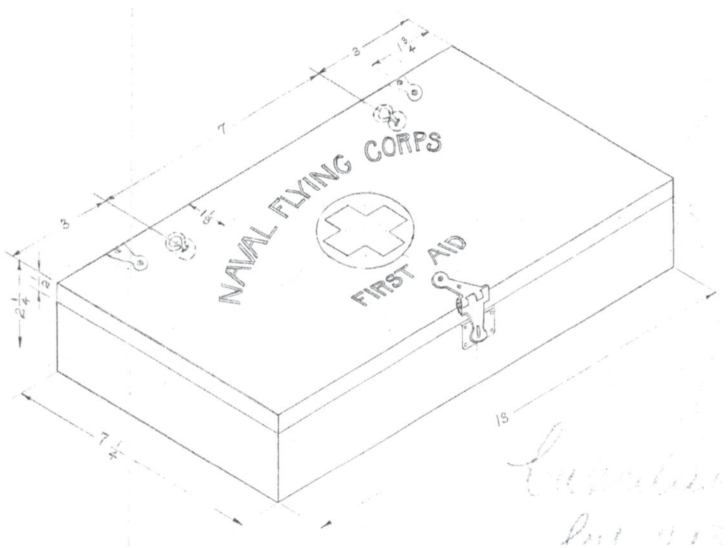

(Left) Drawing of the first aid kit for Naval aircraft, 1923, measuring 13" x 7-1/4" x 2-1/4"; unclear if fabricated of metal or wood. Top was stenciled with the red cross and labeled Naval Flying Corps, First Aid. Components included spool of adhesive plaster 2" x 10 yards., 2 1" compressed gauze bandages, six 2½" compressed gauze bandages, triangular compressed bandage, two1 oz. tubes of Carron oil, three 2 oz. packages of cotton gauze, one 3 yd. compressed absorbent gauze, head dressing, six tubes tincture of iodine, 2 oz. package of compressed lint, 12 safety pins, scissors, two vials spirit of ammonia, tourniquet and wire mesh for a splint.

(Left and below) Coinciding with adoption of USAAC first aid kit, the BuM&S began revision for a new type of unit. Adopted as standard on 5 August 1936 was the Aircraft, First Aid Package, US Navy detailed under NAF Dwg. No. 213886. Components were contained in a tray/drawer that slid into the outer container and secured shut by two camlocks on each side. Exterior was red for visibility. Unit measured 4-1/2" x 7-1/2" x 2 3/4". Kit was contracted prior to being officially adopted; on some examples contract number and date are stamped on outer container. This example has contract award of 7 November 1935 to Mine Safety Appliances Co., the only known contractor. There were slight variations of the components.

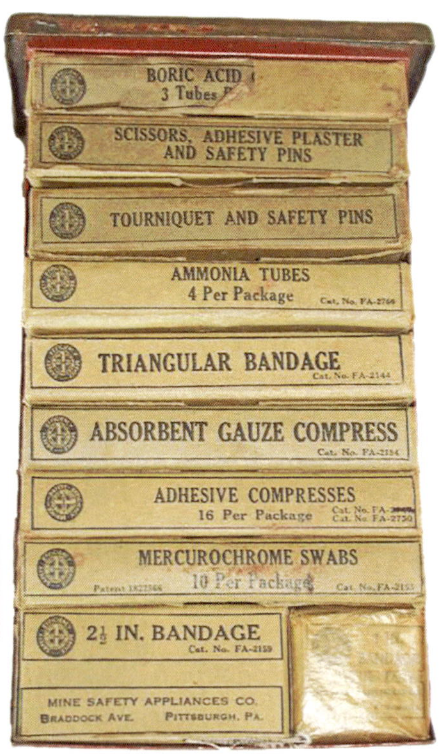

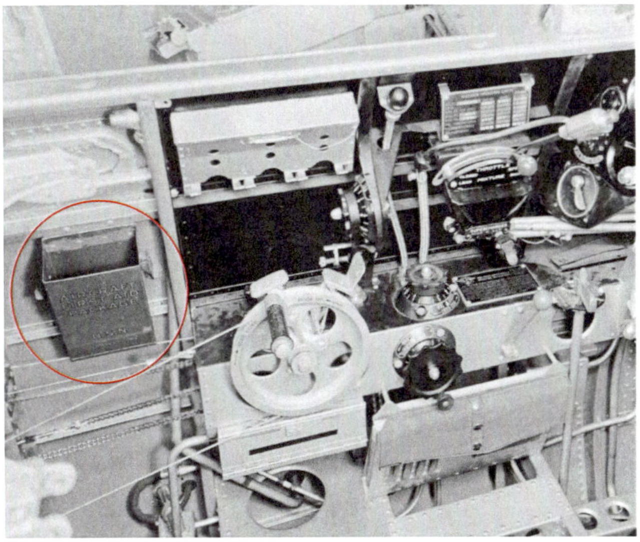

(Above) NAF-213886 first aid package. First aid kits were also standard for VF class or single-seater aircraft in the prewar era. Brewster F2A-2. Note 10 gauge flare rack, above right of kit. (National Archives)

(Above) The NAF-213886 first aid package attached to bulkhead in Martin PBM Mariner, c.1940. (National Archives)

(Below) NAF-213886 first aid package in PBY, c.1940. Triangular strap for securing parachute pack of QAC parachute pack. (National Archives)

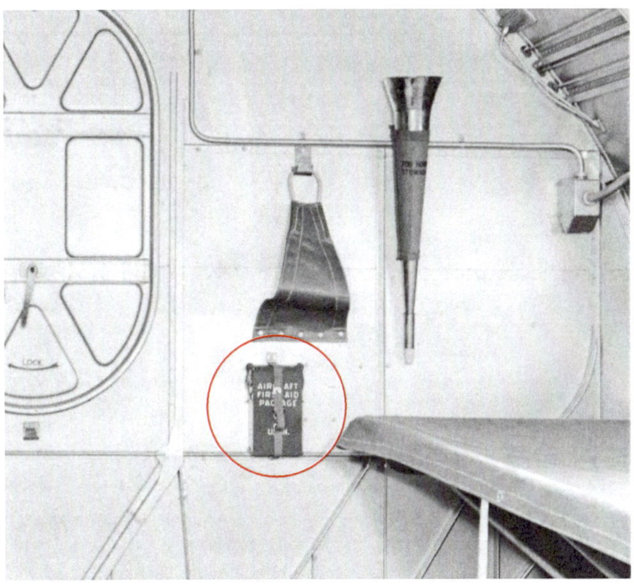

(Below) NAF-213886 first aid package as part of display of standard equipment for PBY-2, c.1937. (National Archives)

In the 1930s, demand arose, and, as mentioned previously, these units were in their evaluation process. During this era the US Army aeronautic and US Navy aircraft package created and shared a basic standard kit. The year 1939 was pivotal in many respects: war had erupted in Europe and US tensions with Japan had increased and standard first aid kits had been in service for several years.

On 24 April 1939 the Materiel Division adopted a revised aeronautical kit it titled Kit, First Aid, Aeronautic, Improved under USMD Specification No. 6-274. This unit was a complete overhaul from the previous 99518 unit and included a duck canvas container, incorporated a better method of attachment to the aircraft fuselage, and was enlarged to accommodate a new set of components. As part of the design, the new container allowed for ease of access to the individual components and applied a painted red cross in combination with the USMD caduceus for increased visibility in an emergency. This new unit was assigned USMD stock number 99516 (complete) and 99517 (container only). As part of this revision, issuance was changed to one per pursuit aircraft and one per three individuals in all other types of airplanes.

It is important to note that stock numbers are not a credible dating system. In most sequential stock series, numerical blocks were intentionally left unassigned to allow for future expansion.

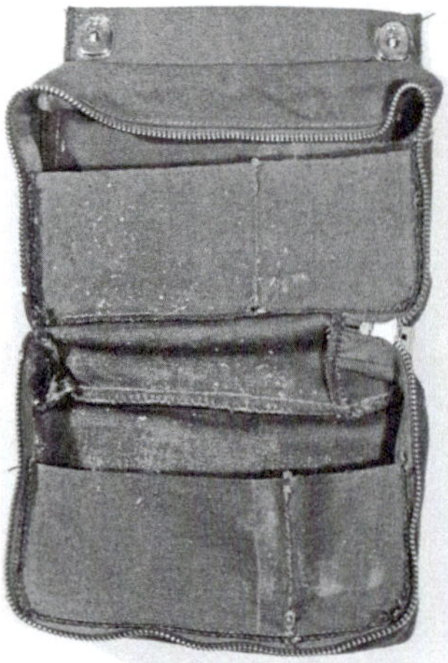

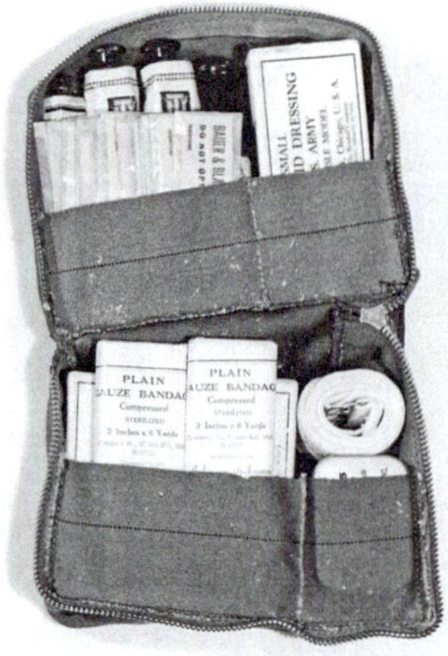

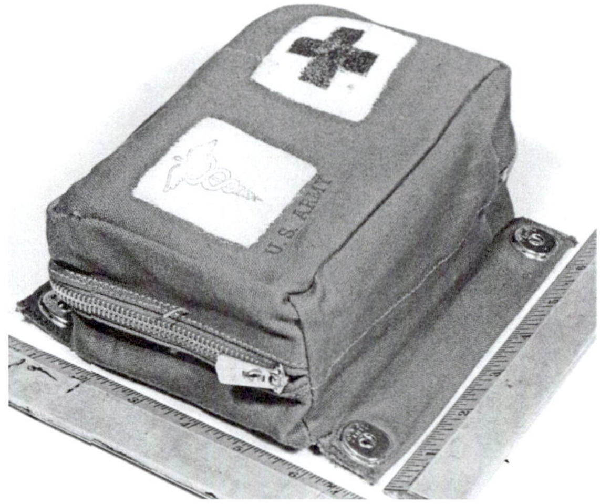

Improved aeronautic first aid kit of 1939. New securing method includes four studs secured to the fuselage, and four Lift-the-dot fasteners on the container for attachment and quick release. Existing examples are often stamped with the USMD stock number 99517, which was for the container only. (National Archives)

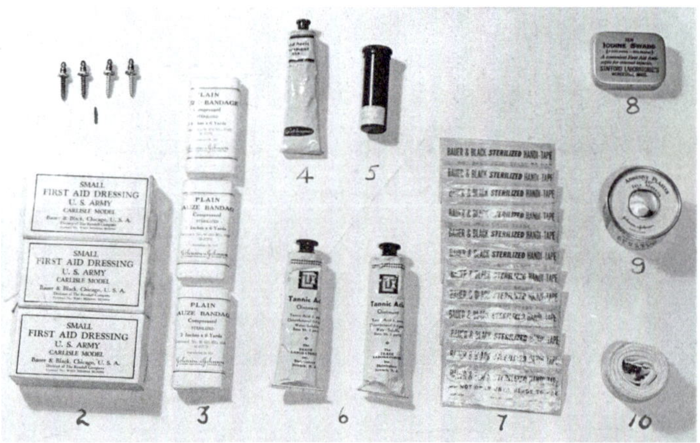

1 - 4 Lift-the-dot fasteners, woodscrew type.
2 - 3 Small first aid dressing
3 - 3 Plain gauze bandages, 3" x 6 yds.
4 - 1 Tube, boric acid ointment, 1 oz.
5 - 2 Doz. Acetylsalicylic, 5 gr.
6 - 2 Tubes, Tannic acid ointment, 1 ½ oz. each
7 - 12 Handi-tape, sterilized, medium
8 - 10 Iodine swabs, with metal container
9 - 1 Adhesive plaster, 1" x 5 yds.
10 - 1 Belt, canvas 1" x 3'

(Left and above) Proposed components for the new aeronautic first aid kit, 99517. (National Archives)

(Below) Kit, 99517, is seen attached to fuselage of transport aircraft.

(Below) Existing example of the 99517 improved aeronautic first aid kit. It was fabricated of duck cotton measuring 6" x 4½" x 3" and red cross and USMD caduceus were painted on front.

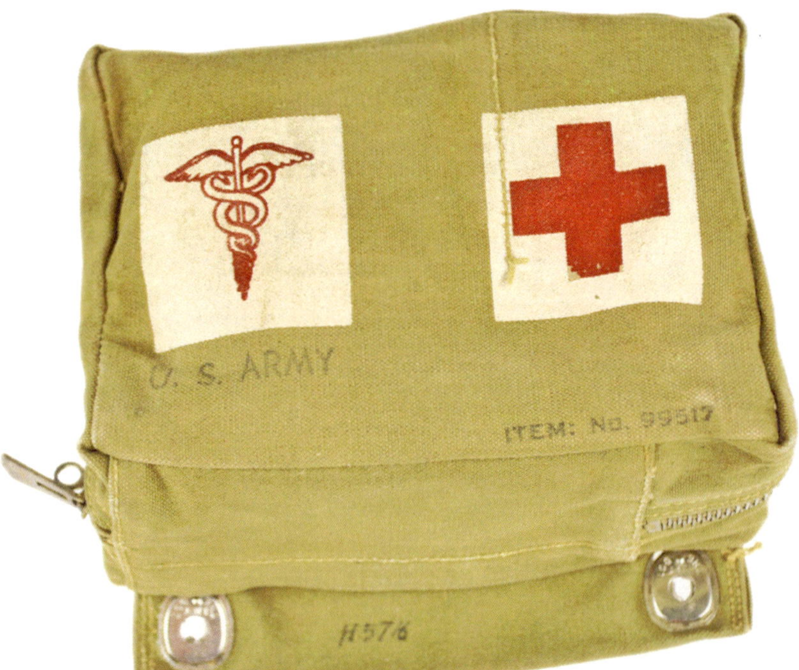

After service testing the US Navy first aid package, the BuAer established its own requirements for a suitable aeronautic first aid kit. This new version is outlined in NavAer Specification M-448. In many ways, it mirrors the USAAC counterpart in that it improved accessibility to the components and the attachment method. The new aluminum container would include a hinged lid and two swivel hangers and utilized a standard 10-unit assembly as outlined in the Navy Department stock class number 2-495. The M-448 was adopted as standard on 15 July 1940 and was requisitioned under ASO stock number R57-C-8630 for the container only. The components were requisitioned as a 10-piece filler unit under ASO stock number R57-F-4005. It could also be requisitioned from the BuM&S under stock number S2-1370. The container was procured by the ASO, with the components requisitioned through the BuM&S. The M-448 officially superseded the NAF 213886.

The components at the time of standardization were: a package of tannic acid, a package of spirit of ammonia, one packet of iodine swabs, one 2" compress bandage, two 4" compress bandages, a triangular compress, a small battle dressing and tourniquet. This assembly was referenced under BuM&S Specification 57-K-032. As a side note, the Navy references the M-448 as a 10 unit kit but there are only nine physical boxes; it is not clear what constitutes the tenth unit.

This aircraft first aid package conforms to original specification M-448 with aluminum container. Label in lower left-hand corner confirms at least one contractor, Davis Emergency Equipment Co. Components updated per amendment to Specification 57-K-032 to include sulfanilamide and sulfadiazine but retain some pre-inter-war packages such as the 2" compresses and tannic acid ointment. However, kit lacks morphine and iodine package. Unit measures 7" x 4" x 2-1/2". (Kriss Usherwood Collection)

The units discussed thus far were comprised of pre-WWII developments and revisions. As the United States entered the war, these were tested under combat conditions, leading to revisions to meet new tactical requirements. Procurement of the kits was made on a limited basis and they were not universally installed in aircraft, presenting an issue to overcome in the first year of the country's involvement in the war. A new emphasis was then placed on all research and development, which continued at a rapid pace throughout the war.

One issue of concern to the BuAer was the manufacture of the M-448 container in aluminum. Due to the War Production Board's restrictions on aluminum, placed in 1942, the BuAer was forced to revise the specification to fabricate the container from a cellulose acetate plastic. Amendment-A to NavAer Specification M-448 was therefore made, effective 1 June 1942, revising the container to be fabricated from plastic and updated to include the components from revised Specification 57-K-032B.

The M-448A container was assigned ASO stock number R57-C-8635. Container and handle were changed to plastic, retaining only the piano hinge, two swivel hangers, and securing latch in metal. Components provided with this unit are of Specification 57-K-032B, retaining filler unit stock number R57-F-4005/S2-1370. Example was assembled by Conray Products Co.

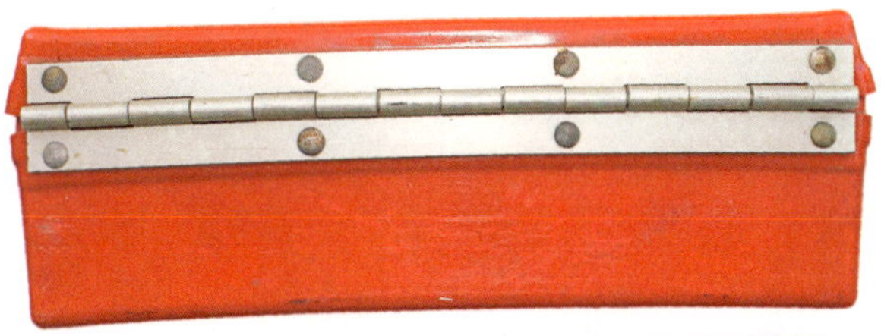

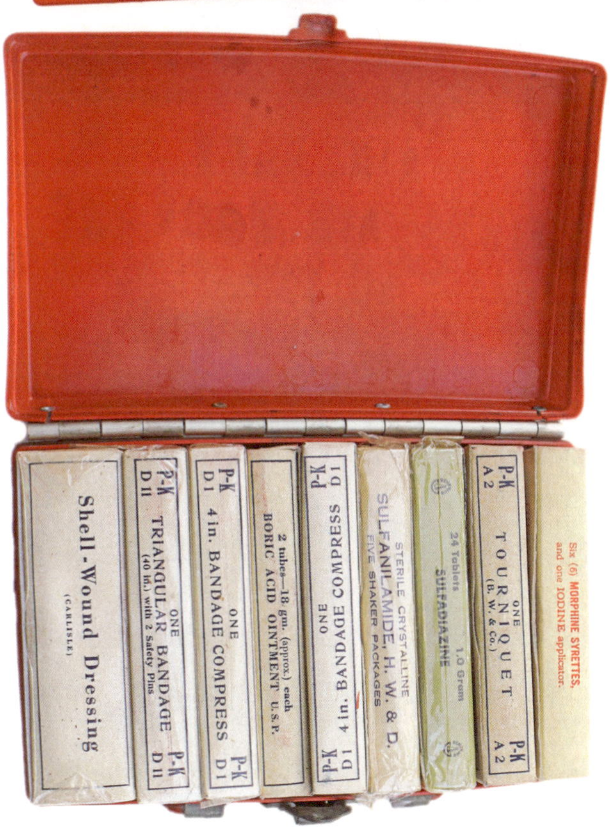

Burroughs Wellcome & Co. was another contractor with variations to M-448A container fabrication. Brackets securing the handle to the bottom were metal and the stud to secure the latch was integrated as part of the mold into the lid. Components for this example meet the requirements of Specification 57-K-032B. Plastic containers were reported to have durability issues and warped under extreme conditions as shown in this example; aluminum container was preferred by operational units. Container provided by Burroughs Wellcome had the nomenclature on the lid in silver; it was white on the Conray unit. (Kriss Usherwood collection)

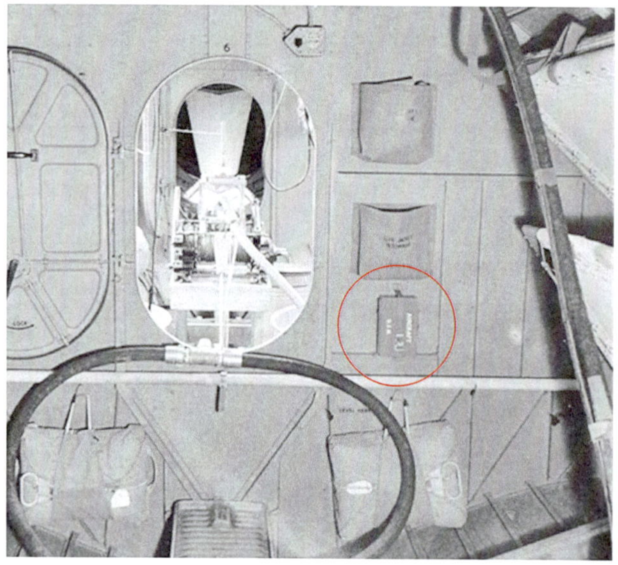

(Above) M-448 first aid package secured to bulkhead of Navy PB2Y-2 Coronado, April 1942. Both models were secured to bulkheads in brackets of designs at the discretion of the aircraft manufacturer. (National Archives)

(Left) Aluminum M-448 first aid kit installed at gunner's position on OS2U Kingfisher, June 1942. Gunner James Elardo, an 18-year naval veteran was formerly a first-class boatswain's mate on the USS Utah. (National Archives)

(Below) M-448A first aid package as part of a complement of provisions for a PBY Catalina. (National Archives)

By the time the United States entered World War II, war had been raging in Europe for more than two years. Within the allied coalition, information was exchanged fairly smoothly. Liaisons from the USAAF and USN had been in Europe, and RAF representatives had been in the United States, as early as 1940. The US-UK partnership was strong throughout the war. USAAF liaisons from Britain consistently submitted reports on activities, observations and recommendations to the Materiel Division. Having seen combat and its effects, several memoranda were submitted in October of 1941 to the Materiel Division that discussed current inadequacies the existing US aeronautic first aid kit would have in treating combat injuries when compared to RAF and Luftwaffe models.

Soon after, the United States entered WWII with survival and emergency kit concepts on the drawing boards. In January of 1942, the newly redesignated Materiel Command presented a list of modifications and improvements for the aeronautic first aid kit. This would mark the first in a rapid series of revisions. At the time, it was already redesignated from stock numbers 99516 and 99517 to new USMD stock class numbers 97765 (complete) and 97770 (container only). The primary revision was the addition of an outer pocket for iodine and band-aids for the treatment of minor injuries. This unit was accepted and designated under USMD Specification No. 753, USAAF Dwg. No. 42D7411.

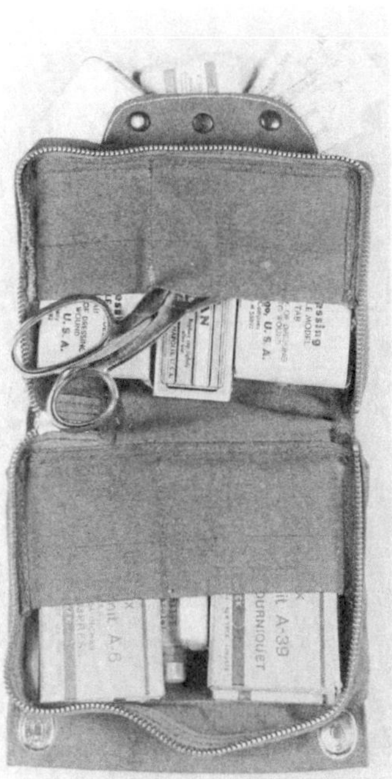

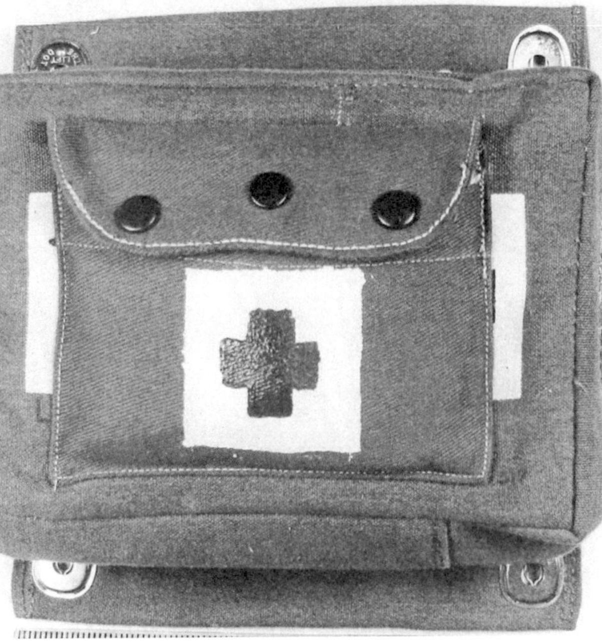

(Left and Below) Concept for aeronautic first aid kit presented in January 1942. Under external pocket were two painted panels from the 99517; the last three digits of the former stock number can be seen at the lower right corner. Late manufactured units of the earlier pattern have stock number 97770 printed at that location. (National Archives)

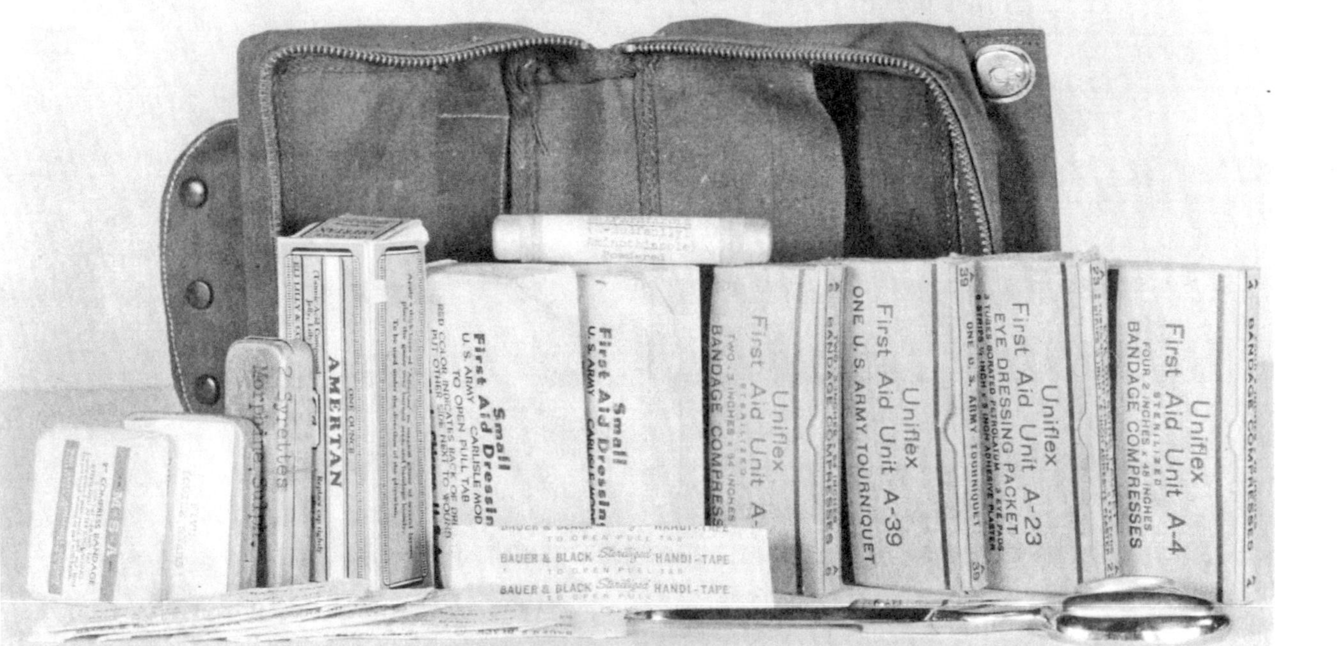

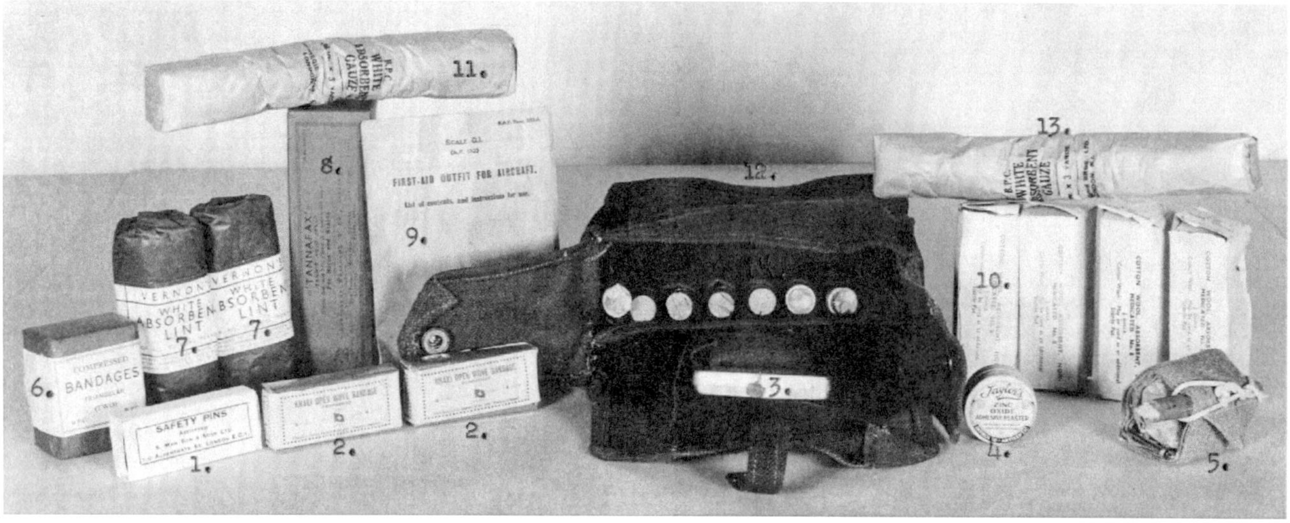

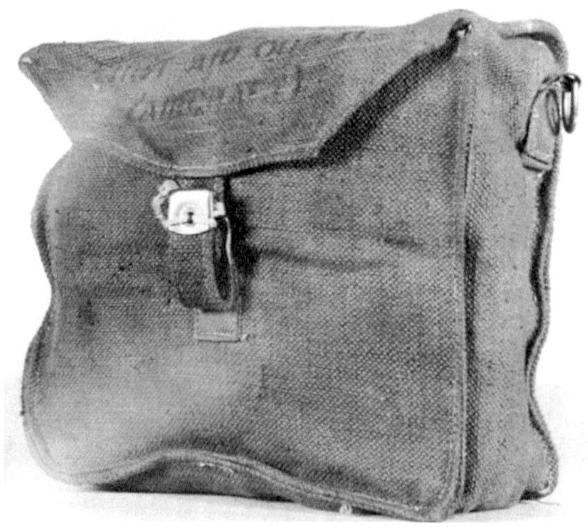

(Above and right) RAF first aid outfit used and evaluated by USAAF units, 1941. (National Archives)

(Below) Display of captured German equipment illustrates the Sanitatspack Fur Flugseuge (Sanitation Pack for Aircraft), manufactured 1941. This and the RAF unit were evaluated by USAAF prior to drafting Dwg. 42D7411. (National Archives)

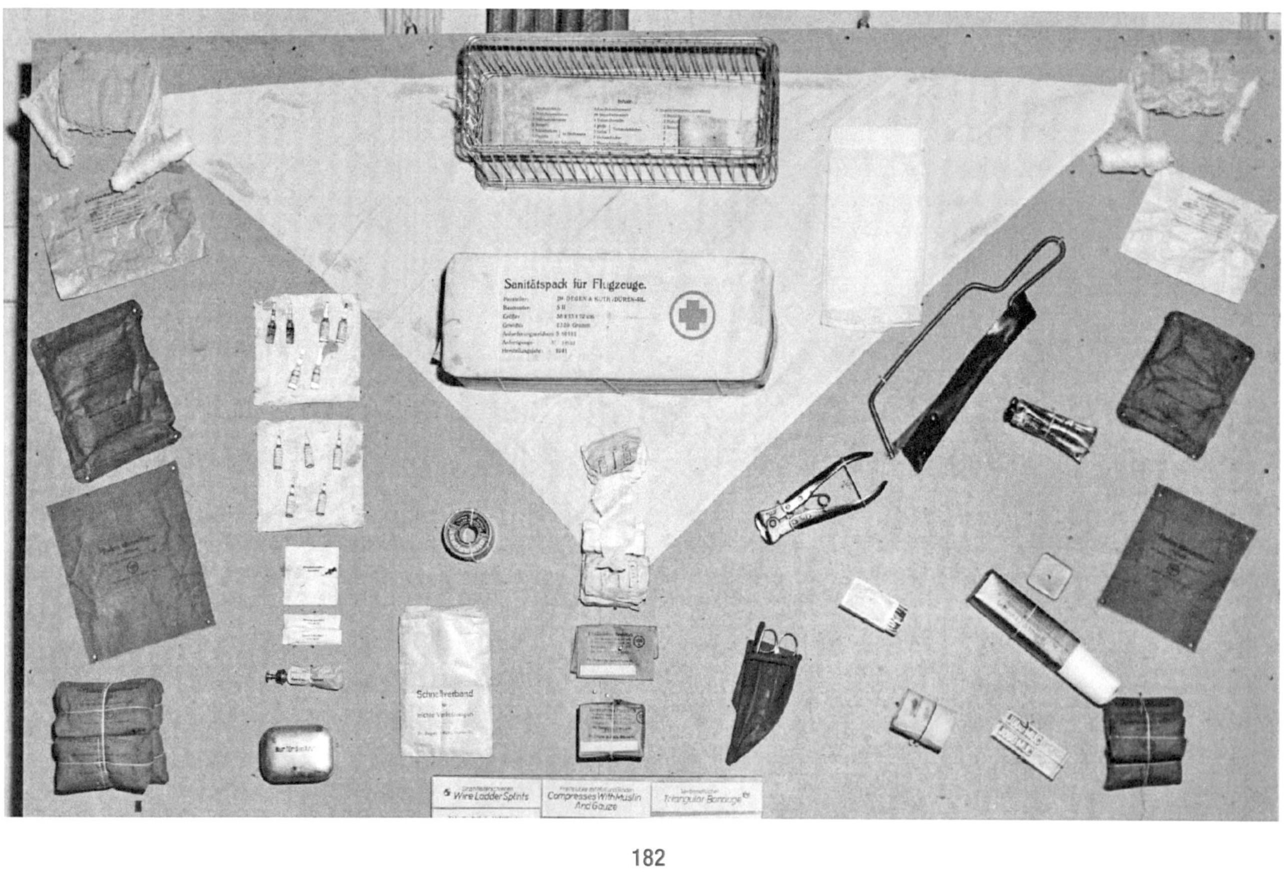

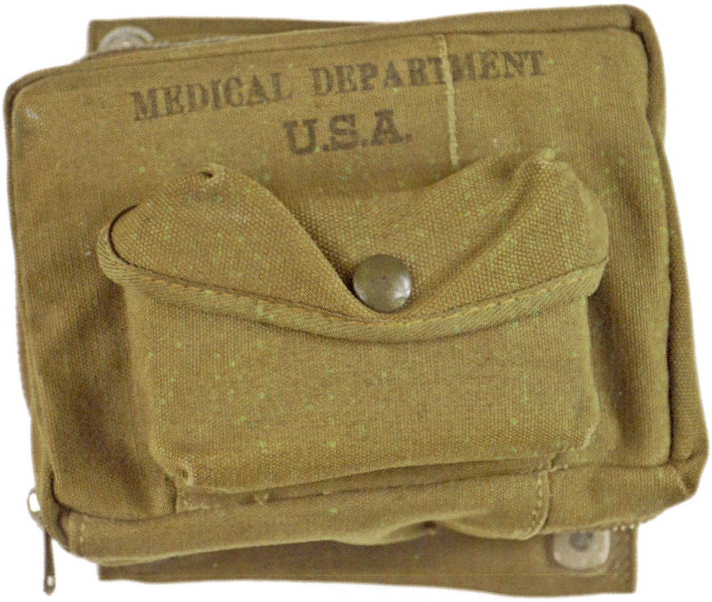

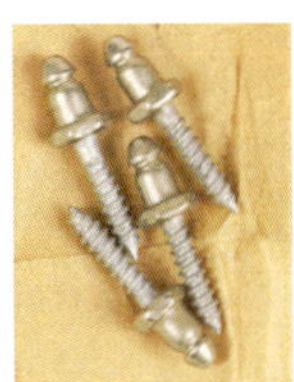

First production aeronautic first aid kit under Specification No.753 and Dwg. No. 42D7411, an expedited version as it did not meet all requirements of the January 1942 memorandum. Container 99770 was used but with an external pocket. This was the first attempt to address injuries peculiar to aerial combat. Additions include an eye dressing set, morphine syrettes, sulfanilamide, sulfadiazine, and scissors for cutting flight clothing.

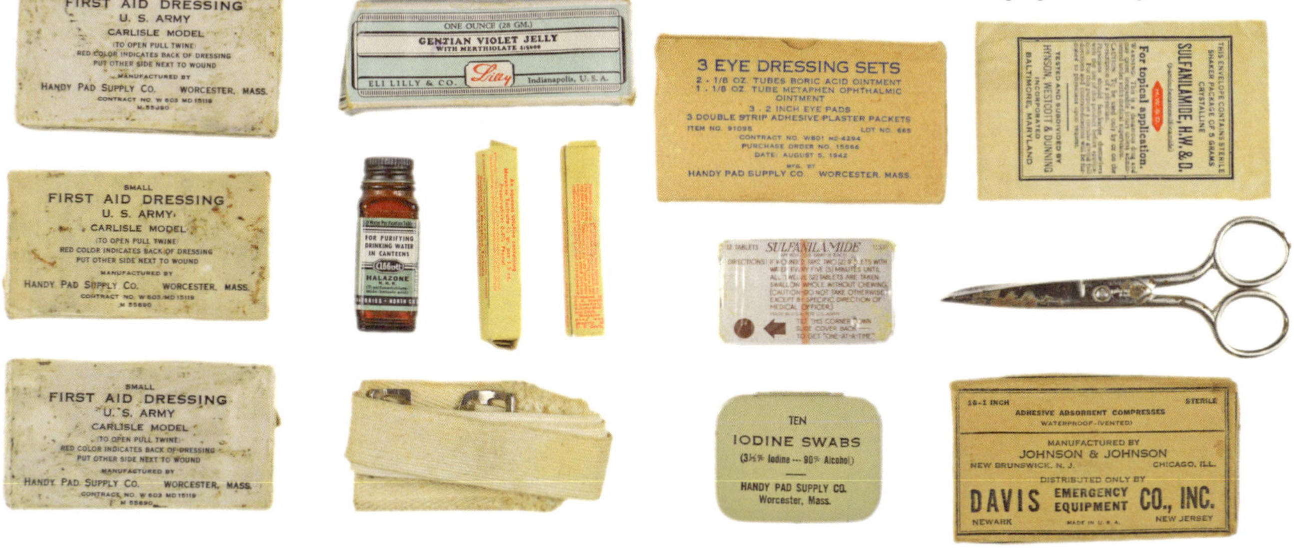

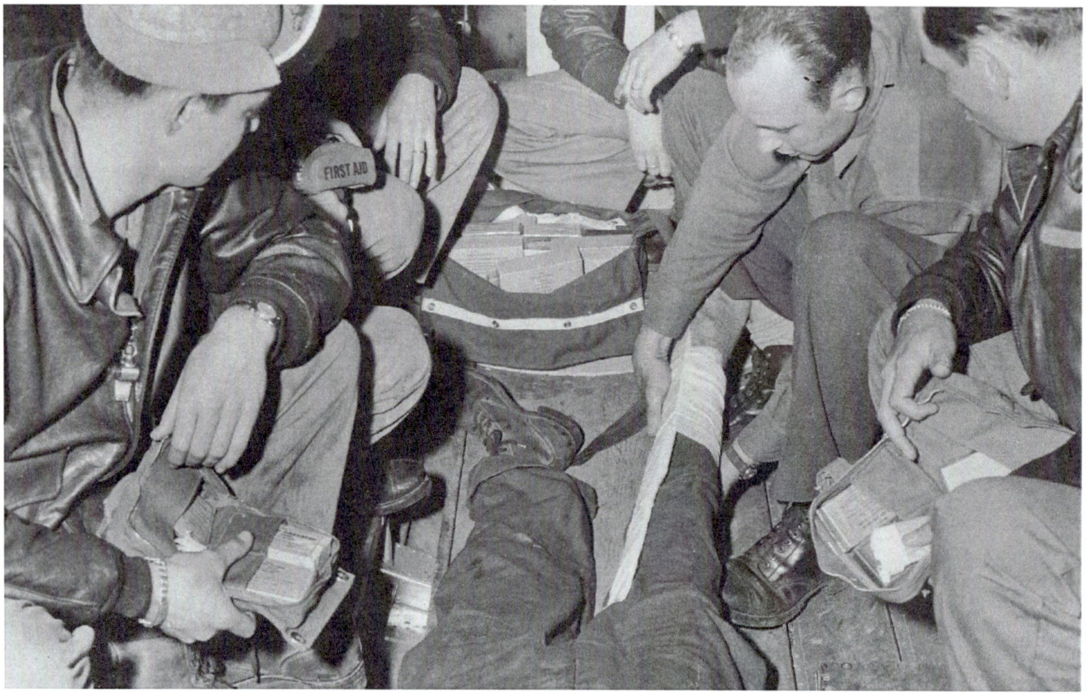

8th Air Force aircrew members undergoing medical aid training simulating flight accidents, c.1944. Two members are holding first generation 42D7411 first aid kits.

The next revision of aeronautic kit, dated 27 January 1942, under USMD Specification No. 753A, slightly enlarging the compartments and adding carrying strap on back. Purpose of carrying strap was for removal of unit and ability to carry while wearing heavy flying gloves. The components remained the same.

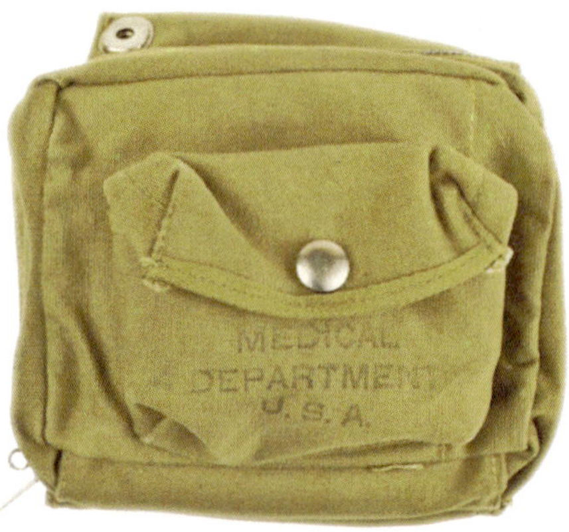

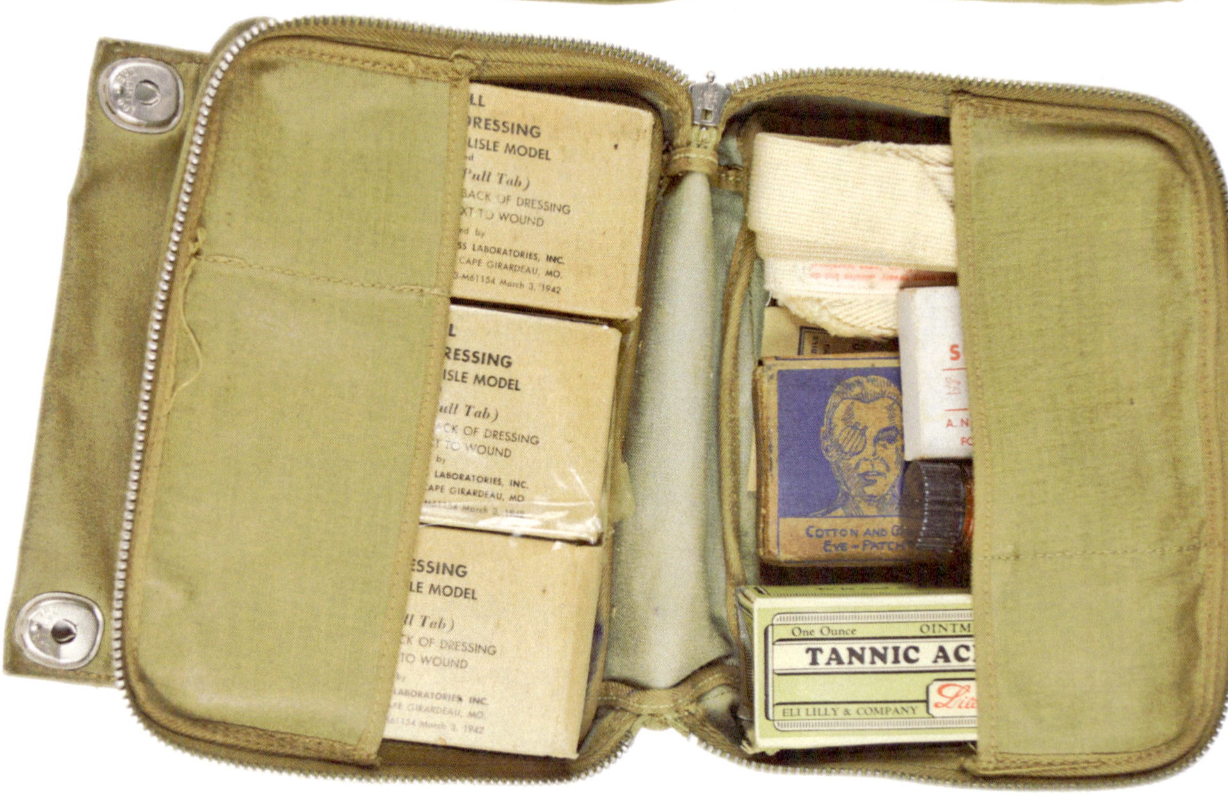

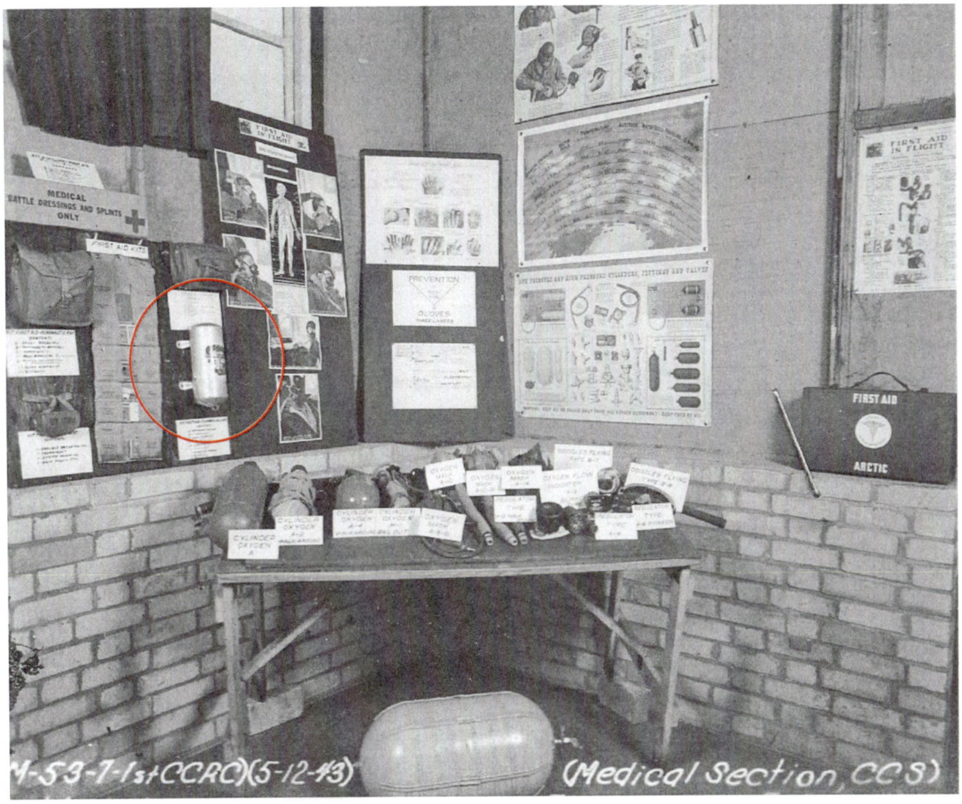

Display board in England shows array of first aid kits provided for USAAF aircraft and flying personnel. Note cylindrical aluminum container, a commercial first aid kit manufactured by E.D. Bullard Company, listed by USAAC in 1938 as authorized supplier of aeronautic first aid kits. (Prior to WWII, first aid kits for aircraft were provided as contractor-furnished equipment when required.

Surviving example of the Bullard first aid kit, conplete with its aluminum container and all contents..

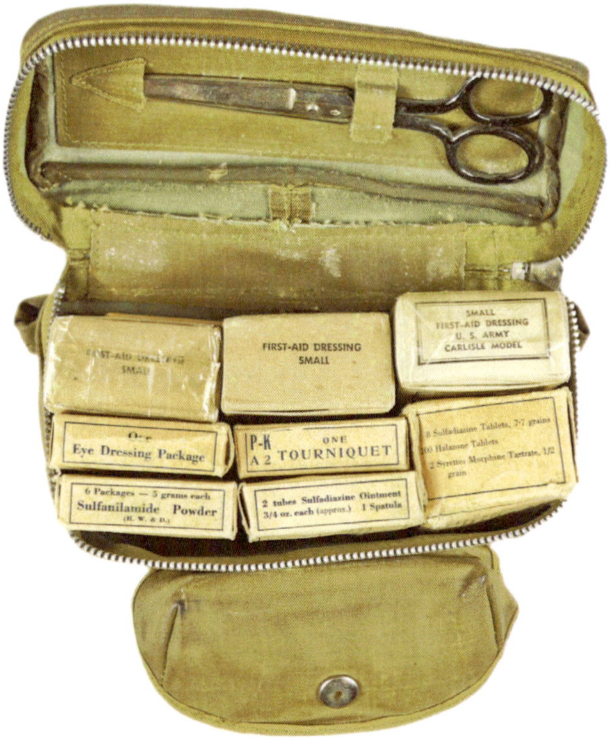

First two production patterns were expedited to meet requirements outlined in January 1942 engineering memorandum, which also directed components be arranged in an orderly fashion and be generally visible. Unit pictured here was the next pattern adopted under USMD Specification No. 1631, 3 October 1942. At this juncture, the three principle variations share stock numbers 97765 and 97770, but the nomenclature for this unit was Kit, First Aid, Aeronautic (New Style). When requisitioned, any of three patterns were provided.

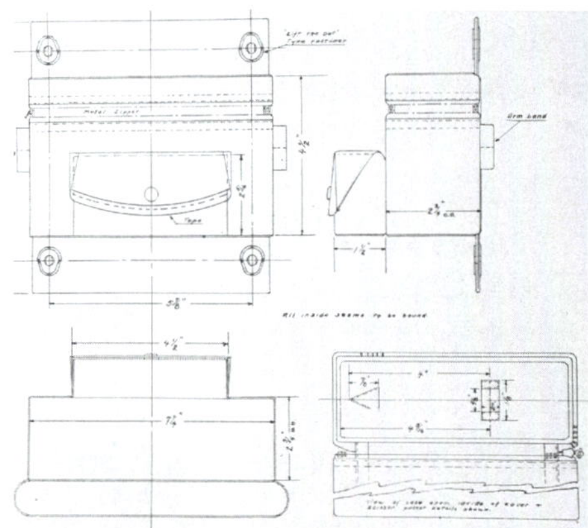

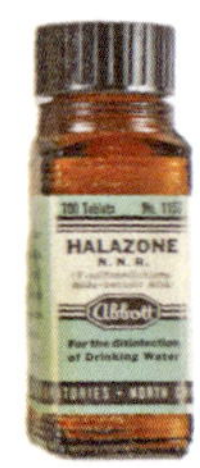

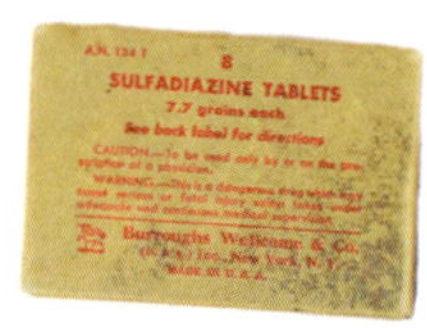

Training aids poster for the aeronautic first aid kit, including parachute first aid packet, October 1942. (USAF)

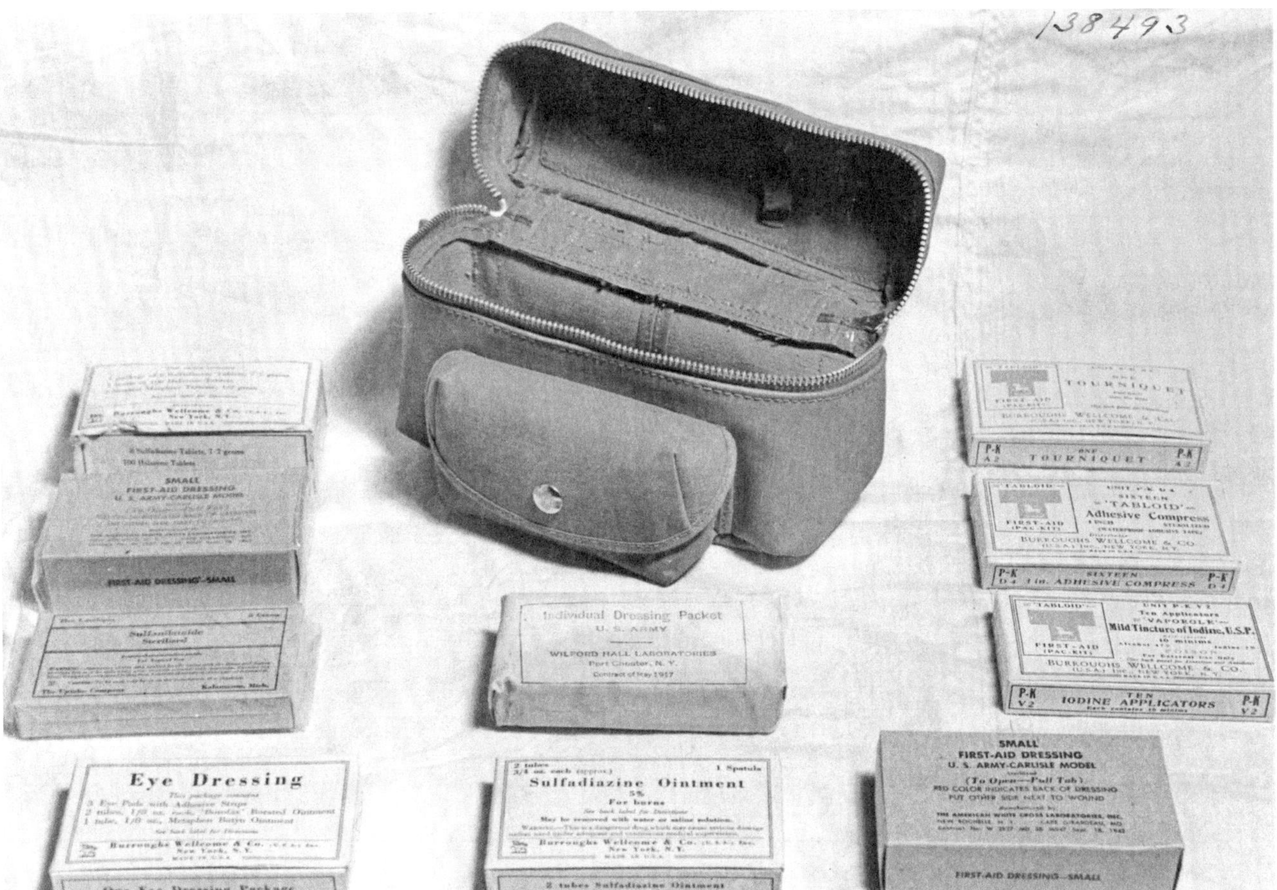

Wartime image of third production aeronautic kit No. 1631, February 1944. (USAF)

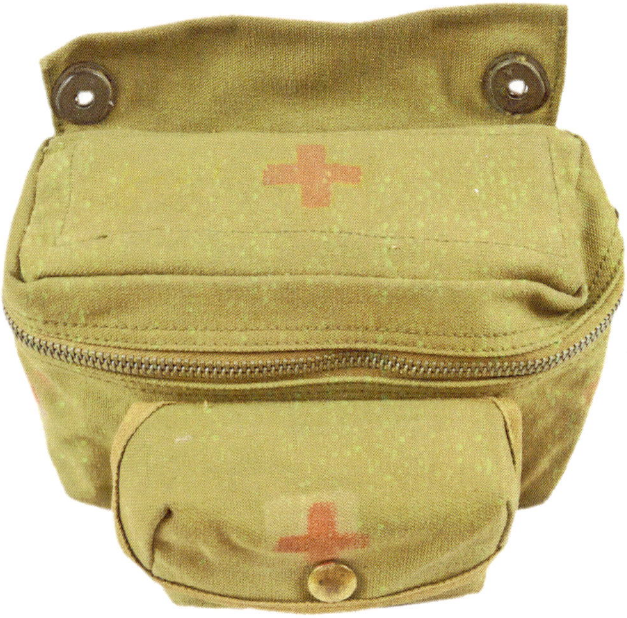

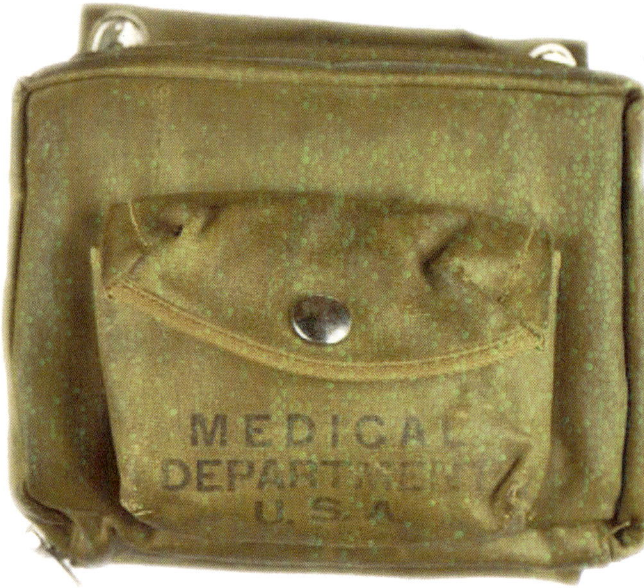

Initial containers were fabricated of duck canvas. By January 1943 standards were established on immersion testing for waterproofing first aid kits. To meet these standards, units were to be manufactured of plasticized vinyl. This was prescribed in Specification 1631. (right) Plasticized 42D7411. (left) Nonplasticized 1631. Manufacture of earlier patterns overlapped production of new style detailed in Specification 1631.

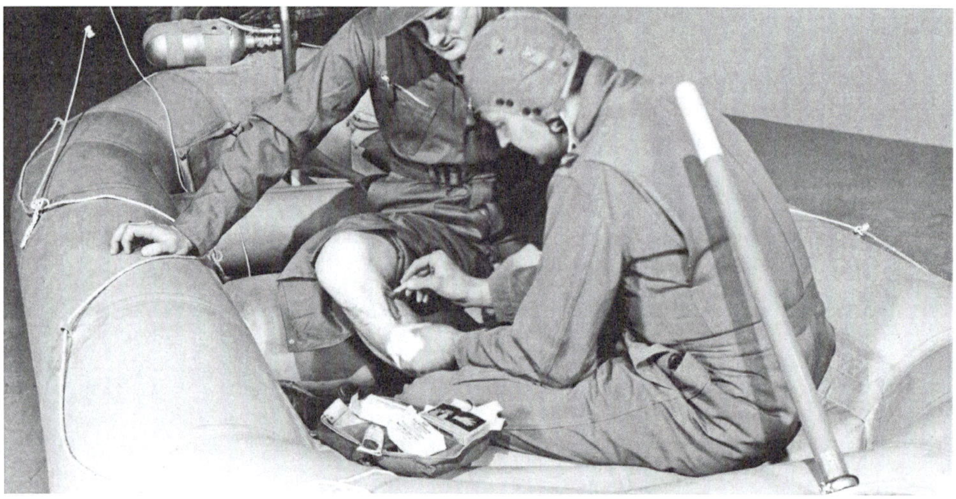

(Left) Aeronautic first aid kit eventually integrated as standard accessory for multi-person life rafts. Two USAAF members simulating first aid treatment with aeronautic first aid kit while in inflated rubber raft. (USAF)

(Right) Crew Chief, Sgt. W.H. Banks in a 9th AF C-53 transport from England to France, c.1944. Above his left shoulder are two aeronautical first aid kits. (National Archives)

The units developed in 1942 were designed to meet initial wartime requirements and were rushed into production. The first deliveries were reserved for aircraft preparing for overseas departure. As stock would allow, they would next be supplied as Government Furnished Equipment (GFE), to aircraft manufacturers to be installed in aircraft before delivery. The new wartime requirements specified a ratio of one unit per fighter plane and one unit per two individuals on all other classes of aircraft. Technical Order 00-30-139 of 15 December 1942 was released for service distribution and described the guidelines for replacements, first of the aeronautic kit then revised to include the parachute first aid packet. This technical order, with its subsequent revisions, was generic in that it listed the classes of aircraft authorized for the installation of the aeronautic first aid kit and advised that the table of quantity allowances be reviewed for specific aircraft as per their T/O&E.

Entering 1943, first aid kits were reviewed by the Army/Navy/British Sub-Committee of Standardization for First Aid Kits. This subcommittee was established to help formalize and streamline the development process. The aeronautic kit and several other units were being monitored. On 27 March 1943, the USAAF drafted a revision to Specification No.1631 as Amendment-C with the principal revision being the inclusion of boric acid ointment. All other changes were technical details for the manufacturing process. The committee recommended a change to require one small and two large Carlisle dressings. In August, based on the acceptance of the change to the dressings, the committee concluded that this new unit be adopted for joint Army/Navy/British use under Specification 1631-C.

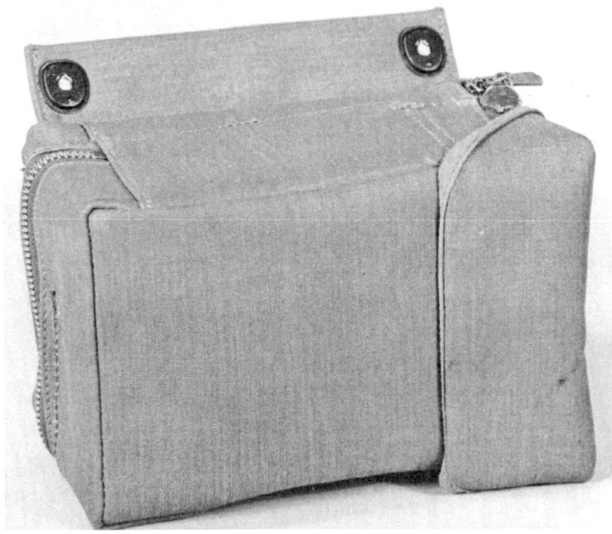

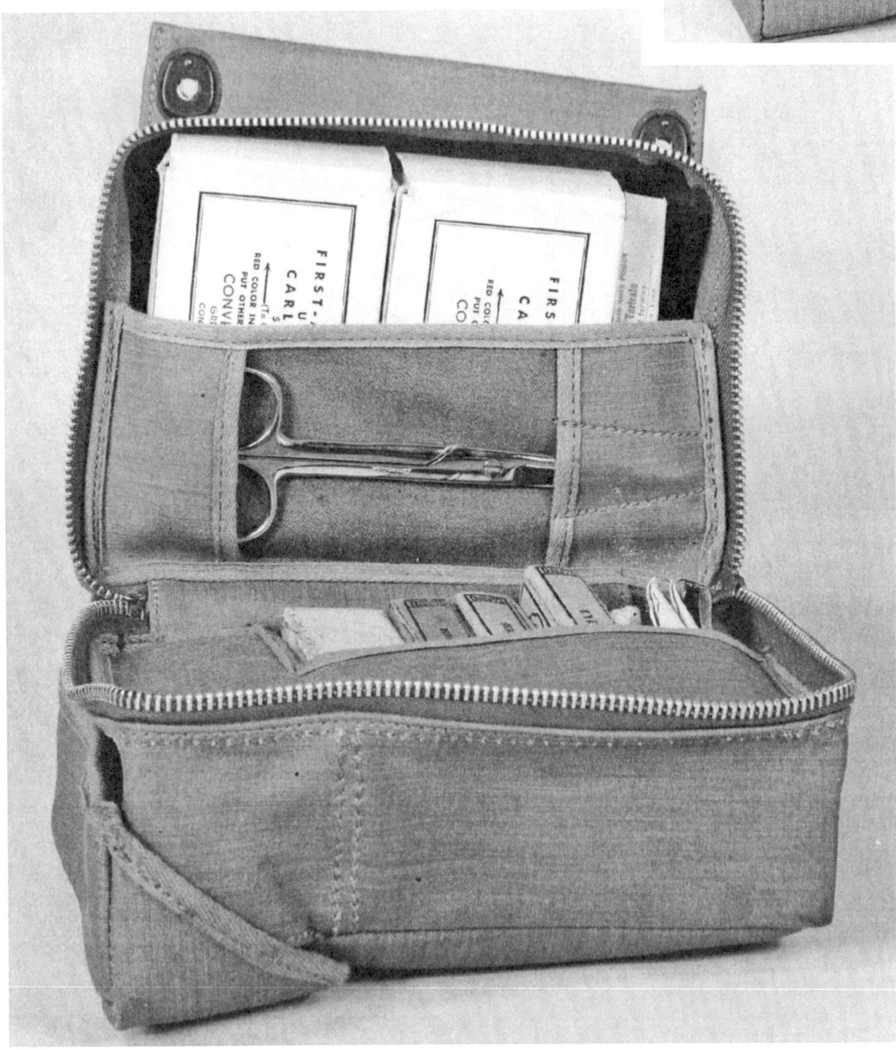

To accept new components, container needed revision. On 19 January 1944, the USMD NY Purchasing Office released Dwg. No. 144 illustrating construction details of new kit. Specification 1631 was revised to reflect new container and its components in Amendment-D 4 March 1944, which became the procurement specification. Container was additionally referenced under USAAF Dwg. No. 44G7956 since the AML performed initial engineering. Stock class number 97765 was retained to reflect the complete unit but stock number 97766 was assigned to new empty container. Preproduction sample of new aeronautic first aid kit 44G7956. (National Archives)

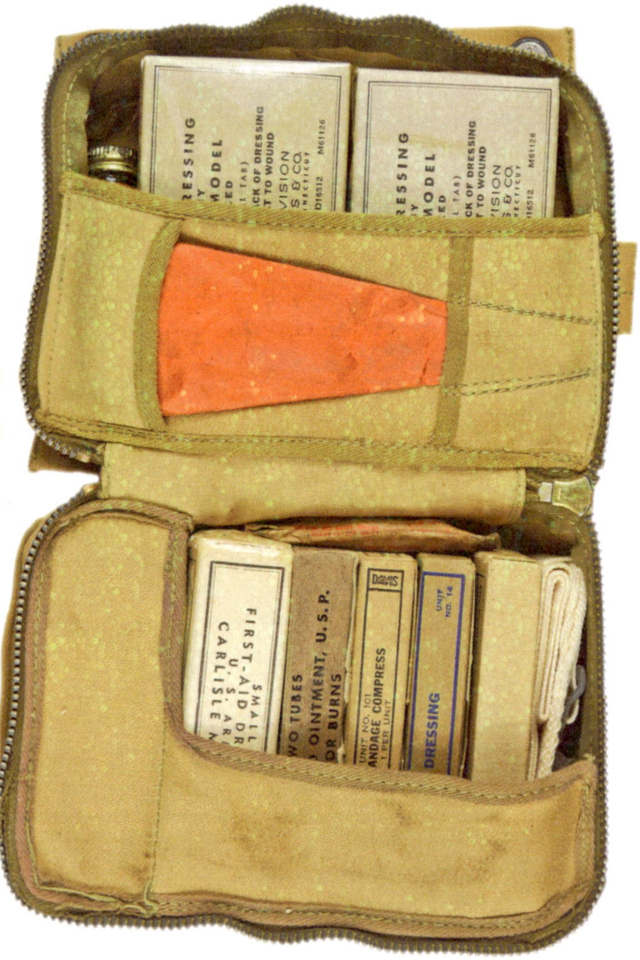

(Left and above) Example of aeronautic first aid kit per USAAF Dwg. No. 44G7956, fabricated of waterproofed fabric and redesignated under USMD stock class number 97766 (container only). Complete unit was still referenced under stock class number 97765 and after March 1944 as 9776500.

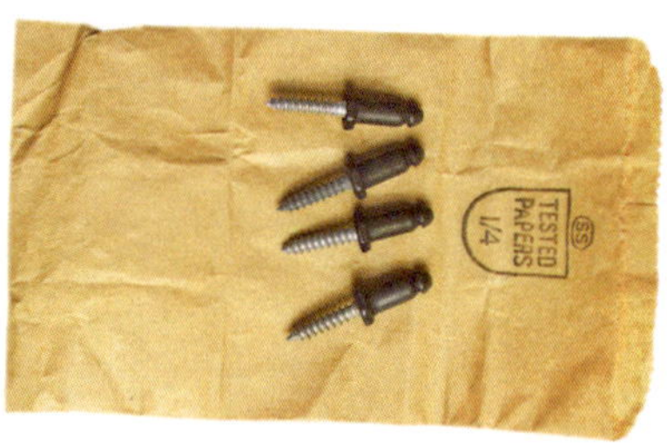

Components indicate early production unit: they are orderly and generally visible. Also, with large dressings, it was more suitable to treat a variety of potential flight injuries.

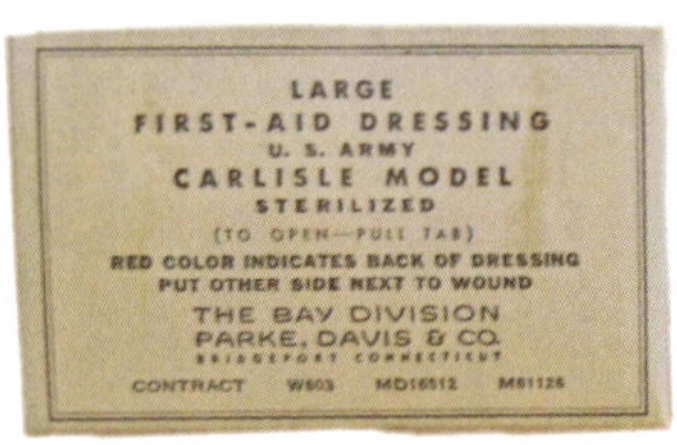

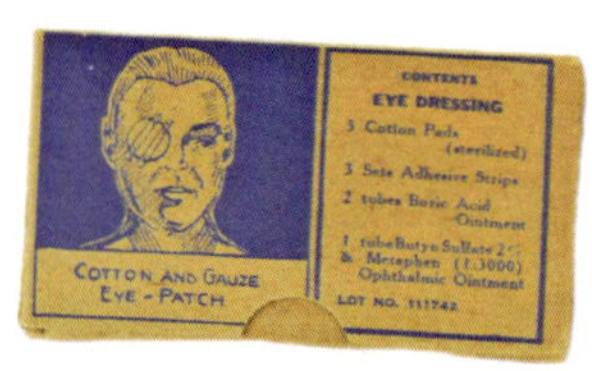

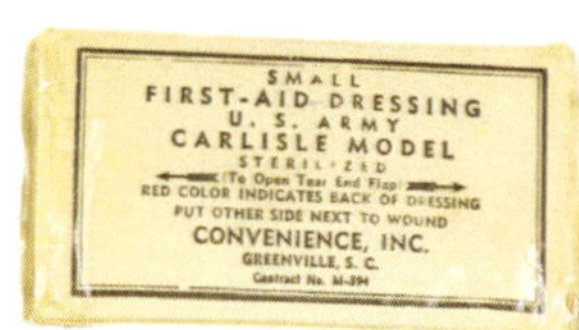

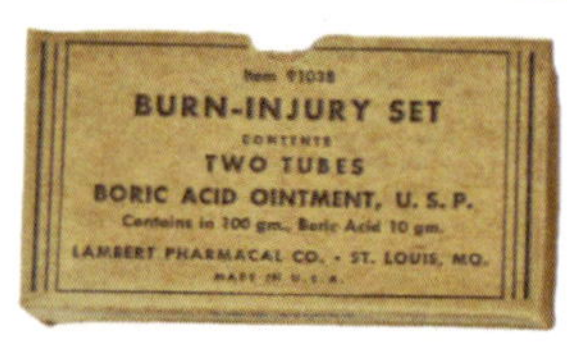

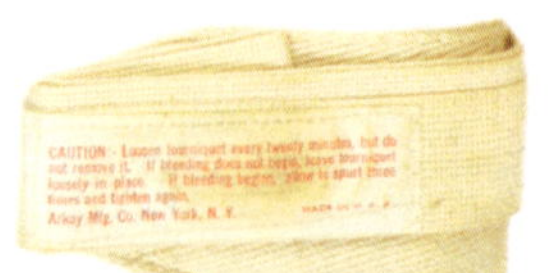

External pocket and prescribed components. Early production runs included components with 5-digit stock numbers; image with components in the pocket best represents a post-March 1944 packing as the 1" bandage box has a seven-digit stock number.

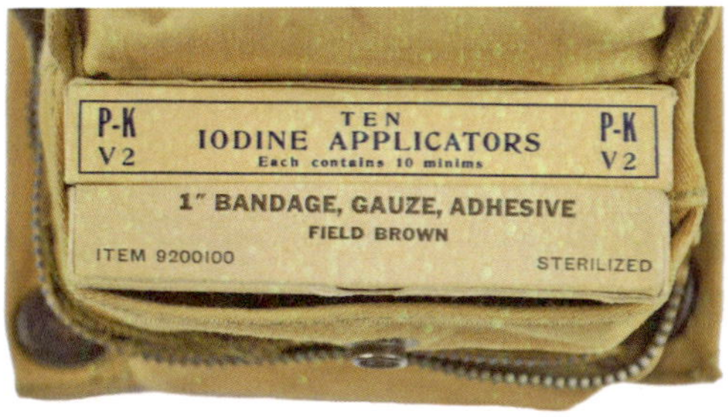

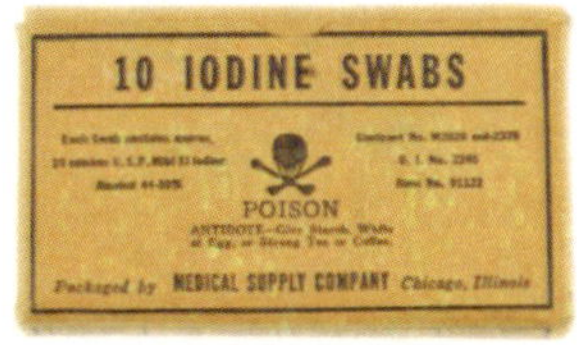

(Below) 1944-1945 packaging with new laminated foil-lined asphalt wrappers containing camouflage Carlisle bandages. Fabric color used in construction of container had no relevance to anything; merely a manufacturer variation. Components packed in these kits varied considerably. Units pictured here are general reference on what is known to be correct. Contracts were awarded to an assembly contractor with components obtained from a variety of manufacturers. Components were delivered to a central location, then units assembled, packaged and shipped to their destinations. Units also receive regular inspection and individual packages could be replaced multiple times during service life of container.

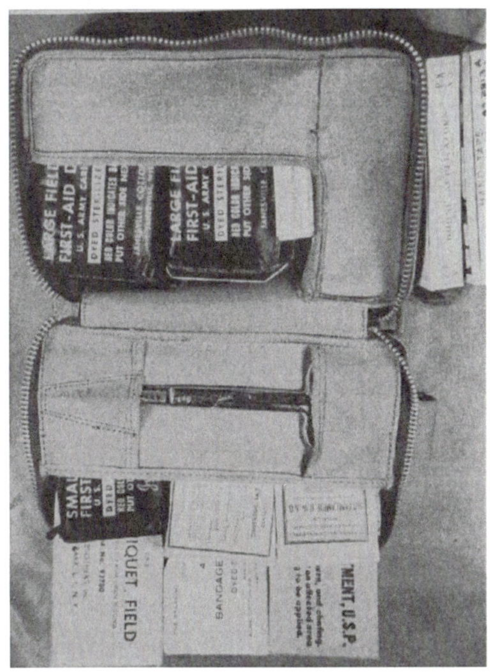

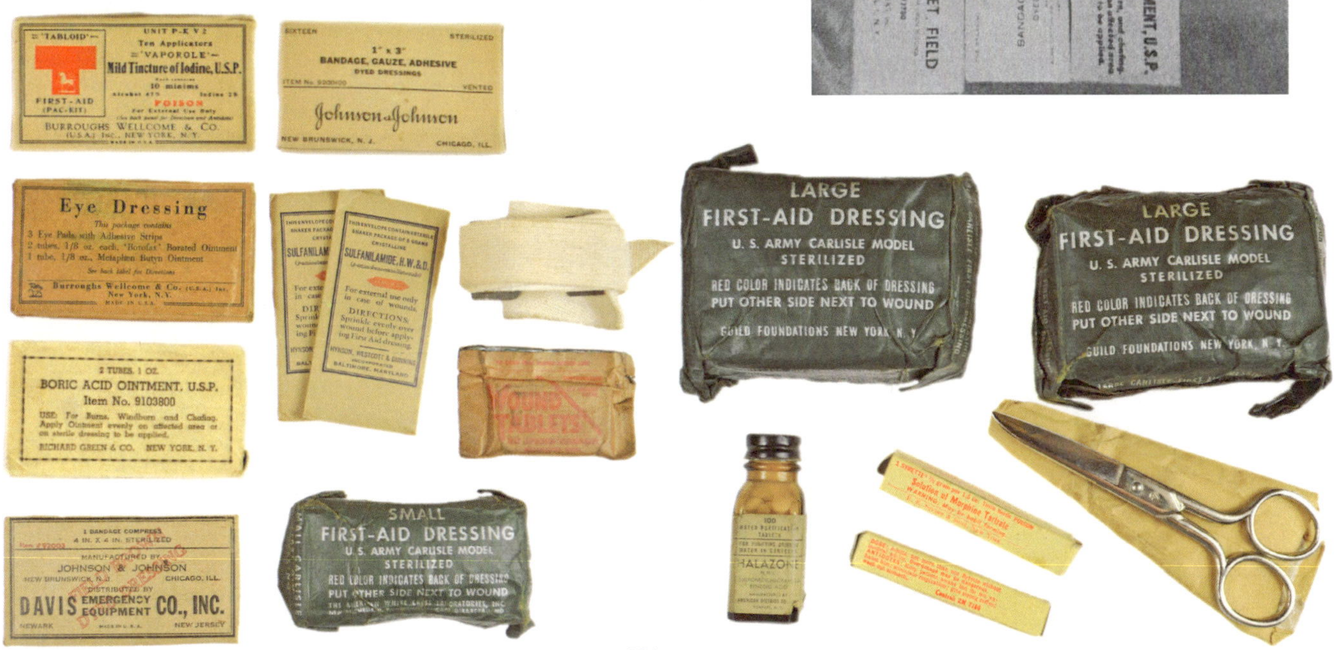

As of August of 1943, the aeronautic first aid kit as detailed in USAAF Dwg. No. 44G7956 was approved for joint Army-Navy standardization. Until the process could be finalized by the Joint Army-Navy Board, however, a series of temporary specifications was assigned to this unit and referenced through the duration of its service life during WWII. The first interim Army specification at the time of approval was D-97765, later changed to US Army Specification No. 10-3022 at an unknown date and referenced under USMD Specification No.1631-D. In December of 1943, the BuAer announced this unit would be made available.

In the following month, January 1944, the BuAer began negotiations for procurement of the aeronautic first aid kits at a projected requirement of 35,000 units for calendar year 1944. This negotiation preceded the involvement of the BuM&S and would be supplied to aircraft manufacturers as GFE. The units on requisition to the BuAer would be shipped complete per US Army specifications and stocked for replacement purposes under ASO stock number R57-K-8524, superseding the M-448. At the close of 1943, the BuAer was using only two standardized types of first aid kits for aircraft: the M-448 (10 unit) and the First Aid Kit for Pneumatic Life Rafts (five-six unit). The unit for pneumatic life rafts was supplied as standard in life rafts and air-droppable and emergency parachute kits, available to all flying personnel.

The integration of the US Army aeronautical kit was a slow process through 1944; it replaced the M-448 when available. By October 1944, the BuM&S was involved, assigning interim Specification 57K8 (INT) and stock number S2-1054. The units were still being assembled under Army specifications. It was established by this time that the aeronautic kit would only be available for multi-engine combat aircraft, on the premise that life rafts and droppable kits were equipped with the S2-1061 6-unit kit and that all flying personnel would be equipped with individual first aid kit S2-1063. When the aeronautic kit became available in quantity, Technical Note 25-45 was released on 31 March 1945 detailing what first aid kits were available for which aircraft and their issuance. The breakdown of quantities per aircraft outlined in the technical note was detailed as to models, types and even whether over land or sea. In sum, transport aircraft were to have four units and heavy to medium bombers were to have two to three units per airplane.

This unit saw service within the US Navy; property marked Carrier Air Service Unit-54 (CASU-54). Another variation within the components that is correct for the USAAF; however, battle dressings are US Navy-marked.

Two patterns of aeronautic kits secured to fuselage of USAAF Douglas C-47, c.1945. (National Archives)

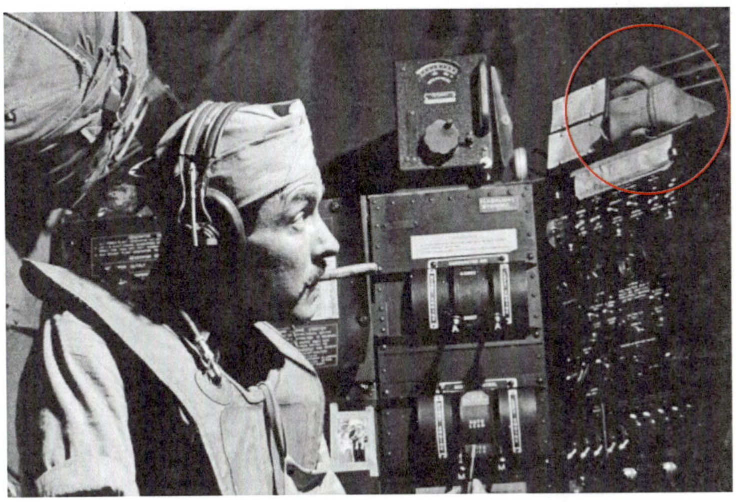

Aeronautical first aid kit resting on radio in engineer's compartment aboard US Navy PBM-5, Tanapag Saipan, April 1945. (National Archives)

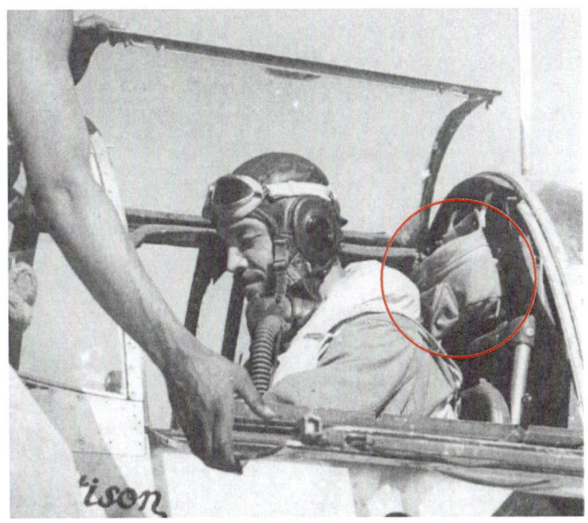

Aeronautic first aid kit secured in USAAF P-51 piloted by Capt. William Mattison of the 332nd FG, Italy, August 1944. (National Archives)

The theft of aeronautic kit components kits was an ongoing problem in both services, principally the morphine syrettes. Surprisingly, scissors were also popular. Initially, the kits were to be delivered complete to the aircraft manufacturer for installation then sealed by the resident inspector by using either a lead or wax seal.

Upon aircraft delivery, the units were often found to be missing the morphine syrettes. Orders were then issued for the kits to be sealed before shipment to the aircraft contractor. Still, many reports of missing morphine and other articles were reported; this was in large part due the use of unmarked seals. Seals could be easily removed, the kits pilfered, then resealed by the enterprising individual. Both services experienced theft at the factory and while in service. The first attempt within the USAAF to remedy the situation was to have the kits shipped to the factory without the morphine, then authorized medical personnel would install the morphine and seal the unit. But this did not solve the issue of theft by service personnel. The problem was addressed in Technical Order 01-1-117 dated 3 December 1943 and called Installation, Use and Inspection of First Aid Aeronautic Kits. It mandated that the kits were to be sealed with an identification stamp with special crimpers, which were a controlled item at the contractor, factory and delivery points. This helped limit the amount of theft but did not eliminate it. For the BuAer, the aeronautic kits would be delivered to supply points with the USMD stamped seal. To address theft, the BuAer assigned a custodian to remove the morphine from the complete first aid units, storing the syrettes in a locker when not in use.

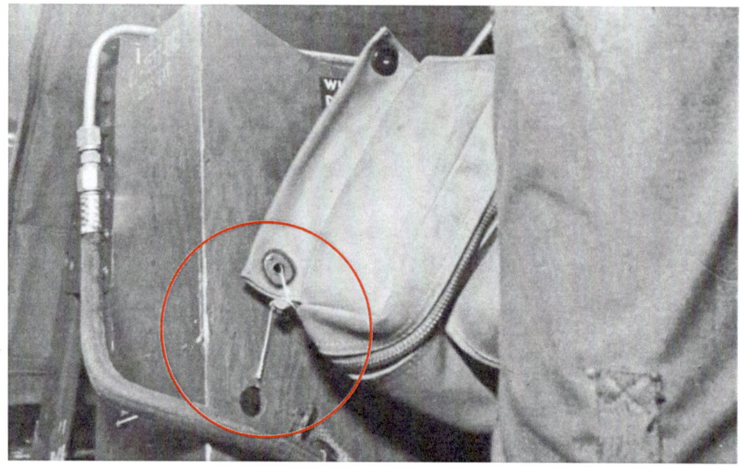

Official Medical Department seal per Technical Order 01-1-117. The correct method of sealing the 42D7411 and 1631 aeronautic kits was to attach wire first at the base of the zipper then to adjacent Lift-the-dot fastener. (National Archives)

(Right and far right) 42D7411 and 1631 aeronautic kits modified with leather tab and eyelet as improvement over Lift-the-dot securing method. Modification originated at Kansas City Medical Depot, documented as early as February 1944.

(Left) Method of sealing aeronautic kit in Dwg. No. 44G7956 was refined by adding small grommet adjacent to zipper to thread the wire.

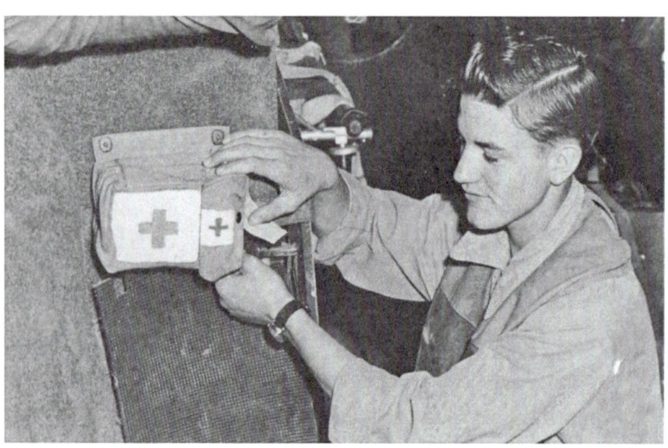

(Left and above) Technical Order 01-1-210, dated 16 January 1945, titled Identification Markings for First Aid Kits, requires first aid kits Aeronautic 97765, Arctic 97762, Jungle 97763, Parachute 97785 and Ambulance 97517 be marked with red cross for easy identification purposes in emergency. Markings were to be placed front and center. Being a late war directive, the practice and application of the red cross is predominately observed in postwar images, such as here, c.1952.

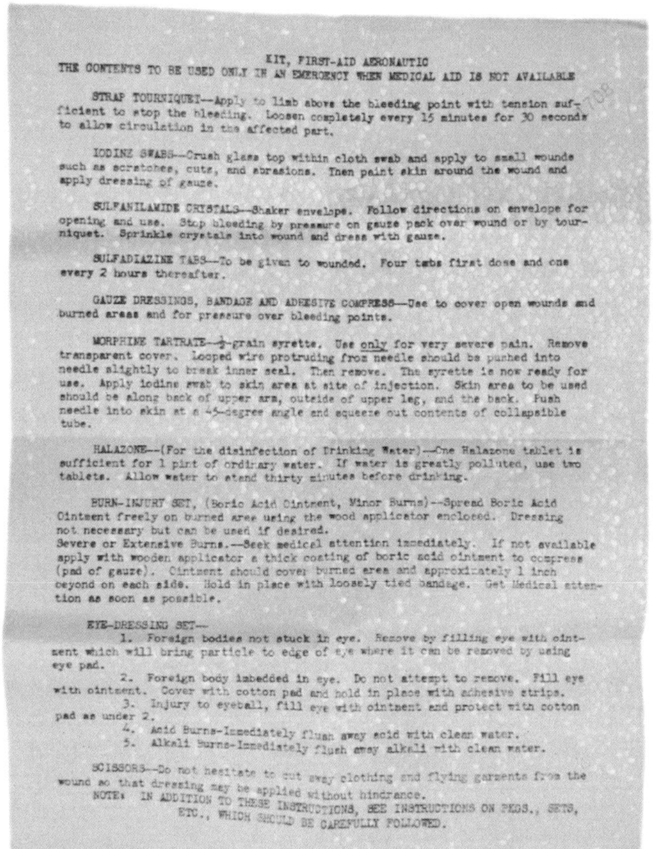

KIT, FIRST-AID AERONAUTIC
THE CONTENTS TO BE USED ONLY IN AN EMERGENCY WHEN MEDICAL AID IS NOT AVAILABLE

STRAP TOURNIQUET—Apply to limb above the bleeding point with tension sufficient to stop the bleeding. Loosen completely every 15 minutes for 30 seconds to allow circulation in the affected part.

IODINE SWABS—Crush glass top within cloth swab and apply to small wounds such as scratches, cuts, and abrasions. Then paint skin around the wound and apply dressing of gauze.

SULFANILAMIDE CRYSTALS—Shaker envelope. Follow directions on envelope for opening and use. Stop bleeding by pressure on gauze pack over wound or by tourniquet. Sprinkle crystals into wound and dress with gauze.

SULFADIAZINE TABS—To be given to wounded. Four tabs first dose and one every 2 hours thereafter.

GAUZE DRESSINGS, BANDAGE AND ADHESIVE COMPRESS—Use to cover open wounds and burned areas and for pressure over bleeding points.

MORPHINE TARTRATE—½-grain syrette. Use only for very severe pain. Remove transparent cover. Looped wire protruding from needle should be pushed into needle slightly to break inner seal. Then remove. The syrette is now ready for use. Apply iodine swab to skin area at site of injection. Skin area to be used should be along back of upper arm, outside of upper leg, and the back. Push needle into skin at a 45-degree angle and squeeze out contents of collapsible tube.

HALAZONE—(For the disinfection of Drinking Water)—One Halazone tablet is sufficient for 1 pint of ordinary water. If water is greatly polluted, use two tablets. Allow water to stand thirty minutes before drinking.

BURN-INJURY SET, (Boric Acid Ointment, Minor Burns)—Spread Boric Acid Ointment freely on burned area using the wood applicator enclosed. Dressing not necessary but can be used if desired.
Severe or Extensive Burns.—Seek medical attention immediately. If not available apply with wooden applicator a thick coating of boric acid ointment to compress (pad of gauze). Ointment should cover burned area and approximately 1 inch beyond on each side. Hold in place with loosely tied bandage. Get Medical attention as soon as possible.

EYE-DRESSING SET—
1. Foreign bodies not stuck in eye. Remove by filling eye with ointment which will bring particle to edge of eye where it can be removed by using eye pad.
2. Foreign body imbedded in eye. Do not attempt to remove. Fill eye with ointment. Cover with cotton pad and hold in place with adhesive strips.
3. Injury to eyeball, fill eye with ointment and protect with cotton pad as under 2.
4. Acid Burns-Immediately flush away acid with clean water.
5. Alkali Burns-Immediately flush away alkali with clean water.

SCISSORS—Do not hesitate to cut away clothing and flying garments from the wound so that dressing may be applied without hindrance.
NOTE: IN ADDITION TO THESE INSTRUCTIONS, SEE INSTRUCTIONS ON PKGS., SETS, ETC., WHICH SHOULD BE CAREFULLY FOLLOWED.

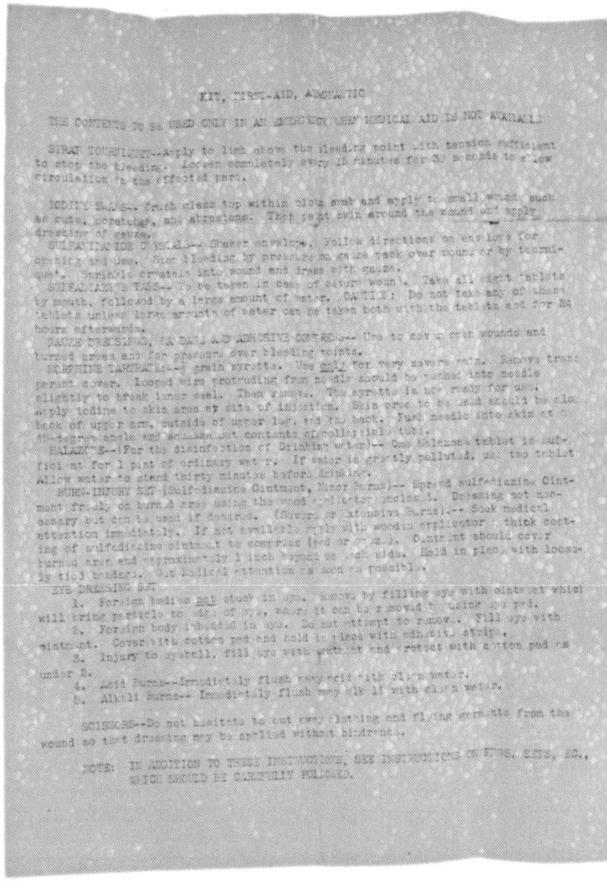

KIT, FIRST-AID, AERONAUTIC
THE CONTENTS TO BE USED ONLY IN AN EMERGENCY WHEN MEDICAL AID IS NOT AVAILABLE

Instruction sheets were provided for basic medical treatment. Instruction sheet provided in the1631 production kits and for 44G7956.

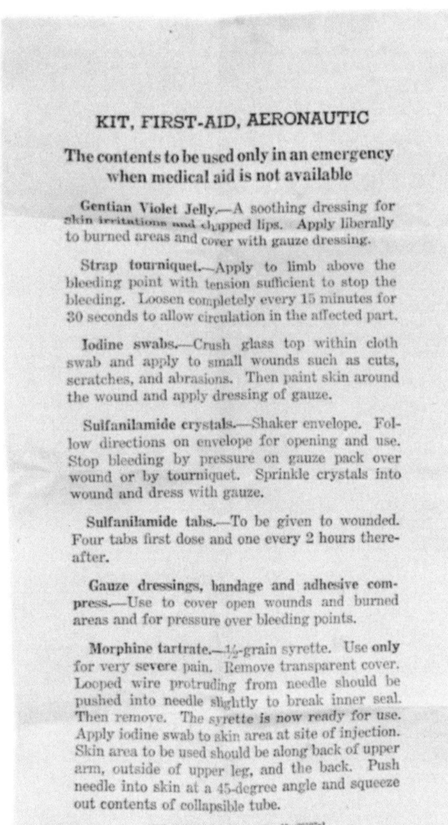

KIT, FIRST-AID, AERONAUTIC

The contents to be used only in an emergency when medical aid is not available

Gentian Violet Jelly.—A soothing dressing for skin irritations and chapped lips. Apply liberally to burned areas and cover with gauze dressing.

Strap tourniquet.—Apply to limb above the bleeding point with tension sufficient to stop the bleeding. Loosen completely every 15 minutes for 30 seconds to allow circulation in the affected part.

Iodine swabs.—Crush glass top within cloth swab and apply to small wounds such as cuts, scratches, and abrasions. Then paint skin around the wound and apply dressing of gauze.

Sulfanilamide crystals.—Shaker envelope. Follow directions on envelope for opening and use. Stop bleeding by pressure on gauze pack over wound or by tourniquet. Sprinkle crystals into wound and dress with gauze.

Sulfanilamide tabs.—To be given to wounded. Four tabs first dose and one every 2 hours thereafter.

Gauze dressings, bandage and adhesive compress.—Use to cover open wounds and burned areas and for pressure over bleeding points.

Morphine tartrate.—½-grain syrette. Use only for very severe pain. Remove transparent cover. Looped wire protruding from needle should be pushed into needle slightly to break inner seal. Then remove. The syrette is now ready for use. Apply iodine swab to skin area at site of injection. Skin area to be used should be along back of upper arm, outside of upper leg, and the back. Push needle into skin at a 45-degree angle and squeeze out contents of collapsible tube.

U. S. GOVERNMENT PRINTING OFFICE 16—26307-1

(Left) Instruction sheet provided in the 753 and 753A.

(Below) Instruction sheets for aeronautics kits 44G7965. 2-sided instruction sheet with contents list on reverse. The listing of interest calls out for the addition of ammonia inhalants for the external pocket.

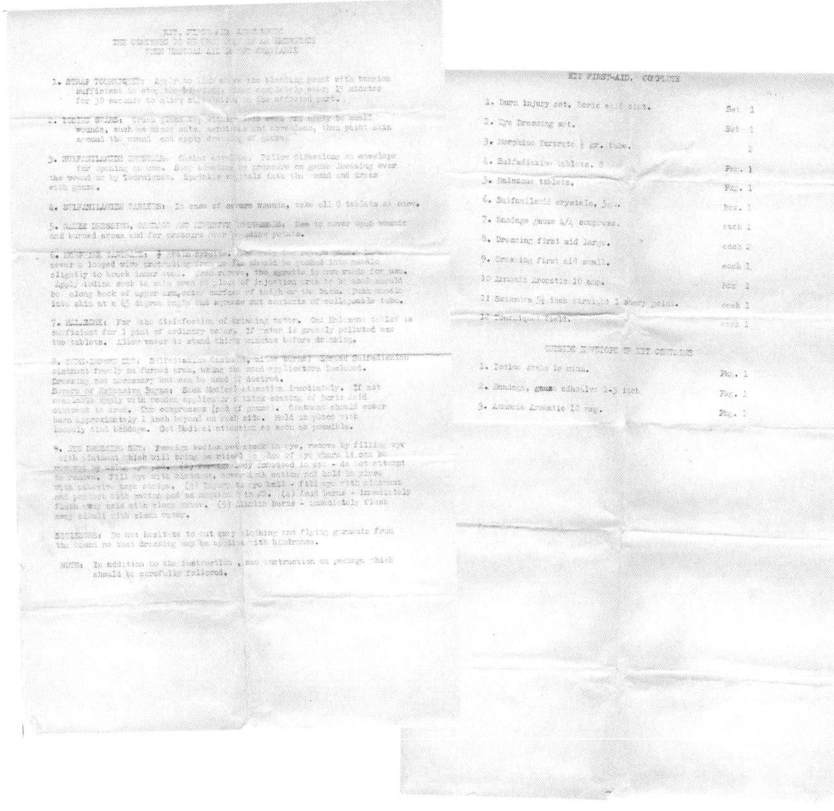

Components listing for aeronautic first aid kit from US Army Medical Department Supply Catalog.
** The June 1943 catalog revised the listing to stock class number 91038 Burn-Injury Set, Boric Acid Ointment.*

NOVEMBER 1942 ARMY MEDICAL SUPPLY CATALOG LISTING

97765 KIT, FIRST AID, AERONAUTIC, COMPLETE

STK. NO.	ITEM	UNIT	QTY.
97770	KIT, FIRST AID, AERONAUTIC, CONTAINER	EACH	1
1K27505	HALAZONE, NNR, 1/16 GR TAB, 100	BOTTLE	1
91042*	BURN-INJURY, SULFADIAZINE, OINTMENT*	SET	1
91095	EYE-DRESSING SET	SET	1
91157	MORPHINE TARTRATE, USP, 1/2 GR SOLUTION	EACH	2
91204	SULFADIAZINE, 0.5 GM (7.7 GR) TAB: 8	PKG.	1
91211	SULFANILAMIDE CRYSTALLINE, 5 GRAMS IN STERILE INDIVIDUAL DOUBLE-WARPPED ENEVELOPE	PKG.	1
92060	DRESSING, FIRST AID, SMALL	PKGS.	3
93692	SCISSORS, 1 POIUNT SHARP, 5 1/2-INCH, STRAIGHT	EACH	1
93780	TOURNIQUET, FIELD	EACH	1
	OUTSIDE ENVELOPE OF KIT		
91122	IODINE SWAB, 10 MIN	PKG.	1
92000	BANDAGE, GAUZE, ADHESIVE, 1 BY 3 INCHES	PKG.	1

*** The January 1944 listing in The Air Surgeons Bulletin shows a quantity of two (2) envelopes for the sulfanilamide using pre-March 1944 five-digit stock class number 91211 and illustrating the term "respectively". This should be taken into consideration when reviewing lists. However, specifications continue to detail a quantity of one (1) box.*

MARCH 1944 ARMY MEDICAL SUPPLY CATALOG LISTING

9776500 KIT, FIRST-AID, AERONAUTIC, COMPLETE

STK. NO.	ITEM	UNIT	QTY.
9776600	KIT, FIRST AID, AERONAUTIC, CONTAINER	EACH	1
9103800	BURN INJURY SET, BORIC ACID OINTMENT	SET	1
9109500	EYE-DRESSING SET	SET	1
9115700	MORPHINE TRATRATE, 1 TUBE	EACH	2
9120400	SULFADIAZINE, 8 TABLETS	PKG.	1
9121100	SULFANILAMIDE, CRYSTALLINE, 5, 5-GRM, ENEVELOPES**	BOX	1
9200300	BANDAGE, GAUZE, COMPRESS, FIELD BROWN, 4-INCH BY 4 INCH	EACH	1
9205100	DRESSING, FIRST AID, LARGE, FIELD BROWN	EACH	2
9206100	DRESSING, FIRST AID, SMALL, FIELD BROWN	EACH	1
9369200	SCISSORS, 5 1/2 INCH, STRAIGHT, 1 POINT SHARP	EACH	1
9378000	TOURNIQUET, FIELD	EACH	1
	OUTSIDE ENVELOPE OF KIT		
9112200	IODINE SWAB, 10-MINIM, 10	PKG.	1
9200100	BANDAGE, GAUZE, ADHESIVE, FIELD BROWN,1-INCH BY 3-INCH 16	PKG.	1

By January 1945 the Kit, First Aid, Aeronautic could be found under US Army Medical Department stock number 9776500; prior to March 1944 it had been 97765, Bureau of Aeronautics Aviation Supply Office under R57-K-8524 and the Bureau of Medicine & Surgery stock number S2-1054. By September 1945, 310,000 aeronautic kits were procured for the USAAF and 77,000 for the US Navy. On 21 March 1946 it was given joint Army-Navy standardization, referenced under Specification JAN-K-303. As part of this Army-Navy specification, the red cross and USMD caduceus were to be factory-applied and other design features of 44G7956 were to carry over. Development of this aeronautic kit was long and then returned to the design of the 1939 unit.

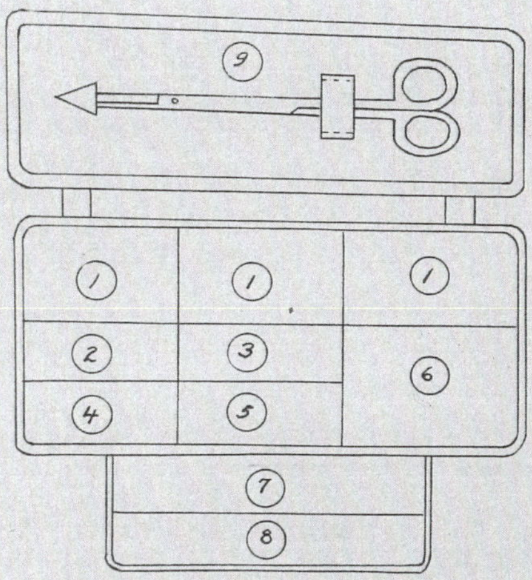

1 - Dressing, First Aid, Carlisle, Small
2 - Sulfanilamide
3 - Tourniquet Field
4 - Sulfadiazine Ointment
5 - Eye Dressing
6 - Halazone 100
Syrettes Morphine Tartrate
Sulfadiazine Tablets
7 - Iodine Swabs
8 - Adhesive Compresses
9 - Scissors

9776500 Kit, First-Aid, Aeronautic, Complete:

LIST-OF-CONTENTS PER W.D. MED. SUPPLY CATALOG, 1944, INCLUDING CHANGE NO. 6

(Items in Parenthesis are Authorized Substutions)

ITEM NO.	NOMENCLATURE	UNIT	QUANTITY
9776680	Kit, First-Aid, Aeronautic, Container:	each	1
9103800	Burn Injury Set, Boric Acid Ointment: (9101000 Acid, Boric, Ointment, 1-oz tube, 1 for 1)	set	1
9109510	Eye Dressing Set, M-2: (9109500 Eye Dressing Set, 1 for 1)	set	1
**9115700	Morphine Tartrate, 1 tube:	each	2
9120400	Sulfadiazine, 8 tablets:	pkg.	1
9121805	Tablet, water purification, individual 100: (9121810 Tablet water purification individual 100 1 for 1)	btl.	1
9200300	Bandage, gauze, compress, field brown 4" by 4" (9200200 Bandage, gauze, compress white, 4" by 4" 1 for 1)	each	1
9205100	Dressing, First-Aid, large brown: (9205000 Dressing First-Aid, large white 1 for 1)	each	2
9206100	Dressing, First-Aid, small field brown: (9206000 Dressing, First-Aid, small white, 1 for 1)	each	1
9369200	Scissors, 5-1/2" straight, 1 point sharp: (9369000 Scissors 4", 1 for 1)	each	1
9377900	Tourniquet, elastic, field: (9378000 Tourniquet, field, 1 for 1)	each	1
	OUTSIDE ENVELOPE OF KIT TO CONTAIN		
9112200	Iodine, swab, 10 minim: (9112000 Iodine swab, 1 1/2 cc., 6 box, 1 for 1)	pkg	1
9200100	Bandage, gauze, adhesive, white 1" by 3", 16: (9200000 Bandage, gauze, adhesive, white 1" by 3", 16, 1 for 1)	pkg.	1

NOTE: The above list to be placed in each item 9776500 Kit, First-Aid, Aeronautic, and this list to be used for requisitioning replishment items.

(Above left) Diagram showing prescribed packing for the 1631 per USMD Dwg. No.57.

(Above) Contents list for late 1945 to immediate postwar 44G7956 unit. (Kriss Usherwood Collection)

9776500 KIT, FIRST AID, AERONAUTIC, COMPLETE

STK. NO.	ITEM	UNIT	QTY.
9776680	KIT, FIRST-AID, AERONAUTIC, CONTAINER	EACH	1
9103800	BURN INJURY SET, BORIC ACID OINTMENT*	SET	1
9109510	EYE DRESSING SET, M-2	SET	1
9115700	MORPHINE TRATRATE, 1 TUBE	EACH	2
9120400	SULFADIAZINE, 8 TABLETS	PKG.	1
9121805	TABLET, WATER PURIFICATION, INDIVIDUAL 100	BTL.	1
9200300	BANDAGE, GAUZE, COMPRESS, FIELD BROWN, 4 INCH BY 4 INCH	EACH	1
9205100	DRESSING, FIRST AID, LARGE FIELD BROWN	EACH	2
9206100	DRESSING, FIRST AID, SMAL FIELD BROWN	EACH	1
9369200	SCISSORS, 5 1/2 INCH, STRAIGHT, 1 POINT SHARP	EACH	1
9377900	TOURNIQUET, ELASTIC, FIELD	EACH	1
	OUTSIDE ENVELOPE OF KIT		
9112200	IODINE, SWAB, 10 MINIM	PKG.	1
9200100	BANDAGE, GAUZE, ADHESIVE, WHITE 1 INCH BY 3 INCH 16	PKG.	1

(Right) By 1946, updates in components had been applied. Contents list for late-war to immediate postwar 44G7956.

The aeronautic first aid kit was designed to be universal for a multitude of emergency situations. There were also units designed for specialty purposes. The parachute first aid packet was designed to meet a special requirement to address wounds in flight as a supplement to the parachute emergency kit. By late 1941 the Materiel Division had standardized the Type B-1 Alaskan emergency kit and was preparing to standardize the Type B-2 jungle kit. In review, it was determined that the first aid items in the parachute kit were not accessible while in flight or during a parachute descent, nor was the aeronautic first aid kit in an easily accessible location. In January of 1942, the Materiel Command initiated development of a packet to contain a basic complement of a dressing, morphine syrette, and tourniquet, making it standard for all parachutes. The emphasis was to treat wounds in flight, during the parachute descent and upon a parachute landing. It would then be incorporated into the parachute emergency kit.

The unit adopted was a container measuring 4" x 2-1/2" x 2-1/2", with a zippered closure and two tie tapes to secure it to the parachute harness, referenced under Dwg. No 42D11238. The packet was released for procurement in April of 1942. The Materiel Command was able to bypass the formalities through the USMD since it was able to procure just the container. Once in supply, the three components were installed from standard medical stock. The zipper would allow quick access and ease of opening with one hand, but also quick access for pilfering. The morphine in the parachute first aid packet suffered the same fate as that in the aeronautic first aid kit: it was often stolen. Though the 42D11238 had a plasticized material, the zipper allowed water seepage. By the summer of 1942, a process to tamper- and waterproof the unit began.

On 11 November 1942, a revised packaging method was accepted titled Packet, First Aid, Parachute referenced under USMD Specification No.1636. It employed an olive drab plasticized container with heat-sealed edges to prevent tampering and to increase waterproofing. The package had to be torn open. In April of 1943, the zippered unit was declared obsolete.

The revised unit was sent to Fort Bragg, N.C. for field testing. Results of the tests were incorporated into the next revision in Technical Order 00-30-139, issued 26 May 1943. It mandated the parachute first aid packet be issued on the basis of one per parachute except parachutes used in primary and basic training. The technical order also stated the unit was to be attached to the right shoulder strap opposite the pull ring and that the printing of First Aid be exposed. By the close of 1942 no guidelines had been established for the attachment to parachutes nor was it included in the November 1942 Medical Supply Catalog. The parachute first aid packet eventually appeared in the revised June 1943 Medical Supply Catalog under stock class number 97785 as Packet, First Aid, Parachute.

(Below) Concept design of new parachute first aid packet 42D11238 prior to approval for production. Packet fabricated of heavy waterproof cotton duck fabric with zippered closure for quick easy access with one hand. (National Archives)

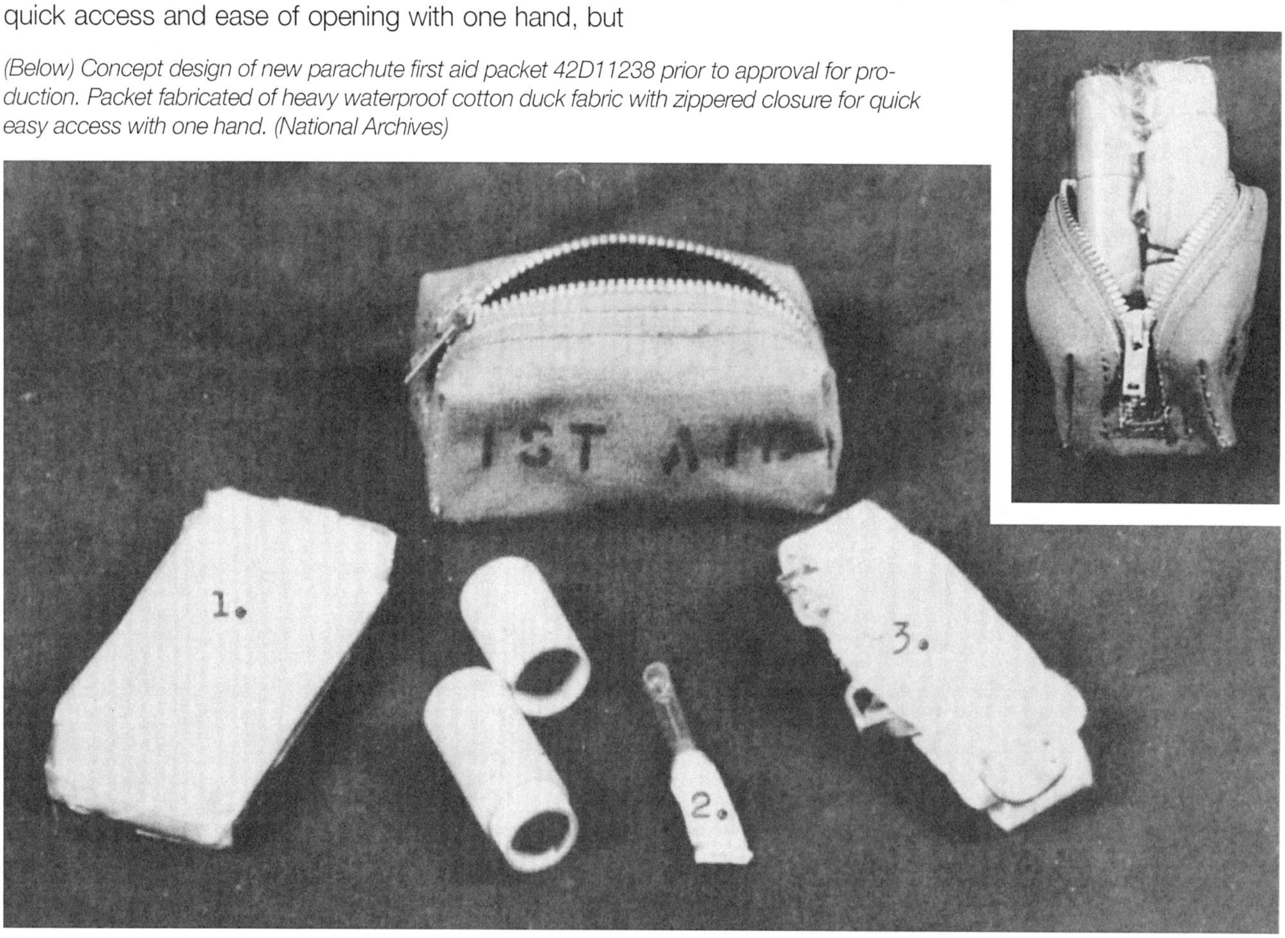

(Above) Wright Field volunteer demonstrating concept of parachute packet 42D11238 attached to eventual prescribed location on parachute harness, March 1942. (USAF)

(Right and below) Description of parachute first aid packet 42D11238, accompanied by illustrations of prescribed components. It was intended for confined or isolated positions where aeronautic kit was not easily accessible; this was prior to authorization to be issued with all parachutes. (USAF)

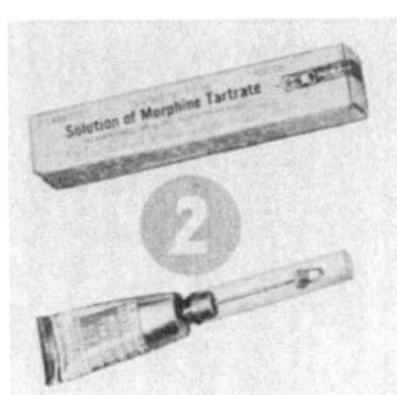

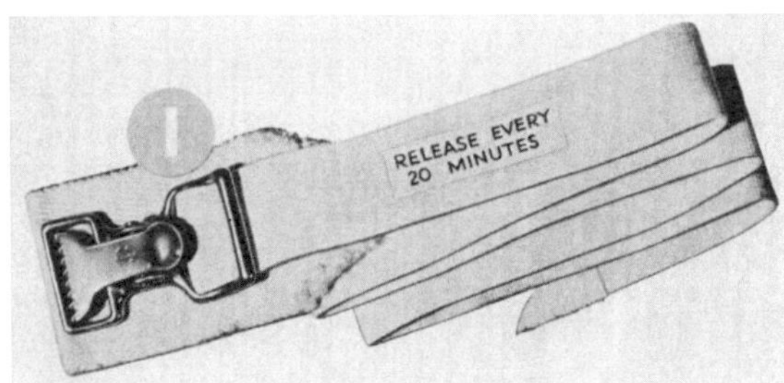

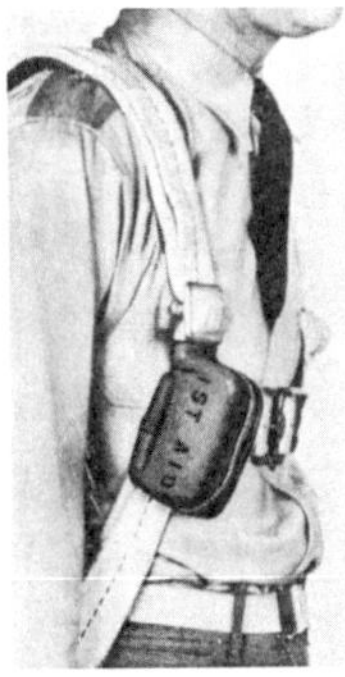

KIT, FIRST-AID PARACHUTE

(Medical Dept. NS-9)

To be attached to the parachute harness for constant availability. Should be carried in Gun Turrets and other cramped spaces where the larger kit is not accessible. Contains tourniquet, morphine and wound dressing as illustrated in above photos, Nos. 1, 2 and 3.

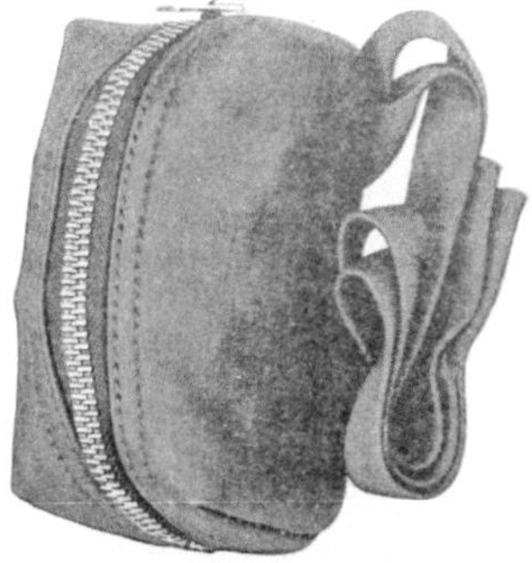

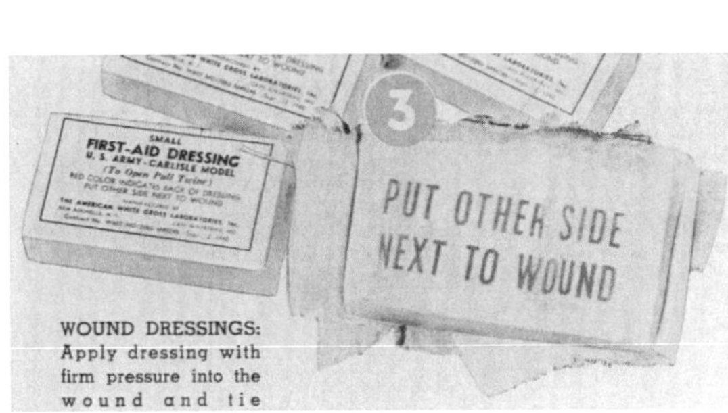

The 42D11238 contained a morphine syrette, small Carlisle field dressing, and tourniquet. Contractors were Bauer & Black Co. and Davis Equipment Co.

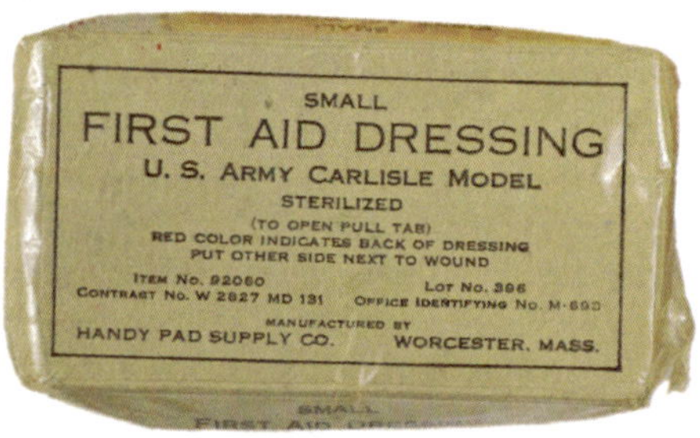

(Left) Waist gunner, S/Sgt. William Fleming preparing for a mission, England, July 1943. Attached to parachute harness was 42D11238 first aid packet. (National Archives)

(Below) B-24 crew member of 376th BG with parachute first aid packet 42D11238 attached to QAC harness, Bengasi, Libya, August 1943. (National Archives)

Revised packet submitted September 1942 with improved packaging method, approved 11 November 1942 by Specification 1636. Drawing shows construction details and packaging method. Packet was comprised of a single sheet of fabric heat-sealed along three edges. Components and attachment method remained the same but items now contained in sealed waterproof container. Design reduced manufacturing cost: zipper alone for the 42D11238 was $0.30. Package change was projected to save $360,000. (USAF)

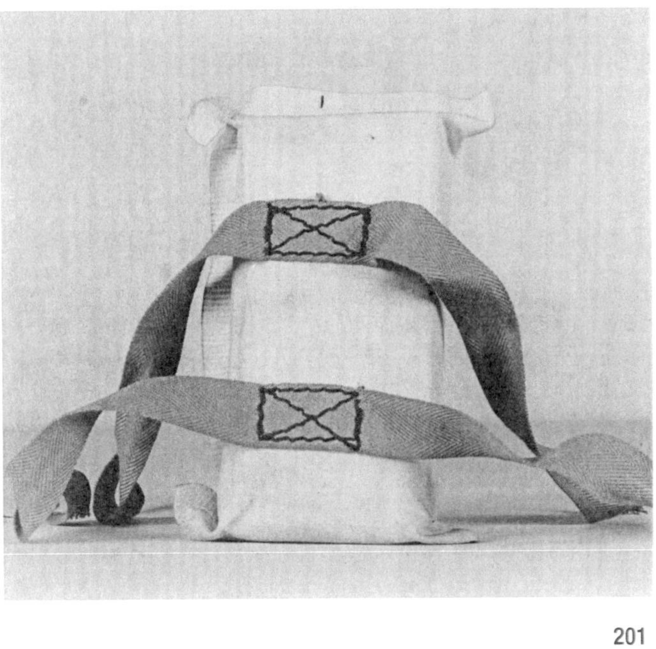

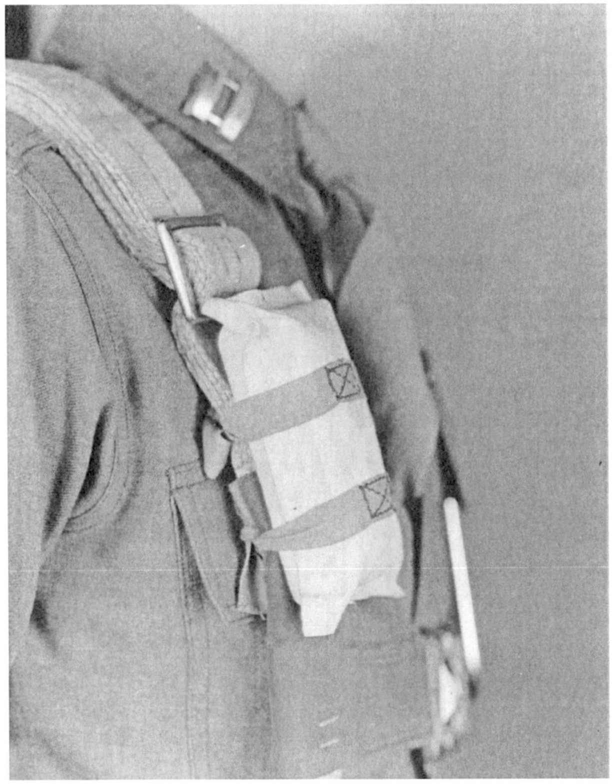

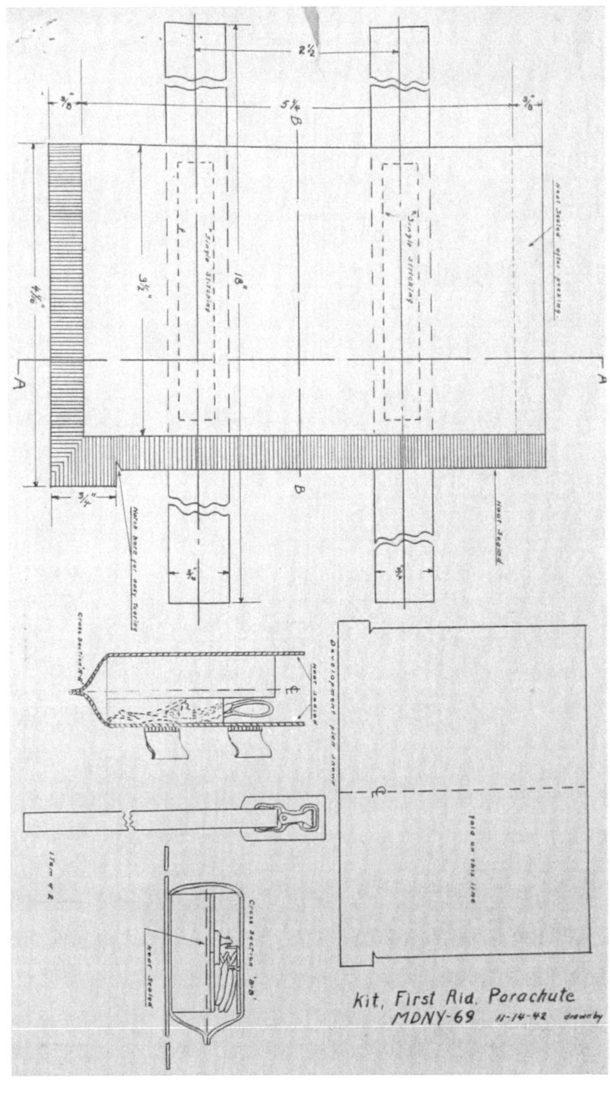

Effective 11 January 1943 Amendment-A was applied to Specification 1636. The container was re-engineered to include two sheets of material, heat-sealed along all four edges, with tearing notches applied near corners. Components continued to be a morphine syrette, small first aid Carlisle dressing, and tourniquet.

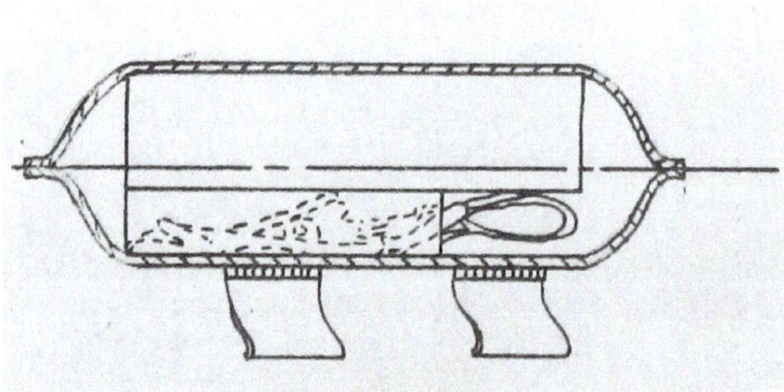

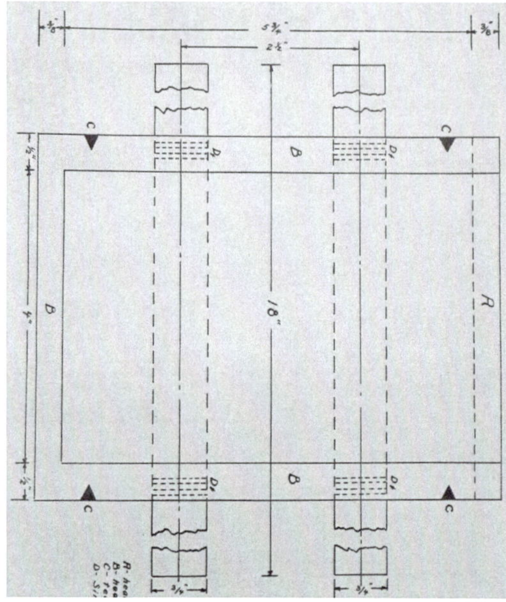

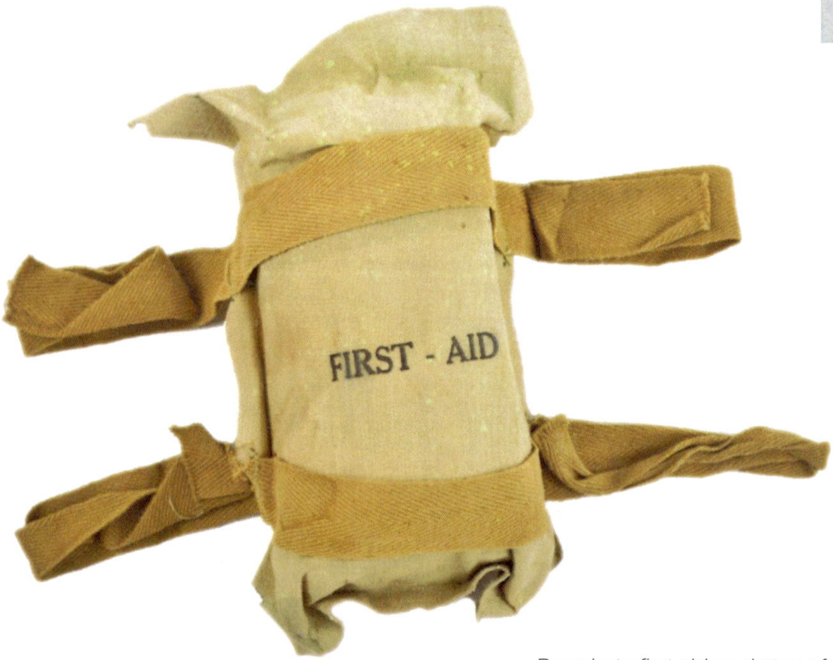

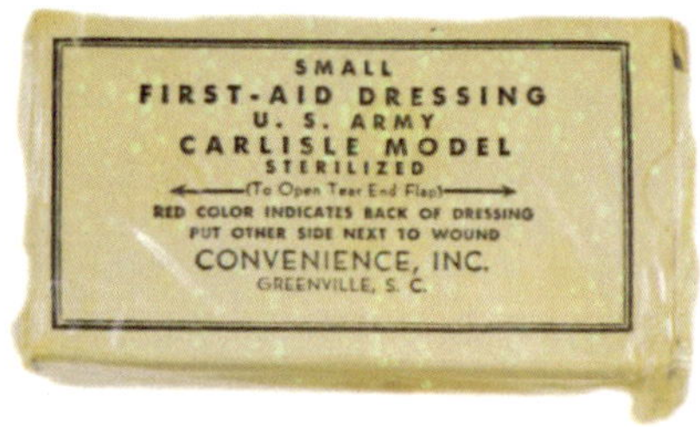

Parachute first aid packet conforming to Specification 1636-A.Font and colors of the container varied, but Olive Drab was specified. Note crushed morphine syrette, a result of flimsy packaging. By the close of FY1944, 543,000 parachute first aid packets had been procured.

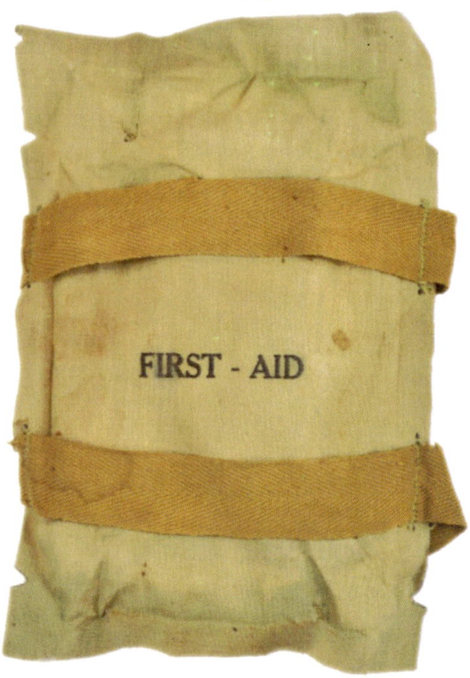

(Right) B-29 aicrew member, Sgt. Leo E. McBride, wears the parachute packet attached to parachute harness as prescirbed by technical order, October 1944. (National Archives)

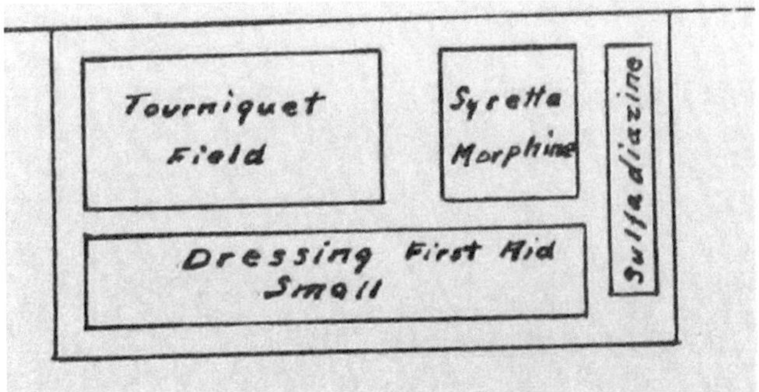

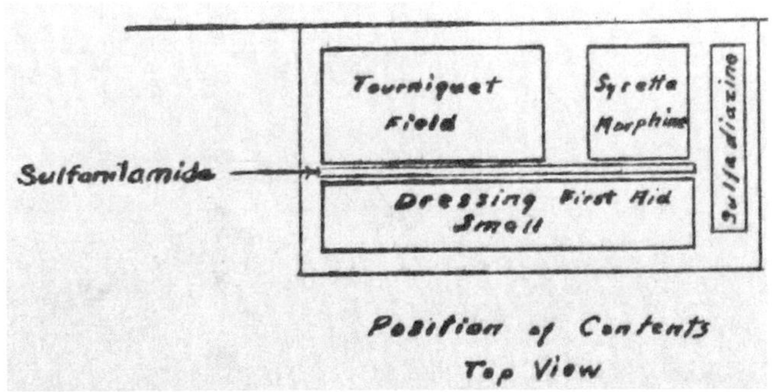

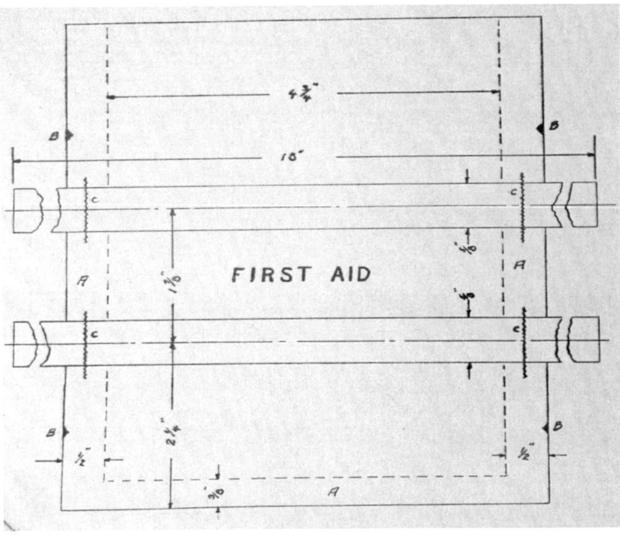

On 6 July 1943 Amendment-B went into effect, adding 8 tablets of sulfadiazine. According to the drawing, container was slightly enlarged at width of ¾", allowing the addition. Cross-section from the top of the packet shows new packaging method. Effective 1 October 1943 Amendment-C added one 5 gm. envelope of sulfanilamide. It also directed that morphine syrettes were to be packed in a paper cylindrical tube made of spiral-wound chipboard to help prevent crushing. Cross-section shows the sulfanilamide between dressing and tourniquet.

(Below) Packets were not always worn as prescribed in Technical Order 00-30-139 as shown by 2nd Lieutenants Earl & Courtney Shonkin of the 450th BG, Italy, December 1944. (National Archives)

(Below) B-29 radar operator, S/Sgt. Stanley Sienkiewicz, of the 45th BS wears the parachute first aid packet on the right shoulder as prescribed, China, c.1944. (National Archives)

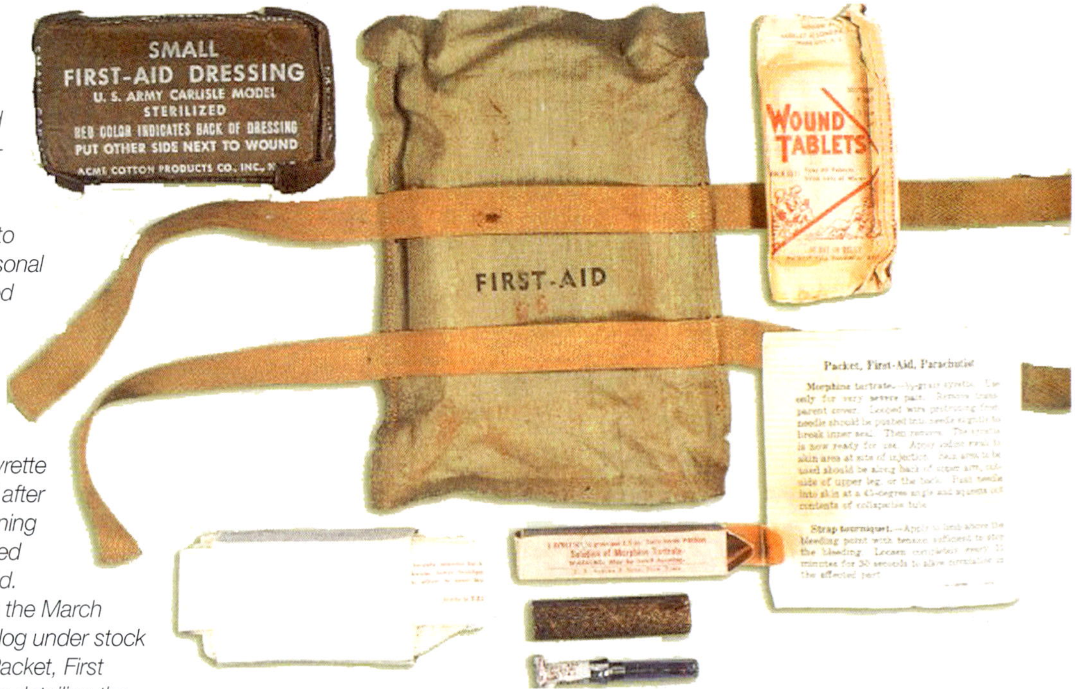

Amendment-D of 7 July 1944 added field brown Carlisle first aid packet with asphalt wrapper and prescribed inclusion of instruction sheet. Amendments of parachute first aid packet were created to remedy deficiency in personal medical provisions noticed in late 1943. Addition of medical instructions had become required by then for all medical kits. Note improved "crushproof" morphine syrette packaging, implemented after a flood of reports concerning crushed syrettes submitted to USMD and Wright Field. The 1636-D was listed in the March 1944 USMD supply catalog under stock number 9778500 titled Packet, First Aid, Parachute, Complete detailing the addition of sulfadiazine tablets, envelope of sulfanilamide (not pictured) and specifying the small field brown dressing. Yardley of London Inc. is only known contractor for this late type of parachute first aid packet. (Kriss Usherwood Collection)

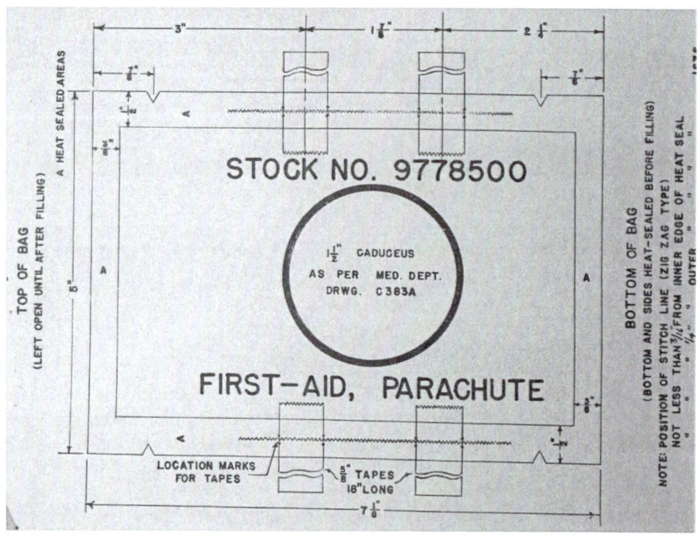

(Right) Final amendment, Amendment-E dated 12 February 1945, detailed new container to have stock number, nomenclature, and USMD caduceus printed on the face. This, however, was never approved since development was proceeding on new individual packet.

(Below) Display board at office of Liaison Committee on Emergency Rescue Equipment, c.1943. (National Archives)

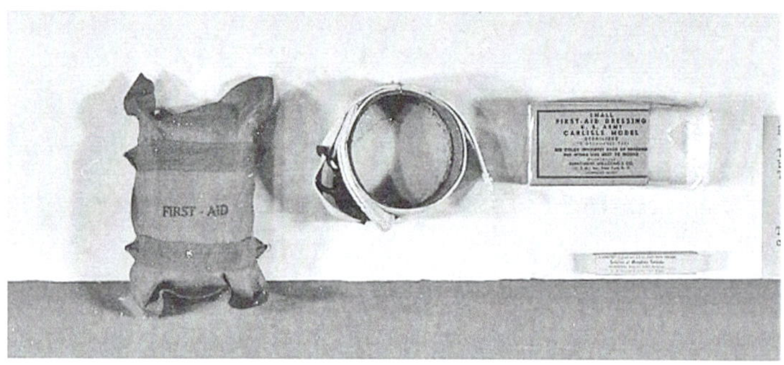

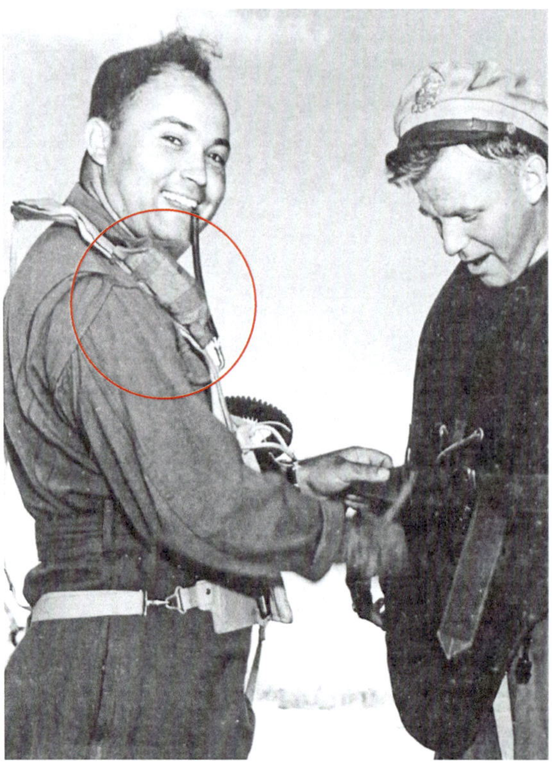

(Left) USAAF Crew members of the 864th BS adjust gear prior to take off for a bombing mission over Palau Islands, November 1944. Parachute packet had been secured to life preserver. This was not an uncommon practice. Materiel Command issued an unofficial directive to discontinue this method as it interferes with the proper functioning of the life preserver. (National Archives)

A frying pan insert was another specialty first aid item developed by Wright Field with a relatively short development history. Like the parachute first aid packet, it was designed for emergency parachute kits, specifically for the Type B-4. In April of 1942, the Materiel Command invited a group of experts in their respective fields of anthropology, geology, and medicine to a series of conferences held in Maryland and New York to evaluate emergency kits. Upon conclusion of these meetings, recommendations were submitted to the Materiel Command on suggested changes. The primary one was to combine the best aspects of both Types B-1 and B-2 emergency parachute kits, prompting the development of what would become the Type-B-4 emergency parachute kit. The three-piece frying pan from the B-1 Alaskan kit was incorporated into this new design. Further, a list of medical provisions determined to be necessary in all climate zones was outlined and, to maximize space, the first aid unit was designed to be stored in the frying pan. The Type B-4 emergency kit was intended to supersede Types B-1 and B-2. The frying pan insert was developed under USAAF Dwg. No. 43B10436 being released for procurement in November of 1942 and assigned USMD stock class number 97106.

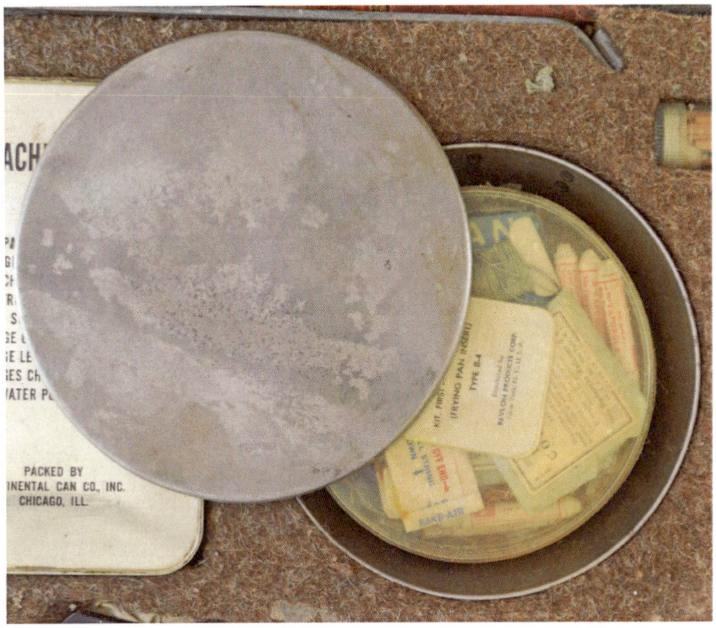

This first aid unit proved superior to medical provisions included in B-1 and B-2 emergency kits. Subsequent technical orders in 1944 included it as standard, making it no longer proprietary to Type B-4. In the development of the Type C-1, the frying pan insert was used in a revised container and many of the principal components adopted for inclusion in the vest carried over from this unit. It was also distributed as a personal provision carried in the flight suit pocket.

(Below) Concept and design prior to release for procurement. (USAF)

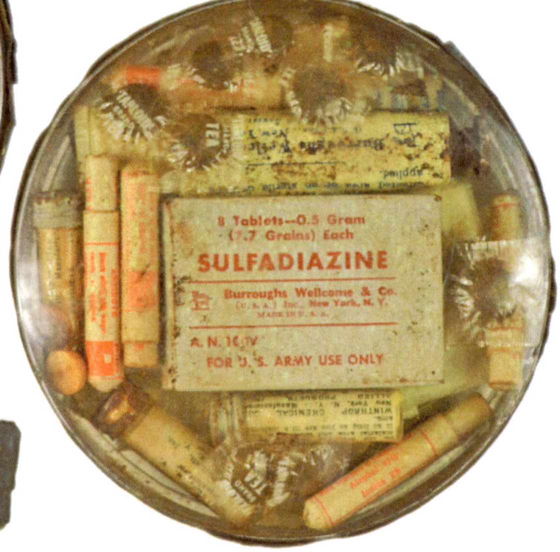

Official nomenclature was Kit, First Aid, Parachute (Frying Pan Insert) Type B-4. Container was fabricated from cellulose acetate, 5" diameter x 1". Two halves were sealed with friction tape and instruction pamphlet was applied to one of the halves. Three known contractors were Allied Products Inc., Revlon Products Corp. and Guild Foundations with 440,000 placed on procurement. Wartime image illustrating the principal components. Multiple subcontractors manufactured individual items resulting in packaging variations.

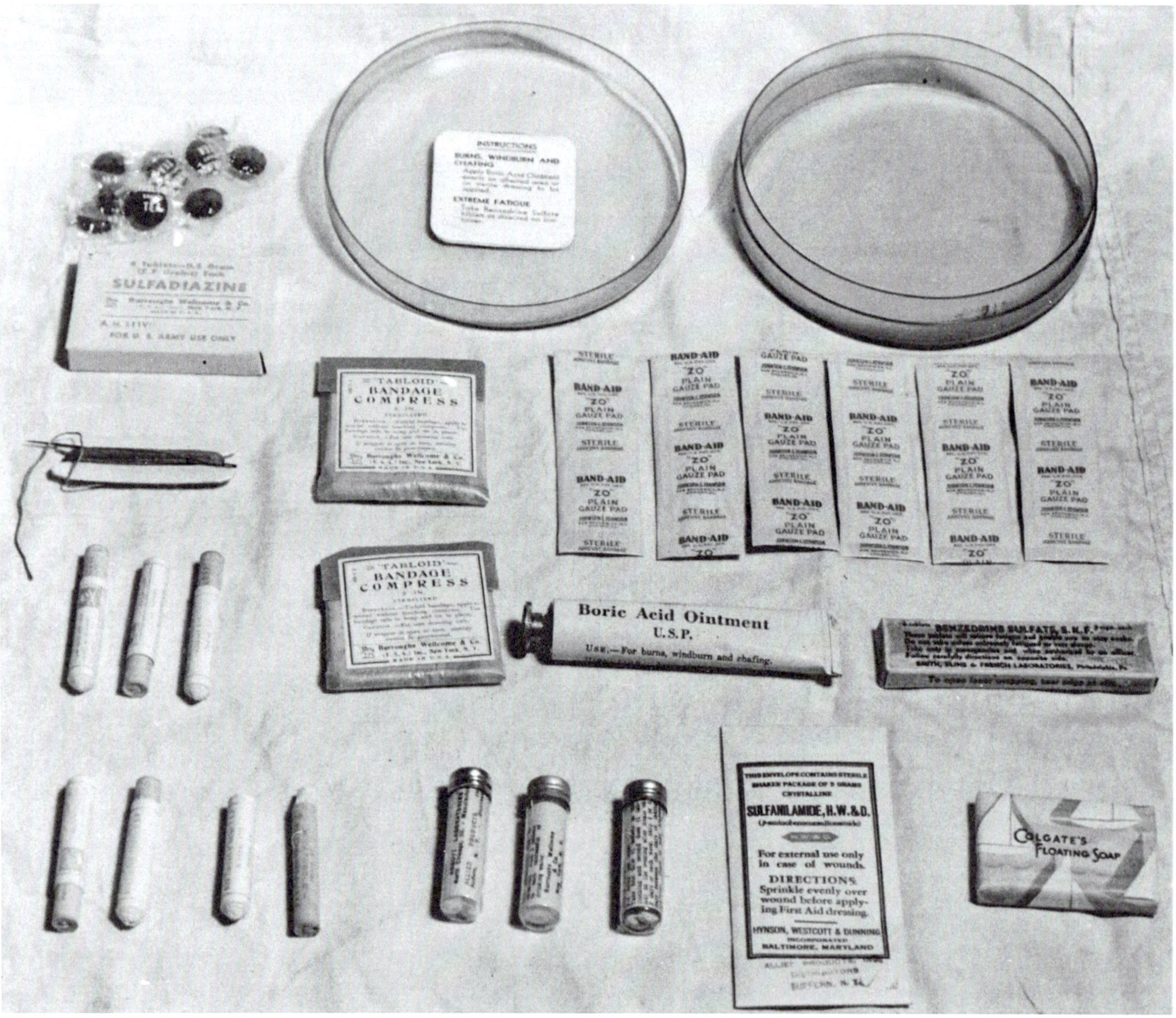

US Army Medical Department supply catalog listings for frying pan insert, June 1943 and March 1944 printings. Several changes were made between the two dates and are reflected in the March 1944 catalog, most significantly, removal of sulfanilamide and the gauze bandages listed as field brown. Examples with these changes have not been found, suggesting that, with the pending standardization of the Type C-1 vest, the insert was no longer needed and no further procurements were made.

JUNE 1943 ARMY MEDICAL SUPPLY CATALOG LISTING

97106 KIT, FIRST AID, PARACHUTE (FRYING PAN INSERT), TYPE B-4

STK. NO.	ITEM	UNIT	QTY.
1K08552	BENZEDRINE SULFATE , NNR, 5 MGM TAB	6	1
1K27507	HALAZONE, NNR, 1/16 GR TAB	EACH`	30
10845	ATABRINE TABLETS, 100 MGM	EACH	12
91038	BURN-INJURY SET, BORIC ACID OINTMENT	TUBE	1
91122	IODINE SWAB, 10 MGM	EACH	6
91202	SODIUM CHLORIDE, 10 GR TABLET	EACH	8
91204	SULFADIAZINE, USP, 0.5 GM (7.7 GR) TAB	PKG.	1
91211	SULFANILAMIDE CRYSTALLINE, USP, 5 GRAMS IN STERILE INDIVIDU-AL DOUBLE-WRAPPED ENVELOPE	EACH	1
92000	BANDAGE, GAUZE, ADHESIVE, 1 BY 3 INCHES	EACH	6
92004	BANDAGE, GAUZE, COMPRESS, 2 BY 2 INCHES	EACH	2
————	NEEDLE, CURVED, 2 1/2" WITH 120 INCHES OF OLIVE DRAB CARPET AND BUTTON THREAD: FOR CLOTHING REPAIR	EACH	1
————	SOAP, CAKE, HOTEL SIZE, SMALL	EACH	1
————	TEA TABLETS, COMPRESSED 8.7 GR	EACH	10

MARCH 1944 ARMY MEDICAL SUPPLY CATALOG LISTING

9710600 KIT, FIRST AID, PARACHUTE (FRYING PAN INSERT), TYPE B-4

STK. NO.	ITEM	UNIT	QTY.
1084500	ATABRINE, 12 TABLETS	TUBE	1
9102925	BENZEDRINE SULFATE, 6 TABLETS	TUBE	1
9103800	BURN-INJURY SET, BORIC ACID OINTMENT	TUBE	1
9112200	IODINE SWAB, 10-MINIM, 10	EACH	6
9120200	SODIUM CHLORIDE, 8 TABLETS	TUBE	1
9120400	SULFADIAZINE, 8 TABLETS	PKG.	1
9121810	TABLET, WATER PURIFICATIONJ, INDIVIDUAL 30 TABLETS	TUBE	1
9200100	BANDAGE, GAUZE, ADHESIVE, FIELD BROWN 1 BY 3 INCH	EACH	6
9200500	BANDAGE, GAUZE, COMPRESS, FIELD BROWN, 2 BY 2 INCH	EACH	2
————	NEEDLE, CURVED, 2 1/2 INCH WITH 120 INCHES OLIVE DRAB CAR-	EACH	1
————	SOAP, CAKE, HOTEL SIZE, SMALL	EACH`	1
————	TEA TABLETS, COMPRESSED, 8.7 GR	EACH	10

The First Aid Kit for Pneumatic Life Rafts was the sole innovation of the US Navy's Bureau of Medicine & Surgery specifically for use in aeronautical pneumatic life rafts. It was part of the progression to include a better complement of emergency equipment for users of the M-3 series.

Medical aid items were first introduced as a standard complement in the 10 April 1941 amendment of NavAer Specification M-3N; This amendment added the emergency equipment container, rations, and signaling devices. Prior to this date, first aid kits NAF 213886 and M-448 were stored aboard the aircraft. The thinking was to have basic medical provisions available to survivors afloat if the packages aboard the aircraft could not be retained. Specification M-3N Amendment- 1 reads as follows: "A first aid kit, containing two syrette units (morphine tube with attached needle), one burn compound unit, one absorbent gauze compress unit, and one 4" bandage compress unit. The burn compound should be 5% tannic acid in a water-soluble jelly and containing a mild antiseptic, such as tricresol or phenol as a bacteric static agent or preservative, put up in a 4 oz. collapsible tube, to be similar in specification to Tannefax or Amertan. The first aid equipment should be packed in a waterproof fabric container, provided with a suitable tear string or 'rip cord' for quick opening."

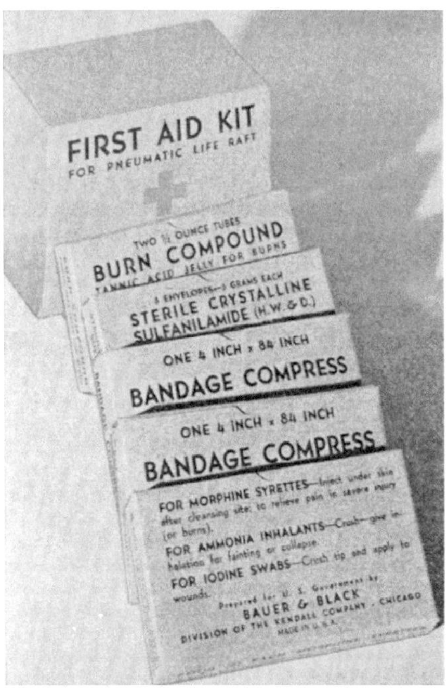

There were two known variations of the 5 unit first aid kit conforming to Specification 57-K-0366A. This kit was assembled by Bauer & Black, displayed on a Liaison Committee on Emergency Rescue Equipment board, c.1943. See page134. (National Archives)

This specification laid out basic guidelines for medical items to be provided by the contractor and requisitioned by the service stations for inclusion. NavAer Specification M-3 Amendment-O dated 11 June 1941 stated: "First aid kit. Furnished and installed by the Government. Kit shall contain: Five 1-1/2 c.c. (.032 gram) tubes morphine sulfate, with attached sterilized needle (one unit), one 4" compress bandage (one unit), one 4" x7" shell wound dressing (two units), and two 1½-ounce tubes of burn compound (two units). The kit will be packed in a waterproof fabric or pliofilm container, provided with a suitable tear string or rip-cord for quick opening, and will be approximately 4" x 2" x 3-1/4" in size." The statement "Furnished and installed by the Government," meant the kits were on separate contracts awarded by the government, and would either be shipped to the life raft contractor or installed after the complete raft assembly was delivered.

The unit in Amendment-O was formalized in January 1942 under BuM&S Specification 57-K-0366 (INT), "INT" signifying interim or temporary. Amendment-A to Specification 57-K-0366, June 1942, incorporated five units from BuM&S Specification 57-K-032B. The five units to be included were two 4" compress bandages, a unit of burn compound, a unit sulfanilamide, and a unit containing six morphine syrettes with an iodine applicator. With this amendment, the five boxes were to be wrapped in pliofilm (cellophane) and secured in a cardboard sleeve. The entire unit was then sealed in a cellophane wrapper for waterproofing. These new units became available by the end of 1942. Prior to their availability, stations assembled first aid supplies from requisitions of the 10-piece filler unit under ASO stock number R57-F-4005. The 57-K-0366A was assembled by commercial contractors, procured by the BuM&S, and shipped complete. The complete units could then be requisitioned from the BuM&S under stock number S2-1055 and requisitioned from the ASO under stock number R57-K-8525. They were still supplied as GFE for life raft contractors. Two further Amendments were made, -B & -C, but are not relevant to physical variations as they are clerical revisions to the preceding amendment being more specific in details for the packaging contractors.

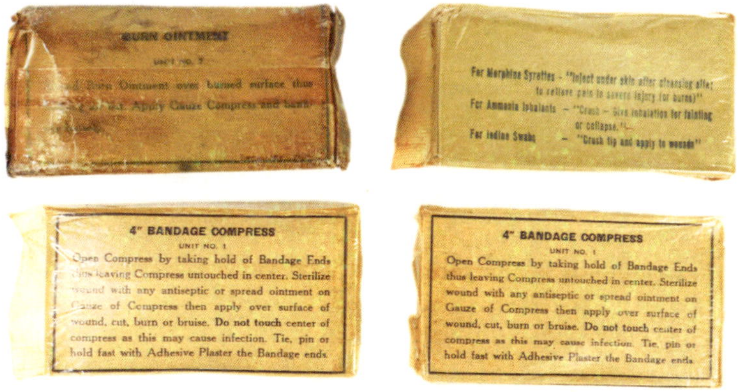

The second version of the 5 unit 57-K-0366A was assembled by A.E. Halperin Co. Sulfanilamide not included in this for unknown reason. Bauer & Black and A.E. Halperin Co. provided boxes of morphine, ammonia inhalants, and iodine. Specifications do not cite inclusion of ammonia inhalant. Each individual component and the complete unit were wrapped in cellophane. Sleeve for this unit measures 4-1/4" x 3" x 2-1/4". Burn ointment is described as a "unit-size" box at 1-3/8" thick, larger than the typical 5/8" thick.

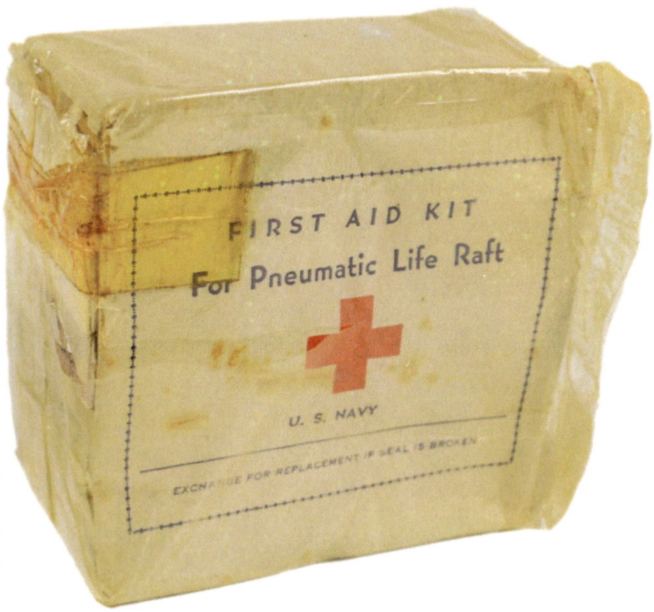

This 5 unit kit conforms to specification. Sleeve measures 4-1/4" x 3-7/8" x 2-1/4".

USAAF requisitioned S2-1055 first aid kits for inlcusion in their multi-person life rafts. One is included in this display for a Type A-3. (National Archives)

To expand medical provisions, the BuM&S determined it could add one more unit with minimal increase in the bulk. In May of 1943, Amendment-D to specification 57-K-0366 added sulfadiazine tablets, increasing the kit to six units. The six-unit kit superseded the five-unit kit only on a procurement basis; both units remained in service and were referenced under the same stock numbers, S2-1055 and R57-K-8525.

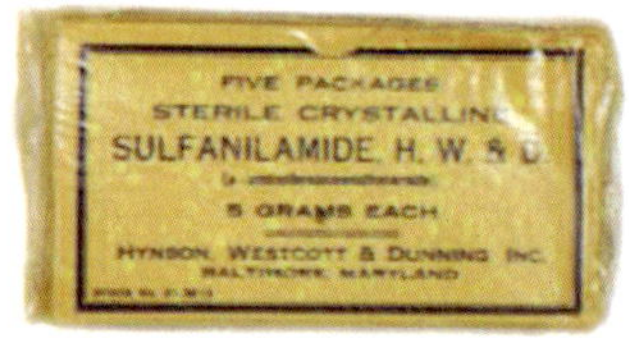

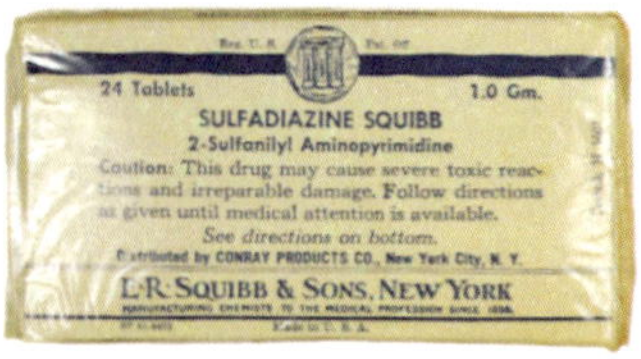

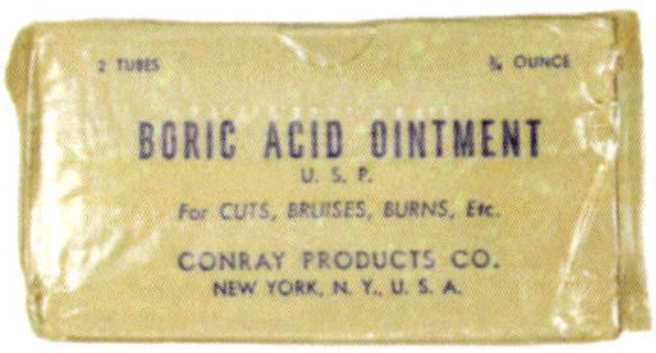

Conray Products Company was the only known contractor for the 6 unit 57-K-0336D first aid kit. Assembly methods matched its 5 unit counterpart. Outer sleeve was white with red printing, like most naval medical items. It measured 4-1/4" x 4-1/2" x 2-1/4".

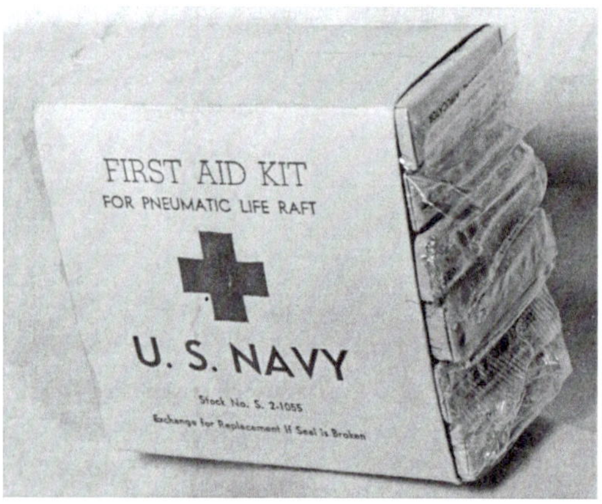

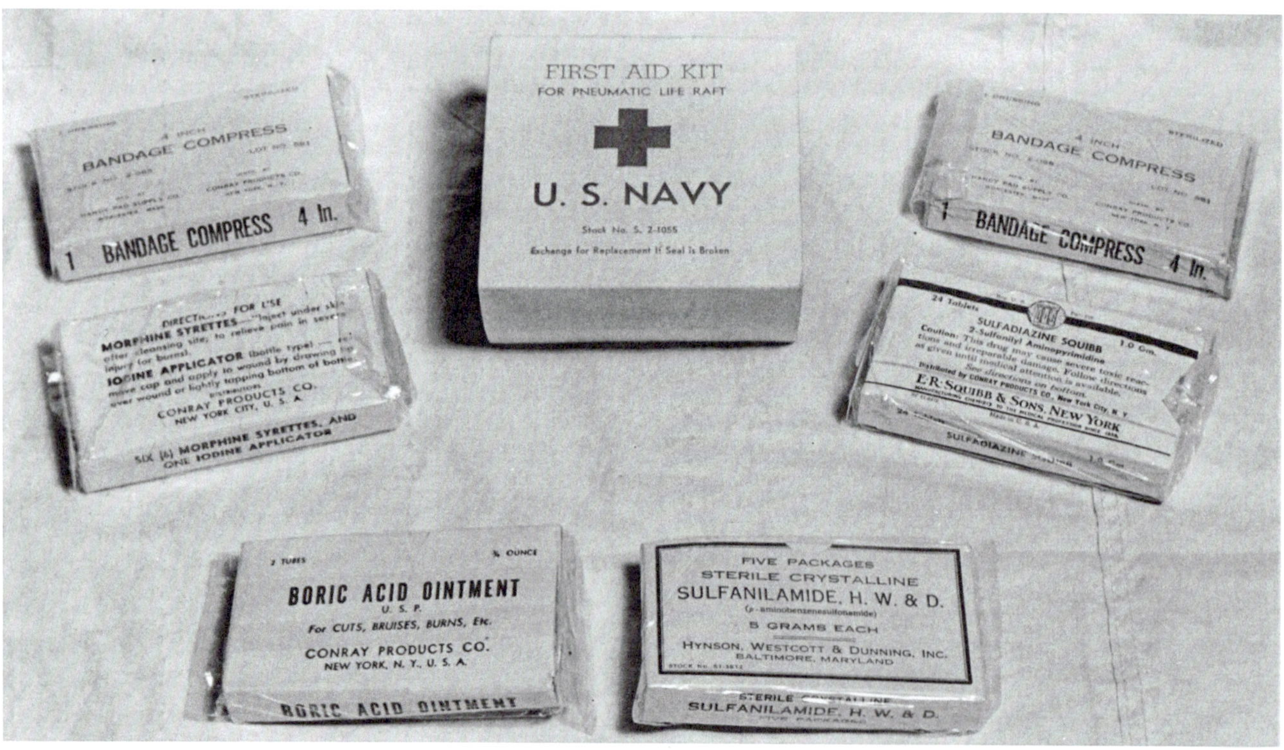

In 1942, BuAer life rafts were equipped with first aid provisions, whereas USAAF life rafts were not. As requirements for lifesaving equipment increased, first aid units were included as standard for USAAF life rafts, integrating both the USMD aeronautic first aid kit and the US Navy pneumatic type. The five-unit first aid kit became the first standard for BuAer multi-person and one-man life rafts by January 1943. The USMD aeronautic first aid kit was the first to be included as a standard accessory in USAAF multi-person life rafts due to its availability, integrating the US Navy type when stocks permitted. Upon adoption of the one-man pararaft, the USAAF included the parachute first aid packet as a standard, later utilizing the US Navy pneumatic type on a limited basis. USAAF units operating in Europe sometimes included a complement of RAF emergency equipment, as its equipment was often found to be superior to what was available from US Government sources.

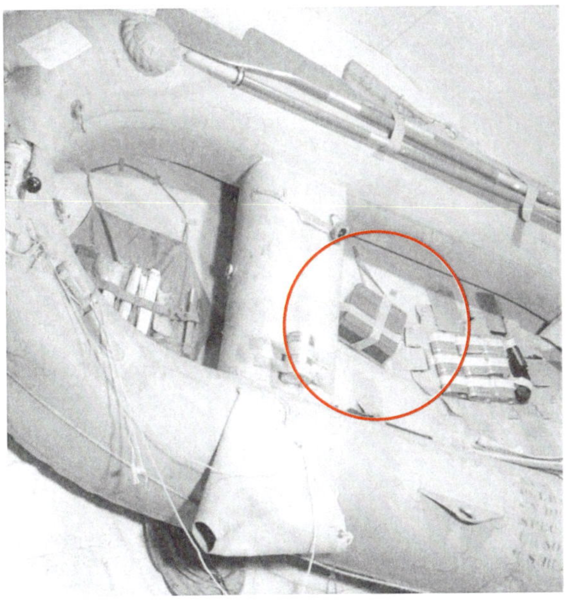

(Above) RAF Large Dinghy First Aid Outfit displayed with a USAAF Type A-2 life raft in the equipment room of the 490th BG, August 1944.

(Above) Large RAF First Aid Outfit measured 7" x 6" x 3-1/2".

(Below) RAF Type-3 First Aid Outfit measured 4-1/4" x 3-3/8" x 3 3/4".

(Below) RAF dinghy first aid outfit Type 3 displayed with USAAF equipment in England. The Type-3 was part of a set of accessories for a USAAF Type A-3 life raft of the 303rd BG, June 1944. The raft had been modified with heavy line attached to bladder. (National Archives)

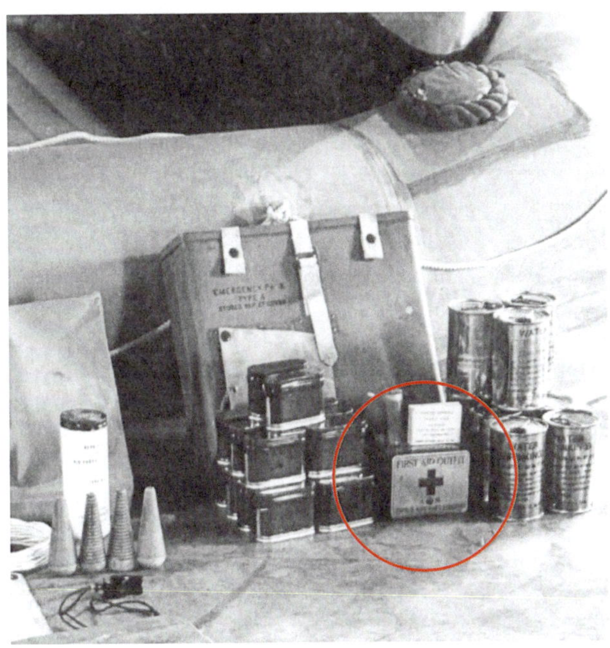

Both the BuAer and the USAAF expressed concern about the waterproofing and durability of first aid kits provided with life rafts. The USMD aeronautic first aid kits would become water-logged and the US Navy pneumatic types were found to be fragile due to the cellophane packaging. Overall, there was dissatisfaction with the lack of a suitable first aid kit for life rafts. In late 1942, both aeronautical services were revising their life raft equipment, and through this process they evaluated each item. Foreseeing an issue with the current status of first aid kits for use in life rafts, the Aero Medical Laboratory began preparing a proposed joint Army-Navy specification for a first aid kit in February 1943. The initial proposals for a kit were based on the Trans World Airlines model used in its standard life raft, which the USAAF was testing by late 1942. The Aero Medical Laboratory considered many recommendations and evaluated proposed units but ultimately decided to repackage the US Navy 57-K-0366 with a durable tear-resistant waterproof container. The Mine Safety Appliances Company provided the two best options for a durable waterproofing package. In April of 1943, Mine Safety Appliances submitted two examples for submersion tests.

Two examples submitted by MSA designated Unit-A and Unit-B. Unit-A consisted of a foil-lined waterproof paper heat-sealed around medical supplies. The package was housed in an outer two-section cardboard box with taped seam. Unit-B was similarly sealed but housed in a single-piece carton then wrapped in waterproof cloth. Both units passed submersion tests.

During the submersion tests, a technical subcommittee on standardization was evaluating aircraft medical kits. This Joint Army-Navy-British Aircraft Sub-Committee met multiple times in the summer of 1943. In 6 August meeting, it was decided to adopt the proposed Army-Navy specification from the AML with the principal design by the Mine Safety Appliances Co. Unit-B. The decision was to amend the existing BuM&S specification 57-K-0366D, merging it, and redesignating it under a joint Army-Navy specification. It was then decided that the BuM&S would have responsibility for the revision until a suitable federal specification could be drafted. The BuM&S released Specification 57-K-1 Type-A (INT) 15 October 1943 to include the new packaging method, which was assigned a new stock number S2-1061.

New first aid kit for pneumatic life rafts per Specification 57-K-1 Type-A with 3-step waterproof packaging including original sleeve, with the 6 medical units retaining laminated cellophane wrapping, inserted into the chipboard sleeve then sealed in foil-lined waterproof asphalt paper. Assembly was secured in a wax-coated chipboard carton and glued closed. Packaging allowed resealing between use, protecting contents from climate and storage deterioration. Following medical department initiatives to camouflage first aid equipment, cartons were a neutral color and marked CAMOUGFLAGED. Note in image (left), both white and neutral brown Sulfanilamide boxes for purpose of comparison. Property mark was also changed from US NAVY to US GOVT because the new kit was a joint Army-Navy item. Units in both images were assembled by Conray Products Co. and measured 4-5/8" x 4-1/2" x 2-1/2".

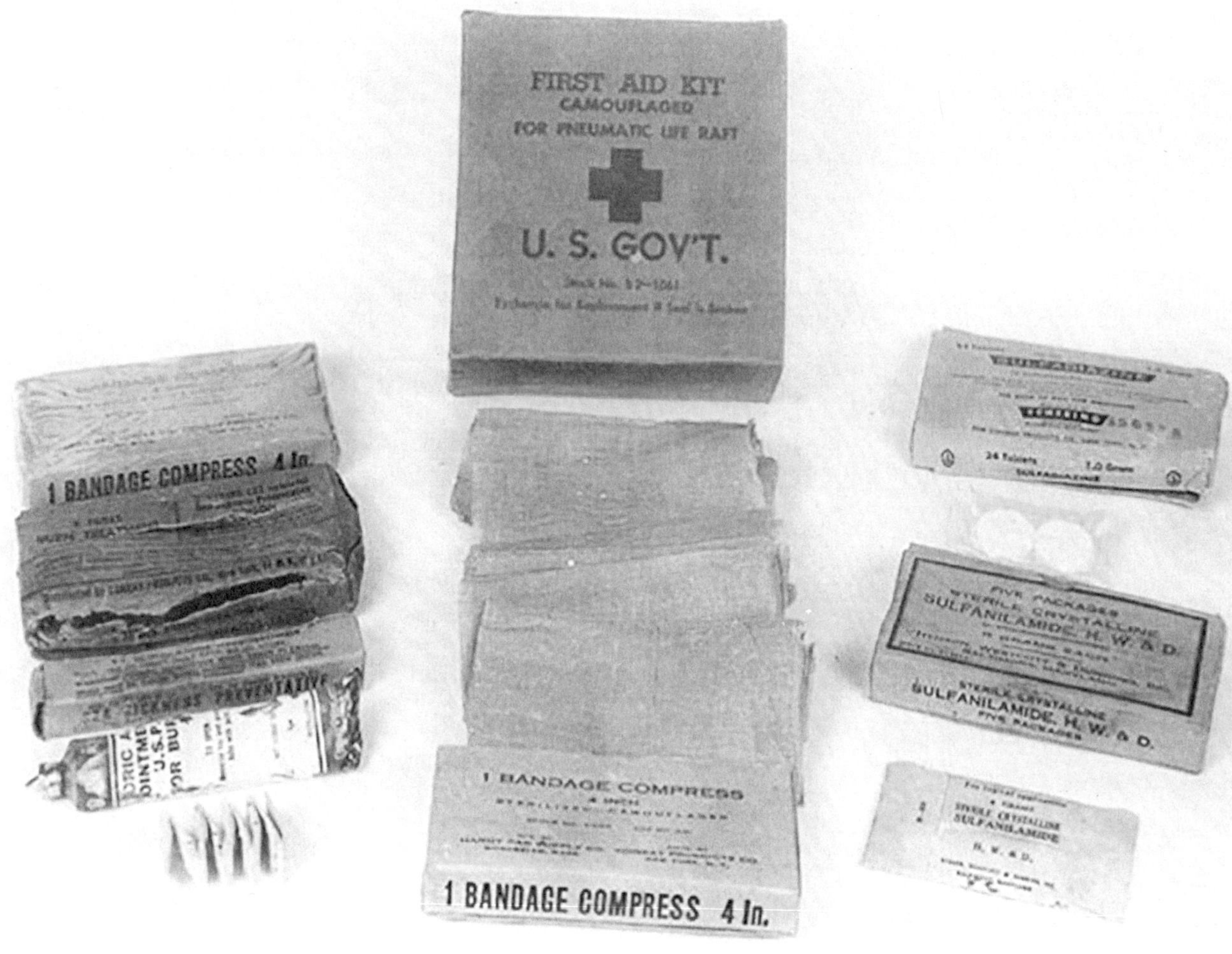

The Conray Products Co. and the Mine Safety Appliances Co. were the only two known contractors for the First Aid Kit for Pneumatic Life Rafts, Camouflaged S2-1061. There were two variations, which can be viewed as the first and second production types of each contractor. One feature discussed by the sub-committee during the meetings in August of 1943 was the inclusion of a seasickness preventive. Scopolamine was approved for standardization, to be packaged with the burn treatment. Procurements of the 57-K-1 were approved in September of 1943, expediting manufacture prior to the availability of the seasickness preventive. The S2-1061 would enter service in the late spring of 1944 and was standardized on 30 September 1944 under Joint Army-Navy Specification JAN-K-87. These were available under BuM&S stock number S2-1061, ASO stock number R57-K-8525, and USMD stock number 9776900. Both the ASO and the USMD did not amend their stock classes to differentiate the new unit for requisitioning. The older types would be used until supplies were exhausted.

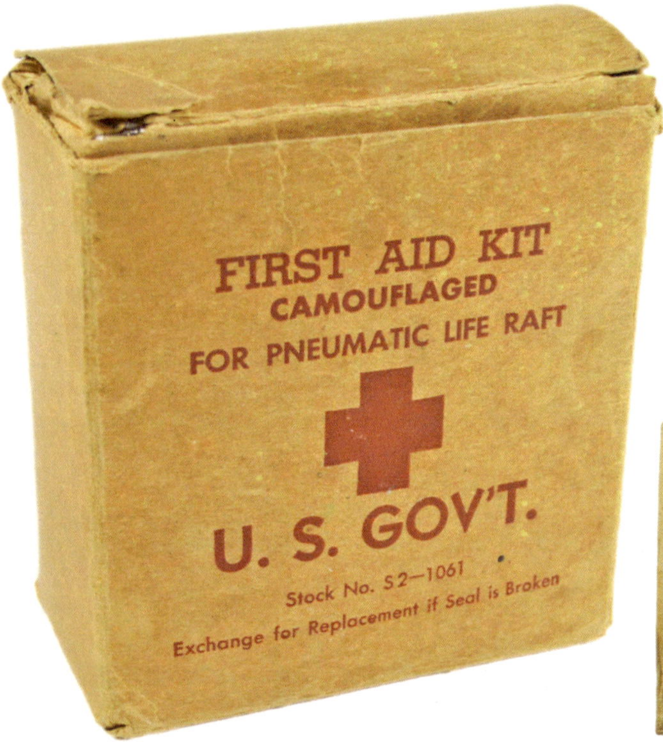

Conray Products Co. S2-1061: packaging and markings remained consistent throughout production. Shown are master carton, sealed insert (opemed to show contents) and back of master carton. Same markings were applied to front and back of insert sleeve.

(Below) Six Components of first expedited production run of the S2-1061 by Conray Products Co.

(Below) Updated six components by Conray Products for the S2-1061 with the inclusion of seasickness preventive in combination with the burn treatment.

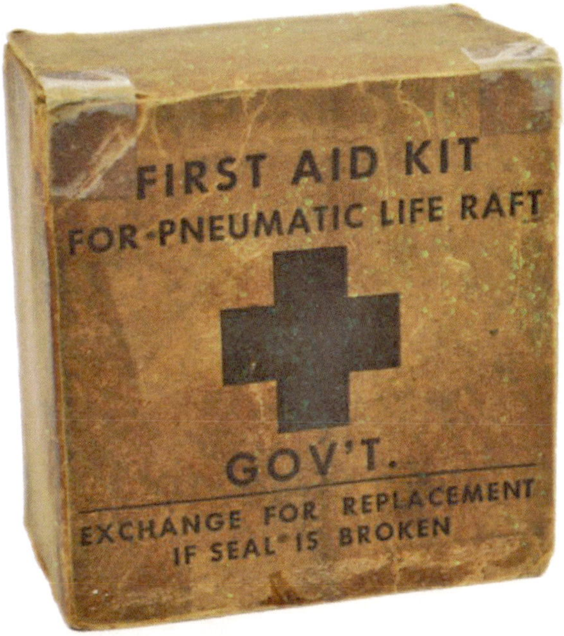

First production of Mine Safety Appliances Co.'s S2-1061 omitted inner sleeve, using only sealed waterproof wrapper. Unit measured 4-5/8" x 4-5/8" x 2-1/2". L-R: Master carton, insert, and back of master carton.

Components included in first production of Mine Safety Appliances Co.'s S2-1061.

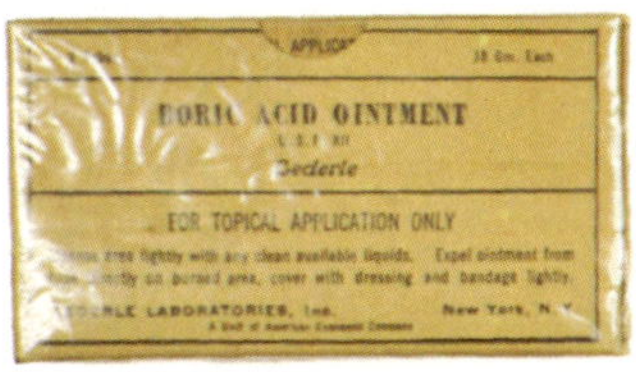

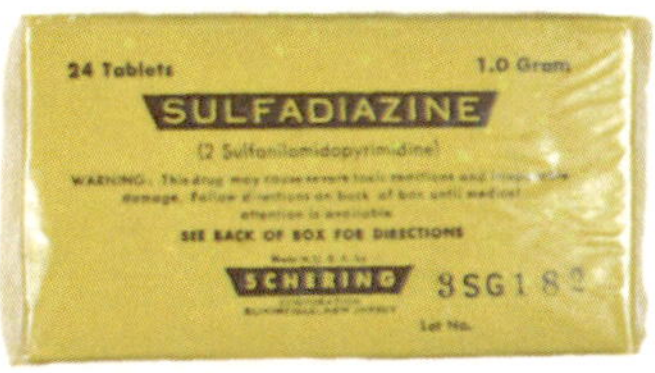

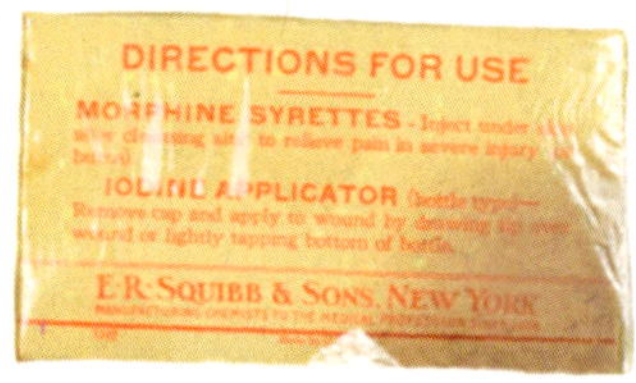

Second production of Mine Safety Appliances Co.'s S2-1061, updated with combination seasickness preventive-burn treatment package. Master carton included the word CAMOUFLAGED and the phrase WITH SCOPOLAMINE. The master carton also included BuM&S stock number S2-1061.

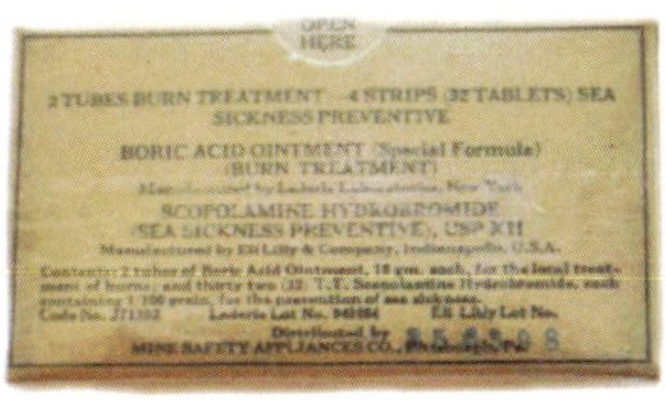

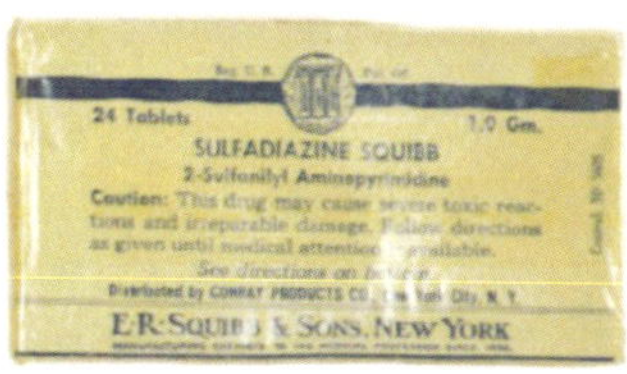

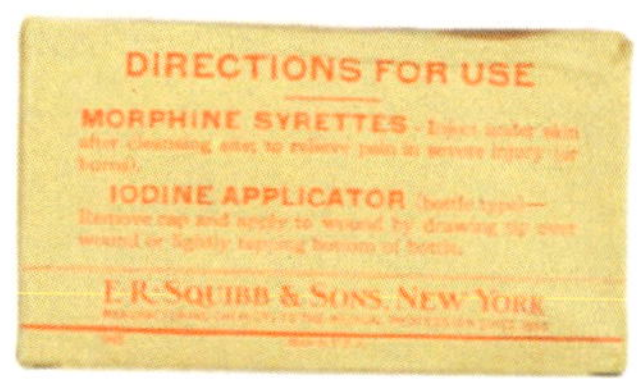

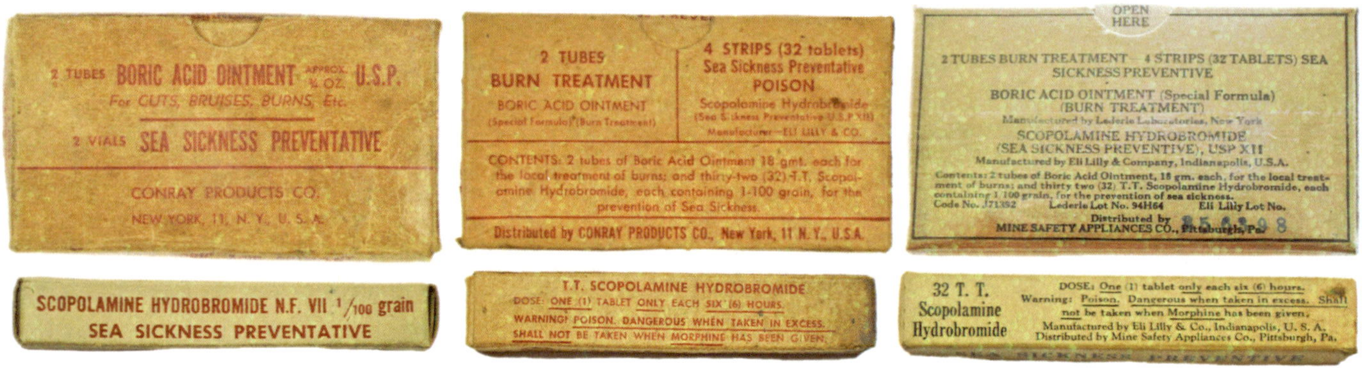

By early 1944, burn ointment renamed treatment because the former term ointment was thought less applicable to saltwater burns, chaffing burns, etc. Three variations of burn treatment and sea sickness preventive units are illustrated here. (left) From 1st production of the Conray Products Co. containing 2 vials of seasickness tablets. (middle) Conray Products Co.'s amended 2nd production unit containing seasickness tablets in 4 laminated asphalt strips with the revised term treatment on the carton. (right) Mine Safety Appliances Co. version with 4 laminated strips. All three contained two tubes of boric acid ointment.

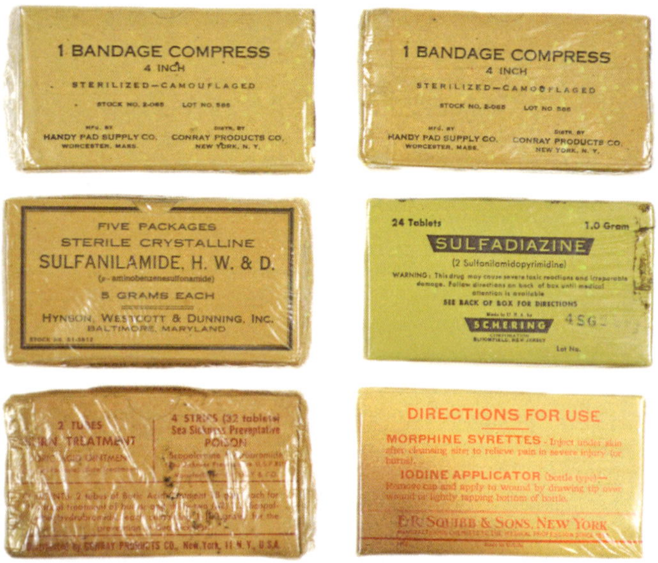

(Above) Second production S2-1061 components including the revised burn treatment-seasickness box. Components were not proprietary to any specific contractor; they used variations.

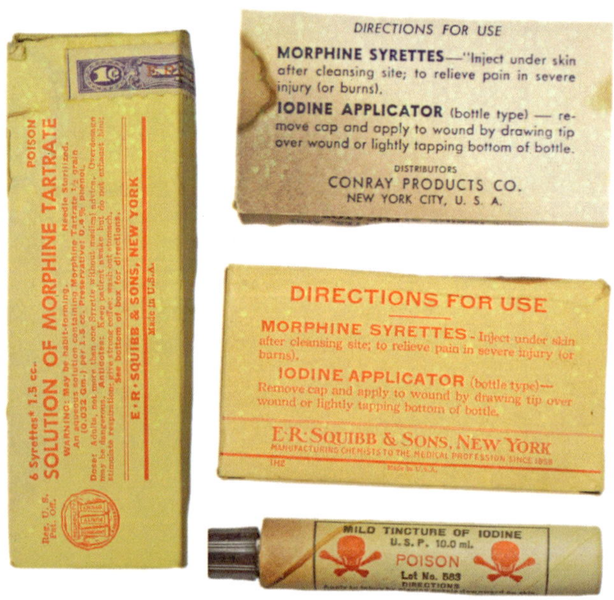

(Above) Morphine syrette and iodine applicator boxes. Both units contained six syrettes and one iodine applicator. Contents were the same for both packaging contractors

Other variations of the bandages, sulfadiazine and sulfanilamide used in S2-1061 for both contractors.

First Aid Kit for Life Rafts of Vessels is a relative of the First Aid Kit for Pneumatic Life Rafts. The difference was the bureau for which they were intended: First Aid Kit for Pneumatic Life Rafts was for the Bureau of Aeronautics and the First Aid Kit for Life Rafts of Vessels was for the Bureau of Ships. The latter was used on Navy life rafts, lifeboats and small watercraft such as amphibious transports. Life Rafts in its title was misleading and does not refer to aeronautical pneumatic types. Emergency life rafts aboard vessels were typically nonpneumatic.

This first aid kit exemplifies two services doing something their own ways. It was referenced under the same BuM&S Specification 57-K-0366 as the pneumatic type but was listed under stock number S2-1056. Though the components were the same, the master carton indicates the differences. Specification 57-K-1 (INT) affected this unit; stock number S2-1062 was assigned. Contract numbers were printed on the unit; the two known contracts were N140-14156, awarded January 1943, and N140-23115, awarded June 1943, both to Conray Products Co. The second contract straddles the issuance of Specification 57-K-1 (INT) providing variation of components like that of the S2-1061. Units under contract 23115 had either stock number S2-1056 or S2-1062.

The S2-1056 included the same components as S2-1055 but omitted the cellophane wrappers. The durable fiber outer container had a telescoping top and sheet metal end caps measuring 4-1/2" x 4-1/2" x 2-1/2". See p.50, VOL-1.

(Left and below) BuShips life raft or more correctly titled life float. Elliptical and rectangular types were available in various sizes from 7' to 13' long. The rectangular type was the new standard at the beginning of WWII. The ring consists of strips of balsawood in varying lengths and assembled in layers joined by waterproof glue. Outside was wrapped with waterproofed coated canvas. Bottom platform sinks in the water supported by a netting.

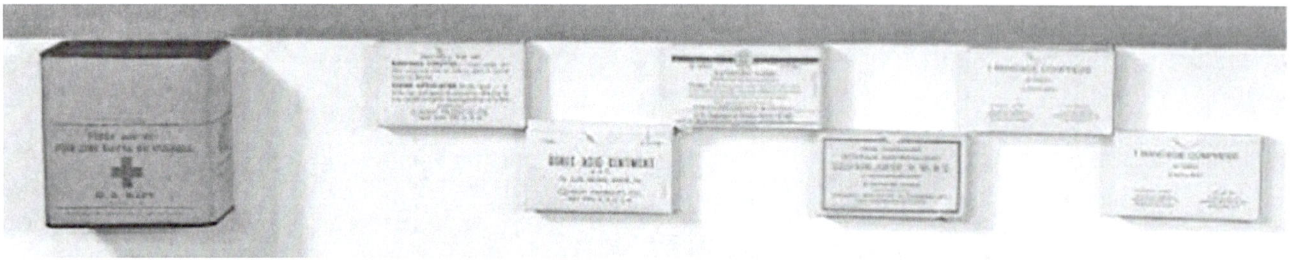

S2-1056 on display board of the Liaison Committee on Emergency Rescue Equipment, c.1943. (National Archives)

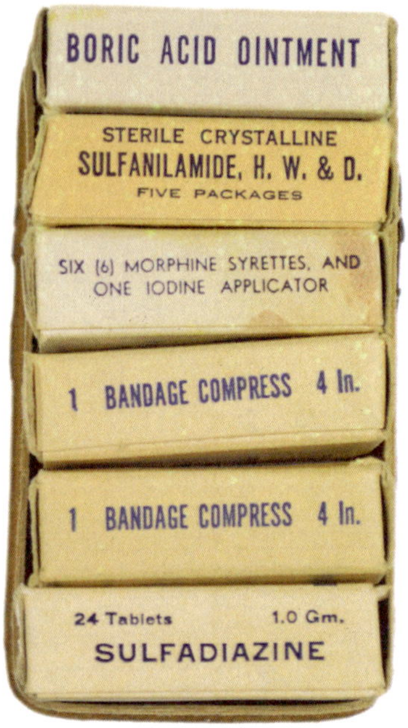

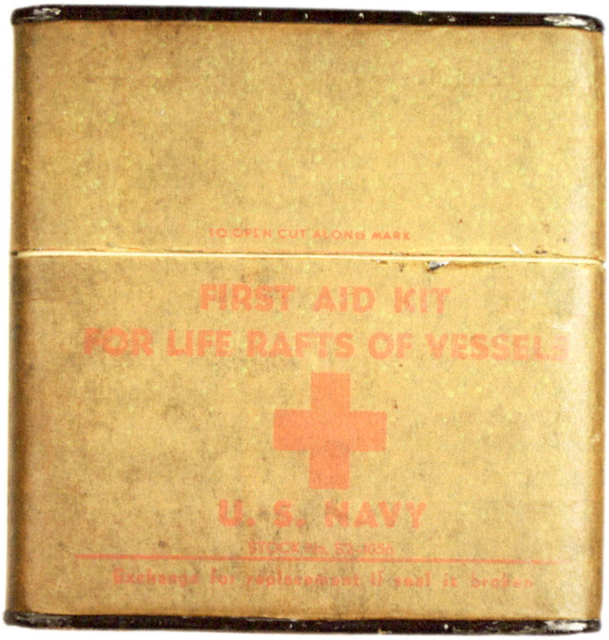

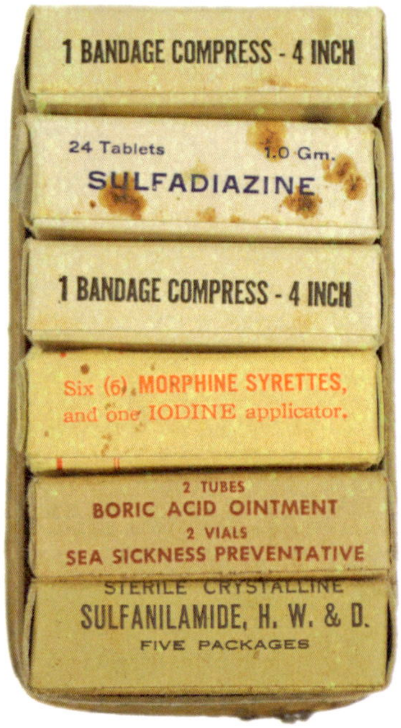

Master container remained unchanged for the S2-1056 and S2-1062. (left) Components included in S2-1056. (right) Examples of components of S2-1062. Outer fiber container was wrapped and sealed with an asphalt-coated paper. A red line around the circumference indicated opening cut location.

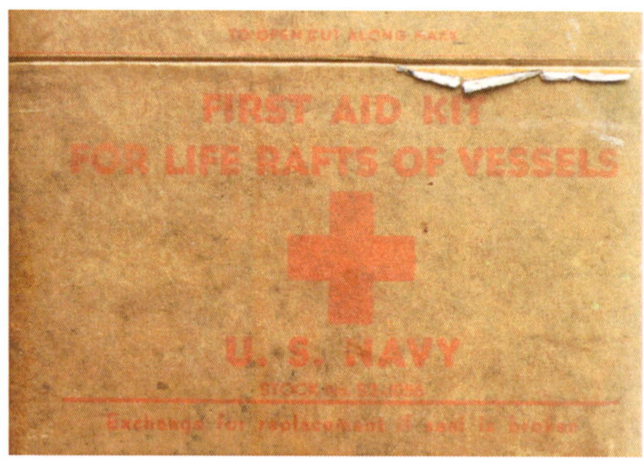

(Below) Stampings provided by the contractor. Nomenclature and stock numbers were provided on the face with the contract number on the side of the container.

The medical kits discussed thus far were units initially developed in a laboratory setting, often reflective of a perfect world scenario. Faults and deficiencies were realized once they were used in real world situations, which was to be expected. Revisions were then applied to improve the kits in an attempt to cater to the needs of the respective services but even then, they were not entirely satisfactory. The next generation of first aid kits was developed almost entirely from service recommendations and streamlined.

Beginning in the spring of 1943, the AAF Materiel Command and the Aero Medical Laboratory took steps to simplify emergency equipment for fliers. On paper, three medical kits were standard for use in fighter aircraft: aeronautic first aid kit 97765, parachute first aid packet 97785 and the frying pan insert 97016. This was thought to be more than adequate to provide basic medical provisions to flight personnel. As per T/O&E, all pilots would have this equipment at their disposal, with the aeronautic kit in the cockpit, frying pan insert in the emergency parachute kit, and the parachute packet secured externally to the parachute harness. However, the reality was that emergency parachute kits were not universally worn and the aeronautic kit was not retained in the event of bailout or ditching. This left only the parachute packet to provide medical provisions, and it too was not always worn. It was concluded that the existing medical supplies were found to be wholly inadequate.

In November of 1943, HQ 8th Air Force submitted a proposal to remedy the deficiency of personal medical provisions of a design by the 67th Reconnaissance Group titled Individual First Aid Kit – Aviators. This unit was similar to the British First Aid Outfit for Air Crews MK-II in its packaging and size at 4"x 5"x 1", but greatly differed in components. This unit could be carried in or sewn on flight clothing, thereby always being with the user during flights, ditching, and bailouts and meeting service requirements. Components included a Carlisle dressing, tourniquet, morphine syrettes, tube of sulfadiazine burn ointment, envelope of sulfanilamide, sulfadiazine tablets, energy tablets, band-aids, halazone and safety pins. Immediate action was taken by the Materiel Command to procure 2,000 kits from the Mine Safety Appliances Co. for service testing, with one change. The British MK-II packet was opened diagonally, destroying the packet for further use. The US service test version was changed to open across each end, much like the parachute packet, enabling suitable storage for the components after opening. These kits were being delivered to the 8th Air Force by the end of February 1944.

By the end of June 1944, the Materiel Command received adequate reports of service tests conducted by the 8th Air Force and the Central Medical Establishment rated the unit as satisfactory. In the reports, the components were found to be adequate but stated the packaging made the removal of the individual components difficult. The primary recommendations were to redesign a case that could be opened with one hand or the teeth by a wounded individual, the case should be reusable, and an individual first aid kit of this design be issued to all flying personnel. In July of 1944, the Materiel Command officially requested that AML develop and standardize an individual first aid kit for air crews. At this point, removal of aeronautic first aid kit 97765 from fighter aircraft was officially considered. It had been proposed as early as October of 1943 but was declined on the opinion that it may be of value in flight or crash landings.

The AML took initial steps to review a design of an individual first aid kit already in development by the Naval Medical Research Institute. The features they were evaluating were the packaging method (utilizing a cellulose plastic container for the components) and a cloth carrying case. Influenced by the Naval design, the AML drafted drawings in July of 1944, assigning Dwg. No. 45B11136 for the plastic container and Dwg. No. 45B11137 for the cloth carrying case titled Packet, First Aid, Individual, Aviator. The prerequisites for the proposed medical kit were that the kit size and weight be minimal, the kit be refillable, shock- and abrasion-proof, and carried in a pocket, which emulated the guidelines of the Naval version. By description of the AML model, it copies the Naval design with slight deviations. Dwg. 45B11136 describes the two plastic shells slipping together then sealed with tape. It measured 4" x 4-7/8" x 1". Also, like the Naval model, an instruction sheet was to be placed in one of the halves so as to be seen from the outside. Dwg. 45B11137 describes a nylon carrying case with a pocket for the plastic container with a 20" securing flap to be wrapped around the unit twice then fastened by two glove snaps. The nylon carrying case was thought to be universal, either carried in the pocket, on a belt, or on a parachute harness. The components were the same as the service test model delivered to the 8th Air Force, but the tourniquet was omitted.

As part of the research process, the AML also investigated metal containers, including stainless steel, manganese, and aluminum. Metal containers were quickly found to be better suited for the task. With the decision to switch to a metal container, the nylon case was eliminated. The size of the container was increased and a minor change was made to the components. The sulfadiazine burn ointment was replaced with a tube of petrolatum and the band-aids were removed and replaced by five yards of adhesive tape. This revision was referenced under Assembly Dwg. No. 45B13082, 14 August 1944.

With its development, and it being a suitable replacement for the aeronautic first aid kit, the AML revised Technical Order 00-30-139 directing the deletion of the aeronautic first aid kit in fighter aircraft effective 13 November 1944. This directive placed pressure to standardize an individual first aid kit for pursuit airplanes. Policy was established that at least a limited procurement of such a kit should be made to supplement the requirements for fighter groups. Further procurements would make them available to aircrew on bombardment aircraft. It was the intention to issue an individual first aid kit at a ratio of one per fighter pilot and one per aircrew member, replacing parachute packet 97785 and making the aeronautic kit 97765 a secondary means for medical treatment and possibly eliminating the need for it altogether. This new unit was being considered for joint Army-Navy standardization as well as the US Navy model, by which time it was already in limited production.

In January of 1945, the decision was made to tentatively standardize the individual packet detailed under USAAF Specification No. 94-40944, Dwg. No. 45B13082. It was given the title Packet, Aircrew, Individual, First Aid, Type A-1 and assigned stock number 8300-623715. An initial 1,000 packets were fabricated for service testing and assembled by the H.L. Burnett Co.; some were assembled by Erle P. Halliburton Inc.

In February, the Materiel Command decided to standardize the packet and begin steps to procure the unit to temporarily satisfy requirements for fighter groups. The next month these limited service test quantities were distributed to the European combat theater, with the bulk being delivered to the 1st, 2nd, 3rd and 4th Air Forces operating in the United States. On 26 March 1945, Type A-1 first aid kit was standardized following issuance of authority to procure 15,000 packets under the assigned title provided in Dwg. 45B13082: Packet, First Aid, Individual, Aircrew Member. A split contract was awarded to the H.L. Burnett Co. and E.P. Halliburton Inc. This contract (number unknown) was expedited prior to the acceptance of the Surgeon General, who had not yet assigned an official USMD stock class number. The limited quantity production of 15,000 packets conformed to the original drawing and specification in that it included sulfanilamide powder and two 1" x 2-1/2 yard strips of adhesive plaster wound on a sheet of clear acetate. Since the USMD stock class number had not yet been assigned, the aluminum packets were stamped with Materiel Command stock class number 8300-623715. Deliveries from this order commenced in May and were completed in June of 1945.

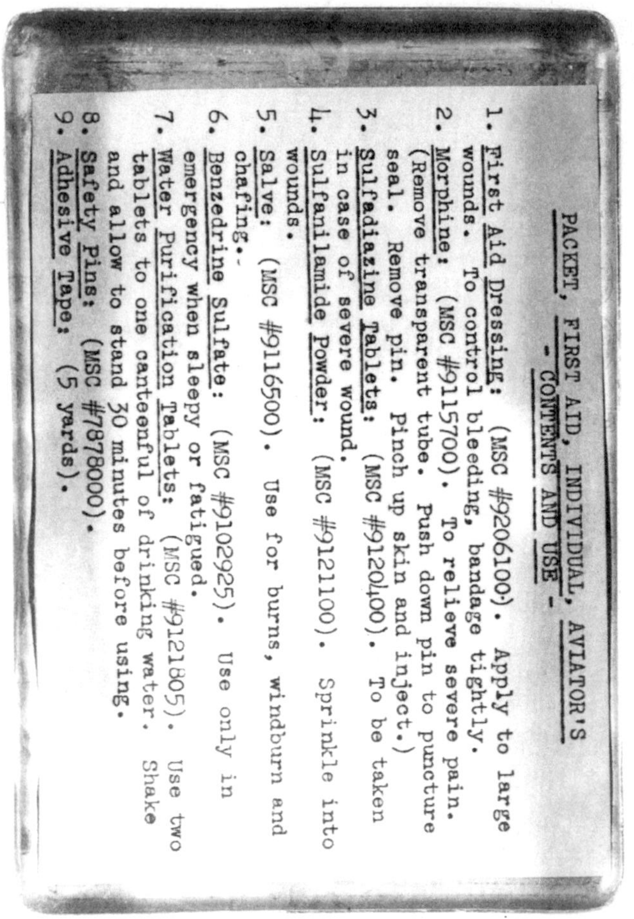

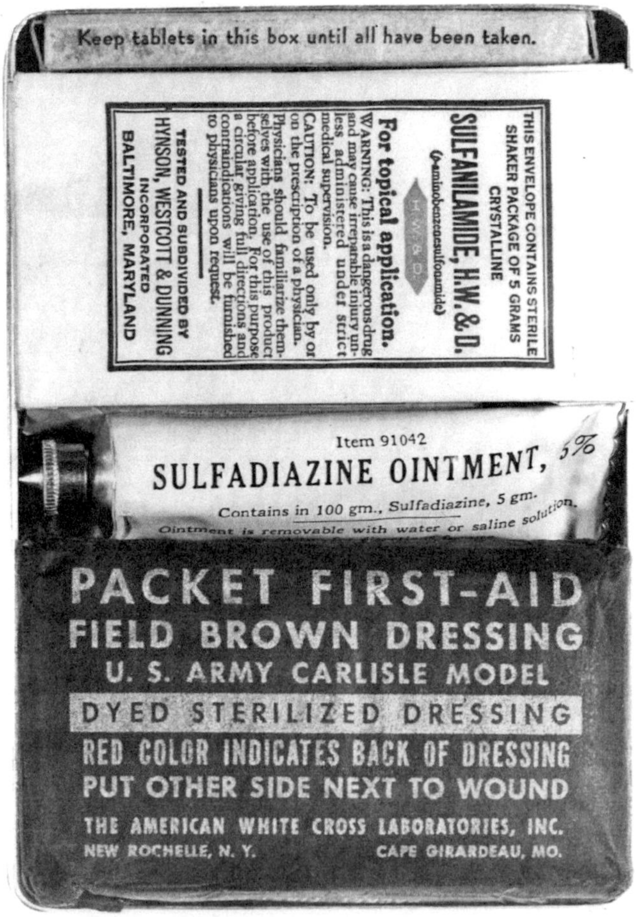

Concept model of transition to aluminum container; nomenclature of Aviator's was retained. (USAF)

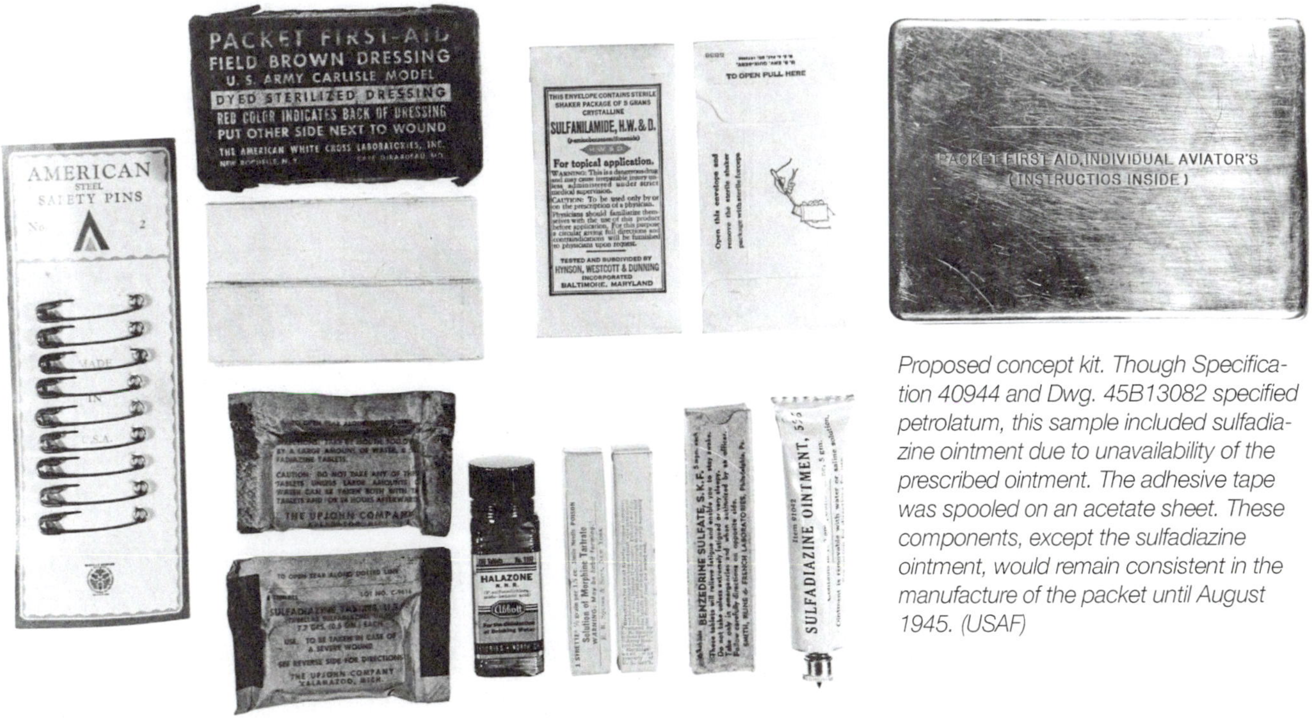

Proposed concept kit. Though Specification 40944 and Dwg. 45B13082 specified petrolatum, this sample included sulfadiazine ointment due to unavailability of the prescribed ointment. The adhesive tape was spooled on an acetate sheet. These components, except the sulfadiazine ointment, would remain consistent in the manufacture of the packet until August 1945. (USAF)

The 1,000 service test samples procured in January 1945 would include amended nomenclature of Aircrew Member.

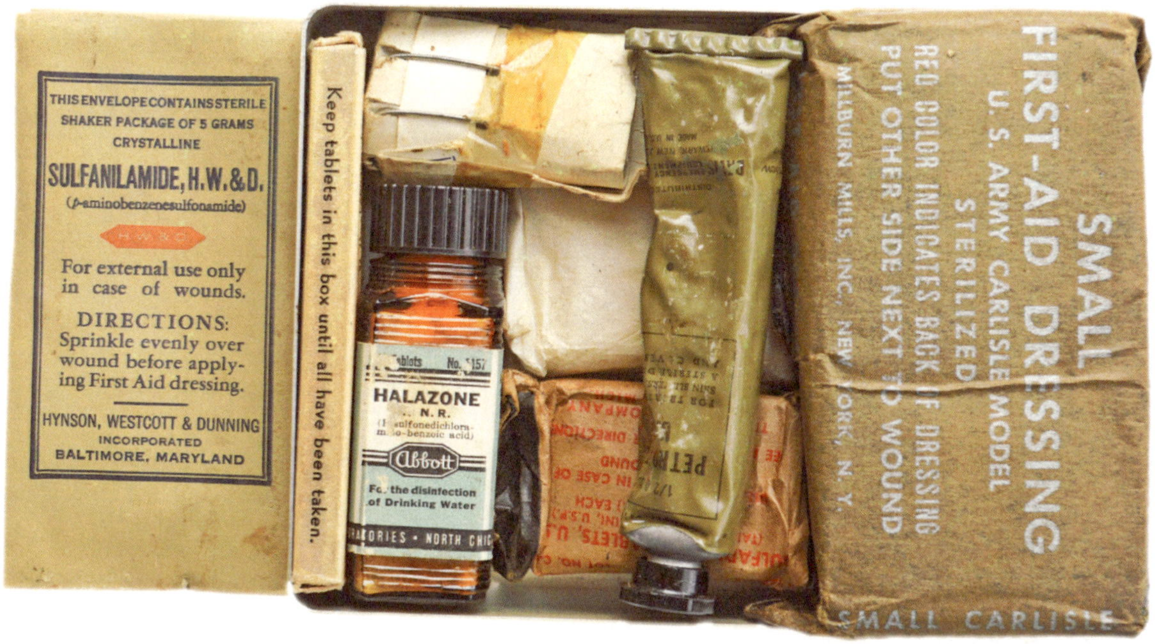

On 6 April 1945, the Surgeon General had approved the unit for standardization. Procurement Specification No. 1753 was released on 4 June 1945 under stock class number 9714000. The Surgeon General recommended the addition of either motion sickness preventive or scopolamine hydrobramide (seasickness preventive) in place of the sulfanilamide but this was not agreed to by the AML for unknown reasons. By this juncture, the packet had been approved for quantity production and suited the needs of the USAAF with emphasis for overland operations. Its counterpart, developed by the US Navy, was oriented toward over water operations and already in quantity production. On 7 July 1945 two contracts were awarded for the aircrew packet to the Mine Safety Appliance Co. under contract 30-084-MD-10673 and the W&K MFG. Co. under contract 30-084-MD-10674, for an unlisted quantity.

Since the latter part of 1943, the development of first aid kits had been monitored by a joint Army/Navy/British Sub-Committee for the standardization of aircraft first aid kits. In this case, the committee had been interested in the development of the USAAF and US Navy units for joint Army-Navy standardization.

The debate continued through 1945 with differing opinions within the Army and Naval aeronautical services. The USAAF wanted one kit for overland and another for over water but ultimately lost the bid for standardization of its packet. This was in part due to not adding the seasickness tablets as recommended by the US Navy and Surgeon General. In September of 1945, the joint subcommittee ruled in favor of a joint Army-Navy standardization of the US Navy's Kit, First Aid, Camouflaged (Aviation Individual), resulting in the cancelation of contracts awarded in July for the USAAF aircrew packet.

The quantity manufactured from July until official joint Army-Navy standardization in October is unknown. However, the contracts were projected to be completed in November, suggesting the bulk of the contract had already been delivered. A month prior, the USMD issued a revised specification under Amendment-A (1753A), 7 August 1945, deleting the sulfanilamide as directed in Section II, War Department Circular 77 dated 10 March 1945, which directed the discontinuation of the self-administering of sulfanilamide and its removal from individual first aid kits.

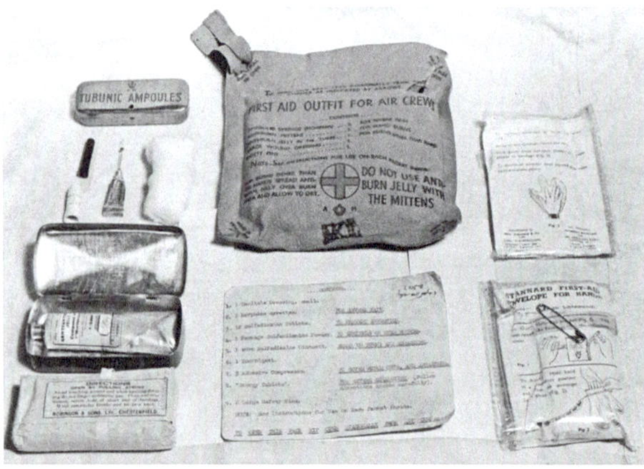

(Left) RAF First Aid Outfit MK-II. Package opened diagonally from one of the four corners. One corner had grommets with the addition of tapes, aiding a user wearing gloves to open the package. (National Archives)

(Right) USAAF P-51 pilot Lt. Merriman of the 20th Fighter Group in England, c.1944. Attached to waist strap of his Type B-3 life preserver was an RAF MK-II First Aid Outfit for Air Crew. Its use within the US Air Force operating in the ETO and MTO was mentioned in various reports on emergency rescue equipment; it was preferred over standard parachute first aid packet 97785. (National Archives

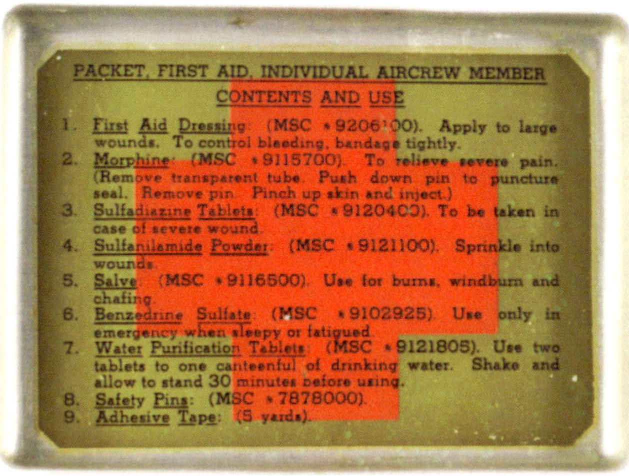

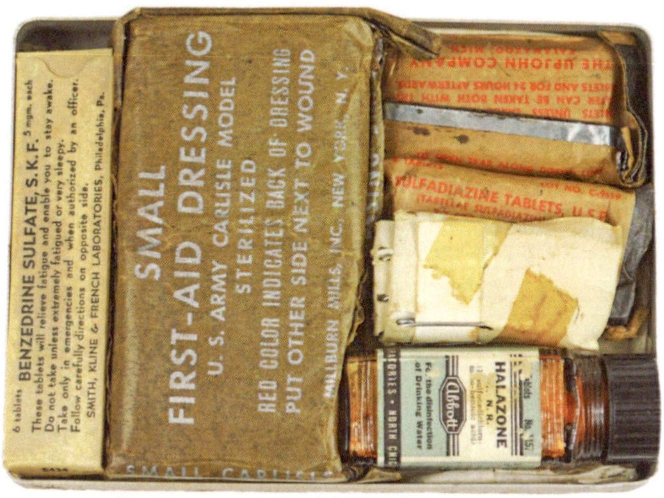

One of the 15,000 first aid packets procured in March 1945. Lid was stamped with Materiel Command stock class number 8300-623715. It consisted of aluminum telescoping halves and instruction label applied to inside of lid. Drawing illustrates packing order for components.

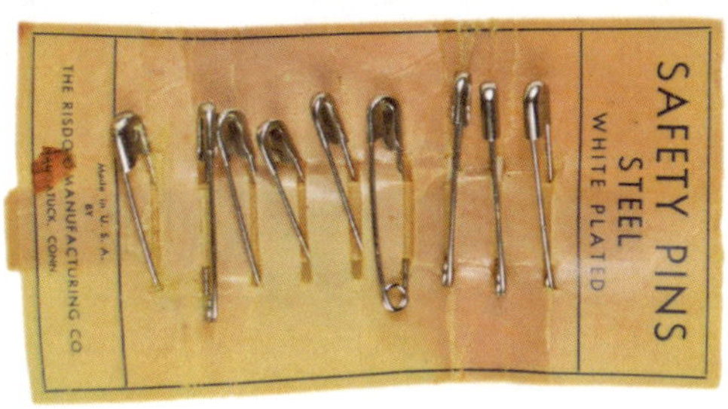

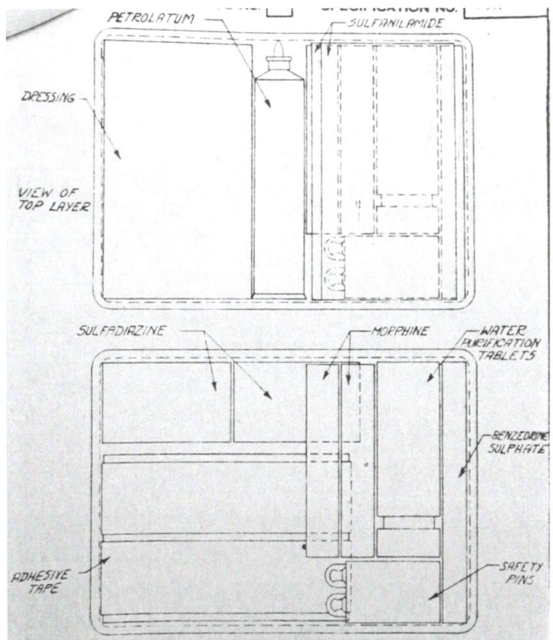

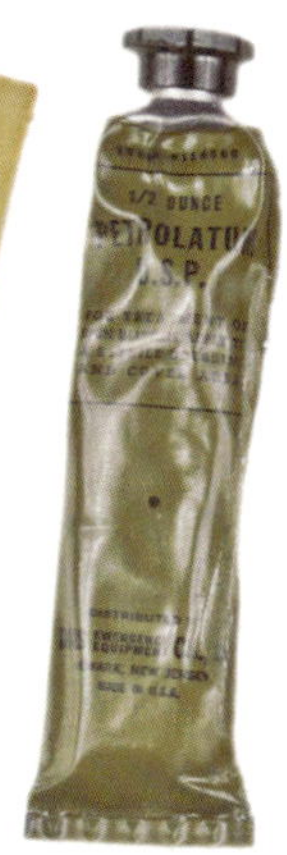

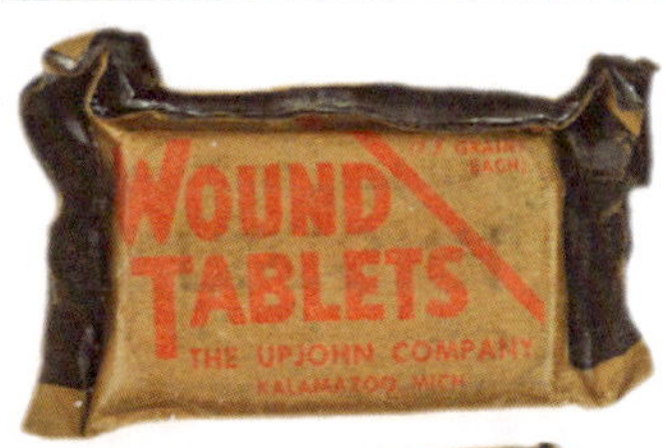

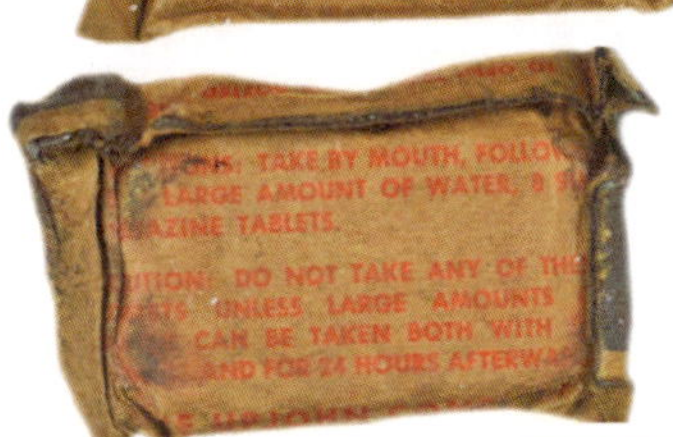

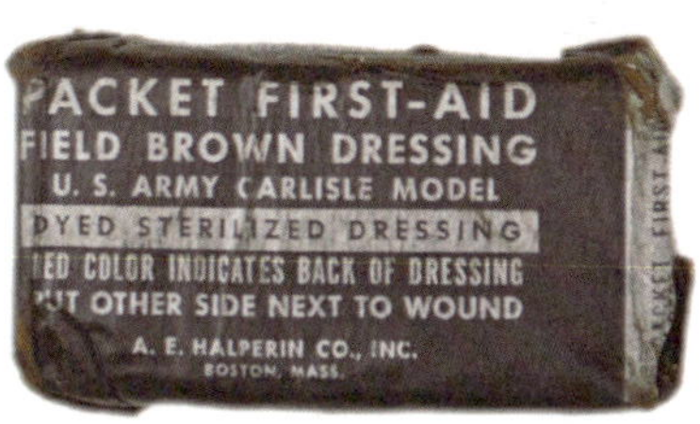

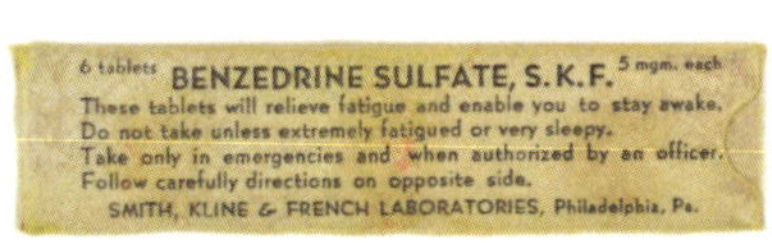

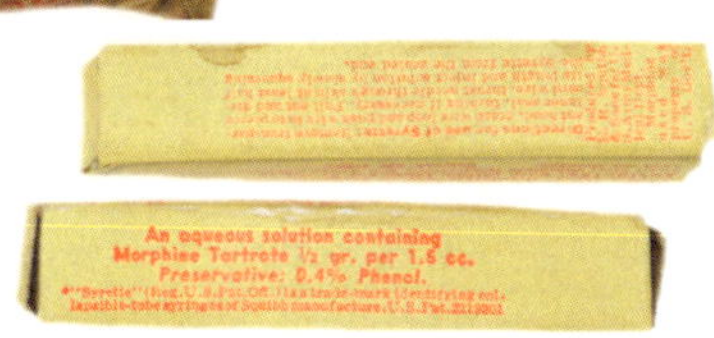

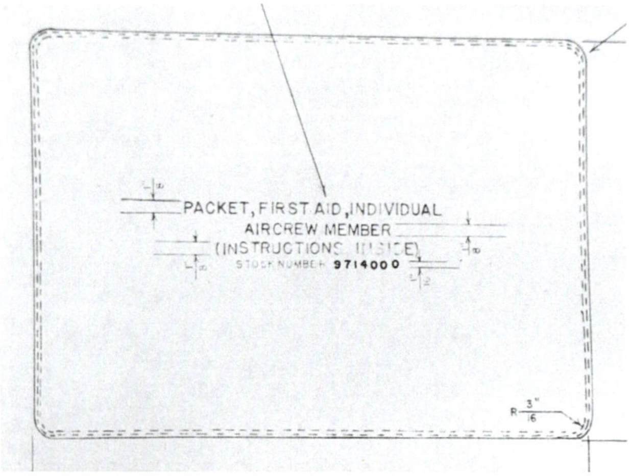

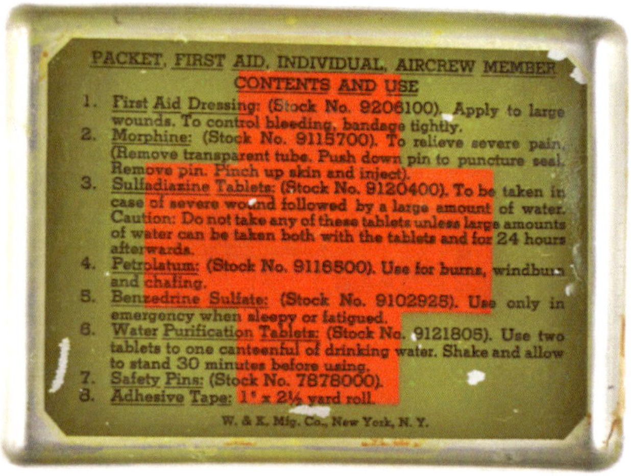

Packet procured in July of 1945 included revised USMD stock number 9714000. Specification 1753 detailed the inclusion of sulfanilamide and adhesive tape 1" x 5 yards. spooled on an acetate sheet. Only change was the inclusion of USMD stock number. Unit here was packed per Amendment-A of August 1945, which removed the sulfanilamide and replaced the spooled adhesive tape on an acetate sheet with spool of tape 1" x 2½ yards. Amendment-A states the adhesive tape shall consist of commercial type spool 1-7/8" in diameter and 1-5/16" H. There was some variance within the components: this unit came with safety pins in a paper envelope but pins on a card were accepted. Aluminum containers were to be sealed with a 3/4" wide friction tape 39-1/2" long wrapping around the container twice. SAVE THIS TAPE - USE TO RESEAL PACKET OR DRESS WOUNDS was stamped on the friction tape.

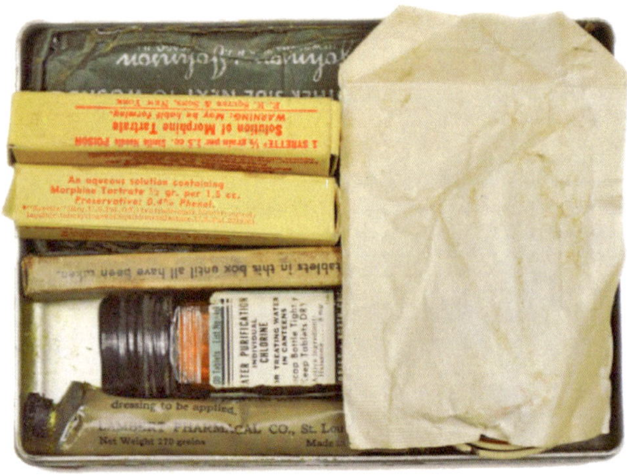

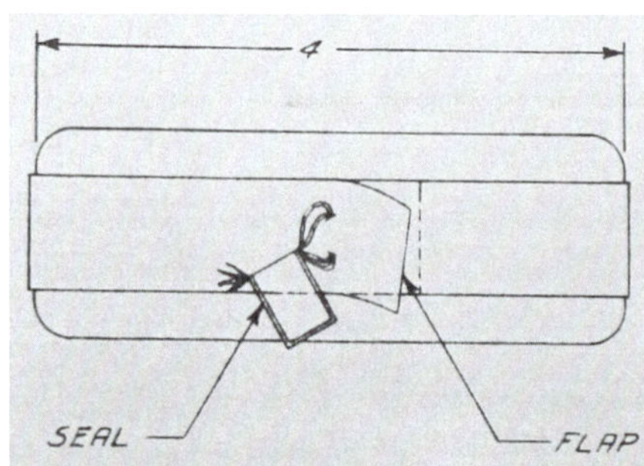

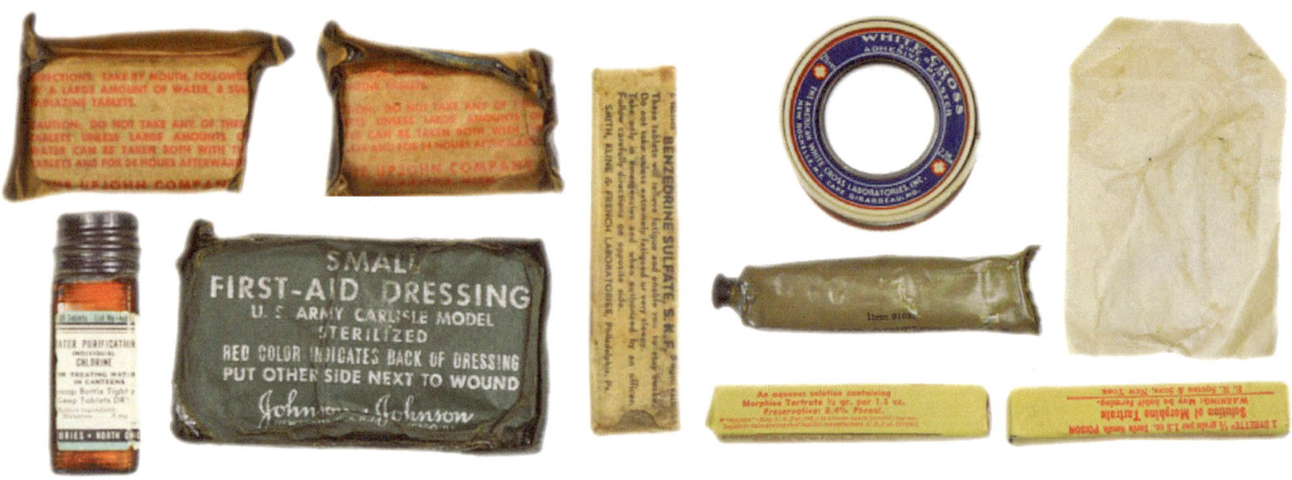

To better understand the development of the US Navy individual first aid kit, it is beneficial to have a working knowledge of the one-man parachute raft and the emergency parachute kit. The kit was implemented in line with the changes of those two units. By late 1943, the pararaft was being used service-wide in conjunction with the parachute emergency kit. Both units had a standard complement of first aid emergency provisions of S2-1055 or S2-1061 types. By the end of 1943, the BuAer had established an unofficial policy to streamline existing and new equipment.

To meet service requirements and recommendations, the BuAer began steps to streamline the pararaft and back pad kit by combining the two into one compact seat kit. To accomplish this goal, certain items had to be dropped or redesigned. The expectation was that all flying personnel would be using the pararaft with a complement of personal equipment emphasizing an individual source of emergency supply. In January 1944, the BuAer requested that the BuM&S develop a small compact individual first aid kit. Service requirements established that a basic kit should be carried by all flying personnel or for use by all pararaft-borne individuals. The objective was to design a compact water-resistant first aid unit that could be carried conveniently by individuals and not impair them in their duties, and to issue it as personal equipment. The established guidelines presented to the BuM&S included: the size and weight be minimal, contents be limited to short-term needs, and it be shock-, abrasion-, and waterproof. The development also supported Air-Sea Rescue activities, which changed the primary purpose of emergency equipment from survival to rescue. It was thought that the unit should contain medical items to treat cuts and abrasions, not major injuries, since rescue and proper medical treatment was thought to be imminent. Tentative specification and research project X-371 was executed 31 May 1944 by the Naval Medical Research Institute for the design of an individual first aid kit for aviation personnel in an attempt to meet the details outlined by the BuAer. The item devised was a two-unit plastic acetate container with individually sealed and waterproofed components provided in a cloth carrying case that could be worn on the belt or in the pocket of a flight suit. The components would include a 4"x 4" compress bandage, a 3" by 6 yard roll of bandage gauze, sulfadiazine tablets, seasickness preventive, two morphine syrettes and a triangular bandage. Instructions were to be printed and placed inside both halves of the container. The basis of issue would be one for each pilot and aircrew member flying outside the continental United States.

In June 1944, BuM&S began negotiations with Conray Products Co. for assembly of first aid kits, accepting a fixed price contract 12 July 1944 for their manufacture. Contract established price per unit with a limited quantity manufactured for field testing. BuM&S assigned stock number S2-1063.

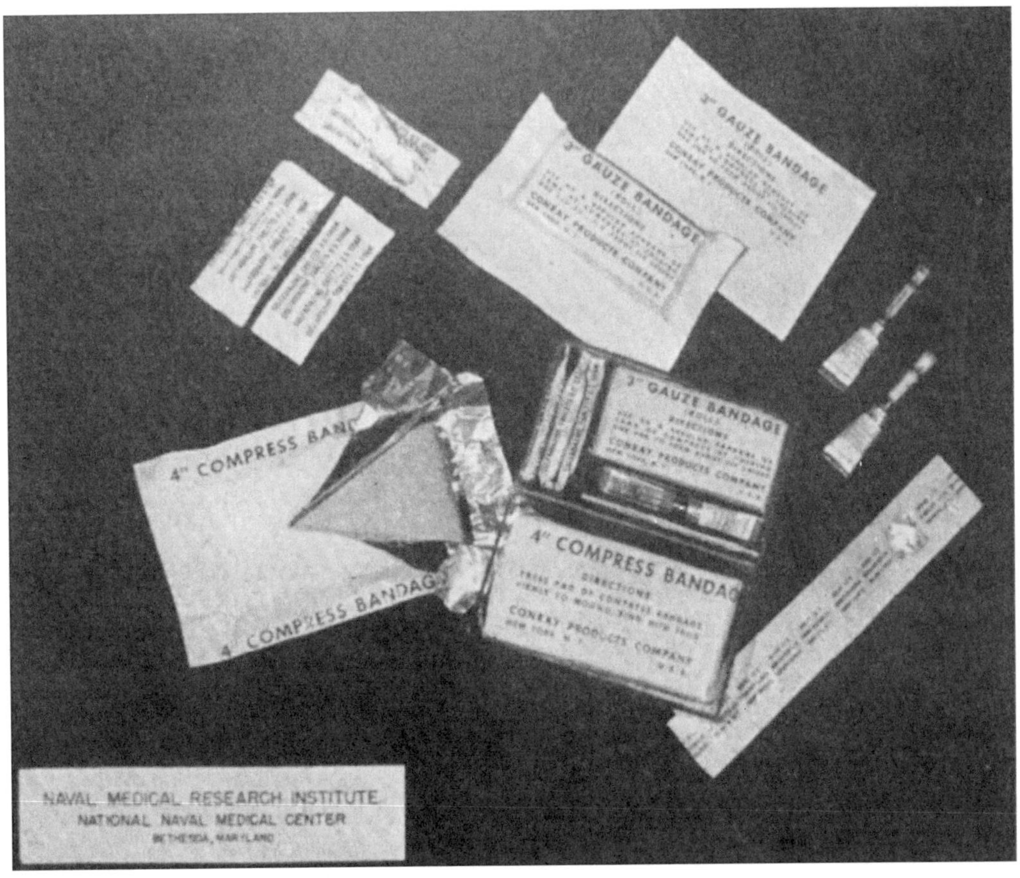

Assembly of US Navy individual first aid kit S2-1063.

The concept was to ultimately eliminate the use of the back pad kit, and, to make room in the new seat kit, the first aid provisions were deleted. The individual first aid kit was designed to be carried by the individual as part of his personal complement to supplement the pararaft. Upon standardization of the Type PK-1 pararaft in November of 1944, production of the individual first aid kit was authorized for a quantity of 100,000 units, awarded to the Conray Products Co, titled Kit, First Aid, Camouflaged (Aviation Individual). Deliveries were being made in limited quantities to the fleet in January 1945, with a reported amount of 5,000 in circulation by February. Due to the limited amounts available, distribution was controlled to carrier-based personnel operating in VF, VB, and VT squadrons or air groups. It was issued as personal equipment to pilots and aircrew available under ASO stock number R57-K-8524-50. After February, an average of 10,000-15,000 units were delivered each month to central supply points. Of the 100,000 placed on procurement, 75,000 were scheduled for west coast stations. In March of 1945, 500 were sent to the USAAF for field testing. The USAAF was contemplating its adoption as an over water first aid kit, and the aircrew member packet was considered an overland supplement. The Kit, First Aid, Camouflaged (Aviation Individual) was officially standardized under joint Army-Navy Specification JAN-K-280 on 31 October 1945.

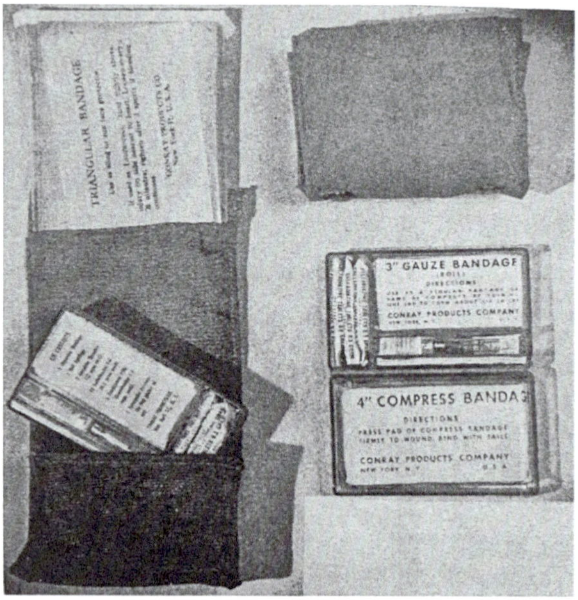

Complete assembly of Kit, First Aid, Camouflaged (Aviation Individual). Acetate container measured 4" x 2-1/2" x 1-1/2". There were two strips of tape, one to secure the two halves together around the full perimeter, and one strip along one long edge to act as a hinge. Cloth carrying case incorporated a pocket to secure the container and another for triangular bandage in flap. Outside of fabric case was stamped with red cross, property mark, and stock number. Complete unit was secured closed by two glove snaps. Inside of flap had brown label with nomenclature and manufacturer.

Two halves of acetate insert included compartments with instruction labels. Triangular bandage was stored separately. Each component was sealed in a waterproof wrapper. The complete unit was designed to be refillable.

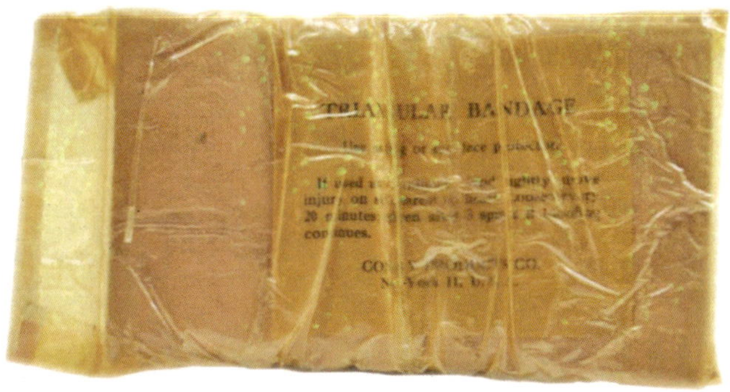

Comparison of materials used in manufacture of the case. Initial production comprised HBT material, meeting requirement of "suitable material" of the tentative specification. Later in production, case was made of mildew-proof OD Army duck cloth incorporated into the JAN-K-280 specification. Cases are identical in all other respects, including brown label, patterned dome snaps, and stamped stock number on reverse.

(Bbelow, left and right) Aircrew of VC-12 using individual aviator's kit. Fleet service testing found the most practical location to wear the first aid kit was on the belt or life preserver. USS Core, April 1945. (National Archives)

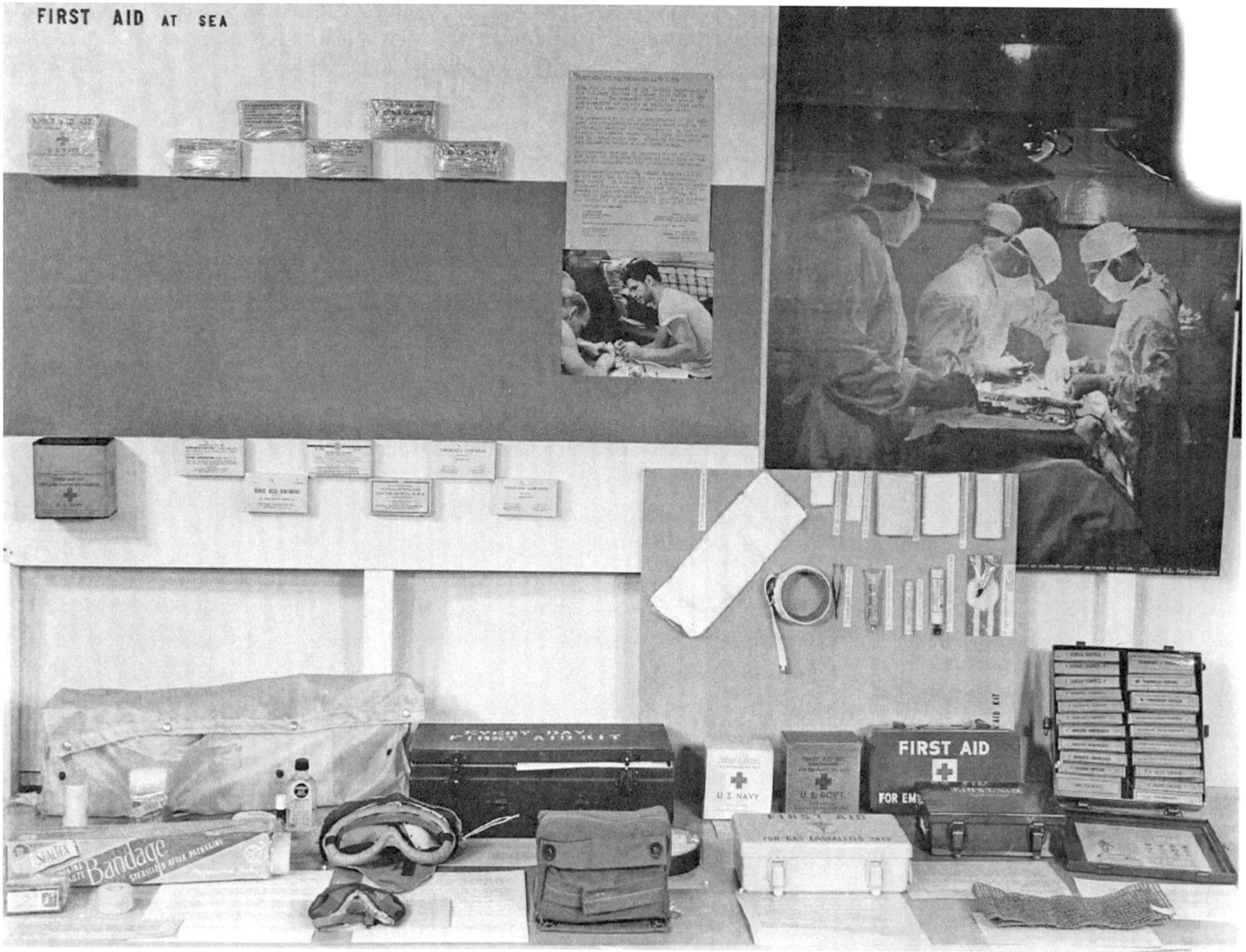

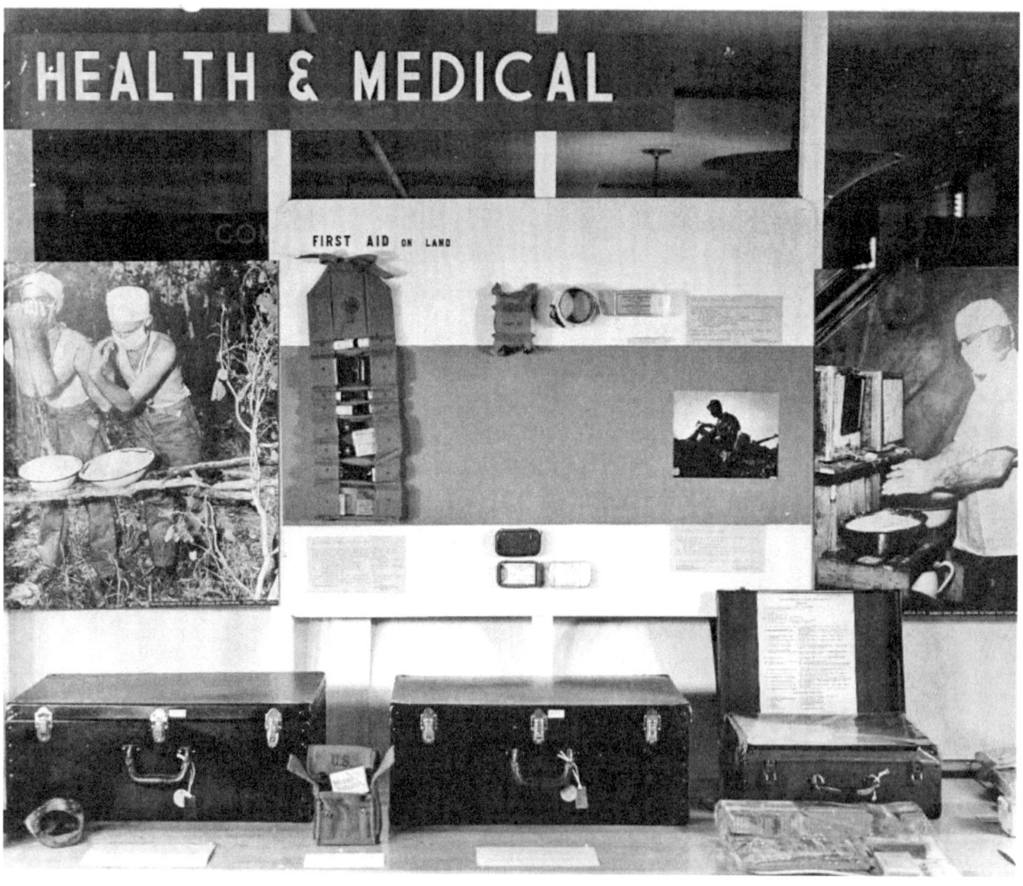

Two photographs from display room at Liaison Committee on Emergency Rescue Equipment, oriented toward medical provisions, c.1943. The articles pictured encompass the services of the US Navy, US Army, USCG, and Merchant Marine. This was representative of what was available for first aid at the close of 1943 and early 1944. (National Archives)

The M-2 jungle medical kit, developed in 1943, was used by aeronautical services of the USAAF, BuAer, and USMC as a supplemental individual first aid provision. Pouch was designated by the USMD as Pouch, First Aid, Jungle to replace roll type jungle first aid kit adopted in 1942. Initial procurements in 1943 were for the pouch only, assigned stock number 97661. On 4 March 1944 the procurement specification was released for complete assembly under USMD Specification No.1696 titled Kit, Jungle, Medical, Individual M-2. Pouch was supplied to contractor by the government, and contractor would assemble the unit as prescribed, shipping it complete. When requisitioned complete, stock number 9710900 was used. It contained field brown Carlisle dressing, 3 band-aids, sulfadiazine tablets, vials of Atabrine and iodine, halazone tablets and insect repellent. In AmendmentA (1696-A) of 30 November 1944 the field brown Carlisle dressing was detailed to include a double sleeve chipboard carton, and the quantity of band-aids was increased to 12. The packing method in the USMD drawing, seen bottom left.

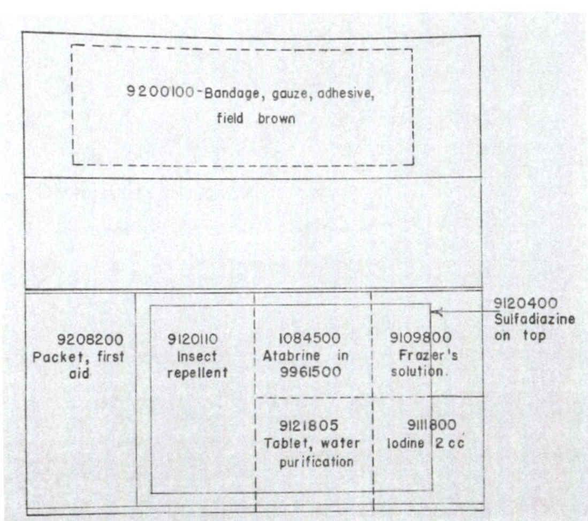

(Right) USAAF B-25 aircrew member of the 81st BS was using M-2 medical kit while stationed in India, c.1945. (National Archives)

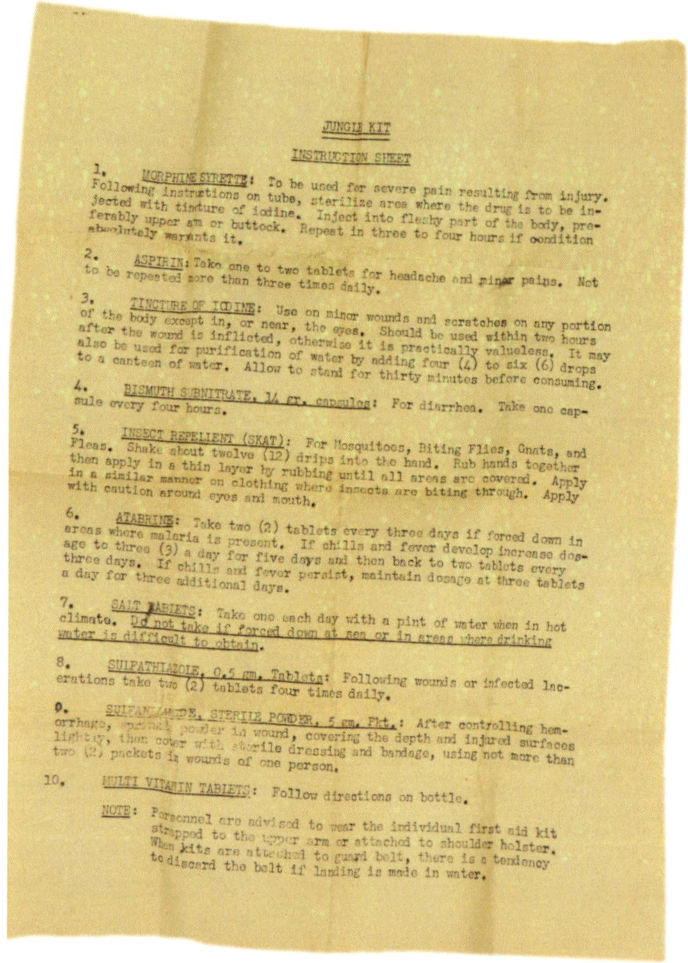

JUNGLE KIT

INSTRUCTION SHEET

1. MORPHINE SYRETTE: To be used for severe pain resulting from injury.
Following instructions on tube, sterilize area where the drug is to be in-
jected with tincture of iodine. Inject into fleshy part of the body, pre-
ferably upper arm or buttock. Repeat in three to four hours if condition
absolutely warrants it.

2. ASPIRIN: Take one to two tablets for headache and minor pains. Not
to be repeated more than three times daily.

3. TINCTURE OF IODINE: Use on minor wounds and scratches on any portion
of the body except in, or near, the eyes. Should be used within two hours
after the wound is inflicted, otherwise it is practically valueless. It may
also be used for purification of water by adding four (4) to six (6) drops
to a canteen of water. Allow to stand for thirty minutes before consuming.

4. BISMUTH SUBNITRATE, 14 gr. capsules: For diarrhea. Take one cap-
sule every four hours.

5. INSECT REPELLENT (SKAT): For Mosquitoes, Biting Flies, Gnats, and
Fleas. Shake about twelve (12) drips into the hand. Rub hands together
then apply in a thin layer by rubbing until all areas are covered. Apply
in a similar manner on clothing where insects are biting through. Apply
with caution around eyes and mouth.

6. ATABRINE: Take two (2) tablets every three days if forced down in
areas where malaria is present. If chills and fever develop increase dos-
age to three (3) a day for five days and then back to two tablets every
three days. If chills and fever persist, maintain dosage at three tablets
a day for three additional days.

7. SALT TABLETS: Take one each day with a pint of water when in hot
climate. Do not take if forced down at sea or in areas where drinking
water is difficult to obtain.

8. SULFATHIAZOLE, 0.5 gm. Tablets: Following wounds or infected lac-
erations take two (2) tablets four times daily.

9. SULFANILAMIDE, STERILE POWDER, 5 gm. Pkt.: After controlling hem-
orrhage, sprinkle powder in wound, covering the depth and injured surfaces
lightly, then cover with sterile dressing and bandage, using not more than
two (2) packets in wounds of one person.

10. MULTI VITAMIN TABLETS: Follow directions on bottle.

NOTE: Personnel are advised to wear the individual first aid kit
strapped to the upper arm or attached to shoulder holster.
When kits are attached to guard belt, there is a tendency
to discard the belt if landing is made in water.

(Left and above) M-2 jungle medical kit used by USMC aviator. Match safe, vial of Benzedrine and tube of ointment are from a personal aids kit. Glass vials were BuM&S standard stock. Unit included instruction sheet, M-1910 first aid pouch, and Carlisle dressing. (Robin Bartel Collection)

(Right) USMC corsair pilot on Kadena, Okinawa airfield wears M-2 jungle medical kit. In June 1943, BuM&S informed Marine Corps it was preparing to requisition a quantity of US Army M-2 jungle medical kits stock number 97661. This requisition was for empty containers, to be filled with medical provisions by local authorities using existing BuM&S stocked items.

(Right) Unit with BuM&S stock number S2-1064 has a partial complement of components. Tourniquet displays BuM&S stock number and labels on vials were marked MEDICAL DEPARTMENT, U.S. NAVY. Stock number S2-1064 is presumed to be assigned by the BuM&S for those M-2 kits maintained by the US Navy Medical Department. (Robin Bartel Collection)

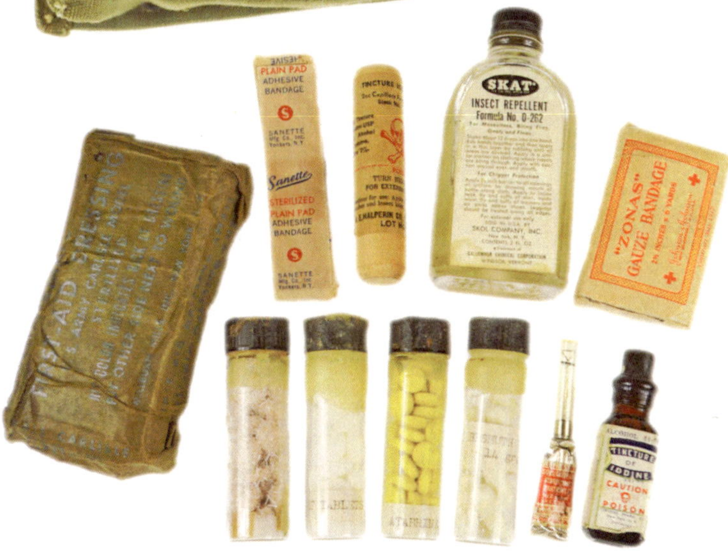

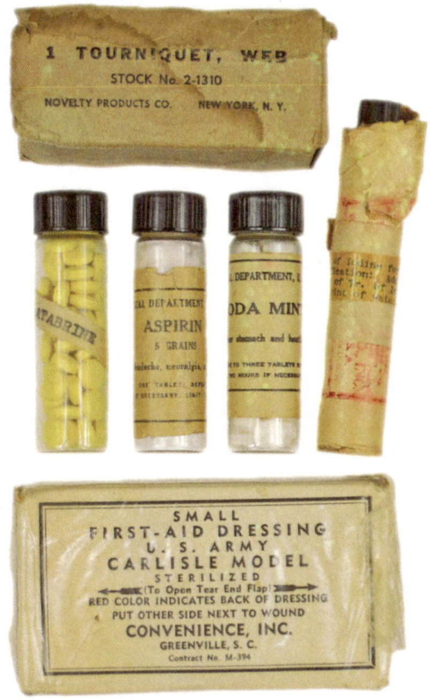

(Above) Bulk of M-2 medical kits procured by BuM&S were of empty container. Requisitions could be submitted to either US Navy or US Army depots, whichever was more convenient. Additional M-2 kit with Naval or Marine origins. The 4 glass vials were proprietary to the US Navy, referenced under stock number 13-065; 6 were included in the officers' Corpsman bag. Glass vials contained tablets of aspirin, salt, Atabrine, and bismuth subnitrate.

(Right) VMF-312 pilot using M-2 medical kit complemented with the addition of the Original back pad kit and pararaft.

(Far right) Lt. William Bennewitz of VMF-213 while aboard aboard the USS Essex, January 1945. M-2 jungle medical kit attached to loose pisltol. (National Archives)

A great deal of development was conducted by BuAer and Marine aviation units at local levels within air groups, and by CASU and Assembly & Repair units. These local commands knew what was best and practical for their operational personnel. Throughout the war, Naval and Marine air units fabricated what was deemed necessary to meet their needs. As the development of emergency equipment progressed, the emphasis gradually moved toward the individual, as indicated by the revisions to pilots' personal equipment and the widespread use of the one-man parachute raft. Featured in the next several pages are first aid units fabricated at local levels for air groups and squadrons. The correlation among these units can be seen in the concept of the standardized aviation individual first aid kit.

(Right) Member of VB-19 demonstrates first aid unit incorporated onto the 23P6 life preserver containing Atabrine, sulfanilamide, sulfadiazine, 4" compress bandage, burn jelly, triangular bandage, and roll of adhesive tape, March 1943. (National Archives)

(Below and below right) Pilots of Carrier Air Group-10 aboard USS Enterprise carry on their belts a locally fabricated first aid general purpose kit, February 1943. These units were described in an article in a BuAer newsletter as containing quinine, salt, sulfanilamide, morphine, a tourniquet, matches, a compass, and rations. L-R: Lt. jg Bruce McGraw of VB-10, Lt. Albert Coffin of VT-10. (National Archives)

C.N. Boder ARM1/c handing Lt. Cmdr. Charles Houston of VB-140 his life preserver prior to taking off from runway in the Russell Islands aboard a PV-1 Ventura, November 1943. The crew's life preservers were outfitted with a first aid packet secured like a sea marker; contents unknown. (National Archives)

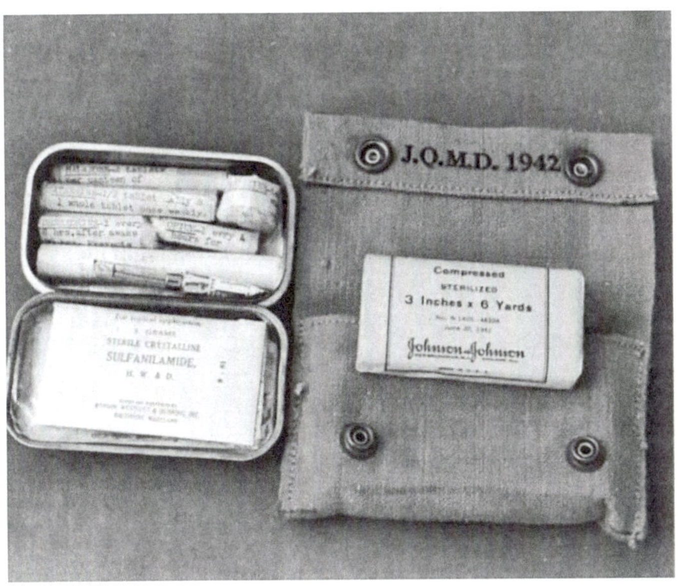

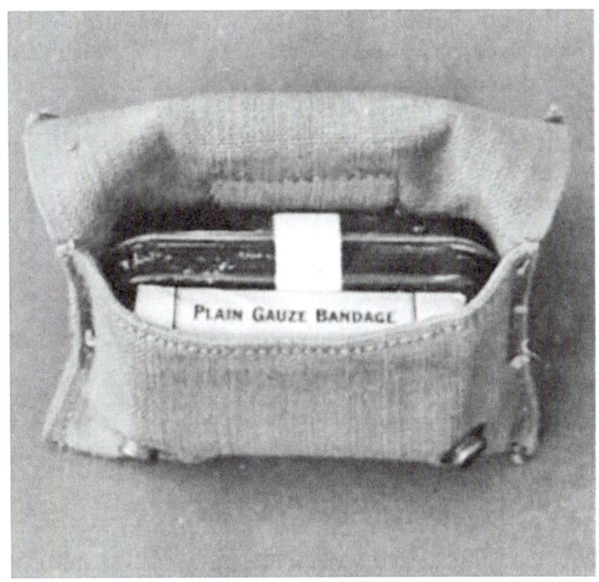

Modified M-1910 first aid pouch demonstrated by unknown Naval or Marine unit. Carlisle bandage was removed and replaced with a 3" x 6 yd. compress with the tin repurposed to contain vials of halazone, Atabrine, Benzedrine sulfate, salt tablets, and morphine syrette, April 1943. (National Archives)

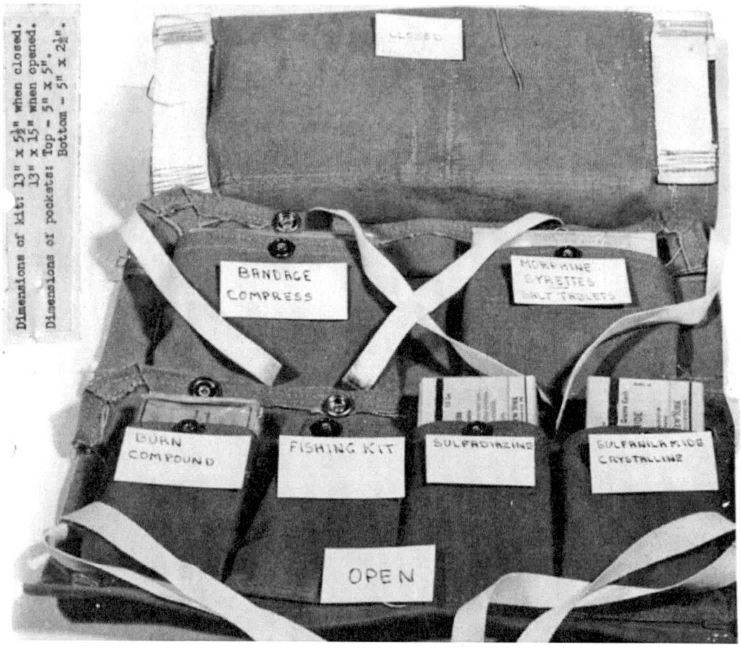

(Above and left, top and below) Prior to the adoption of the individual first aid kit, many air units developed concepts to meet the desire to provide first aid provisions for all flying personnel. Shpwn is a belt kit fabricated by VC-65 in 1944 while conducting operations from the USS Midway. It could to be folded and secured as a waist belt, or be secured to the life preserver as a vest via tie tapes provided. Completed assembly included first aid kit from pararaft and fishing kit. (National Archives)

(Above right) Compact medical shoulder holster devised at Marine Corps Air Station El Toro, featured in BuAer newsletter, August 1943.

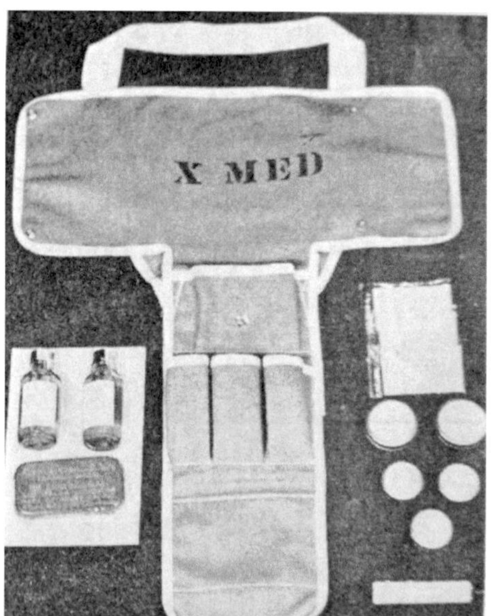

(Right) Medical provisions in the El Toro holster included, aspirin, codeine, sulfathiazole ointment, sulfanilamide power, iodine, Carlisle dressing, and small bottle of brandy.

Marine VMTB squadron member wearing shoulder holster medical kit at operations tent on Peleliu, October 1944. (National Archives)

(Above and below) Another BuAer example of a locally fabricated first aid kit worn as belt. These kits were not proprietary to a single unit but were adopted across the fleet.

(Right)Wartime color image shows VB-5 pilot Lt. jg Fred Joyce wearing a similar medical belt. (Paul Faulkner Collection)

Both United States air forces operated extensive air transport and air evacuation operations, shuttling supplies and personnel, and facilitating Air-Sea Rescue activities. These organizations made additional requirements. BuAer had to meet the needs aboard medical air evacuation aircraft, so personnel used standard BuM&S kits such as the corpsman and field medical units that were readily available or assembled locally. No large medical units were developed as a standard for aeronautical use. The USAAF developed a series of first aid kits to meet a variety of needs, such as special kits for regions and special chests for air evacuation, all of which contained standard USMD items.

The origins for two of the kits date back to the 1930s and were based on a unit designed by Gen. Malcom Grow, which used a metal fishing tackle box to use emergency items while he flew operations over Alaska in 1934. Later, a plywood box was made by the Medical Department Equipment Laboratory, standardized as the Chest, Flight Service on 10 April 1939. Adopted on 23 November 1942 were two regional first aid kits designated as Kit, First Aid, Arctic and Kit, First Aid, Jungle for aircraft operating in remote regions. Each kit consisted of a hardwood veneer or aircraft plywood case and was 17-1/2" x 11-1/4" x 4-1/2" divided into compartments designated as A, B, C, D and E. They were issued one per bombardment and transport aircraft on cross-country flights in jungle and arctic territories.

When Air Sea Rescue aircraft contingents and units began to be deployed, they carried a medical officer and/or Corpsman. Most incidents occurred far from shore, so it was beneficial to carry these trained medical personnel to treat survivors on scene. Use of the hospital corps pouch and aeronautic first aid kits carried in the aircraft was standard but this was considered inadequate by some. The issue was that the contents needed to be dumped out for access – inefficient in cramped conditions aboard the aircraft. Supplies were also limited.

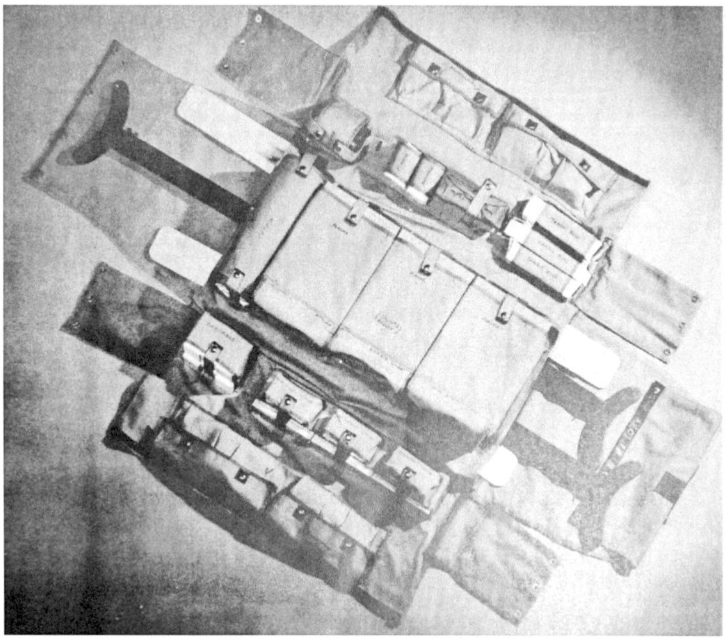

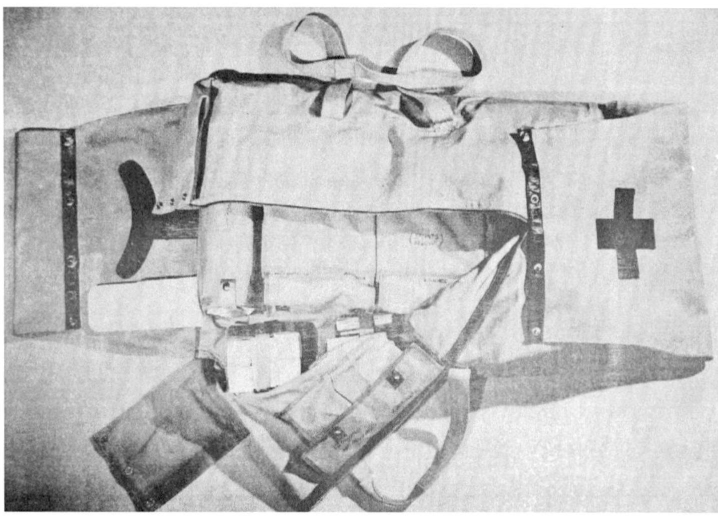

Developed at NAS Pensacola were three kits for use in Air-Sea Rescue operations consisting of a large air-sea rescue kit, small first aid kit, and blanket kit. All were constructed of fabric carried in stock by the parachute loft. Large kit was filled with supplies to treat most medical emergencies and designed for large seaplanes such as the PBY, PBM, and PB2Y. Components were the same as in the corpsman pouch but included plasma, alcohol, and a larger quantity and variety of bandages to address surgical needs. When fully assembled the unit weighed 35 lbs. and measured 30" x 13-1/2" x 4-1/2".

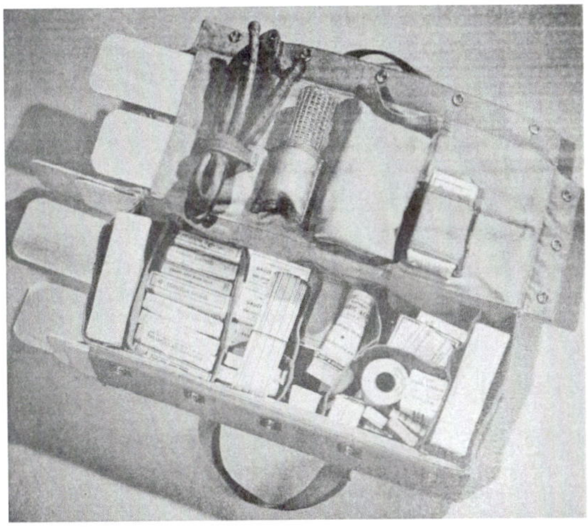

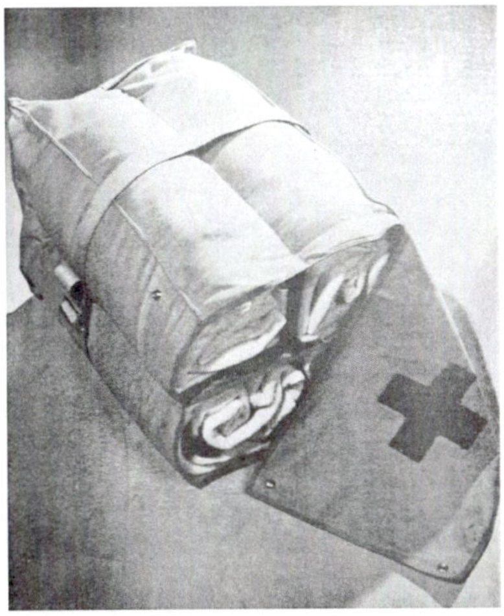

Small first aid kit contained standard components in large hospital corps pouch but in a new container of 18-1/2" x 4" x 8" weighing 8 lbs. Only addition was basswood splints in the walls of the bag. Both large and small units could be conveniently carried by a shoulder strap, but, most importantly, all the medical supplies were easily accessible. (middle) Blanket bag containing 4 standard wool blankets for exposure cases.

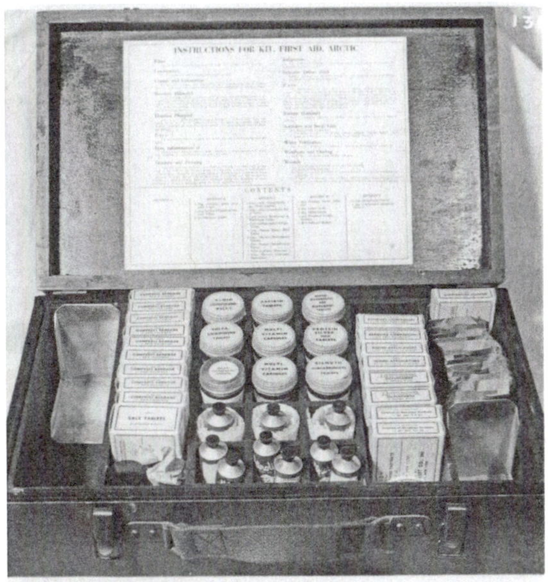

(Above) The Kit, First Aid, Arctic was referenced under stock number 97762 (complete) and stock number 97662 for the empty case. Construction and assembly were referenced under USMD procurement Specification No. 1637. Outside of plywood container included a red caduceus with white lettering.

(Right) Amendment-G of 20 May 1944 revised the case to be made of laminated fiberglass, which reduced the weight and addressed the temperature, termite or rodent damage that occurred with the wooden version. Unit entered production prior to release of the amendment. Early production of the fiberglass models reflected five digit USMD stock number; later production cases included seven digit number. Markings on fiberglass case were black.

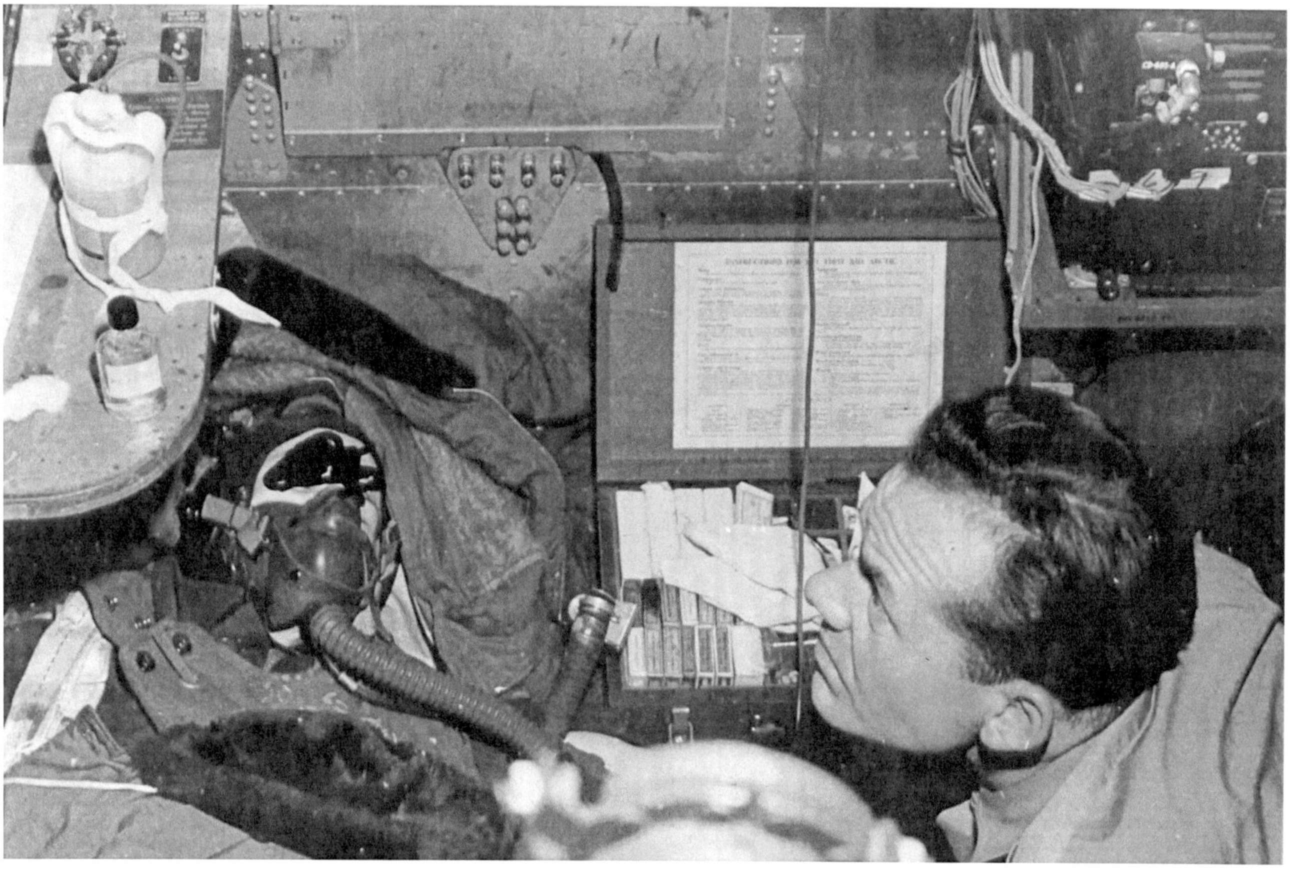

Flight Surgeon Capt. Bryce Welby administers aid to gunner of a B-17 hit by flak on raid over Eisenach, Germany, September 1944. Note open Arctic first aid unit. (National Archives)

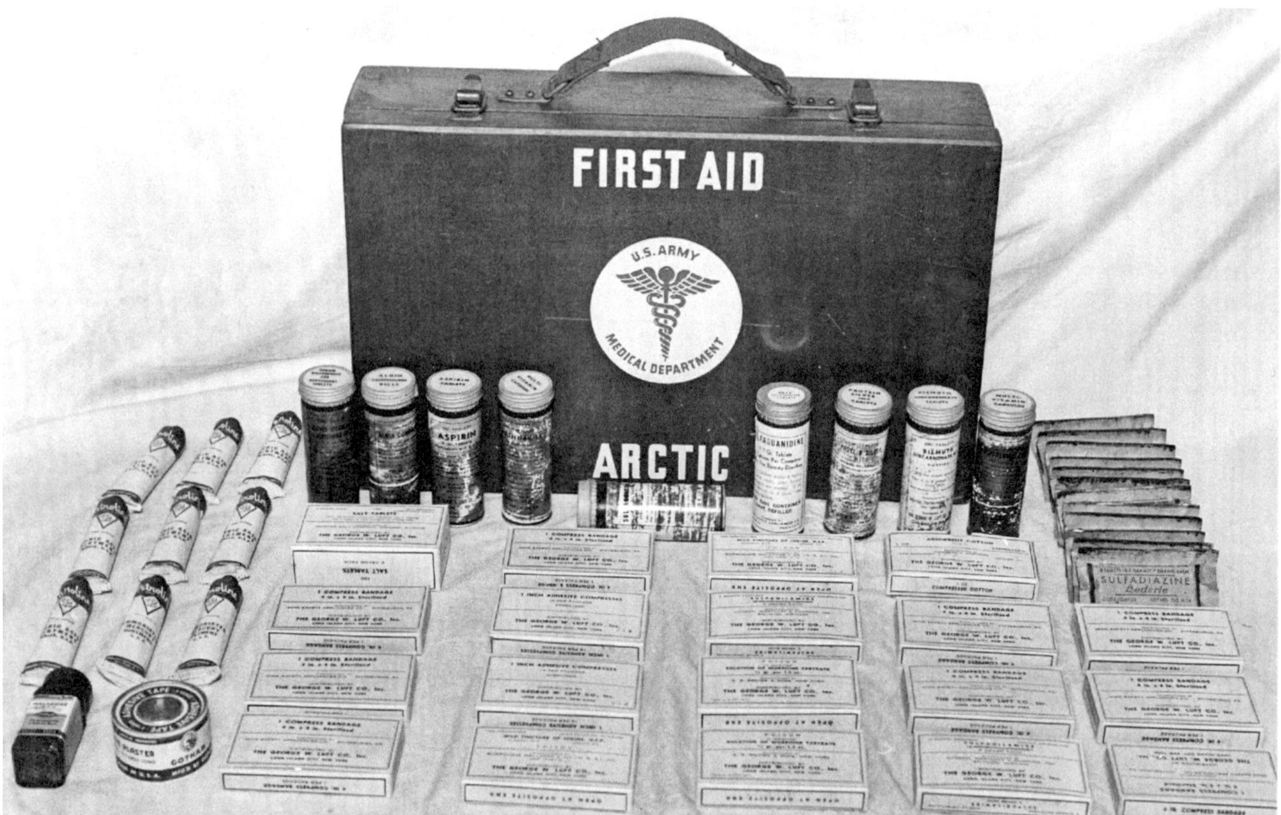

Exploded view of Arctic first aid kit components. Changes to components occurred in its service life to stock numbers, quantities and types of components and were reflected on label pasted to inside of lid. (USAF)

Kit, First Aid, Jungle stock number 97763 (complete) and 97664 (empty). It was referenced under USMD Procurement Specification No.1619. Jungle unit was used under the same conditions as arctic version but included medical provisions peculiar to Asian ailments. It too would undergo revisions to components. The fiberglass conversion was detailed in Amendment-D of 10 June 1944; this example reflects seven digit stock number. (USAF)

Exploded view of jungle first aid kit components. It and arctic kit were superseded by medical kit Type TT-1A and do not appear in 1947 medical supply catalog. (USAF)

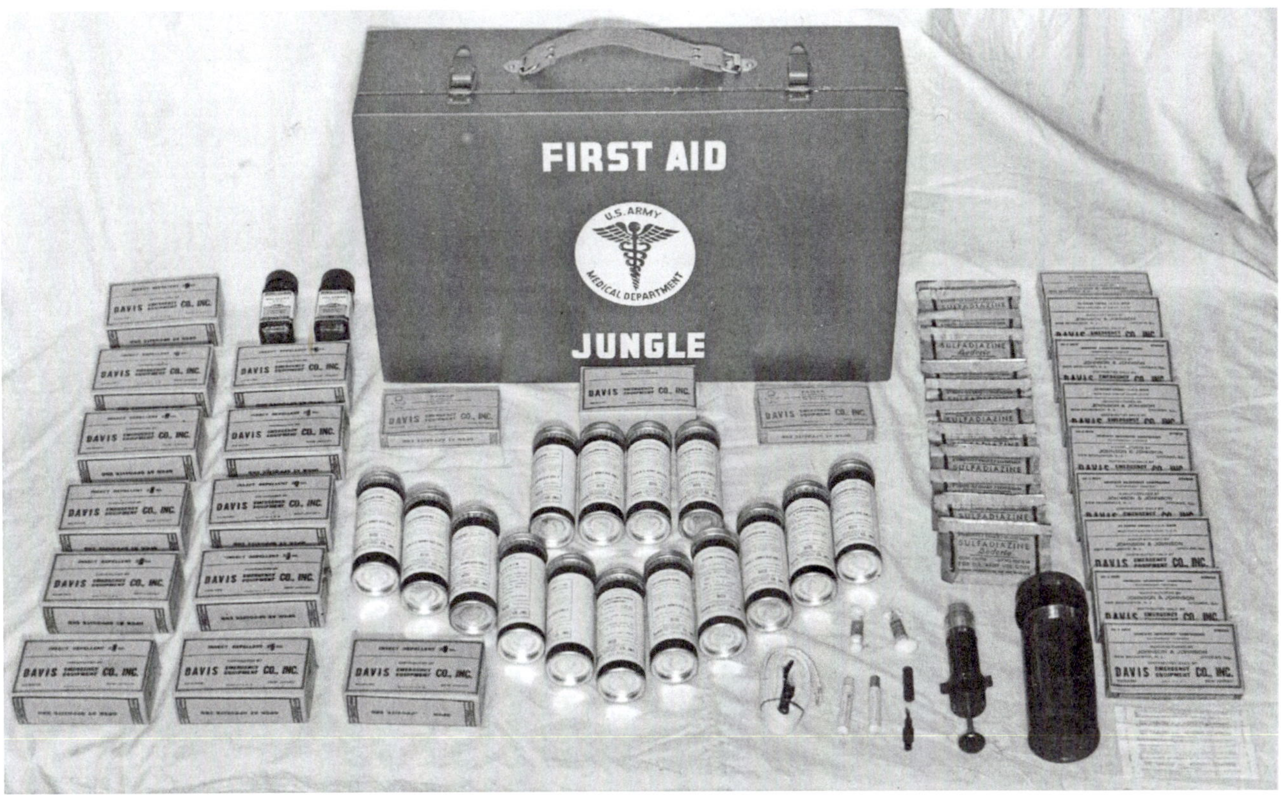

Jungle kit being examined in a USAAF medical supply facility somewhere in Asia-Pacific.

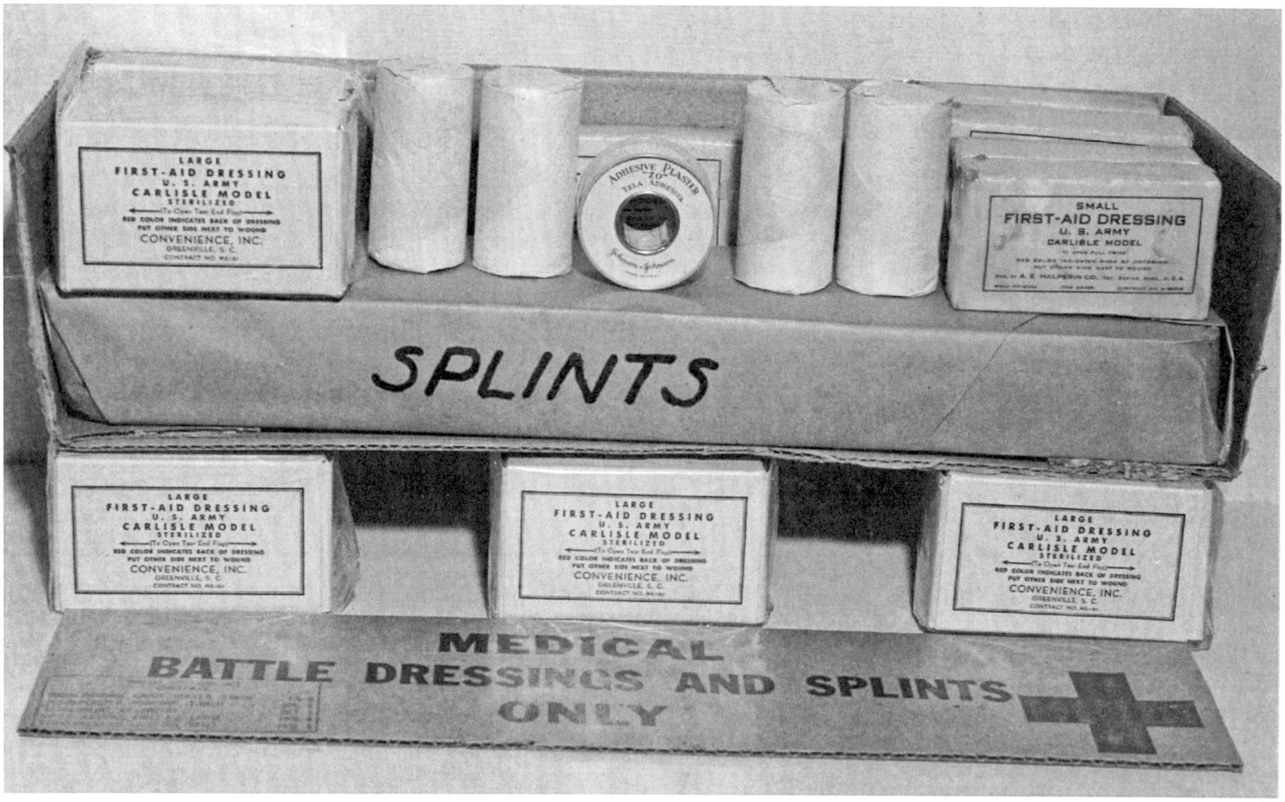

Adopted on 15 February 1945 as supplemental item for bombardment, cargo, and transport aircraft was the Kit, First Aid, Battle Dressings, Complete. It consisted of a corrugated fiberboard container measuring 18" x 5 3/4" x 4-1/4" containing splints, roller bandages, large and small field dressings, and spool of adhesive tape, available under stock number 9776210. Its distribution is under the same conditions as arctic and jungle first aid kits. Outer carton marked on two sides as detailed in Specification No.1732. (USAF)

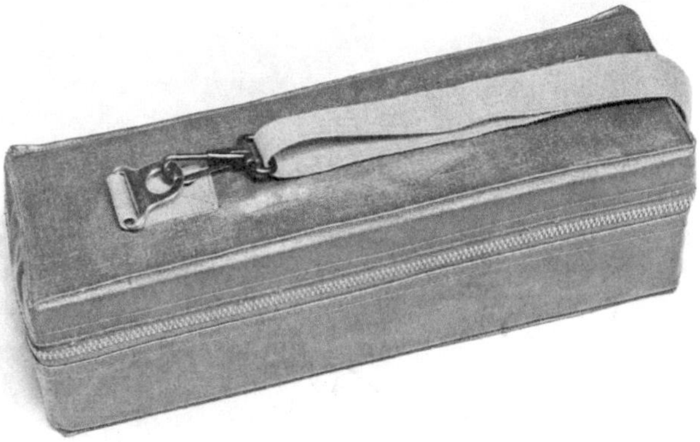

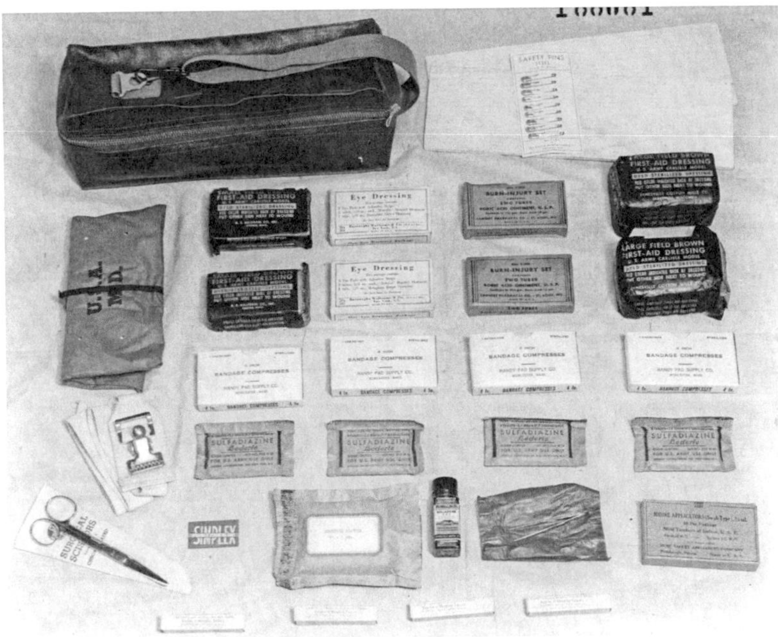

Kit, Medical, Emergency, Type TT-1 was designed for inclusion in the E-18 sustenance kit under Specification No. 94-41036. Container was referenced under USAAF Dwg. No. 45G18837. Design was for arctic and jungle types but was modified to be universal or All-Area. Concept originated just prior to the spring of 1945, but formal specifications were not issued until 29 October 1945. It became a component of Type E-18 and was standardized in conjunction with its assembly on 20 August 1946. Amendment-A of TT-1 became effective 1 December 1947. The kit remained in service until 1 October 1952 when it was redesignated as Medical Emergency Kit, Aeronautical. (USAF)

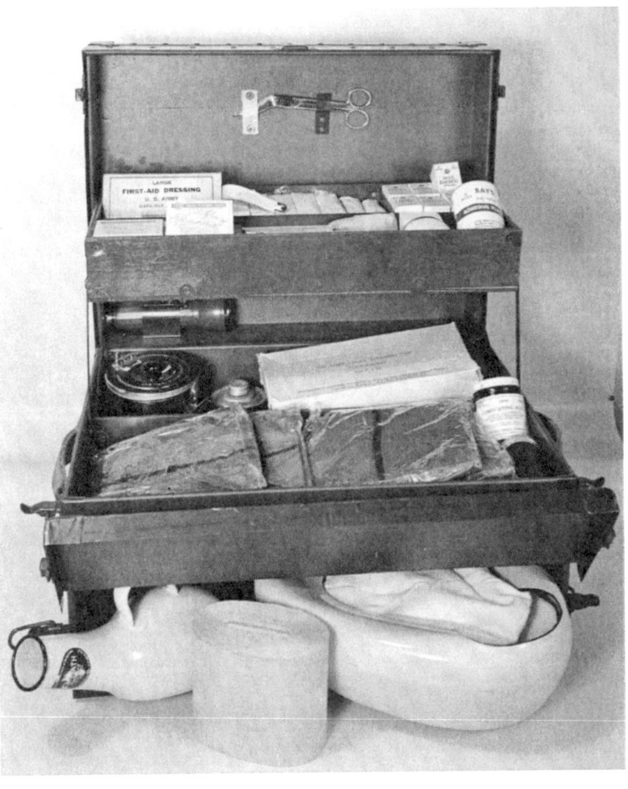

Chest, Flight Service, combined features of USMD medical chests Nos. 1 and 2 being light, compact, and made of plywood covered with OD canvas. It measured 25-1/4" x 12-1/2" x 13-1/2 " and was referenced under stock number 97545. Two trays were built into the chest; when opened, the trays, containing medical provisions. (National Archives)

Chest, Ambulance, Airplane was developed to meet requirements for medical air evacuation units. It was derivative of the flight service chest, available under stock number 97517. These chests were used by Flight Nurses and Flight Surgeons and contained supplies and bedside equipment for long transoceanic or cross-country flights. Most supplies were packed in transparent plastic drawers for quick identification. (National Archives)

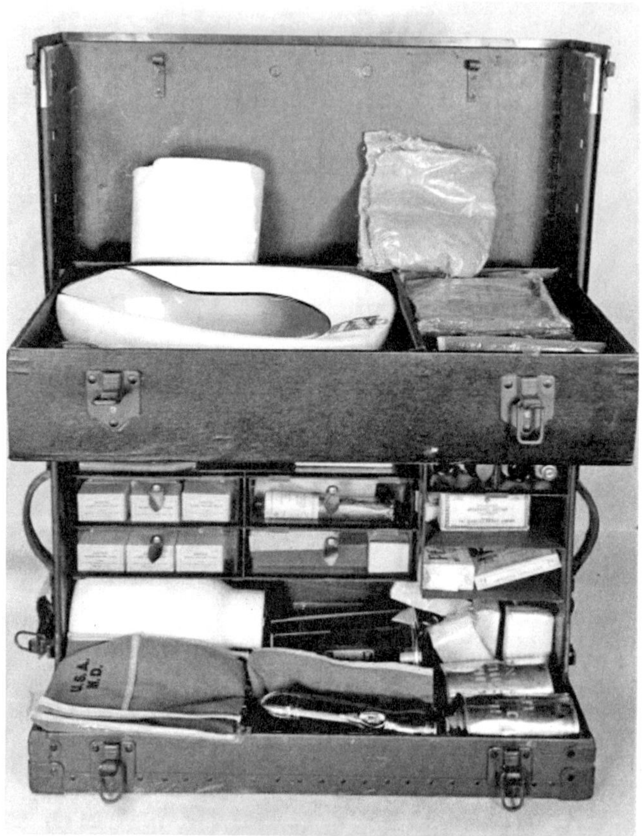

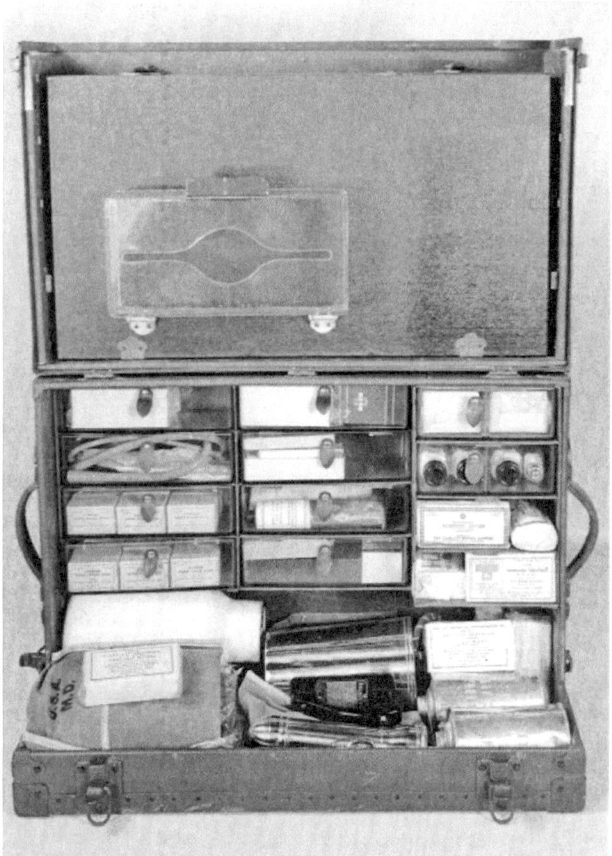

(Right) Flight Nurse Lt. Lucille Tarney inspecting an ambulance chest aboard transport with the 9th AF, March 1944. (National Archives)

Medical transport inbound to Australia from New Guinea, November 1943. Preparing to use the electrically-heated cup provided in the ambulance chest. (National Archives)

(Below) The Gallagher Kit was developed in 1944 by Colonel John Gallagher, USAAF. It was designed to meet service requirements of a compact, highly visible first aid unit combining the aeronautic, jungle, and arctic first aid kits. Designed specifically for B-29 bombardment aircraft, 500 were procured for field testing in September 1944. (National Archives)

(Below) Suitcase roll developed by Capt. Molomut of the Far East Air Forces in December 1944. It was intended to replace the standard ambulance chests, allowing the components to be more visible, orderly, and compact. (National Archives)

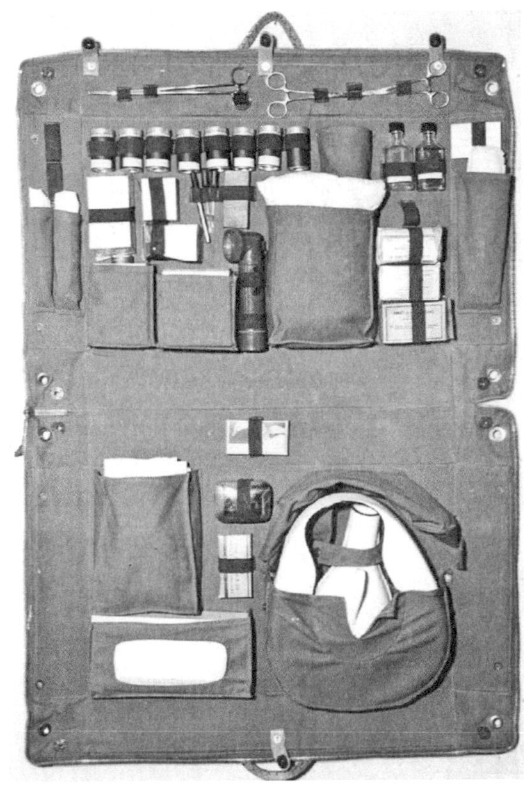

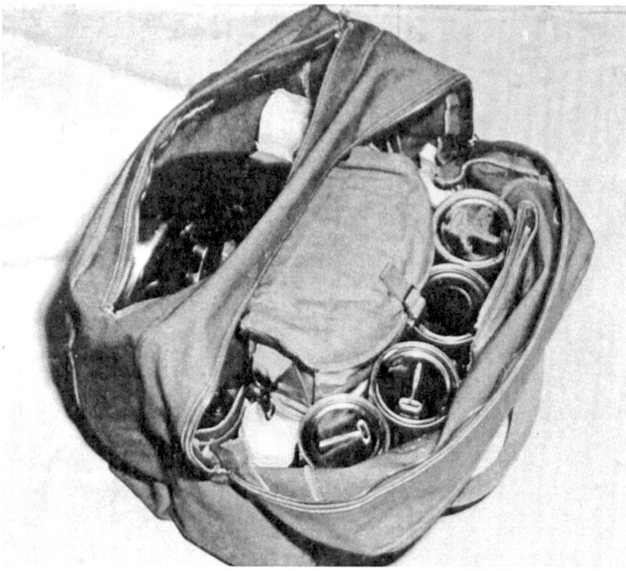

Whether Air-Sea Rescue or air evacuation, medical units were continually adapted to maximize efficiency. Air-droppable unit created by the USAAF 2nd Central Medical Establishment at Guadalcanal for downed aircrews. Unit could be dropped to stranded crews in isolated jungle terrain. It contained supplies to help survivors on their trek out of the jungle. (National Archives)

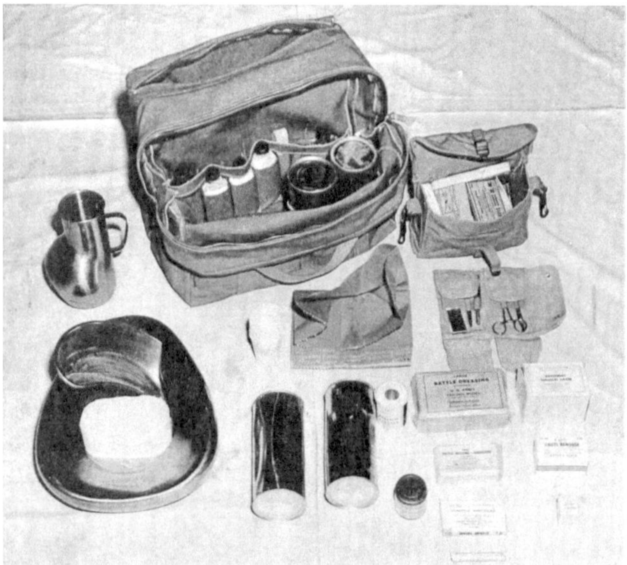

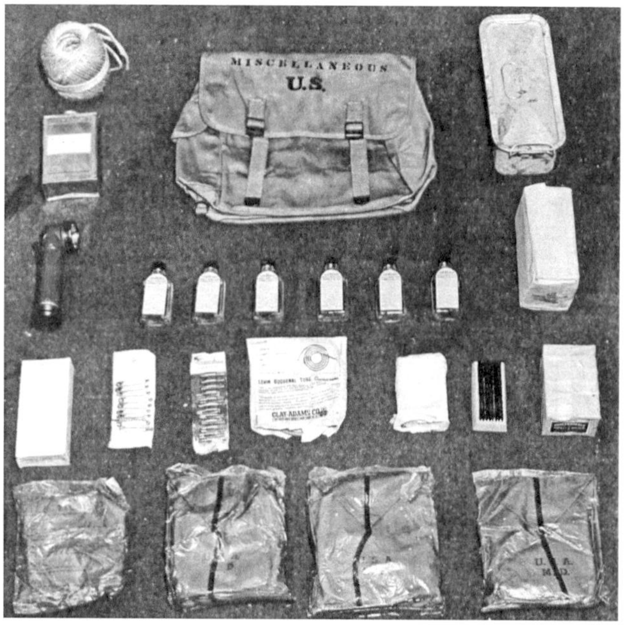

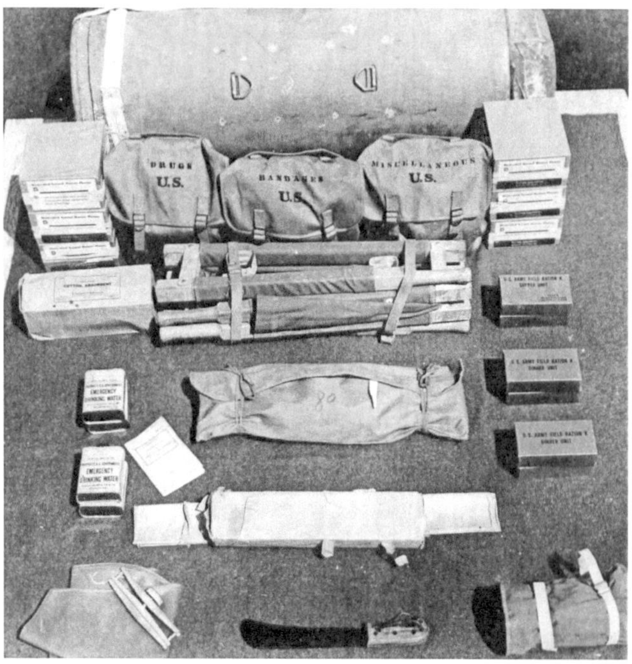

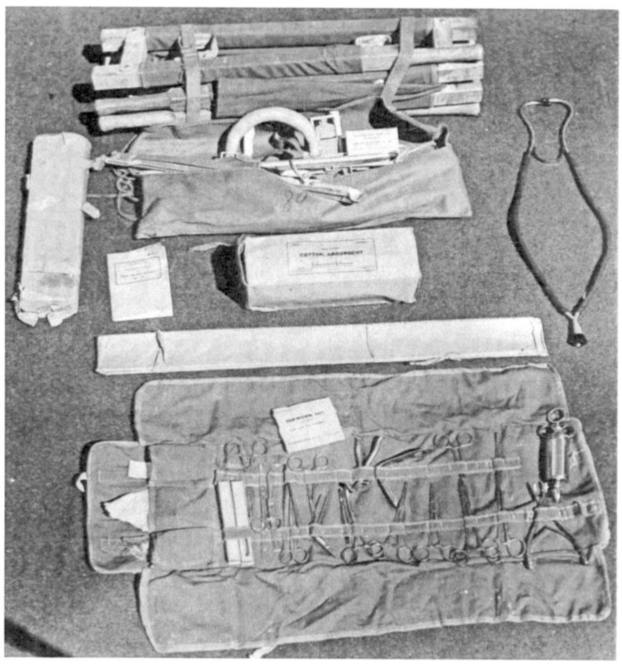

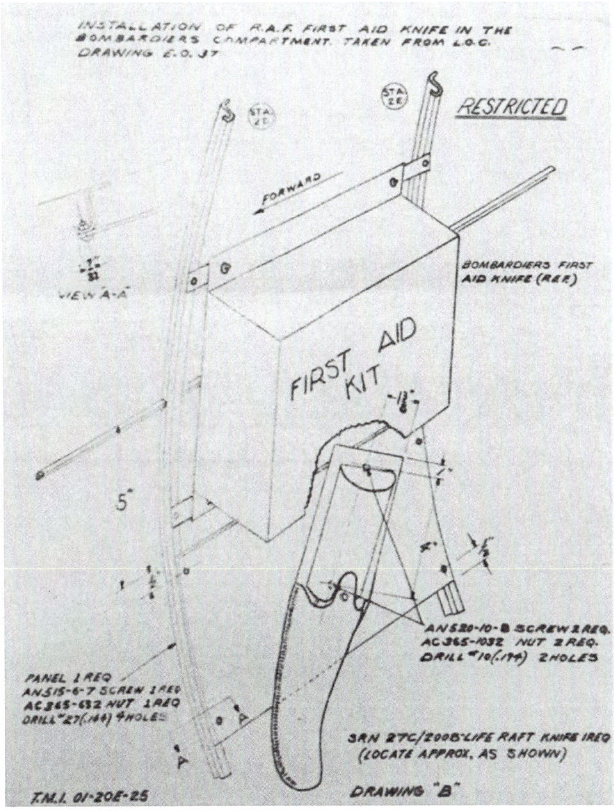

USAAF Knife, Emergency, Curved stock number 7900-470950 was developed by the Aero Medical Laboratory in 1943 for cutting clothing off wounded airmen. Design was based on existing RAF model referenced under stock number 27C/2003, identical to RAF's life raft knife. The RAF first aid knife was service-tested by the 8th AF. In July 1943 8th Air Force directed that two knives be installed in bombardment aircraft, two per heavy and one per medium aircraft. Drawing is part of Technical Maintenance Instruction issued 22 December 1943 by HQ 8th AF Service Command for installation of RAF first aid knife in bombardment aircraft. AML approved this requirement and redesigned the knife, reducing the size and incorporating the same attachment method as the aeronautic first aid kit under Assembly Dwg. No. 44B26161, Knife and sheath were referenced under drawings 44B2662 and 44B2663. In May 1944 the knife was officially required to be installed at staging areas before overseas departures. It was to be installed adjacent to first aid kits in the nose and radio compartments.

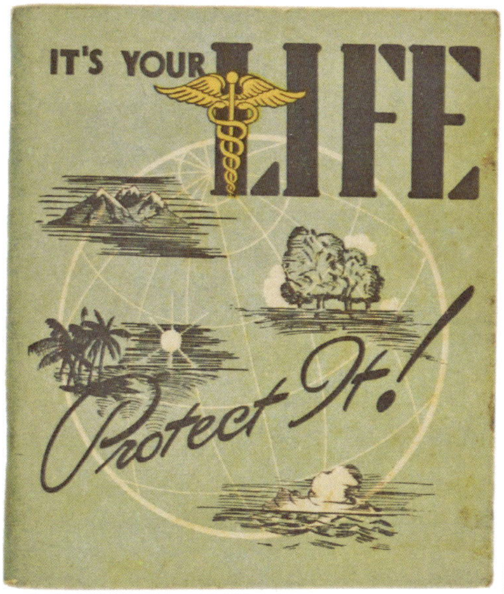

Examples of literature prepared by medical departments for use by aircrew members. (left) 9-page booklet AF Manual No.45, printed November 1944, for inclusion in aeronautic first aid kit with endorsements from USAAF and BuM&S. It is a reprint of information in the Pilots Information File. (middle) 26-page booklet compiled by the Flight Surgeon of 3rd Fighter Command, September 1943. It included information on survival techniques and treatments of ailments. (right) 105-page booklet prepared by medical staff at Will Rogers Field, Oklahoma, March 1944, which covered medical matters in all climate regions, including information on edible plants and animals.

EMERGENCY RATIONS

The development of rations was like any other item included in this study in that rations were (and still are) designed to meet service requirements. The responsibility formally standardized in 1936 for the development of special rations for the US Army was conducted by the Quartermaster Subsistence Research & Development Laboratory (SR&DL) located at the Chicago Quartermaster Depot. Prior to this and dating back to World War I, research and development was conducted by the QMC Subsistence School. In tracing the lineage of emergency rations, it is therefore practical to start from World War I. From experiences gained in the Great War, the QMC developed the Reserve Ration, Trench Ration and the Emergency Ration.

By World War II, as activities increased and, with them, the need to coordinate in the development of special rations, the SR&DL created the Ration Development Branch, which was divided into three sections: Army Ground Forces Section, Army Air Forces Section, and the Nutrition Section.

(Left) The Emergency Ration or Armour Ration was designed during WWI to be an emergency ration for ground soldiers to sustain life when no other food was available. Air services at the time were small and had no requirement for emergency subsistence. This would influence future development of special rations for aeronautical services. It consisted of 3 oz. cakes of a mixture of a beef powder and cooked wheat and 3 oz. chocolate bars. Manufacturing of the Armour ration was discontinued in 1922.

(Above right) Trench Ration gained no traction and was quickly discontinued, but development progressed for the Reserve Ration. Through the 1920s and 1930s, Reserve Ration underwent various revisions; in late 1930s it became the more familiar Type-C ration, a standard for US Armed Forces during WWII. Reserve Ration as it was in the early- to mid-1930s; it consisted of hard tack, chocolate, sugar, coffee and meat.

Requirements for subsistence for the USAAC were minimal from WWI through into the 1930s. During this era, rations were only required on cross-country flights. These were provided with locally available provisions, such as sandwiches, fruits, and vegetables or as classified under the US Army QMC designation of the Type-A Ration. Formal development of an emergency ration commenced in 1933, with the intent for it to be used with a proposed individual parachute emergency kit for cross-country flights. The Reserve Ration was evaluated and determined adaptable for Air Corps use, but not in its current configuration. The packaging was not optimal due to its bulk and the inability to sustain long-term storage. It also contained what could be viewed as luxuries such as sugar and coffee, which are not necessary for an emergency ration. On 28 March 1934, the USAAC issued tentative specifications for an emergency Air Corps ration titled Rations, Emergency, Air Corps. The components were of a streamlined Reserve Ration comprised of eight ounces of hard tack (hard bread) and six ounces of sweet chocolate, packaged in a Kraft carton with the dimensions of 2-1/4" x 3-3/8" x 11". This constituted one ration for one day. Initially, one type of ration was specified for the pilot's emergency kit, referred to as Ration, Cross Country, Emergency, Air Corps. It consisted of three rations in separate cardboard containers wrapped in cellophane placed in one master carton. In June of 1935 this ration was released for use in the parachute kit and authorized for distribution in large bombardment and transport aircraft.

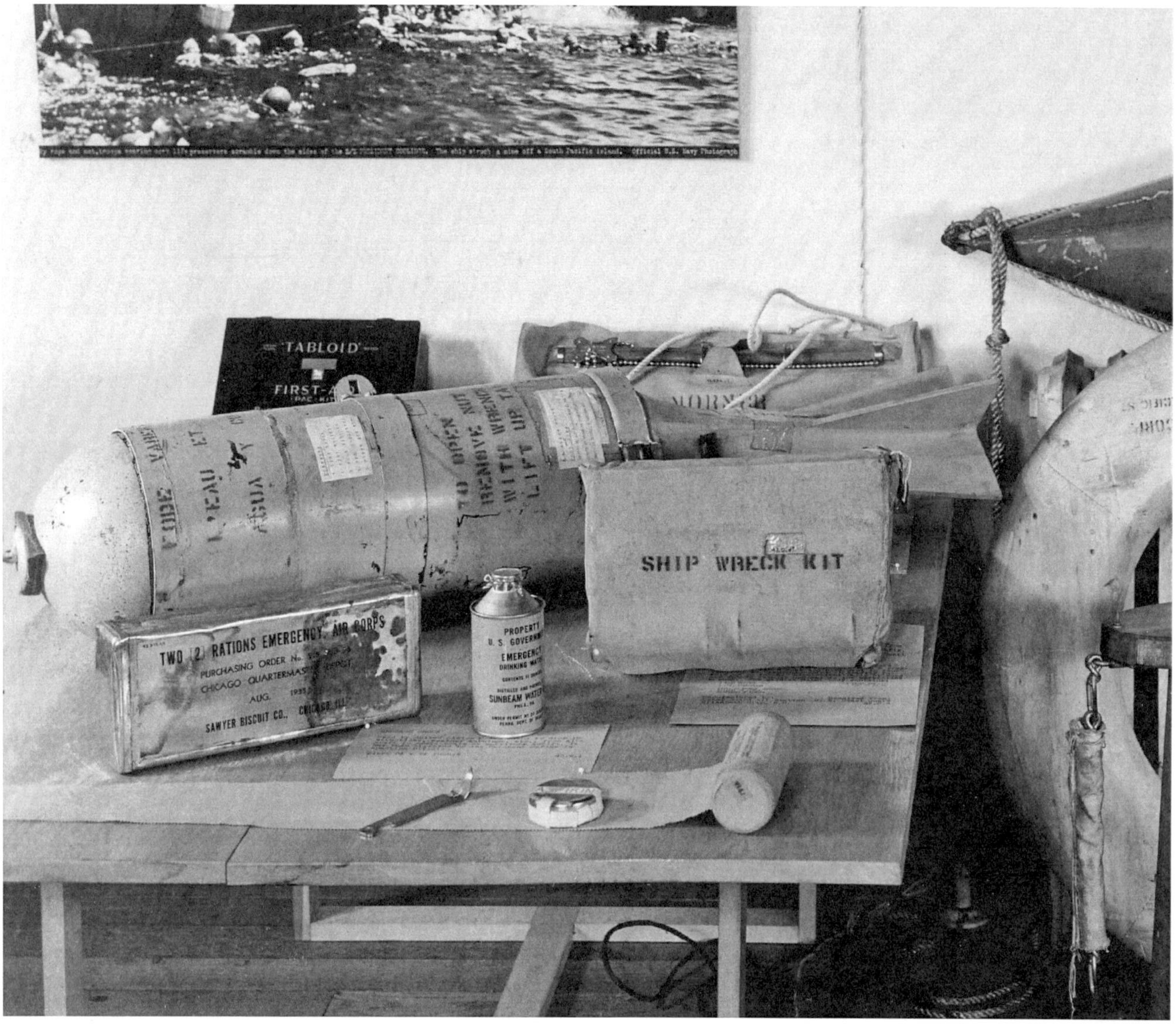

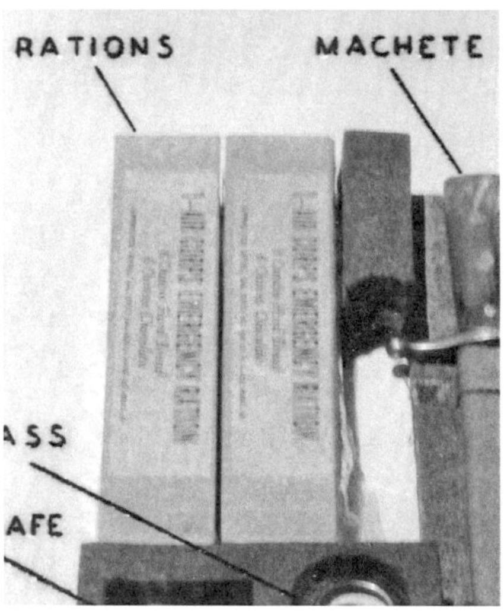

Development of emergency ration immediately preceded development of the emergency parachute kit. The parachute seat kit was designed to accommodate the ration and additional proposed components. For the emergency jungle seat kit, the ration was streamlined to two rations stowed in a sealed tear-strip tin container that prevented deterioration in jungle conditions. Each ration was individually packaged in the cardboard container and wrapped in cellophane. The first procurement of the Ration, Emergency, Air Corps for the jungle emergency kit was awarded in August 1935. Emergency ration the USAAC used in jungle emergency seat kit. Complete assembly measured 2-1/2" x 6 3/4" x 11", available under stock number 8300-680000. One image has the two Air Corps emergency rations removed from the master carton, c.1936. (National Archives)

By the close of 1939 the Type-D ration and the 38D2666 were adopted as standard along with a new standard back-type jungle emergency parachute kit. The tin containers were contracted through the Materiel Division and the rations through the OQMG but were not assembled by a contractor. Jungle kits were ordered empty with the components requisitioned separately. A depot would requisition the tin from the Middletown Air Depot and the D-Ration from the Chicago QM Depot. Once in stock, assembly of the ration would be conducted in-house. This method exposed the chocolate to contamination. Eventual reports stated that assembled units were found with failed soldered seams and infested with parasites. It was then decided to have the complete assembly processed through a food handling contractor to assure proper sterilization.

Following almost immediate disapproval of the seat type emergency kit, both the Materiel Division and overseas stations began work on a back-type emergency kit in late 1936. By converting the standard back pad on parachutes, the ration became exceedingly uncomfortable to wear. It was deemed necessary to reduce its thickness. The change coincides with work of Col. Paul Logan, the inventor of the D-Ration. His innovation was to provide chocolate in 4 oz. blocks. The Ration, Emergency, Air Corps was revised in August 1937: hard bread was removed and replaced with three 4 oz. chocolate cakes. This constituted a ration for one day; six cakes were packaged in a sealed key-opening tin container 9" x 6" x 3/4". Empty container was referenced under Dwg. No. 38D2666-5, stock number 8300-126900. After container was sealed with the chocolate, ration was referenced under the Assembly Dwg. No. 38D2666, available under stock number 8300-126800. The cross-country emergency ration was revised to include 12 4 oz. cakes per tin container similar in appearance to 38D2666, available under stock number 8300-679900. (above) 38D2666 emergency ration. (National Archives)

Packaging of Air Corps emergency rations was transferred to commercial contractors in February 1941. They used new vacuum-sealed tin container referenced under Dwg. No. 41D4293. New container was a thinner gauge key-opening tin but only included 1 emergency ration of three 4 oz. cakes. It also included the instructions WHEN EMPTY USE CONTAINER FOR WATER. It was initially available under stock number 8300-126320, later revised to stock number 8300-679930.

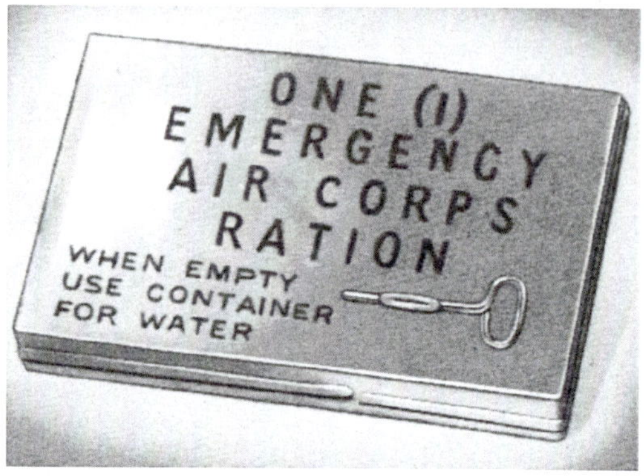

(Above) The 41D4293 emergency ration measures 6-1/4" x 4-1/4" x 7/8". To open, a key-opening tear strip was provided along the outside perimeter. Key was soldered to container face. Packaging was like the 38D2666 but conducted by a professional food handling contractor.

(Right) Upon entering WWII many raw materials were in critical shortage in the US, including tin. In February 1942, emergency Air Corps ration began to be packaged in a fiber chipboard box as detailed in revised QMC specifications. The new box consisted of a wax-coated chipboard carton containing 3 4 oz. chocolate bars. It also reflected new nomenclature of US Army Field Ration D and included instructions. Carton was detailed under USAAF Dwg. No. 42B10959. The new assembly method consisted of placing bars in a sealed fiber master carton, then dipped in wax. In 1942 the USAAF added this ration to the complement of accessories for multi-person sustenance kits and life rafts at the ratio of 5 for Types B-3 & B-4 and 7 for Types A-2 & A-3. Dimensions of this ration were the same as the 41D4293.

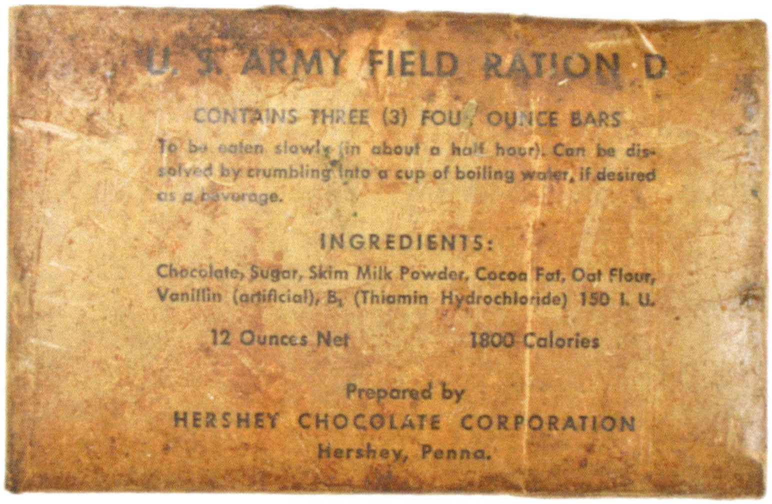

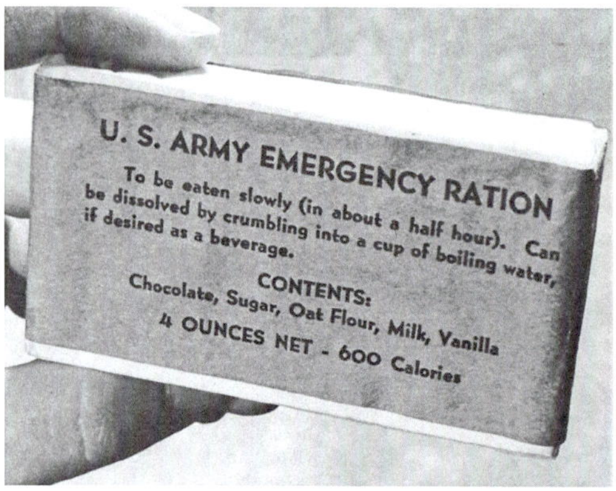

(Left and below) 4 oz. cakes included in the 38D2666 and 41D4293 containers. First tentative specification for Type-D ration were drafted 29 July 1940, with no CQD number. Specification indicated the bar was to be wrapped in aluminum foil then overwrapped in vegetable parchment paper with Kraft paper wrap-around label. These press release images illustrate the first generation of the emergency ration Type-D. The ration directly belows differs in packaging with a single wrap of parchement paper.

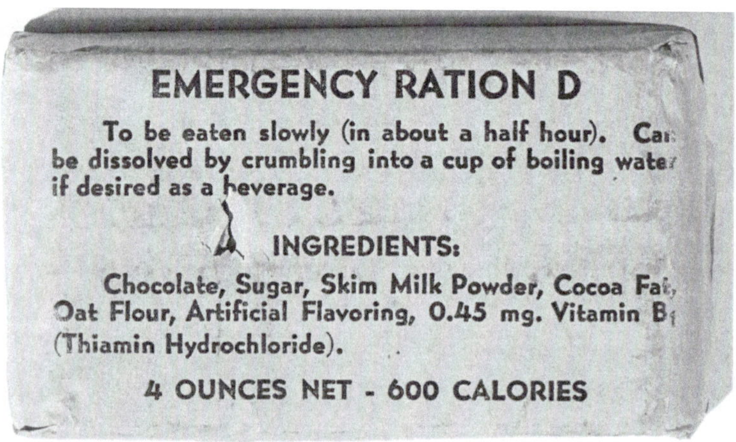

The Type-D ration was the work of Colonel Paul Logan, who began development of an emergency ration while at the QMC Subsistence School in 1935. In 1937, he presented a chocolate bar that would withstand extreme climate changes, give the greatest number of calories in the smallest weight and space (with 600 calories in a four-ounce cake), and assure that the ration was not palatable enough for an individual to consume until an actual emergency arose. Three bars for a total of 1800 calories were determined to be enough sustenance for an individual for one day. In the same year, 90,000 of the bars were procured for field testing purposes. Regular requisitions from the USAAC supplied emergency parachute kits. In October of 1939, the Logan Bar was designated as the US Army Emergency Field Ration D for use by both the Army's Ground and Air Forces. QMC Tentative Specification CQD No.19 (Chicago Quartermaster Depot) was officially assigned 2 October 1941, redesignated as United States Army Field Ration, Type-D. The emergency nomenclature was omitted with the issuance of the CQD specification. The premise was that, like that of the WWI emergency ration, it was intended for use when the ordinary supply of field rations had been interrupted. It was considered a supplemental ration for mobile troops.

The same packaging method was still used in July of 1940. A report submitted in December 1941 indicated the individual packaging method failed a series of tests conducted with chemical warfare gases. For remediation, revised Specification CQD No. 19B was approved on 30 January 1942. CQD No. 19B outlined that the bars were to be sealed in cellophane then contained in a sealed wax-dipped chipboard carton to prevent contamination from chemical agents. This further added water vapor and waterproofing protection. In line with this amendment, the use of tin containers was eliminated for Air Corps emergency rations and replaced by this new method, as detailed in USAAF Dwg. 42B10959.

Further amendments of the D-Ration did occur, but this was the last amendment affecting the procurement for the USAAF. The Type-D ration would quickly give way to the Bail-Out Ration,Type-K ration, and Parachute Ration and its eventual replacement by sweet chocolate bars. By 1943, the use of Type-D rations was discouraged for use in life rafts, as they were found to be thirst-provoking and were officially superseded by the Type-K.

The ration developments prior to 1940 had not changed much since WWI. The Reserve Ration was repackaged into cylindrical tins and adopted as the Type-C ration for field consumption and the Armour ration was reinstituted based on new service requirements as the Type-D ration, for emergency use only. With the introduction of new military tactics, it was believed that a ration should be developed for mobile troops such as paratroopers, tank corps, and other mobile assault units. The concept was to develop a ration that could be conveniently carried in the pocket since the Type-C was too bulky and heavy for the task. With the assistance of the University of Minnesota's Laboratory of Physiological Hygiene Department, the requirements of the new ration were established in early 1941. The ration was to be the opposite of the Type-D in that it was to be palatable, compact, light, require no cooking, and provide vitamins and nutrients. The concept was to have one universal ration but it evolved into three meals: Breakfast, Dinner, and Supper, constituting one day's ration for one person. Though it was intended for a variety of mobile troops it was quickly oriented toward paratroopers and dubbed the Parachute Ration, also known as the Blitz Ration by the University of Minnesota; other unofficial names were applied. In August of 1941, the new ration was ready for field testing. The QMC conducted field tests at Fort Snelling, Minnesota and the USAAF conducted field tests in Albuquerque, New Mexico. Each service praised the new ration and the USAAF immediately requested 30,000 rations in October of 1941, with the Office of the Quartermaster General (OMCG) designating it as the Field Ration Type-K the same month. The Type-K ration was officially adopted 25 May 1942. The USAAF would adopt and integrate the Type-K for use in emergency sustenance kits and life rafts.

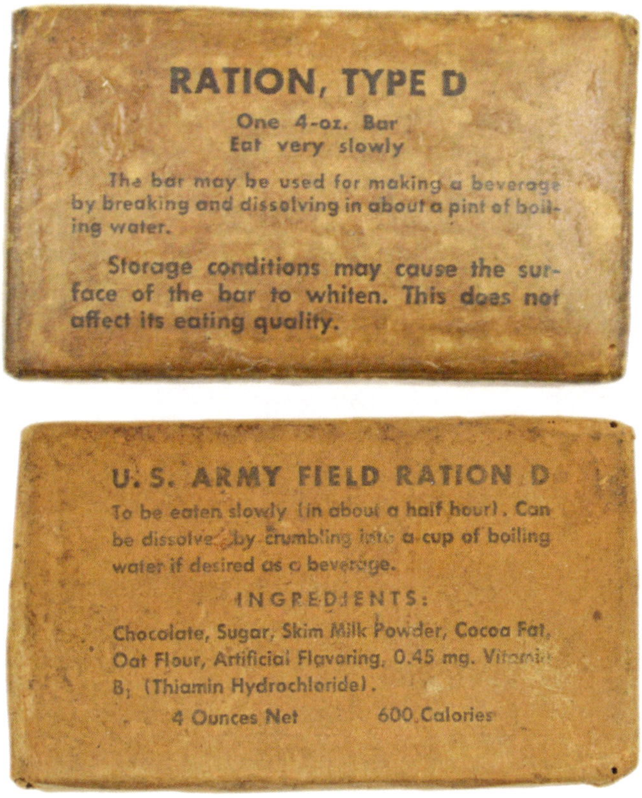

(Left) Two of many variations of the US Army Field Ration D. These were packaged with a single 4 oz. bar and remained standard field ration for the US Army and Marine service forces.

(Above and right) Mobile assault ration field-tested in August 1941 as featured in Newsweek. Each unit was packaged in compact rectangular boxes weighing 1¾ lbs. Three units met the 3,200 calorie daily requirement for one person. For comparison, a person would need to carry 7 lbs. of Type-C rations to obtain 3,600 calories.

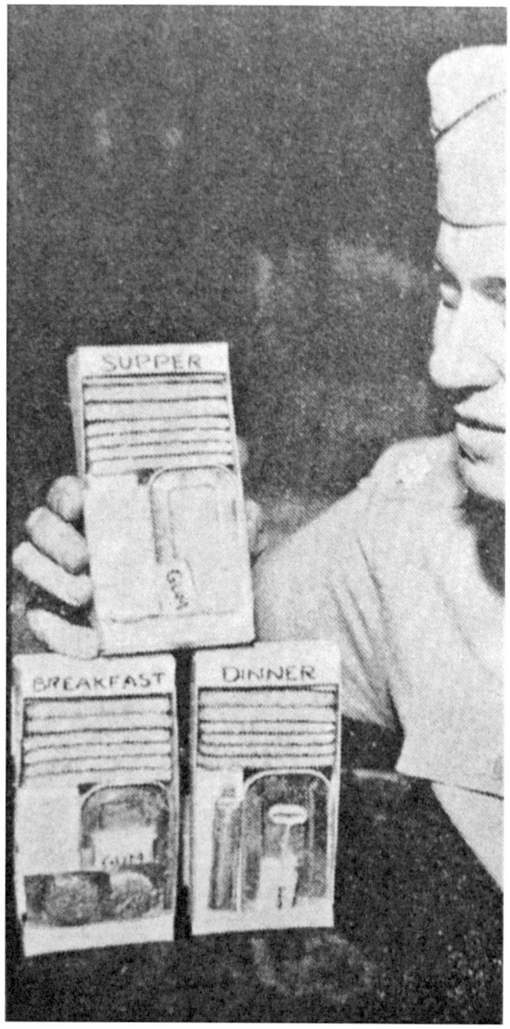

(Left and below) USAAF volunteer demonstrating Type-K rations during New Mexico field tests, August 1941. (National Archives)

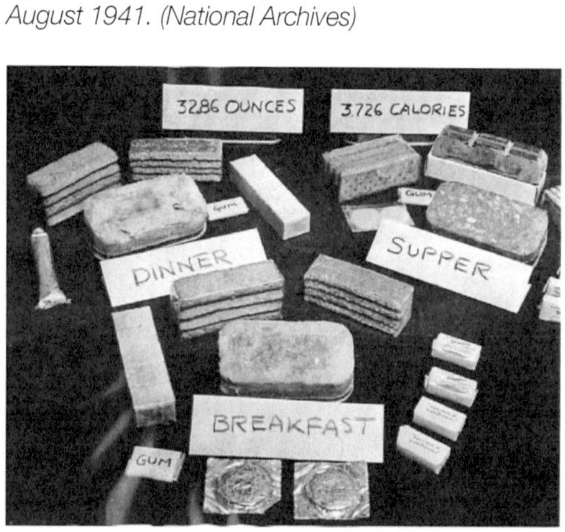

(Above and right) QMC Tentative Specification CQD. No. 28, released 16 December 1941, for new Type-K ration. It detailed two methods of packaging: 1) Components placed in a carton waxed inside and outside. Filled carton then sealed by heating its waxed flaps. 2) One end of the carton was glue-sealed; the individually packaged components were inserted, and the other end of carton was sealed with glue. The filled and sealed carton was then dipped in wax. Upon adoption, the packaging method was refined and carton slightly increased to 6" x 4" x 2". Photographs illustrate first Type-K ration adopted and placed into production. Samples were used in field tests conducted by 29th Infantry Division at Fort Benning, Georgia from February to March 1942. Also included in the tests were C- & D-Rations. (National Archives)

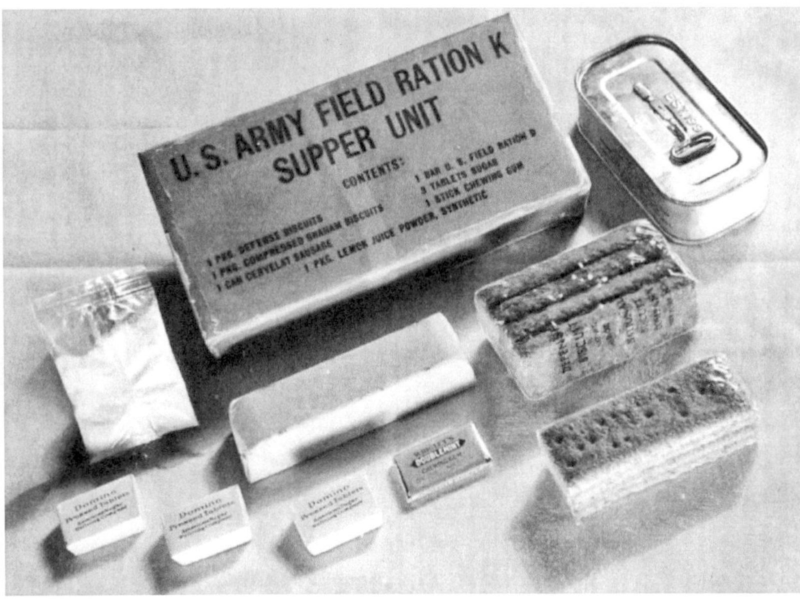

To facilitate production, the SR&DL desired to increase the methods of packaging for the Type-K, allowing procurement from a wider variety of potential contractors. CQD No. 28A became effective on 27 April 1942 to facilitate this. It outlined six procedures for packaging: 1) the packaged components were placed in a carton that had been waxed inside and out; the ends of the carton were sealed by reheating the waxed flaps. 2) The packaged components were assembled in a carton, the flaps of the carton were sealed, and the carton was dipped in a micro-crystalline bath. The waxed carton was then wrapped in a waxed Kraft paper and again dipped in a micro-crystalline wax. 3) The filled and sealed carton was dipped in micro-crystalline wax, inserted in a second carton, and again dipped in micro-crystalline wax. 4) The components were to be placed in a full telescoping carton which was wrapped in Kraft paper, dipped in micro-crystalline wax, and finally overwrapped with waxed Kraft Paper. 5) The components were placed in a lined, unsealed carton double-wrapped with waxed paper. 6) The components were placed in a thread-opening fiber can. The fiber body of the can was to be coated both inside and out with wax.

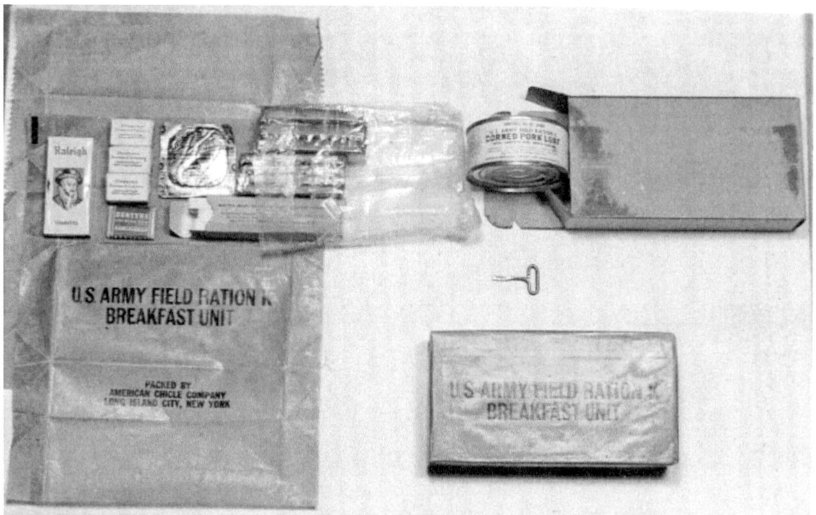

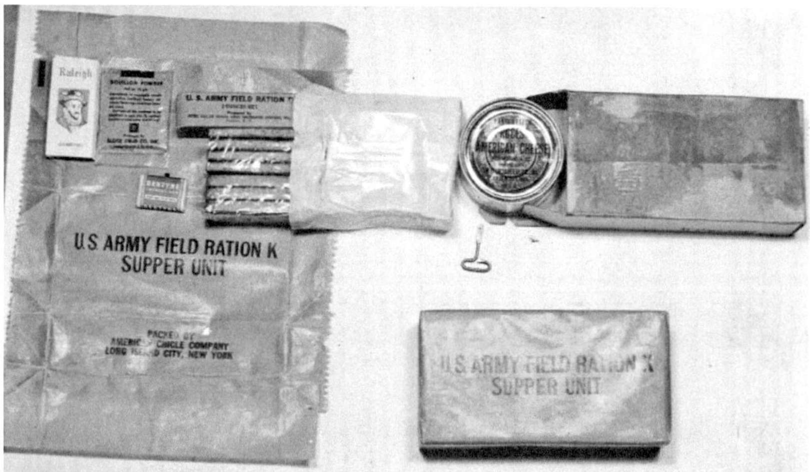

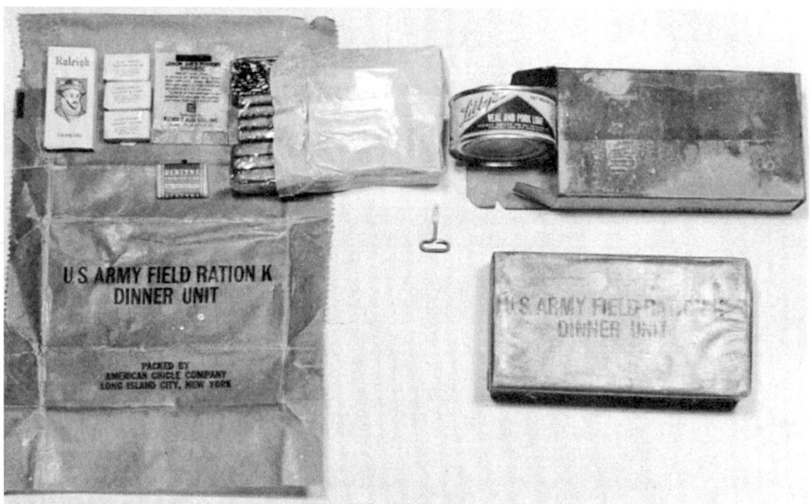

The Type-K ration underwent subsequent packaging revisions, but principally conformed to one of the six methods as detailed in CQD 28A. CQD No. 28B of 20 May 1942 made no changes to the packing. CQD No. 28C 3 August 1942 reduced methods of packaging to 2: wax-dipped carton over-wrapped in Kraft paper, or waxed-dipped carton over-packaged in another carton. CQD No. 28D of 12 January 1943 reduced the packing method to 1: components were placed in specified laminated bags, then inserted in a sealed-end carton, which was then dipped in wax and placed in an outer chipboard carton. CQD No. 28E of 24 February 1943 required the canned meat to include a chipboard sleeve. CQD No. 28F of 22 September 1943 directed that only a cellophane bag be used for packing the components before being inserted into the outer cartons. CQD No. 28G of 31 October 1944 refined the specifications for components. Last revision occurred 31 August 1945 under CQD No. 28H and detailed 2 methods of packaging. Either the filled and sealed carton was to be wrapped in Kraft paper laminated with aluminum foil and waxed tissue, or it was additionally wax-dipped and overpackaged in a chipboard carton. The Breakfast, Dinner, and Supper units illustrated here conform to CQD No. 28A. (National Archives)

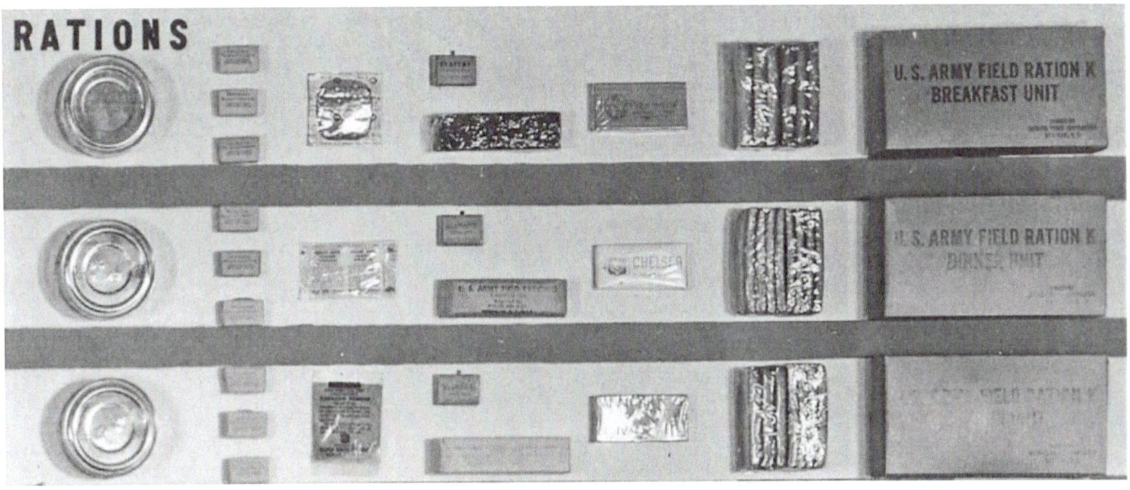

(Above) On a display board were Type-K rations conforming to CQD No.28D. Components were contained in an inner wax-coated and an outer chipboard carton. Standard for USAAF life rafts was nine Type-K rations (5 Dinner, 4 Breakfast), available under USAAF stock number 8300-669200. (National Archives)

(Left and below)) To distinguish rations from munitions and spare parts, rations were packaged in colored cartons. Each unit was assigned a specific color and pattern, which also helped the packing contractor from mistakenly placing rations in the wrong cartons. Three units and their respective colors and patterns. Type-K remained a standard subsistence item in emergency sustenance kits for the USAAF overland but was replaced by Life Raft Ration Type-A in life rafts. BuAer also used the Type-K, if available, for aircrew meals. (below) Colored rations were part of an assembly of Type E-14 sustenance kit. (Robin Bartel Collection)

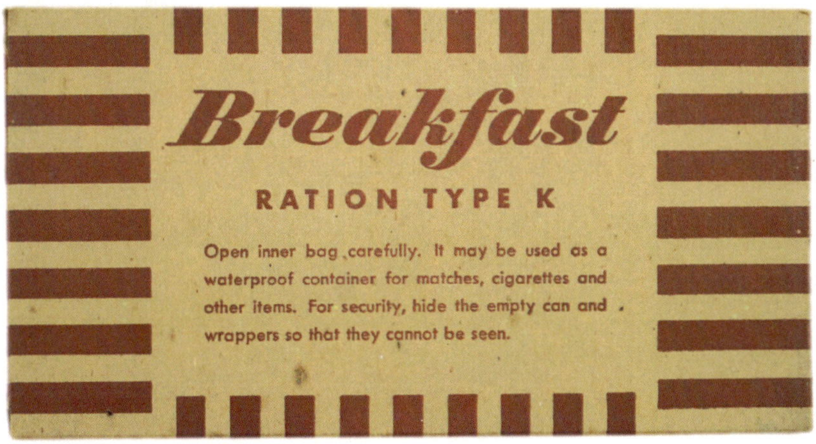

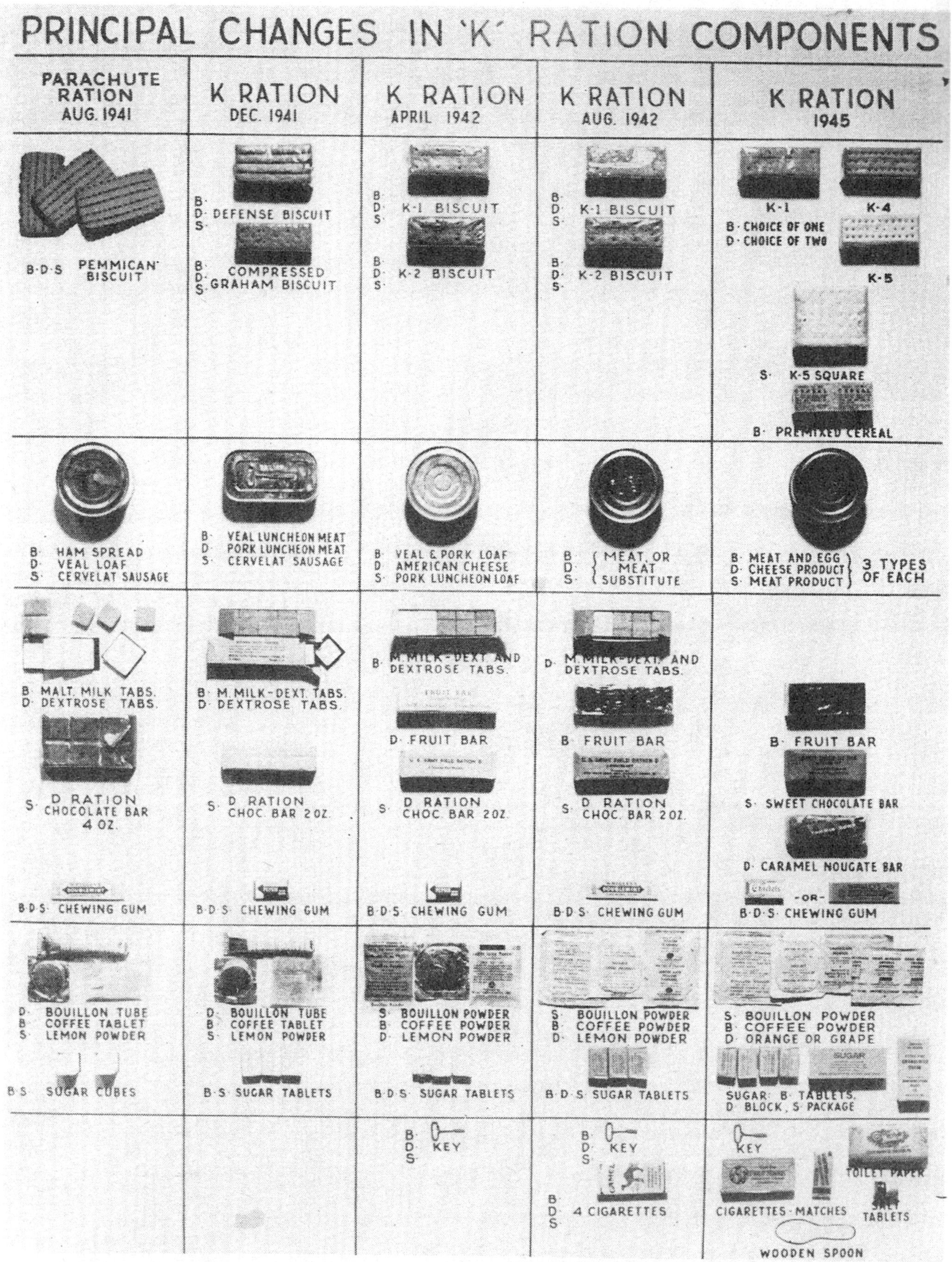

(Below) Chart outlines the changes in the Type-K ration components from its inception to the end of WWII. There were many variations of the components and packing methods; only a few examples are illustrated in this chapter.

With the adoption of the Type-K ration, the USAAF began to evaluate existing emergency rations included in parachute kits. It was found that the Type-D provided no real nutritional value and the USAAF required a ration like that of the Type-K for emergency parachute kits. The SR&DL began development of a suitable ration for the USAAF in the early spring of 1942; however, the USAAF expedited development. The origins for the new ration derived from the development of the Type E-3 sustenance kit copied from the British ration boxes and requirements for in-flight consumption from the 8th Service Command in the ETO with its proposed carbohydrate ration. Compiling the features from those rations, a basic ration for inclusion in parachute emergency kits was developed titled as the Bail-Out Ration.

This ration was a derivative of the Type E-3 but included only food. The foodstuffs of the new ration, from a nutritional standpoint, would give fliers forced to bail out a maximum chance of maintaining their morale and physical condition until they could find other supplies. The Bail-Out Ration was intended to supersede the Type-D complement of three four-ounce cakes in parachute kits and was available for distribution on an individual basis, and was carried in flight clothing. It was also included in life rafts on a limited basis. The rations were soon found to be unsatisfactory: they failed to withstand rough treatment, the package could not withstand pressure at high altitudes inside the carton by allowing air to escape and allowing new moisture-laden air in, which accelerating deterioration.

In July 1942, the USAAF authorized procurement of 100,000 Bail-Out Rations designed for incorporation in parachute emergency kits for use overland as an individual survival food unit. The ration contained two 2oz. D-bars, two boxes malted milk and lemon dextrose tablets, two sticks of chewing gum and a package of bouillon powder. These were in a heat-sealed chipboard craton dipped in wax. By the close of 1942, 460,000 Bail-Out Rations were approved for procurement, available under stock number 8300-669050. The cracker Jack Co. and William K. Wrigley Co. were approved contractors. The Wrigley Co. began production in January 1943.Thier production included the title: US ARMY SPECIAL AIR CORPS RATION.

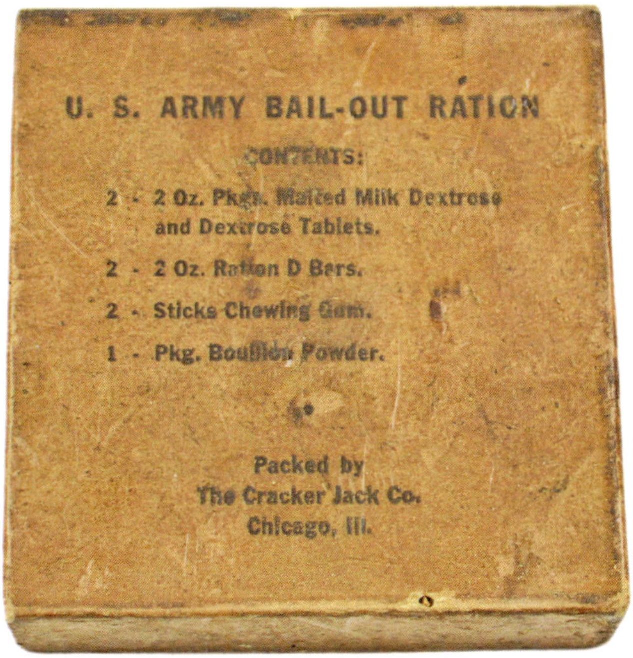

Bail-Out Ration assembled by The Cracker Jack Co., package measures 4" x 4" x 1-1/8".

(Upper right) Wartime photo of the Bail-Out Ration and its components.

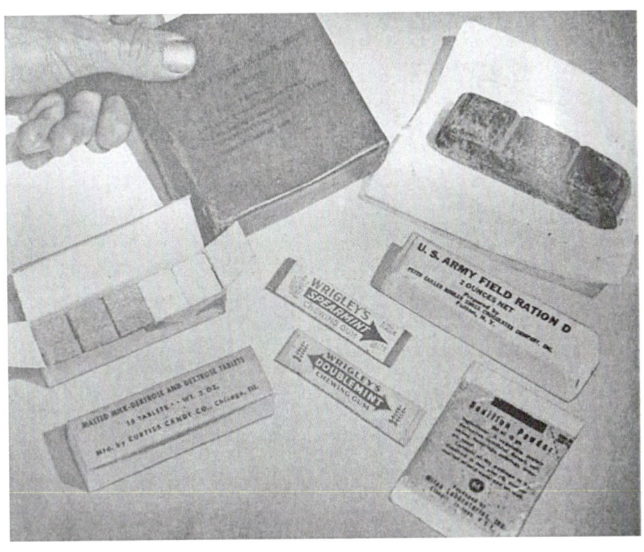

(Right) Bail-Out Ration packaged by Wrigley Co. with nomenclature US ARMY SPECIAL AIR CORPS RATION.

Tentative specifications for the Bail-Out Ration were drafted and submitted 9 October 1942. The SR&DL did not approve of the existing configuration but procurement by the USAAF continued without approval as an interim. By early 1943, the Bail-Out Ration was found to be universally unsatisfactory, failing to meet the rigid requirements of a survival ration. The USAAF did formally establish a requirement for a new type of ration for the Type B-4 parachute emergency kit. First step was a revision to the Bail-Out Ration to include: two 2oz. D-bars, a 2oz. fruit bar, package of hard candy, package of lemon beverage powder and several packages of K-ration biscuits.

To meet this demand, specifications were drafted by the SR&DL to include those proposed additions, thereby increasing the caloric count. The new ration was to be assembled in a key-opening vacuum-sealed tin container, which remained sealed and could withstand extreme temperature and altitude changes. This ration was titled as the Parachute Ration and expedited procurement was authorized without a CQD specification.

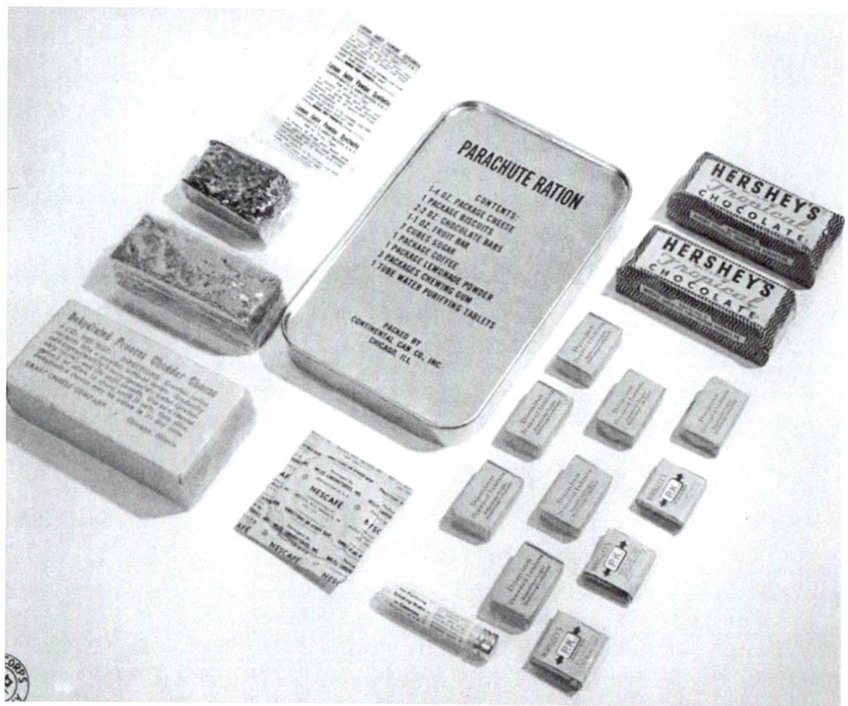

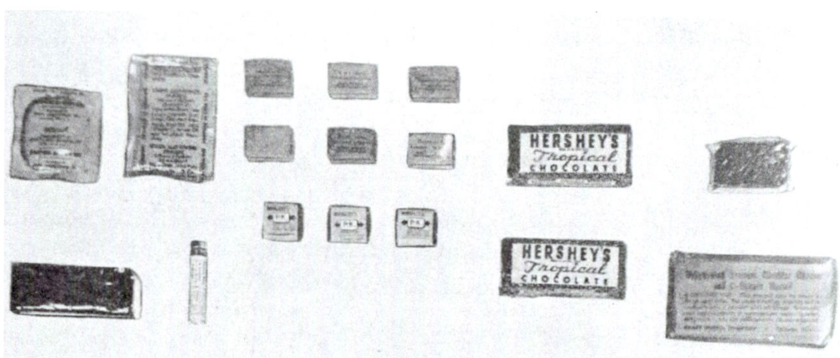

Contracts for the Bail-Out Ration were cancelled in favor of the new Parachute Ration per specifications by the SR&DL. The new ration was packed in a key-opening tin container 4-1/8" x 6-5/8" x 1". This modified Type-K ration included a brick of cheese, biscuits, chocolate bars, a fruit bar, sugar, coffee, lemon powder, chewing gum and halazone. It was designed for use in the Type B-4 parachute kit but was available for integration into the preceding types. Only one procurement was made for a quantity of 85,000, entering production in August of 1943, available under stock number 8300-680650. (National Archives)

In October of 1943, the USAAF requested that a ration in a smaller tin be created for inclusion in an emergency vest then being developed. The SR&DL, in coordination with the Continental Can Co., were quickly able to revise the existing Parachute Ration to a smaller tin, tentatively titled Ration, Parachute, Emergency, Vest Pocket Type. The nomenclature was intended to distinguish it from the larger Parachute Ration. The new ration was officially standardized on 17 July 1944 under Specification CQD No.302 and was titled Ration, Parachute, Emergency, Vest Pocket Type. It immediately entered mass production to supply recently released expedited contracts for the Type C-1 emergency sustenance vest. It would officially supersede the Parachute Ration with the understanding that the Type C-1 would supersede all previous parachute emergency kits, making future procurement of the older ration unnecessary.

In October of 1944, the title was revised to Ration, Parachute, Emergency. Detailed in the original specification, the ration tin was to be stenciled EMERGENCY PARACHUTE RATION. The components of this ration were four one-ounce sweet chocolate bars, a package of hard candy, one cheese and cracker bar, two bouillon cubes, two tablets of sugar, two vials of halazone, two packets of coffee, fourpieces of chewing gum, a cellophane bag (for storing components after opening) and a can opener. The complete assembled ration measured 3-1/8 x 5-1/2" x by 1-1/4".

In January of 1945, the contents were revised under Specification CQD No.302A altering the components to one vial of halazone, one coffee packet, six pieces of chewing gum, two paper clips and one packet of cigarettes. This ration would be available under QMC stock number 56-R-8600 and stock number 8300-680825 from USAAF supply.

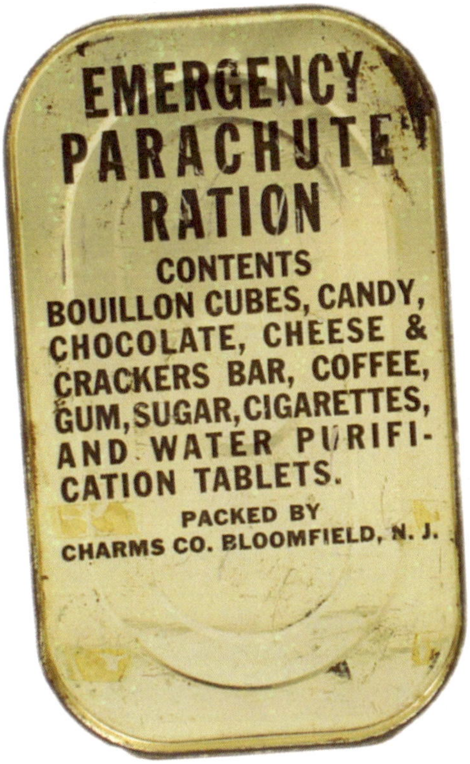

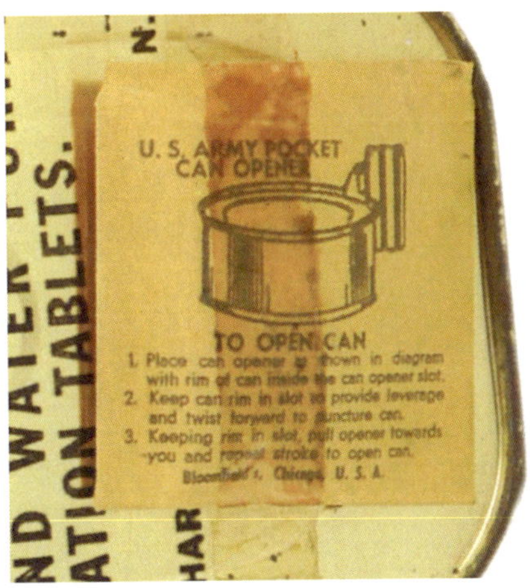

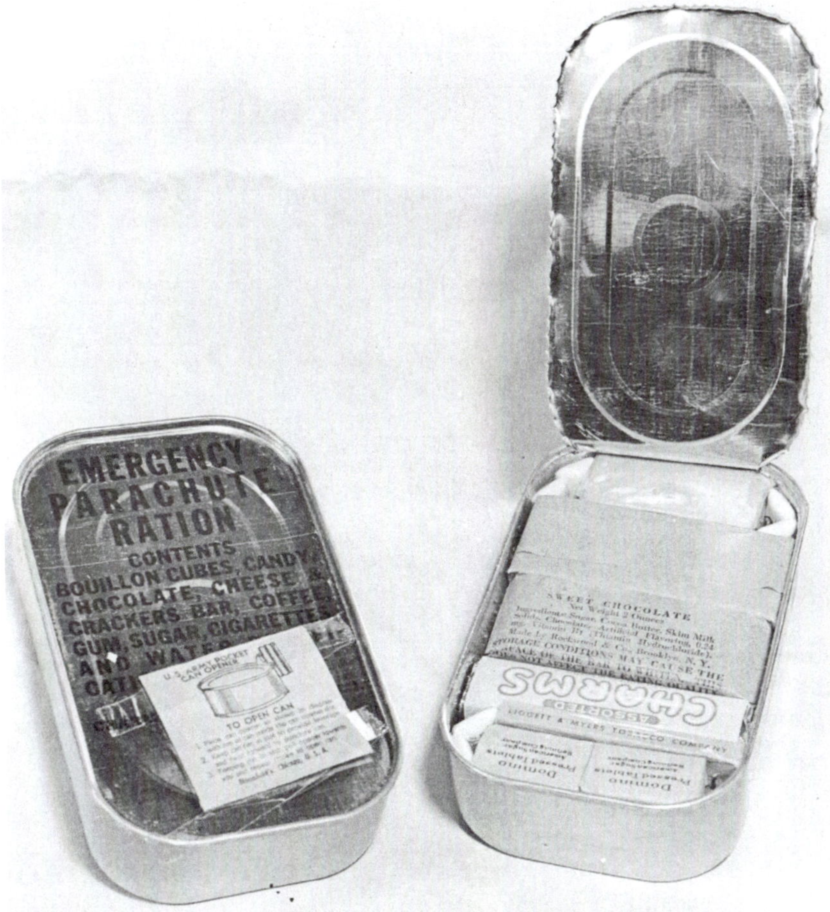

Emergency Parachute Ration CQD 302. Can opener was taped to outside of tin. Charms Co. was the only assembly contractor, producing 1,557,020 rations. Wartime image of ration sealed and exhibiting packing method. (USAF)

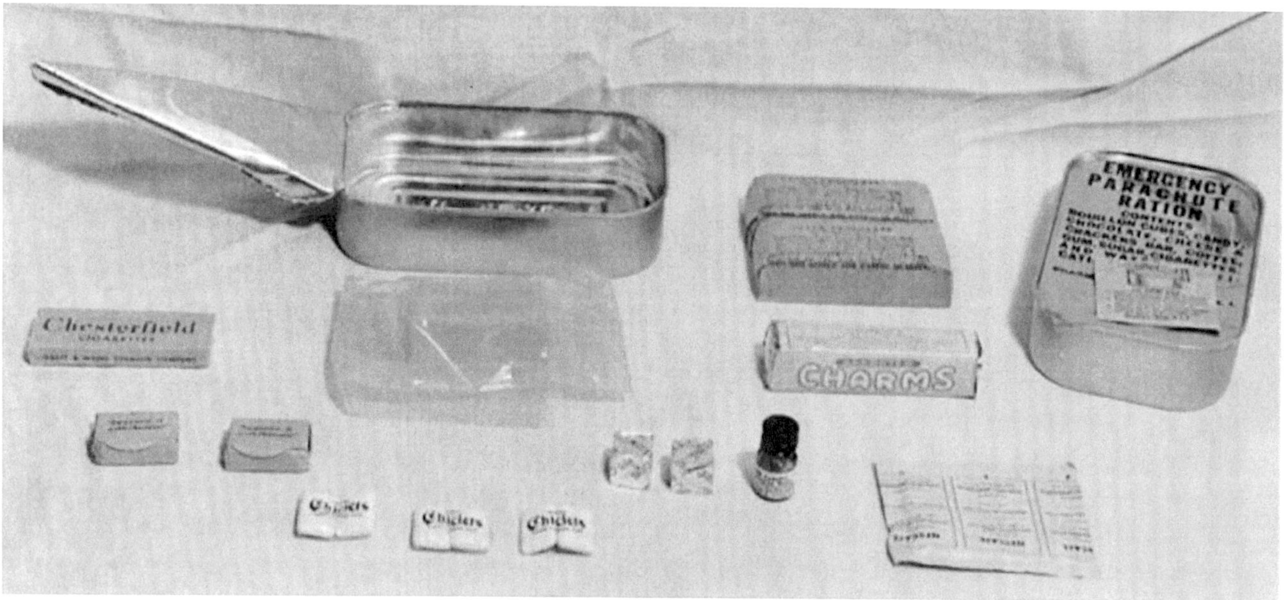

Displays of components of Emergency Parachute Ration. Assembly per January specification (above) and per July specification (right). Note omission of "Cigarettes" on face (below). Cigarettes eased hunger and relieved boredom. (National Archives)

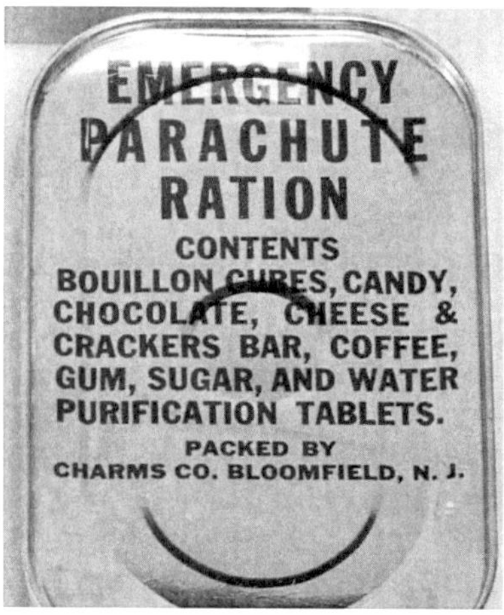

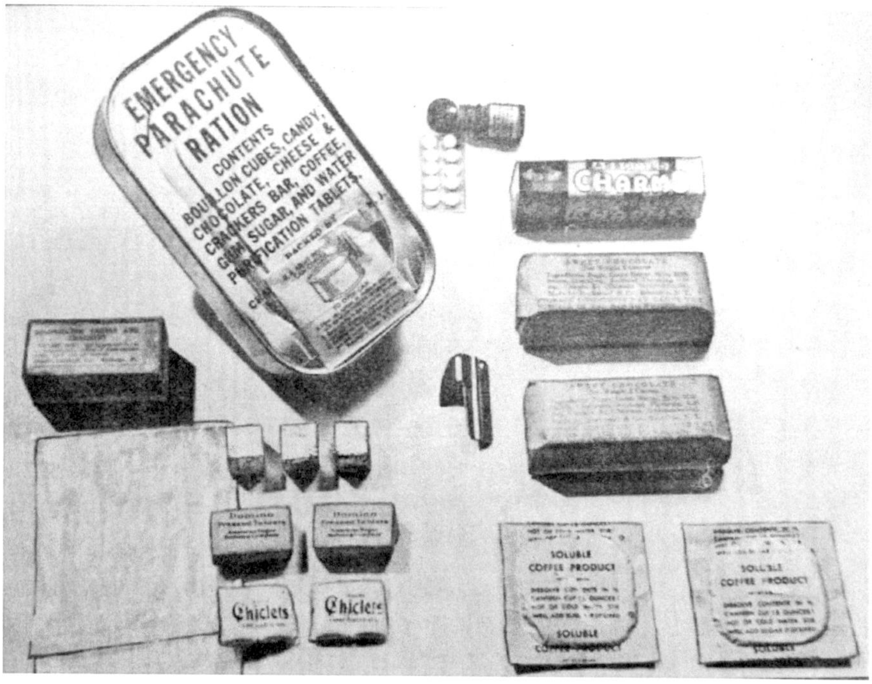

To recap, upon entering WWII the US Army Field Ration D, Dwg. No. 42B10959, was included in USAAF life raft accessories when supply became available. The Type-K replaced the Type-D when it became available by late 1942 and early 1943. In February of 1943, the USAAF informed the OQMG that the Type-K were unsatisfactory for use in life rafts but were still determined satisfactory for over-land sustenance kits. The notable issue with the Type-K was the deterioration of the packaging; when stored with the life raft it was subjected to high heat from being stored in closed metal compartments on aircraft in the tropical sun, and the ration deteriorated when subjected to long periods of immersion in salt water. The Aero Medical Laboratory, in conjunction with the SR&DL, worked on a new ration for life rafts.

Through nutritional studies, a diet of carbohydrates was found to be less thirst-provoking, which led to the consideration that a ration for life rafts include candy, gum and vitamins. By April of 1943, a suitable life raft ration had been developed comprised of candies. On 18 May 1943 the new ration was titled as the Life Raft Ration Type-A and approved for standard inclusion in life rafts and over water sustenance kits. This new ration would include 12 packages of Charms candies, 12 loose Charms candies, six packages of sugar-coated gum, and six vitamin tablets, packed in a square vacuum-sealed key-opening tin. The approved specification was released 4 December 1943 under Specification CQD No.164. Expedited procurement was authorized upon its approval in May of 1943.

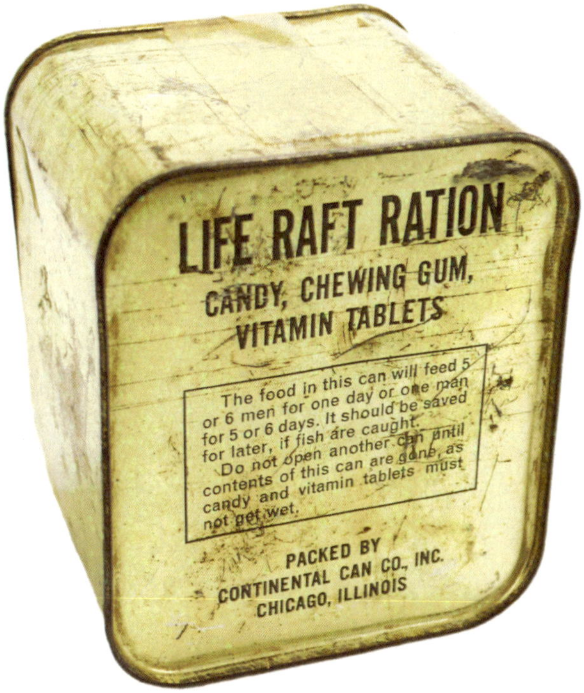

(Above) Continental Can Co. was awarded first contract for assembly of Life Raft Ration Type-A, May 1943. A consumption instruction sheet was included. Specifications were amended to require supply of 7 cans for each life raft or 1 per man. Example from first production, Life Raft Ration Type-A.

(Above) Specification CQD 164, December 1943, dictated that instructions were also to be printed on can, as shown in this post-December 1943 Life Raft Ration Type-A packaged by Continental Can Co.

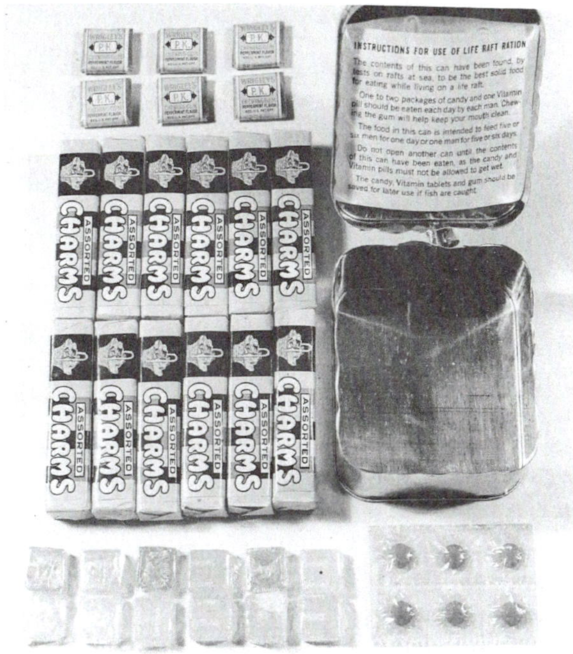

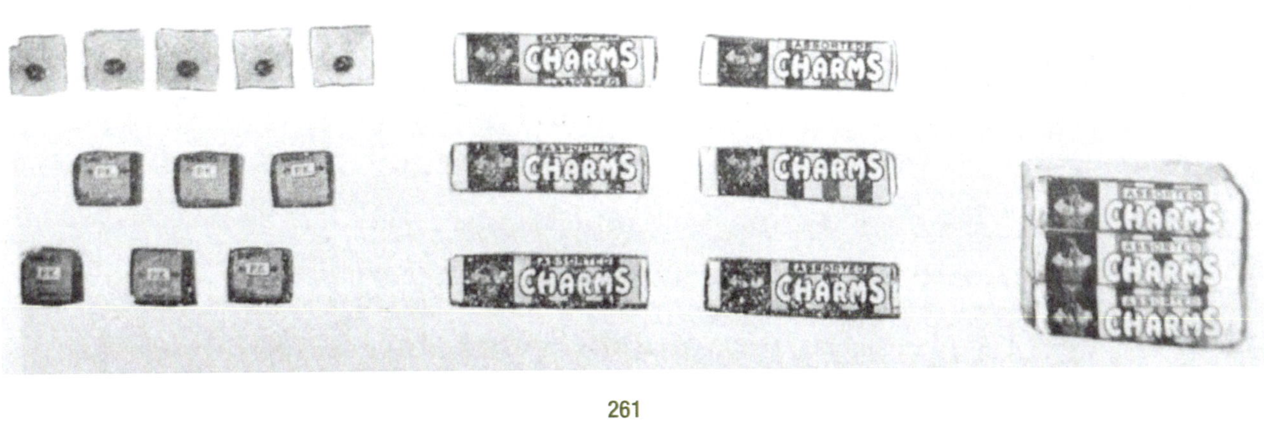

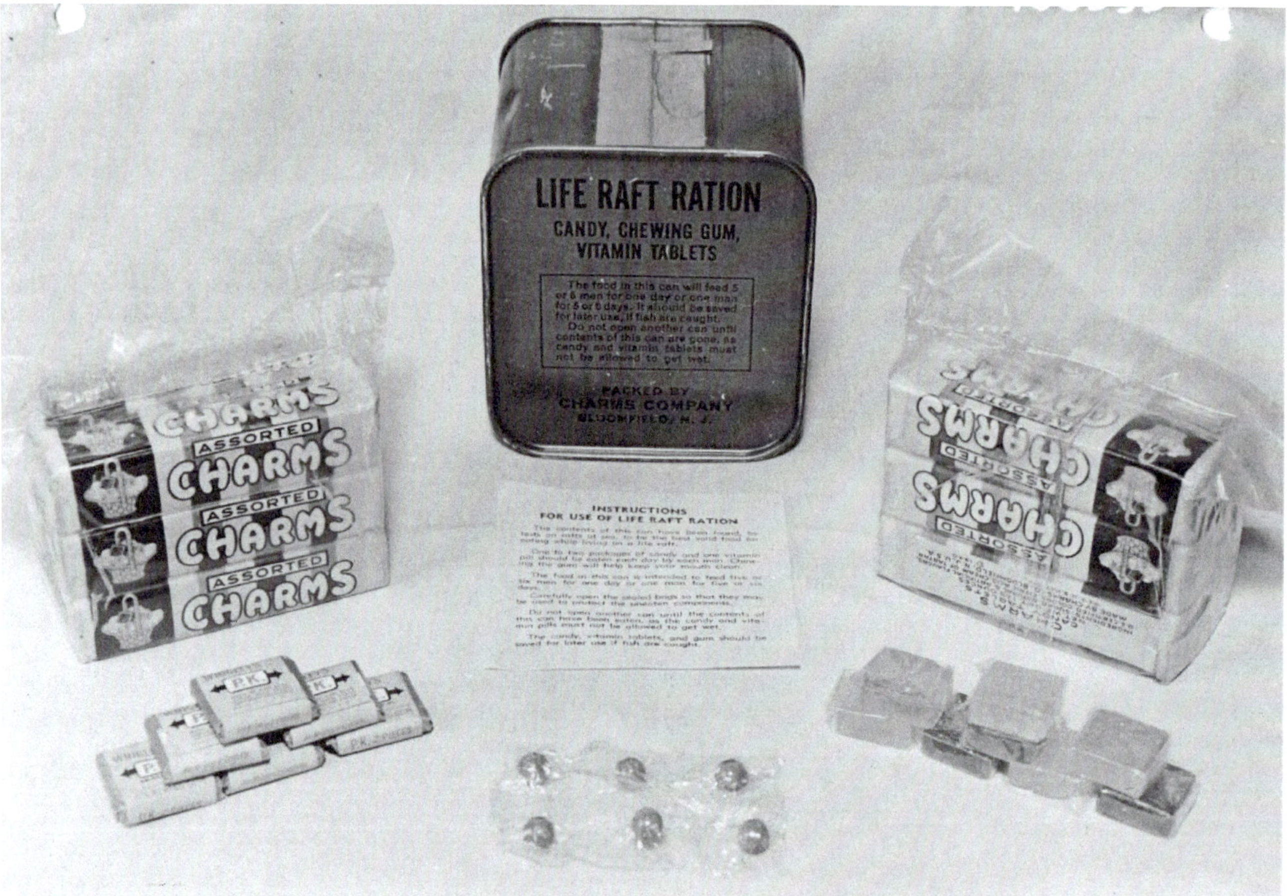

Life Raft Ration Type-A was revised 7 June 1944 under Specification CQD No. 164A; candies were to be overpackaged in heat-sealed laminated cellophane envelope.12 packages of Charms candies sealed in cellophane, six per package. Charms Co. was awarded assembly contracts of the ration. Production of Life Raft Ration Type-A reached 1,753,012 units and superseded Types -D & -K rations for life rafts and over water sustenance kits in August 1943. It was available under QMC stock number 56-R-7225 and stock number 8300-680300 from USAAF supply. Through Spring 1945, USAAF contemplated a universal ration for inclusion in the type C-2 sustenance vest, other sustenance kits, and life rafts. The discussion was to repackage it in a tin of same dimensions as existing Emergency Parachute Ration. Detailed under Specification CQD 164B dated 11 September 1945, new components were four packages of hard candy holding 10 tablets each, package of hard candy holding nine tablets, four penny packages of sugar-coated gum, seven malted milk tablets and an empty plastic bag. This ration never entered production. (National Archives)

In 1944, the USAAF was developing an air-droppable boat, for which it requested the development of a ration large enough to feed ten men for twelve days. The concept presented by the Materiel Command was that it wanted to provide meals to feed survivors, designed to be stored in watertight bulkheads on the boat and accessed through hand holes about 4-1/2" x 5-1/2". To meet this request, SR&DL began development in March of 1944, working with the Higgins Boat Co., to help design a suitable ration. It was standardized on 27 September 1944 and titled Ration, Life Boat, Airborne, referenced under CQD Specification No. 320. The ration was provided in two menus, each consisting of two meals of breakfast and supper. Each menu provided food for two men for one day with 120 rations provided in the lifeboat and one Life Raft Ration Type-A per six lifeboat rations; 20 were supplied. The breakfast menu was in a 4" x 3" x 8-1/2" package and the supper unit in a 3-1/2" x 3" x 9" pack. Only one procurement was made for 96,000 rations, assembled by the General Foods Corp.

(Rught and below) Breakfast unit in Menu #1 consisted of a can of chicken soup, can of beef & pork, can of chopped eggs, B-Unit of coffee and package of hard candy. Supper unit of Menu #1 consisted of C-Rations: Can of frankfurters & beans, can of meat & noodles, one beverage can and package of hard candy. The Breakfast unit in Menu #2 was a can of beef & noodle soup, can of chopped ham & eggs, can of pork lunch, canned beverage and package of hard candy. Supper unit of Menu #2 consisted of a can of meat & spaghetti, can vegetable stew, beverage can and package of hard candy. These rations comprised standard QMC ration components. Hard candies listed were provided in life raft ration. Upon consumption, a life raft ration tin was directed to be opened. (National Archives)

The rations described thus far were oriented toward the requirements for an emergency, to provide basic sustenance to individuals in survival situations. As aviation units began operating in combat theaters, new requirements were established for the ability to feed flight personnel while on missions. To accomplish this, many factors had to be considered, such as what ground facilities were available, length of the missions, and the type of aircraft. These factors affected space limitations. The situation that arose was very much like that of the requirements of cross-country flights encountered by the USAAC in the 1920s and 1930s but in wartime conditions, cross-country flights had become a daily occurrence with flight times averaging from three to eight hours and potentially up to 30 hours. Providing food for air crews was left to the discretion of the local commands supplying fresh foods or Type-A rations.

(Above) Hot and cold foods were prepared for flight crews at local mess facilities. In this series of photos, sack lunches are assembled containing donuts, sandwiches, candy, fruit and more. Provisions were then transported to the flight line and distributed to aircraft. Water, coffee, and other drinks were supplied inflight in vacuumed-sealed jugs and Thermos flasks.

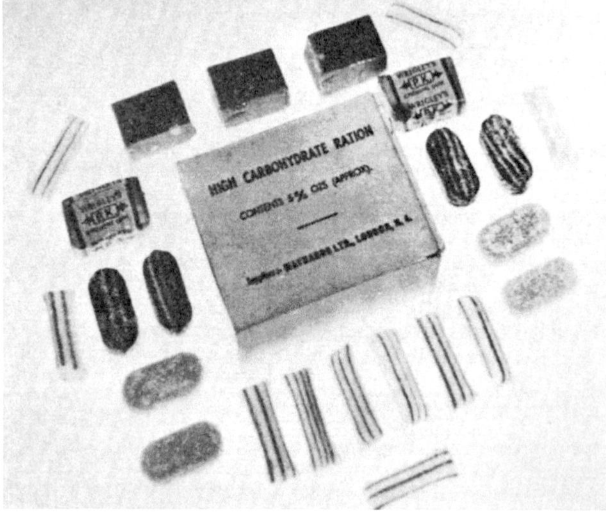

(Left) The first formal attempt to provide a standard ration for inflight meals was made by the 8th Air Force in 1942; it was known as the High Carbohydrate Ration. The priority was not to provide a balanced diet but food that provided energy, helped alleviate hunger, and supplemented Type-A rations. This ration comprised easily obtainable British foods. It was the foundation of the Bail-Out Ration.

In October of 1942, the USAAF requested that SR&DL develop a ration for in-flight consumption with a package holding one meal for three men. In March of 1943, a design was agreed upon and dubbed the Lunch, Combat, AAF. A limited procurement was issued on 19 June 1943 of 6,000 units for field testing. A wax-coated fiber carton measuring 8-1/2" x 7" x 6" contained dehydrated ground beef, precooked rice, coffee, tea tablets, dry milk, bouillon cubes, lemon powder, sugar, salt, butter spread, chili powder or tomato paste, biscuits, hard candy, dried fruit, fruit bars, gum and a can opener. The ration was designed to be prepared utilizing galley kits Types C-1 or C-2. This proved problematic since it had already been determined to be impractical to prepare food while in flight. Additionally, there was limited space aboard bombardment aircraft and galley kits were being considered for discontinuation. However, further procurement was initiated after the ration was officially detailed under Specification CQD No.156 dated 4 October 1943 for a quantity 250,000 units. This specification redesignated the ration to Lunch, Flight, AAF since it was felt the word Combat gave an improper overtone to an inflight meal. Even as the USAAF was contemplating the discontinuation of the flight lunch, minor revisions were still being made. Revised Specification CQD No.156A effectively changed the name back to Lunch, Combat, AAF with some changes in the components. A further procurement of 225,000 was authorized upon this revision. Ultimately, by the close of 1944, the combat lunch was discontinued in favor of finding a better means of providing meals for aircrew members.

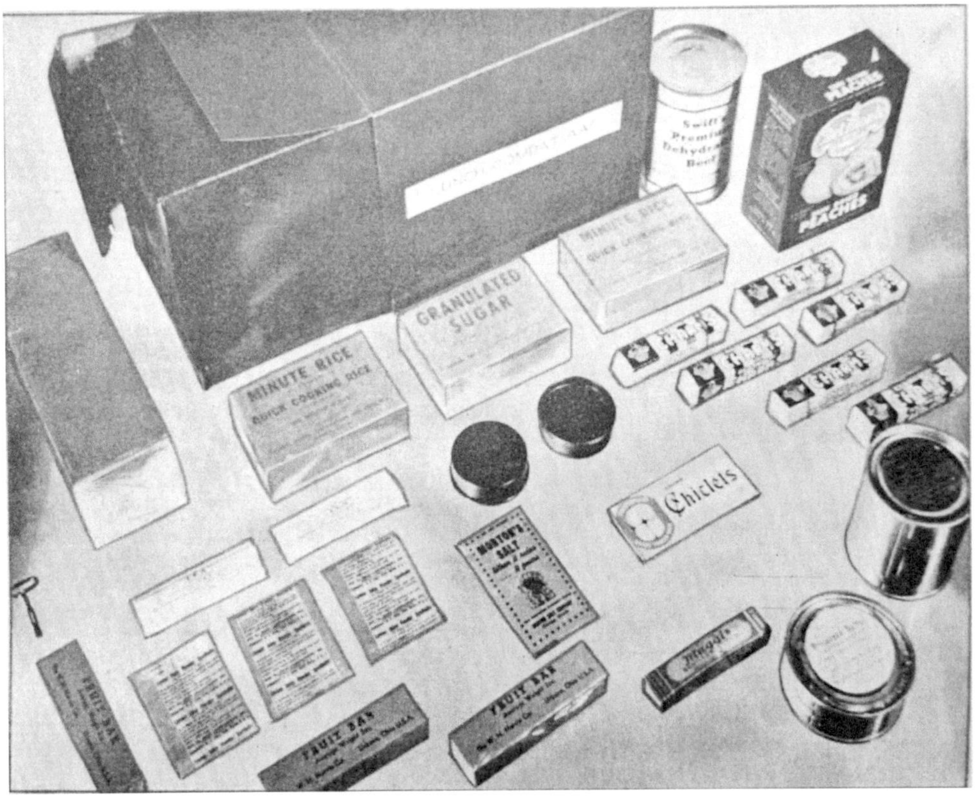

Lunch, Combat, AAF. After its discontinuation, the USAAF attempted to accommodate service requests for meals ready-to-eat by developing a Sandwich Beverage Pack containing ingredients for sandwiches for 12 aircrew members but soon dropped the concept.

The 8th Air Force's High Carbohydrate Ration further influenced rations for USAAF personnel, as its development continued at the Aero Medical Laboratory. In April 1943 they proposed a supplement containing candy be developed to maintain aircrew energy levels on missions lasting longer than an hour. The development process continued through the Fall of 1943, with the ration having to meet the requirement of a pocket-sized package that could be opened with one hand while wearing flight gloves. In November of 1943, 10,000 samples were procured from the Beech-Nut Packing Co. for field testing. This company helped design the ration and had given it the name Strata Sweets; the USAAF designated it as the Air Forces Pocket Lunch. It remained in development until the summer of 1944, when it was officially standardized as the Lunch, Air Crew, and detailed under Specification CQD. No. 319, issued on 15 September 1944. This was followed by a procurement of 15,828,000 units. It was the only USAAF ration maintained strictly through its standard supply channels and was available under stock number 8300-680475. Technical Order 00-35A-36 dated 11 November 1944 states: "Issued, in addition to any other ration, to all Air Crew members on missions three or more hours in duration." The purpose of this ration was to relieve hunger, alleviate fatigue, and improve efficiency of combat crews.

Concept pocket lunch for the USAAF as manufactured by Beech-Nut Packing Co. (USAF)

The Air Crew Lunch.

Contents included fudge and vanilla bars, candy-coated gum, and a candy mixture of peanuts, chocolate drops, cream centers, gum drops, licorice drops, and fondant creams in a tray and sleeved container of 3 3/4" x 3" x 1". (National Archives)

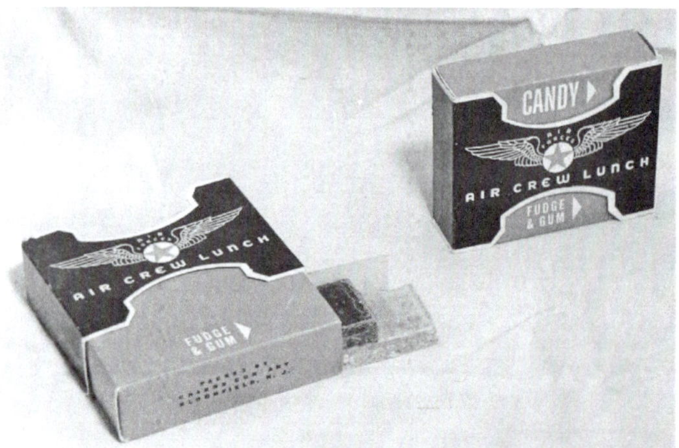

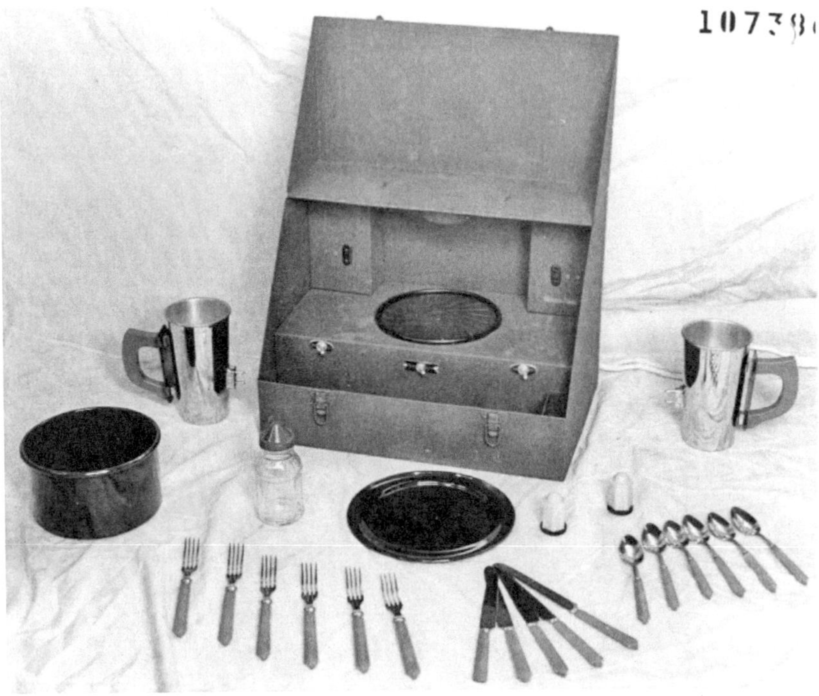

The USAAF developed, standardized and in January 1942 procured 3,024 units of the Kit, Galley, Type C-1 for preparing food inflight. It was designed to cook Type-A rations, measuring 16" x 12" x 12", and contained utensils, a hot plate and two heating cups for meal preparation. It used the aircraft's 24 volt electrical system. The combat lunch was designed to be prepared in Type C-1 or C-2 galley kits. (National Archives)

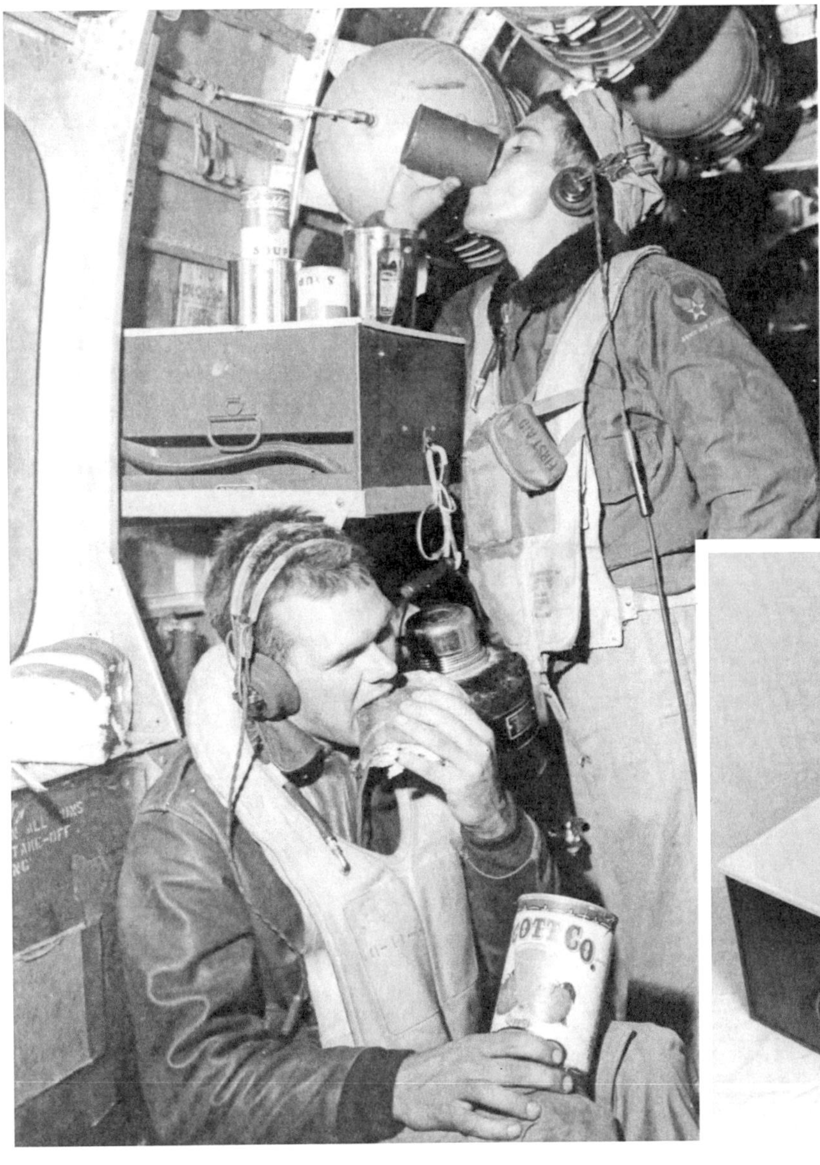

(Left) USAAF B-24 crew members take a moment for an inflight meal. Tray secured to fuselage holds a Type C-2 galley kit, Mariana Islands, c.1944. A liquid jug can be seen below the galley unit. (National Archives)

(Below) The Kit, Galley, Type C-2 was standardized on 2 March 1942. It had a grill, heating cup, two warming wells or heated storage cups and utensils necessary for cooking and eating. It was 30 -1/2" x 13-1/2" x 9-1/2" and 12,684 units were produced. Both galley kits were used on a limited basis as they were too bulky for most airplanes. (National Archives)

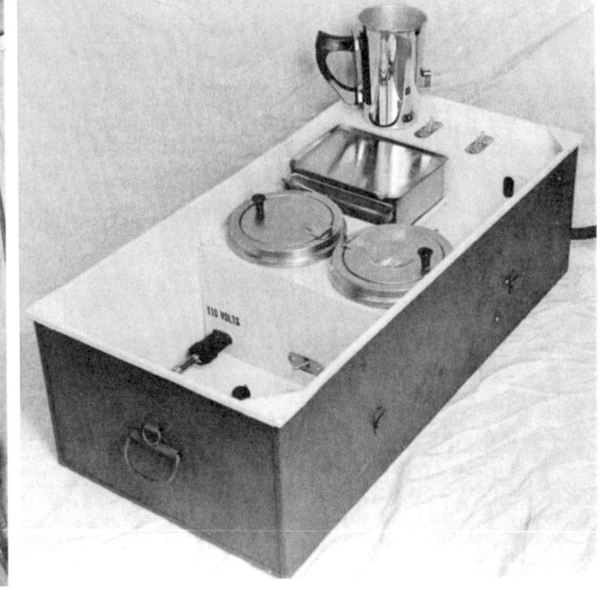

Long-range fighters were provided a Type-K ration lunch packet and hot or cold drink, typically a thermos of hot or cold water for coffee or lemon powder to be mixed in. Galley units were intended for long-range bombardment and ferrying missions and were considered essential for morale and efficiency. AAF Memorandum No. 65-15 dated 3 May 1943 defines the distribution of this equipment. It states that galley kits Type C-1 or C-2 will be provided for all multi-place aircraft. All aircraft not supplied with the galley kit would be supplied a 1- or 2-burner gasoline stove if desired. The purpose was to provide cooking facilities in a forced landing, or at rest-refueling stops. Aircraft could also carry Type-K rations and canned foods and beverages. The gasoline stoves detailed were the Coleman #520 1-burner and the Coleman #523 2-burner. The Type E-4 sustenance kit was designed for this contingency.

Due to inconvenient features of the galley kits and combat lunch, it was decided to provide aircrews meals pre-prepared and in serving-sized portions. In Spring 1944 the USAAF adopted the B-2 Food Warmer, which measured 18" x 24" x 14". It was manufactured by Tappan Stove Company, which called it the FTG-3. The unit held 2 compartments that could maintain 160° F and provide portioned food for 6. Space was provided for 12 cups of hot soup or beverage. The top section held an unheated drawer for utensils, pastries, and fruit. The food was prepared in a kitchen, then stored, and kept warm and ready to serve. Type B-2 was specially designed for use with B-29 aircraft but used in other aircraft as well. Materiel Command Technical Instruction No.1455 dated 10 May 1944 outlines that the B-2 Food Warmer was to be supplied to heavy and very heavy bombardment aircraft or supplied 1 unit per 6 passengers. This instruction also directs that Types C-1 and C-2 galley kits be removed and their use be discontinued. (left) Two aircrew aboard a bombardment aircraft enjoy a warm meal from the B-2 food warmer, November 1944. (National Archives)

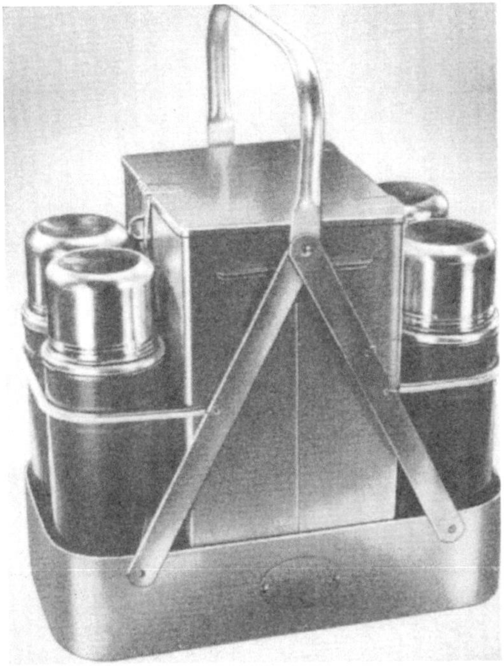

To supplement inflight feeding for cross-country fights the USAAC adopted the Food (Aircraft) Type A-1 food container, detailed under Dwg. No. 41G10440, Specification No. 94-40361. It was designed to carry food and liquids in an aircraft and consisted of a sheet metal container and tray that held four 1 pt. vacuum bottles. (above) On 9 April 1931 the USAAC adopted the Kit, Emergency Cooking and Messing, Type A-1. It comprised a wooden storage box with a two burner stove and utensils. (USAF)

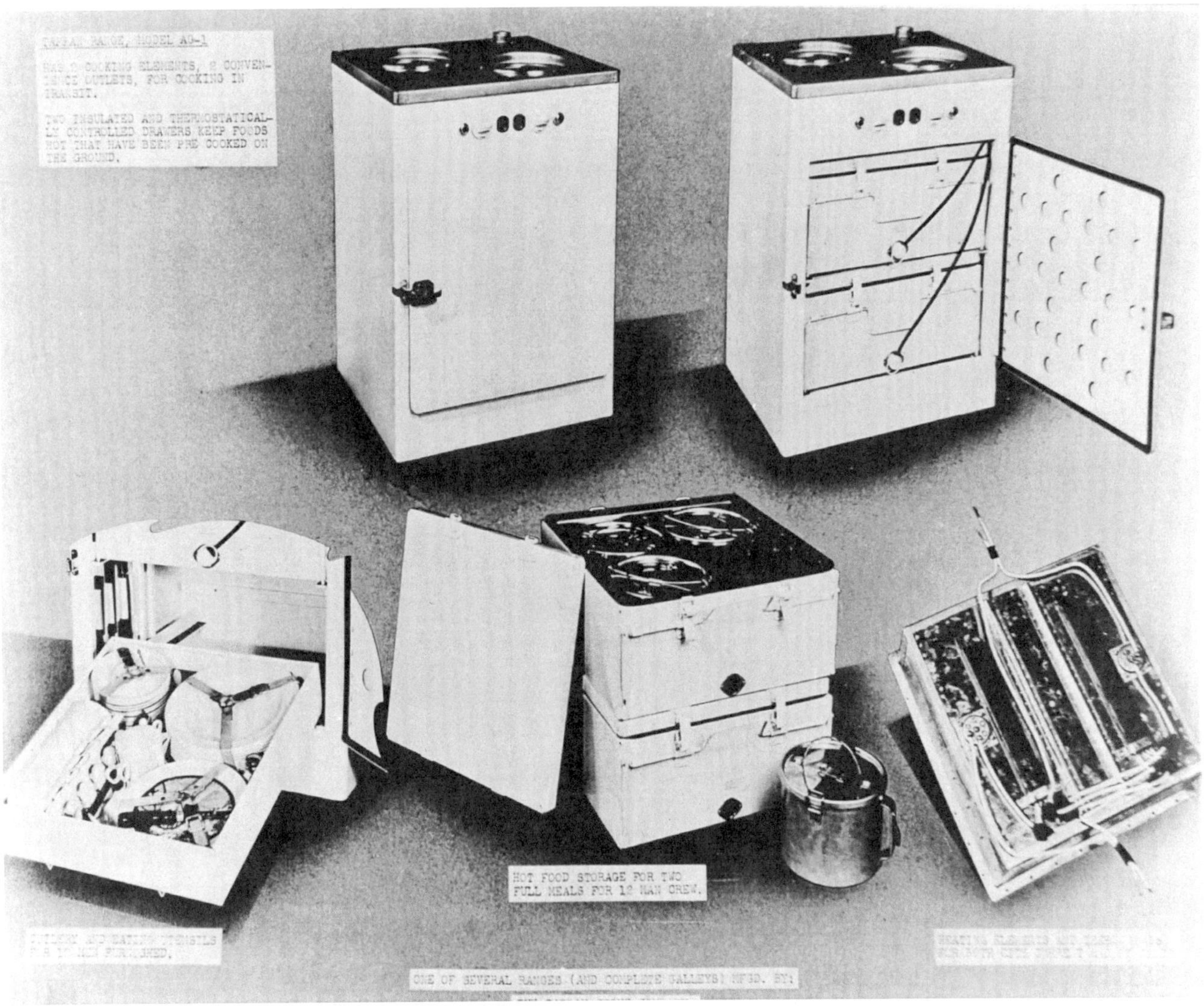

The BuAer also adopted a series of galley kits for aircraft ranging from portable gasoline stoves to electric hotplates. Range unit manufactured by Tappan Stove Co. referenced as Model AG-1. It included two cooking elements, two plugins, and two insulated thermostatically controlled drawers to keep precooked food warm. Cutlery was included. (National Archives)

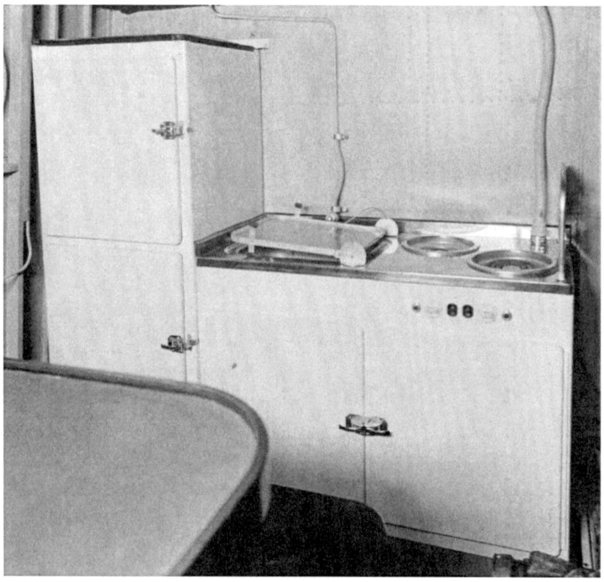

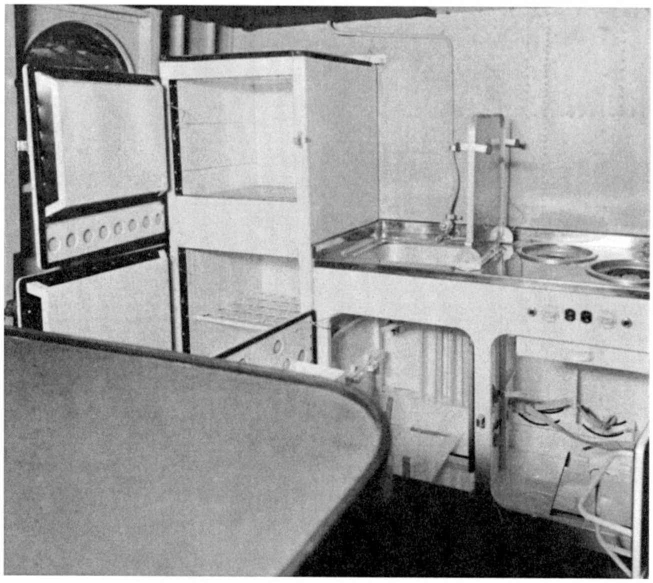

Flying boats such as the PB2Y Coronado and Martin PBM Mariner could accommodate a large galley, having all the amenities of a typical household, including range, sink, cooktop, and refrigerator. The area also included a mess table. These aircraft were designed to transport personnel or VIPs but, more importantly, also to stay aloft for long durations. It was not uncommon for them to remain in flight 8-17 hours. These two aircraft, and PBY Catalina, also contained a toilet. For almost complete self-containment, water storage containers were secured to the fuselage at various locations. Fuel stills were also a standard feature. (National Archives)

(Left) As airplanes flew farther and longer, inflight feeding became a more serious consideration. For the US Navy, most long flights were conducted by flying boats or seaplanes. Here, portable gasoline cooking stove provided in a PBY, December 1939. It includes a large variety of cooking utensils, dinnerware, and jugs and bottles for stowing liquids. (National Archives)

(Right) A mess and cooking unit were adopted as standard under NavAer Specification M-419 15 March 1940, titled Seaplane Ration Equipment for Patrol Airplanes. A new assembly was adopted under NavAer Specification M-560 titled as Flight Mess Gear, which revised cups, bowls and plates to be plastic or Bakelite.

Galley stoves were eventually integrated as a permanent feature in large US Navy patrol airplanes with the same requirements as the USAAF. Mess facilities are shown in PBY Catalina. Hot meals were provided from an electric stove in the plane's galley to sustain crew during long flight. Near the bulkhead door is a container to hold hot or cold drinks. The utensils are as detailed by NavAer Specifications M-419 and M-560. (National Archives)

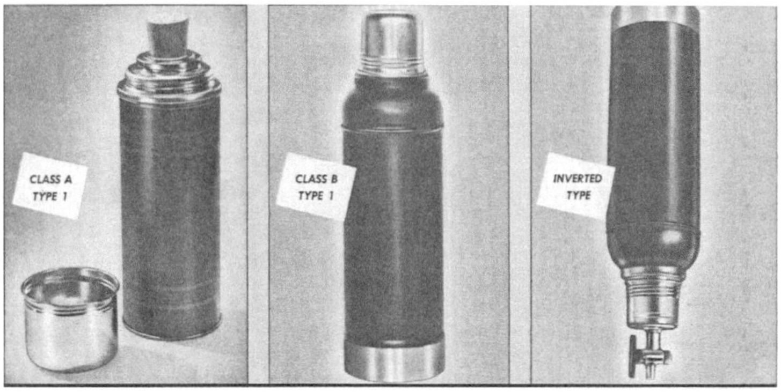

(Above) "Bottles", better recognized as Thermoses, were used by both aeronautical services for storing hot and cold drinks during flights. Specialized vacuum bottles were adopted to accommodate pressurization at high altitudes. Three types of vacuum bottles are shown. Class-A Type-I available in 1 pt. and 1 qt. capacities, Class-B Type-I available in 1 pt., 1 qt., and 2 qt. capacities, and an inverted type available in a 2 qt. version. These were procured by the USAAF; the BuAer procured the Class-A and -B models in the 1 pt. and 2 qt. capacities. The manufacturer Stanley or equal was specified, with an additional manufacturer from the BuAer to be Landers, Frary & Clark.

(Below) Larger vacuum-sealed jugs were available for storage of liquids in 1 gal. and 2 gal. capacities detailed under USAAF Specification No. 94-40297. The features are identical and included a spigot. The BuAer procured the 1 gal. model under USAAF specifications.

USAAF aircrew members enjoying a beverage from a vacuumed-sealed jug. AAF Memorandum No. 65-15 states that everyone who might be required to subsist on special rations would provide themselves with a full canteen of water and mess kit. Additional water supply was included in thermos bottles and jugs, and canned water was in emergency kits.

The bulk of the rations described thus far were developed by the USAAF but did not cover the entire line developed by the QMC. Ration development for the USAAF was complex, focusing on specific aspects for overland, over water and inflight consumption. The US Navy's development of rations was more streamlined as a universal, one-size-fits-all endeavor. The BuShips had a complement of provisions for lifeboats and floats well before WWII but did adopt a standard ration prior to entering the war, as explained later in this chapter. The BuAer contemplated adding rations to its pneumatic life rafts based on recommendations made in January of 1939. These recommendations stem from the developments at the time of the Type-D and new developments of pemmican.

Work had been done in refining production of pemmican to make it more suitable as an emergency ration. It was included in the initial concept of the Type-K for the QMC by the subsistence laboratory in Chicago. In April of 1941, the BuAer adopted rations for its life rafts consisting of a ratio of a can of pemmican, a quarter-pound of sweet chocolate and one ounce of malted milk tablets per man. They were to be stowed in the emergency equipment container. In short, the BuAer loosely adopted the original partial components of the QMC Type-K but gave no consideration to packaging.

Upon entering the war, the nonexistent guidelines for packaging proved problematic, prompting the

BuAer to issue formal specifications for a life raft ration. On 11 May 1942 it was standardized as the US Navy Life Raft Emergency Ration under NavAer Specification M-539. The contents and packaging method were revised to be three 3-1/2 ounce key-opening round tin cans of pemmican, three one-ounce rolls of malted milk tablets packed in a rectangular cardboard container then wrapped with heat-sealed wax paper, and one four-ounce cake of sweet chocolate, with the complete assembly packaged in what is simply described as a commercial container. The specification prescribes that each individual content to be marked US NAVY LIFE RAFT EMERGENCY RATION.

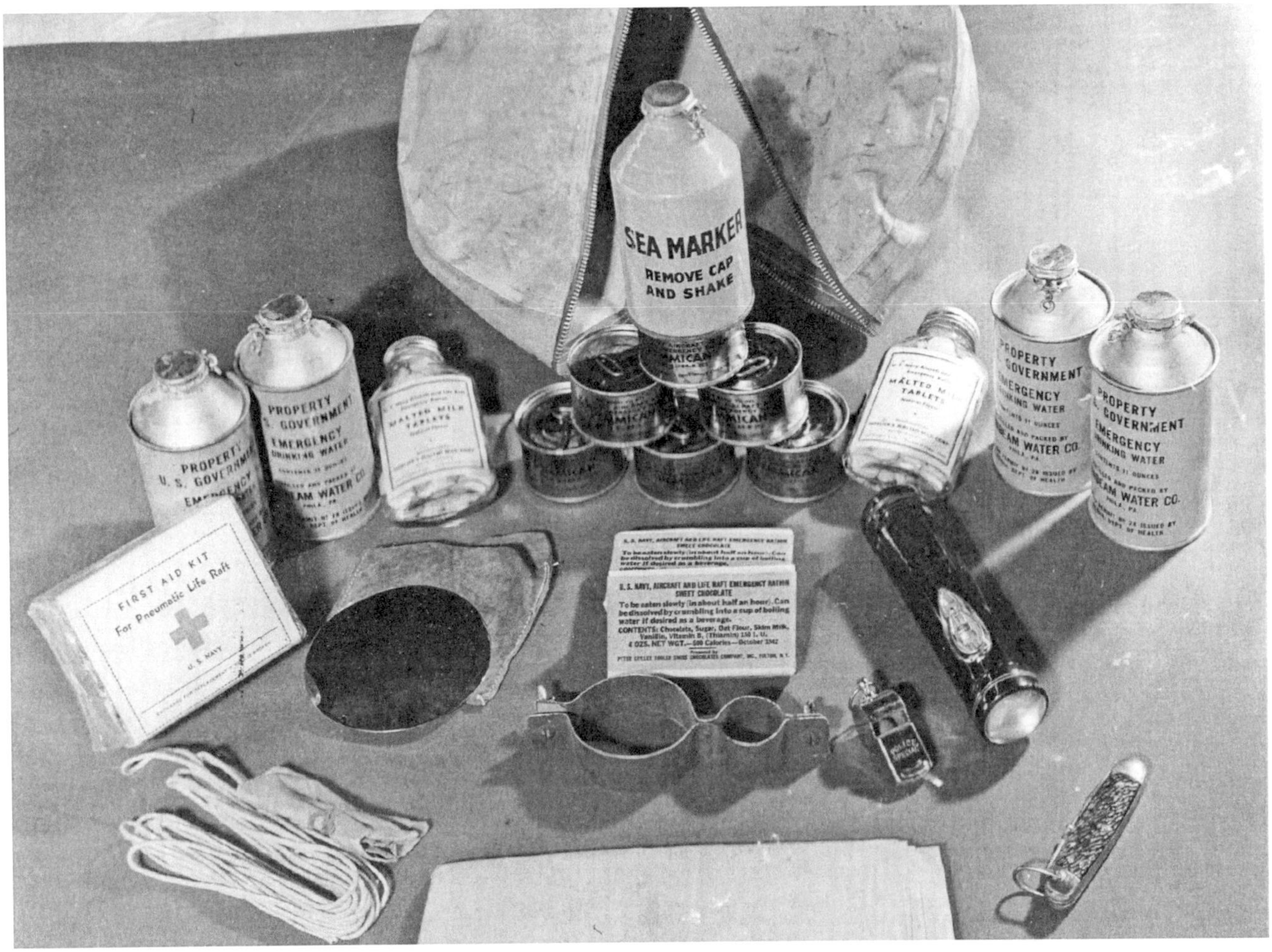

Provisions for MK-II life raft. Supply of a M-539 ration and 2 cans of drinking water were provided for each occupant. Other articles include first aid kit, signaling mirror, whistle, smoke grenade clamp, jackknife, sail, line, and flashlight (TB-8).

(Left) Life raft equipment assmebled by local unit. Note: two variations in the pemmican can labeling. (National Archives)

In July 1942 Specification M-539 was amended for the 3 oz. of malted milk tablets to be contained in a screw-top glass bottle and the name to be changed to US NAVY AIRCRAFT & LIFE RAFT EMERGENCY RATION. This ration was packaged in June of 1942 just prior to the July change. It does not include LIFE RAFT in the nomenclature on the outer container and would remain that way through its production life. The outer container is a standard commercial fiberboard box. (National Archives)

(Below) This M-539 exhibits the early production features of the ration. The box and the individual food components do not include "Life Raft" in their nomenclature. Commercial container measured 6-1/2" x 4-1/2" x 3". Pemmican cans have a printed label applied. The Type-D ration was a 4 oz. cake wrapped in parchment paper and contained in a cardboard container, an authorized deviation.

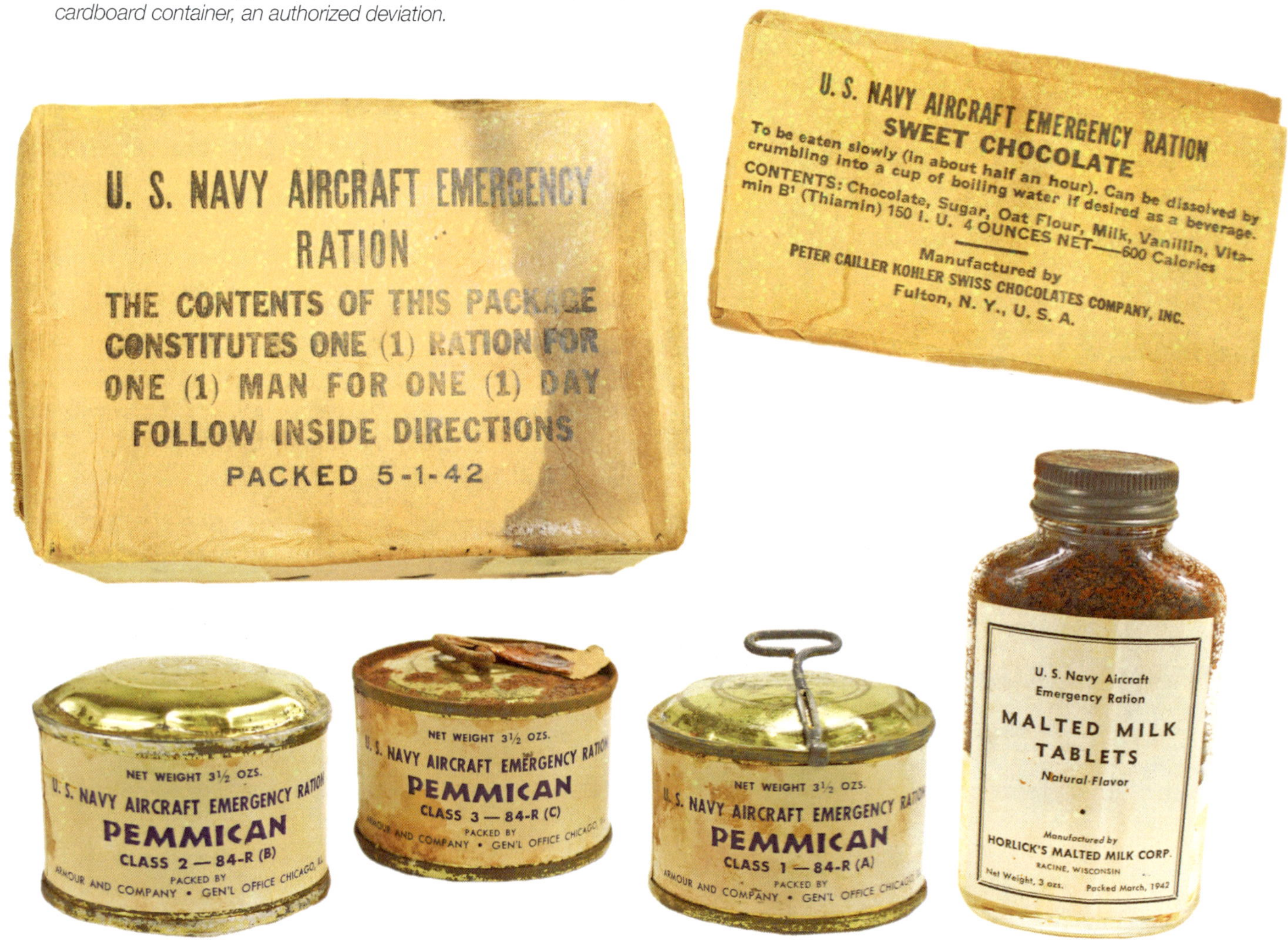

U.S. Navy Aircraft & Life Raft Emergency Ration Specification M-539; updated version had outer wax-coated chipboard carton. Nomenclature US NAVY AIRCRAFT EMERGENCY RATION K reflected its origins. The components were marked as detailed by specification: U.S. Navy Aircraft & Life Raft Emergency Ration. In March 1943, the ration was revised per Specification M-539A; outer carton was then wax-dipped and Type-D sealed in cellophane in outer wax-coated chipboard carton. The complete assembly measured 5" x 9" x 3-1/8". Number produced is unknown; it was discontinued in summer of 1943 and superseded by the tablet type. The M-539 was available from the ASO under stock number R56-R-6125, distributed at a ratio of one per man.

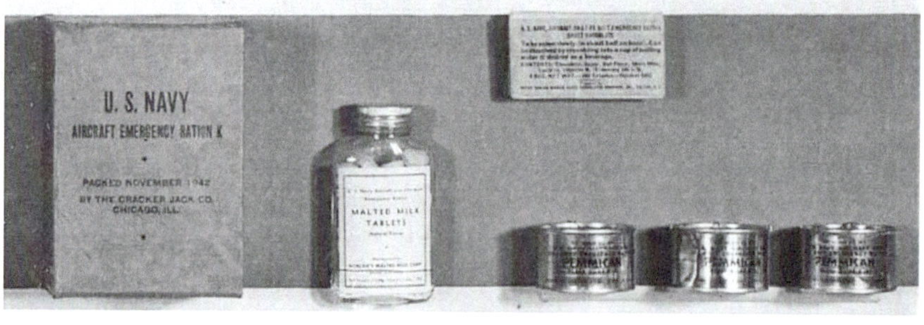

U.S. Navy Aircraft & Life Raft Emergency Ration Specification M-539 with updated components. Two variations of D-Ration and pemmican cans are pictured. No variations in the bottle of malted milk tablets are known other than nomenclature printed on the label. Detailed in the specification, each unit contained three 3.5 oz. cans of pemmican although a 4 oz. variation is illustrated. One chocolate bar is packaged per tentative specifications of the Type-D, with the other packaged per updated specifications of a wax-dipped chipboard carton.

In October of 1942, the Naval Medical Research Institute began coordinating with the BuM&S to evaluate the supply of water and rations provided for lifeboats and life rafts for the BuShips, the BuAer, and the Merchant Marine. The consolidated consensus found it nearly impossible to supply sufficient food and water in emergency equipment. In the evaluation process, the group first had to consider what, in the existing supplies, was most vital to sustain a shipwrecked survivor. A series of controlled tests determined that water was the most vital element and that a castaway could survive up to 10-14 days without food. Limiting the amount of food to allow the maximum amount of water for stowage was therefore emphasized. Field tests determined that a small amount of carbohydrates and fat would maintain an individual just as well as larger amounts of food in that time period.

Based on these findings, a liquid ration of a specially canned salt-free tomato juice and beef consommé was developed. It contained 10 percent solids. To enhance the calories in the liquid rations, hard glucose citric acid candies were provided. This combination of a liquid ration and candy supplement proved satisfactory. The liquid ration was considered for standardization, but by the early spring of 1943 several methods of chemical demineralization were developed that could make seawater potable. Since demineralization provided the means to obtain water without having to store it, efforts shifted toward the development of a solid food ration.

In June of 1943, the NMRI recommended a concentrated solid food ration be developed building on the use of candies. In its opinion, a limited amount of food that furnished the largest number of calories should be supplied, but also and most importantly so that it would not provoke thirst like pemmican and crackers. According to focus groups, food was found to raise morale, and providing a variety of candy increased palatability, energy, and contributed to the conservation of water. Finally, it was determined the ration should be packed in a container that would resist battering and stay sealed when immersed in water. The new ration would include sucrose citric acid tablets to increase saliva output. Sucrose lipid citric acid tablets were added for their fat content, malted milk tablets with added sugar were provided for variety, and chewing gum to moisten and clean the mouth. Multivitamins were included primarily for morale and a plastic bag was added to store the components after the ration had been opened. Samples of a compact tablet ration type were prepared and field tested in July of 1943. The test garnered positive results, so production of the new ration commenced in August of 1943.

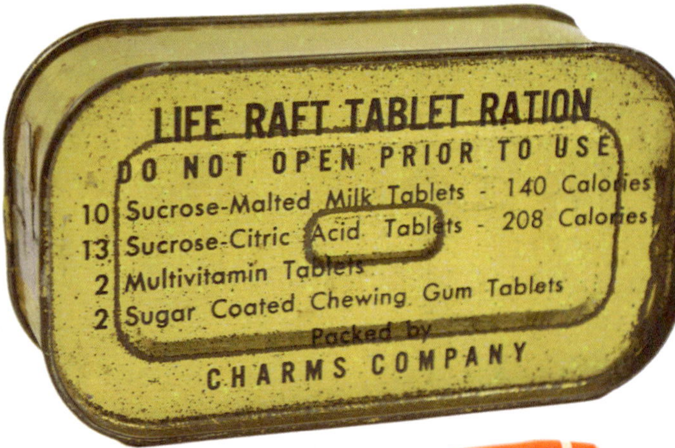

All components for new emergency tablet ration were commercially available except sucrose lipid citric acid tablets. The ration that entered production was expedited prior to drafting formal specifications and availability of the lipid citric tablets; these were not included in initial production. Tablets rations were detailed to be supplied 3 per man, measuring 3-7/8" x 2-1/8" x 1-1/8", with the allotment for 1 man consuming the space of 3-7/8" x 2-1/8" x 3-3/8". Production of the M-539 was discontinued upon award of contract for the tablet type. Tablet ration opened; 2 packages of candies were labeled Tablet Life Ration. These were available for the ASO under stock number R56-R-6300. Expedited procurement of tablet ration occurred August 1943.

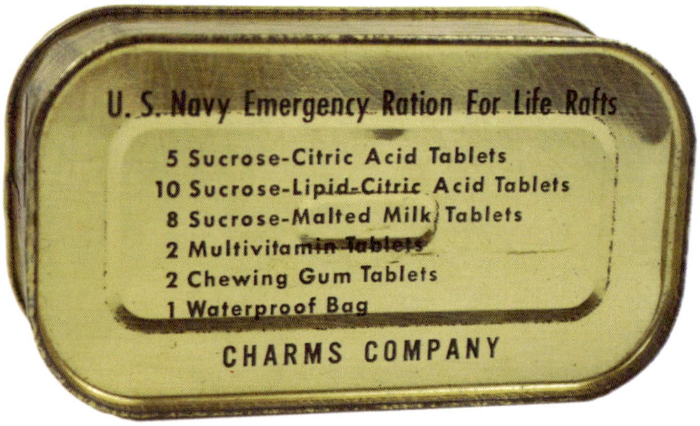

Tablet ration was standardized 4 October 1943 as the M-539B, Life Raft Tablet Ration. Later procurements included the lipid acid tablets, which became available Spring 1944. Emergency tablet ration known officially as M-539B procured after standardization. It remained BuAer standard for several more decades, redesignated under MILSPECS in 1950 and retitled as Food Packet, Life Raft, Aircraft in 1953. For these procurements, 2 tablets packages were labeled U.S. Navy Emergency Ration.

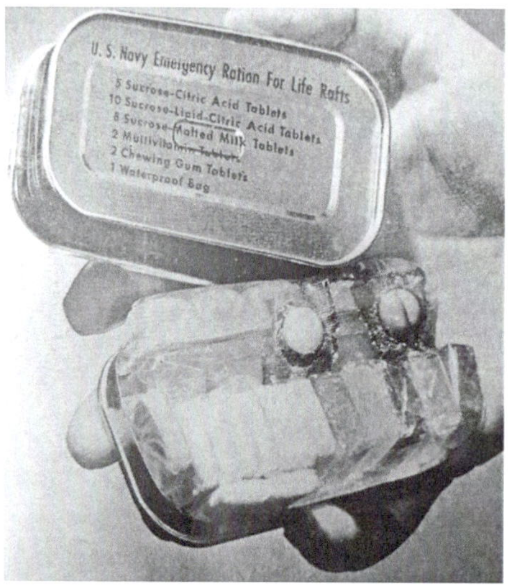

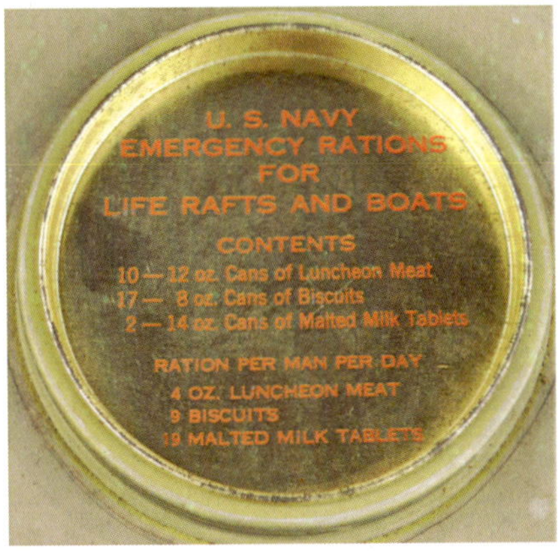

Bureau of Ships US Navy Emergency Rations for Life Rafts and Boats, a.k.a. Abandon Ship Ration. It contained pork luncheon meat, biscuits, and malted milk tablets individually packed in tin containers stored in a waterproof metal cylindrical tin. This ration was designed to supply 1,000 calories per man per day for 3 days and was issued according to the capacity rating of the boat or raft. The master metal container was 12" diameter x 12" H and had a screw-top cap. Emergency water was contained in wooden water breakers. This ration was under evaluation by the NMRI and BuM&S, late 1942-early 1943. Adoption date is unknown. Ration illustrated here for comparison to BuAer M-539 life raft ration. Displays of the ration and some of its components from 1943 to 1945. By 1950, this ration complement would be replaced by the tablet ration as adopted by the BuAer packaged in similar container.

In 1942, with the dramatic increase in wartime shipping, the USCG approved a new emergency ration, the Life Boat and Life Raft Ration. It was based on the original Naval Aeronautical Specification M-539, which consisted of seven ounces each of pemmican, malted milk tablets and chocolate, and the US Army Type-C Biscuit. The ration was packaged in a rectangular vacuum-sealed tin 4-1/2" x 8" x 3-1/4". Later, a tin of water of the same size was also provided. Distribution of the new rations was at a ratio of six cans of water to two cans of food and a can opener. Production began in late 1942 and became regulation for lifeboats on Merchant Marine vessels over 3,000 gross tons in January 1943.

USCG approved and used rations during WWII, starting with adoption of USAAC emergency ration in early 1937. The ration supplied to USCG was marked Aircraft Emergency Rations USCG. Prior to WWII ration was repackaged to include 2 oz. Logan bars sealed in parchment paper in fiber outer carton.

1942 Food ration assembled by L.A. Young & Wire Corp. Contents list indicated US Army Type-C biscuit and US Navy aircraft specification for the pemmican.

Food ration assembled by Multiple Breaker Co. in August 1942 with contents label on top. Water breakers used until packaging transitioned to use of equal-sized water container.

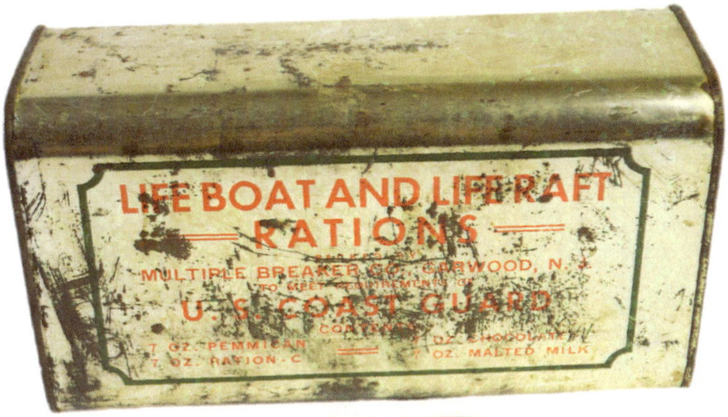

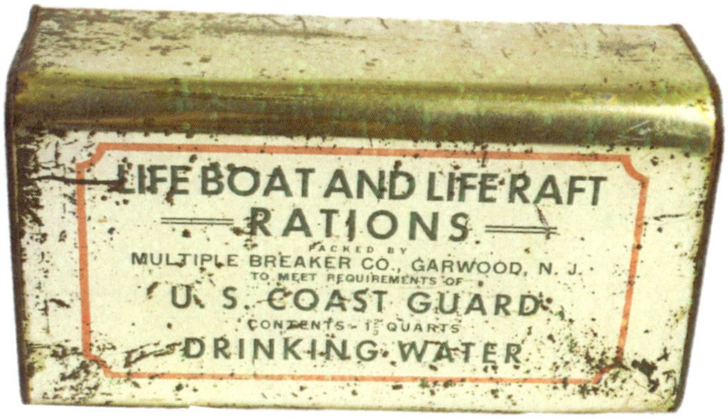

The Multiple Breaker Co. continued production through 1943 and into 1944, changing can labels and producing a companion water tin.

(Below) Components of USCG-approved emergency rations.

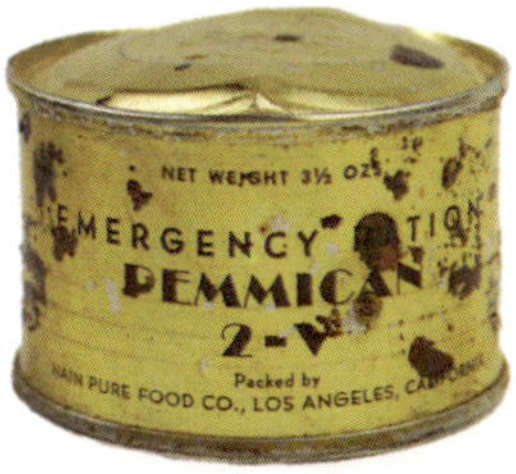

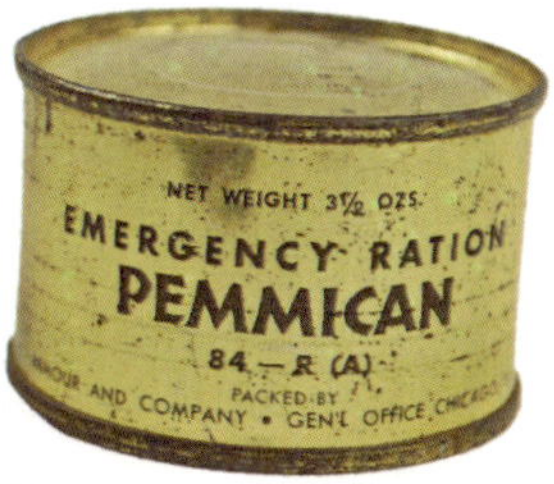

An additional contractor, Chemical Service Co., assembled these 1944-1945 rations in grey cans. The same ration, with slight modifications, continued to be standard into the 1970s. In the immediate post WWII years, BuShips would replace 3 gal. water breakers with USCG approved water tins, pictured here. Production of USCG rations exceeded 1,750,000 by end of WWII.

In the effort to improve provisions for survivors aboard life rafts, the BuAer standardized a can of drinking water distributed at two cans per person for life rafts. Standardized on 19 July 1941 and detailed under NavAer Specification M-504. Upon entering the war, immediate demands arose for the inclusion of water in emergency rescue equipment for the USAAF. The can of drinking water M-504 was adopted as a joint Army-Navy standard on 7 April 1942 and referenced under Specification AN-W-5.

The BuAer M-504 drinking water crown top can contained 11 oz. of distilled drinking water. Per specification, the property mark USN was stenciled on the can. Upon joint Army-Navy standardization in April, the USN was deleted. (National Archives)

Formal specification drafted and released 2 July 1942 as AN-W-5A (Amendment-A) outlined that the can be labeled with the property mark, nomenclature and manufacturer. Like the sea marker can production, it changed to a more cost-effective cone-top can. 11 oz. of distilled water was provided in a can 2-5/8" diameter x 5-1/2" H, available from the ASO under stock number R51-W-135.

When the joint Army-Navy drinking water was standardized, the one-man parachute life raft had just been adopted. It then underwent a total reconstruction by the USAAF, which found the round cans to be uncomfortable when used in the pararaft seat kit and did not store well in multi-person life rafts. In August 1942, the USAAF issued a change order for contractors to switch to a rectangular or flat can of a design to be supplied. The rectangular shape improved comfort when a person was seated on the kit and could be packed more efficiently inside other emergency kits.

By the close of 1942, the BuAer was developing parachute emergency kits and had concluded that the rectangular cans were more efficient. It recommended joint Army-Navy standardization. Expedited procurement of the new flat design was authorized to supply all emergency equipment.

In considering standardization, the flat-water can was in evaluation by the NMRI and BuM&S. The two organizations also evaluated the tablet ration. The flat-water can would hold 14 fluid ounces but contained only 11 ounces of distilled water. This was to prevent can distortion when expanding or contracting in extreme climate changes. The study evaluated whether or not the three ounces should be filled with seawater, with the liquid ration being salt-free and candies providing carbohydrates, the diluted saltwater was thought to help with sodium intake. British studies had found there were minimal benefits to diluted saltwater ingestion and laboratory and field testing on human specimens found the addition of diluted saltwater proved to have no discernible effect on the health of individuals. The conclusion was that it was more beneficial to package the can to the full 14 ounce capacity. The rectangular flat can was officially standardized as the AN-W-5B (Amendment-B) on 31 August 1943 and remained the standard drinking water can until cancellation in June 1950.

(Far left) Expedited production of flat AN-W-5A water can. Reference line served as measuring gauge when refilling it. Can was procured by both aeronautical services prior to formal specifications, with Specification AN-W-5A being referenced at the time of these procurements.

(Left) Standardized AN-W-5B emergency drinking water. Can measured 3-7/8" x 5-1/4" x 1 3/4" .

Packaging dates or date codes were indicated on bottom of cans (middle pictures). Through 1943 date codes were utilized. To avoid confusion, actual dates were provided on perishable items by the close of 1943. The AN-W-5B was provided with the month and year of packaging. Lower images illustrate variations of the cap. The initial cap design (left) and improved knurled cap provided in June 1943 (right).

AN-W-5B was available from ASO under stock number R51-W-145 and from USAAF under stock number 6700-171480.

Drinking water was the most valuable asset for survivors. Emphasis was placed on water production and storage. The BuM&S determined from field tests that individuals would sometimes consume more water than necessary. Since it was never known when rescue would come, water conservation and consumption regulation were critical. BuM&S therefore requested Cruver MFG. Co. to fabricate a measuring cup, marked at ½ oz. intervals, which held 11 oz. It was slightly larger than the AN-W-5B water can and slipped on the can for storage, conserving space. Samples were submitted to the BuAer in January 1944 and, by March, BuAer placed a large order for equipping multi-person life rafts. It became available from the ASO under stock number R83-C-81525, distributed one per multi-person life raft and one for the M-594B Shipwreck kit.

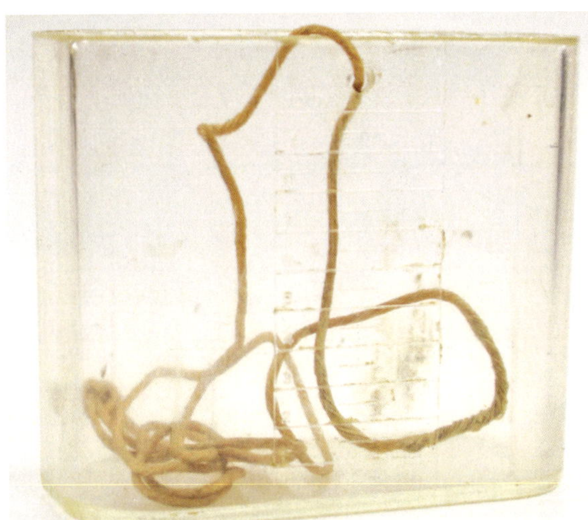

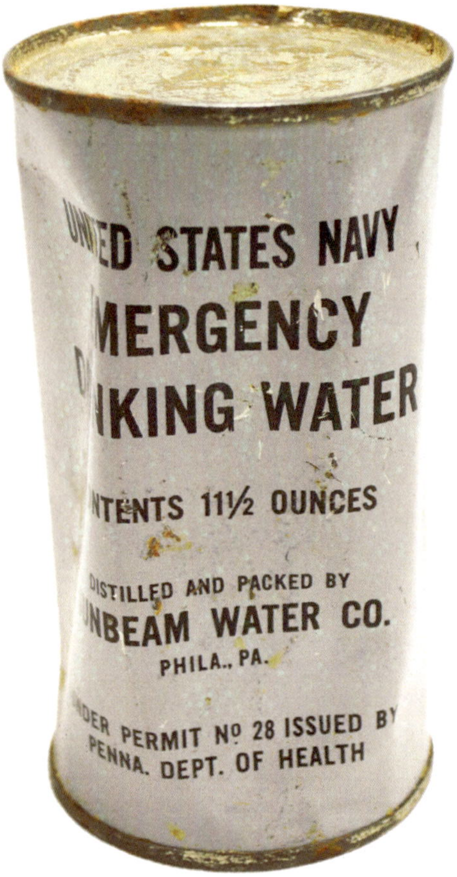

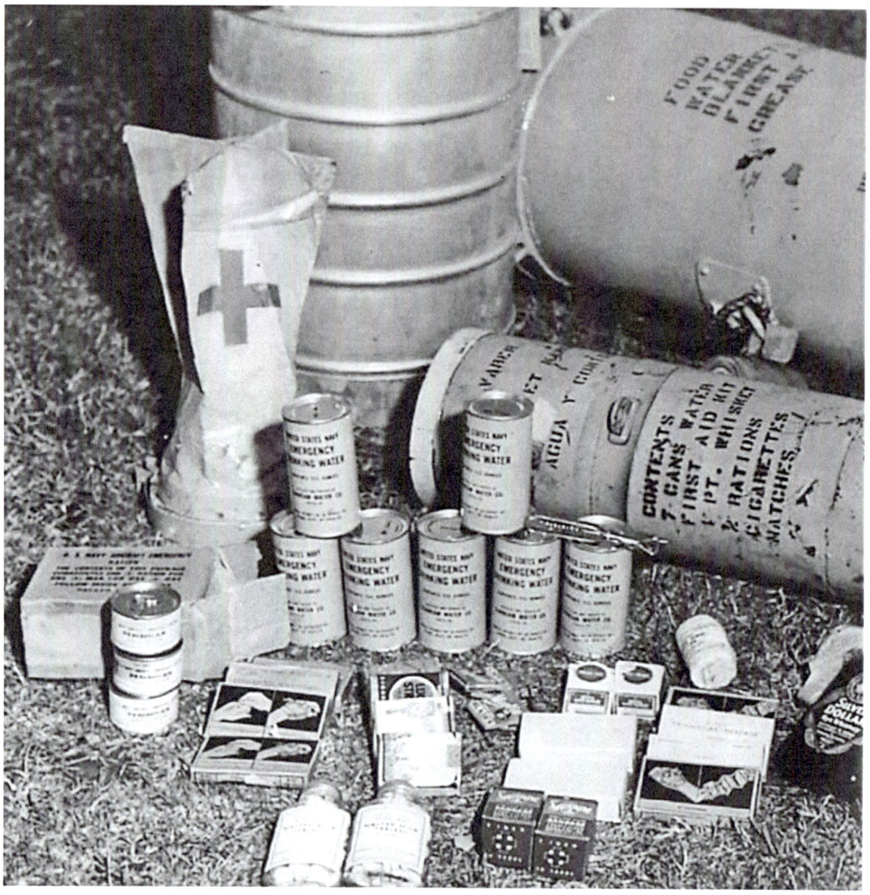

The round drinking water can was similar to the AN-W-5 but lacked the cone top, requiring a can opener. It was labeled UNITED STATES NAVY EMERGENCY DRINKING WATER and contained 11½ oz. The USCG provision bomb used these cans in its complement. Its origins are unknown. Those surviving examples obeserved have 1942 dates. This example is dated May 1942.

In late 1942, the NMRI became increasingly interested in making seawater potable and investigated many proposed methods, but none was found suitable for life raft use. Many of the devices, such as fuel stills and solar stills, were too bulky. In early January of 1943, two naval officers, Lt. Claire Spealman and Lt. William Consolazio, made a breakthrough in the development of a chemical desalting method. Spealman had worked on the process as a teacher at the Medical College of Virginia, and he was motivated after reading an article stating that it was impossible to remove the salt. He took this as a personal challenge and initiated an investigation. Soon after, he was drafted into the US Navy and assigned to the NMRI in October of 1942. While at the NMRI, he befriended Consolazio, who was researching life on a rubber raft for other studies. Together, and in less than three months, they completed the research necessary on a chemical that could remove the salt.

The principle feature was that it was specially designed with field applications in mind. The method they devised consisted of two plastic bags with integrated filters. One bag of seawater would be treated with silver oxide, breaking down the chlorides and magnesium, then filtered into the second bag. The second bag was then treated with uric acid, which precipitated the remaining sodium, dispensing potable water. The process only required 30 minutes.

In February of 1943, the BuM&S authorized the NMRI to conduct trials in the Gulf of Mexico to test various methods of obtaining drinking water from the sea. Prior to testing, the NMRI consulted the Permutit Co. to incorporate its chemicals into the Permutit single-step processing method. The Permutit Co. had been working on the chemical method for some time, but the amount of chemical-to-water-produced ratio was found inefficient. The apparatus was too complicated and expensive compared to the cost of canned water. Together, the NMRI and the Permutit Co. continued to refine the system to meet the service requirements of a compact, rugged, easy-to-use sea water desalting kit.

In July of 1943, field tests were again conducted in the Gulf of Mexico. The resulting revised NMRI-Permutit kit outperformed the Delano solar still by a considerable margin. The tests included the tablet ration and other equipment, such as signaling devices, fishing gear, and clothing. Upon their conclusion, the desalting kit proved it could produce four to six times the amount of water more than the capacity of one water can, requiring only 30-45 minutes to process each batch.

(Above) Compact packaging for the kit Consolazio and Spealman devised with assistance from Permutit Co. It contained four vinylite bags and six briquettes.

(Right, top and bottom) Lt. William Consolazio demonstrating the chemical desalting method that Spealman and he developed, June 1943. (National Archives)

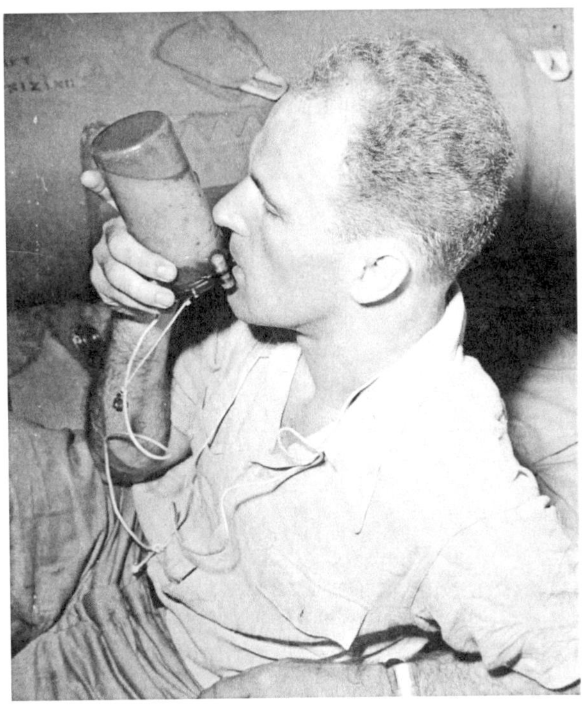

(Above) Permutit chemical desalting kit tested in a life raft at Cape Fear, NC, June 1943, but the device was found unsatisfactory. Consolazio and Spealman refined the system to a single step with bags for efficient storage. (National Archives)

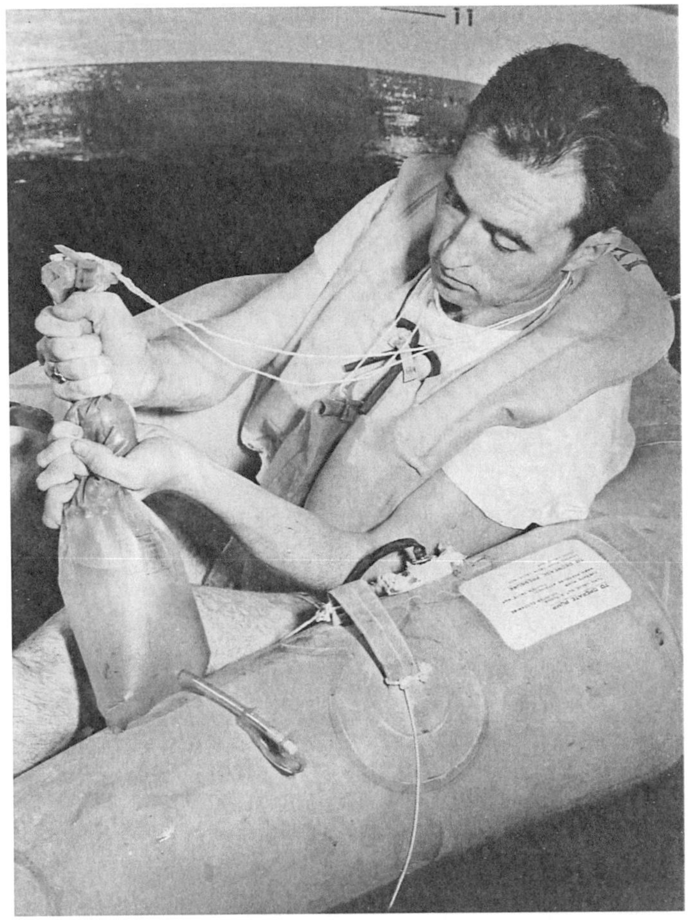

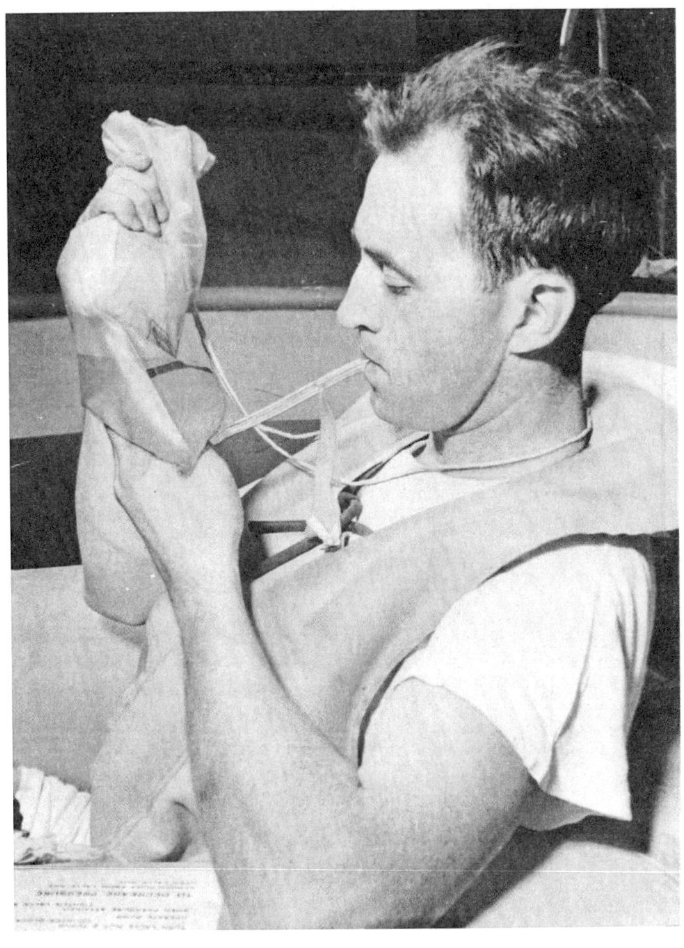

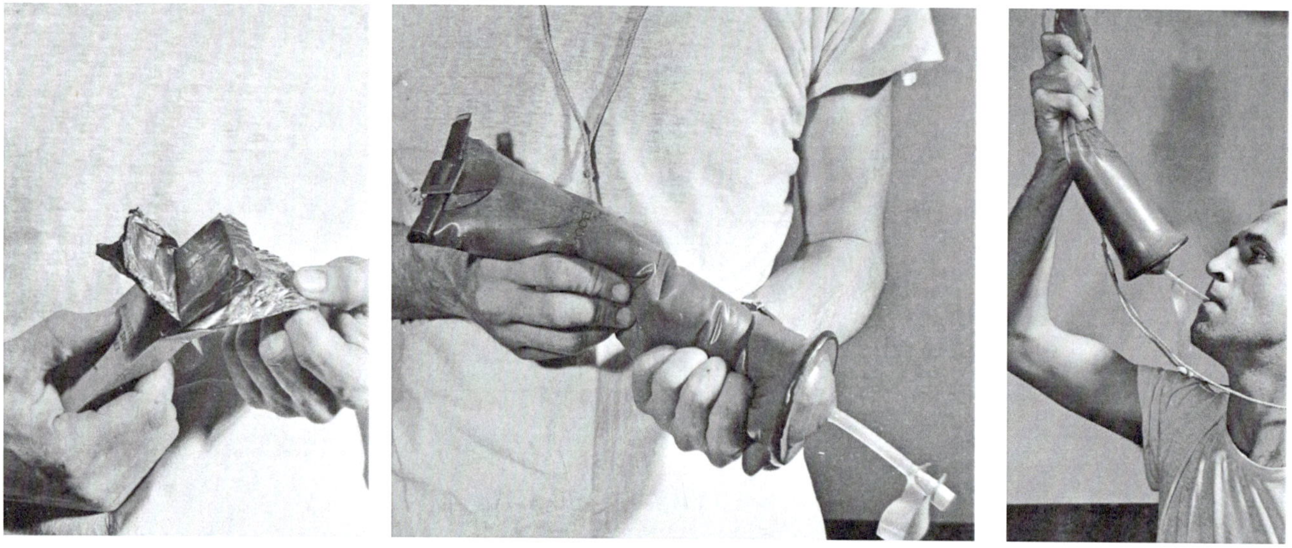

With overwhelmingly positive results of field tests conducted in July, revised NMRI/Permutit method was adopted but needed to be optimized. The BuAer requested kit be the same size as the flat AN-W-5A water can for incorporation into parachute kits, pararafts and life raft equipment kits. By late 1943, 1-step chemical desalting method had been streamlined to include one plastic filtered bag with six chemical briquettes, sealed in tin container only slightly larger than standard water can AN-W-5. (National Archives)

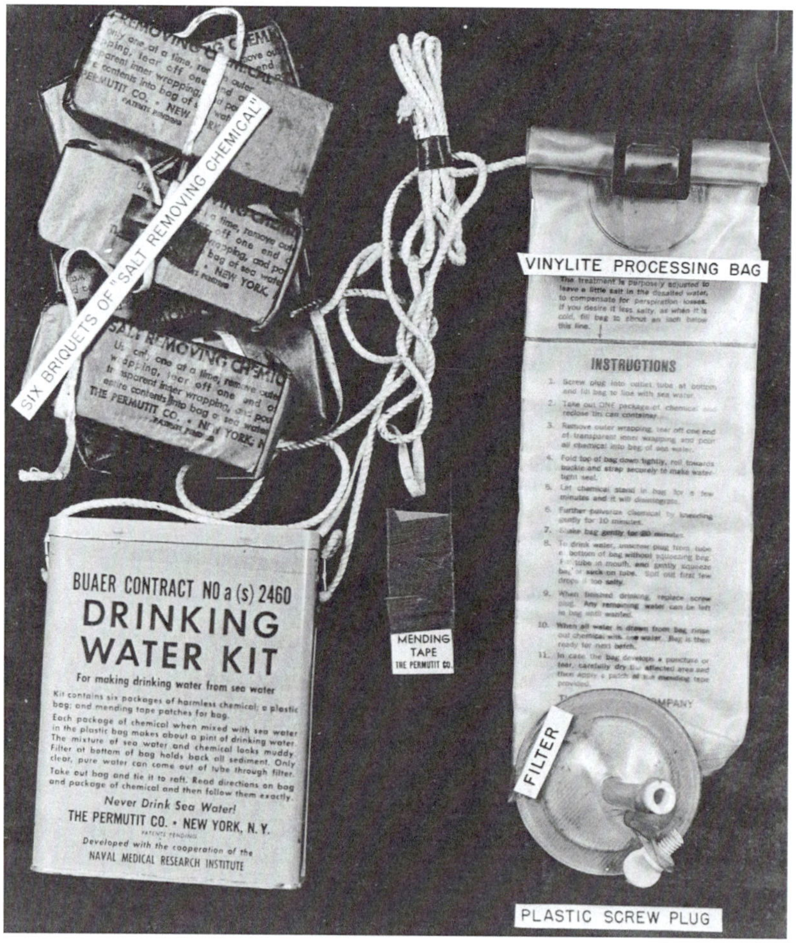

The BuAer awarded an expedited contract in December 1943 for 10,000 units under contract number NOA-2460; first deliveries were expected at supply points March 1944. Additional contracts include NOA-3124, March 1944; 288-28784, February 1945; and 288-32168, June 1945. Drinking water kit was standardized 12 September 1944 under NavAer Specification M-613, available under ASO stock number R83-K-511000. Tin was sealed closed by friction tape and provided with cotton cord as lashing lanyard to prevent loss. Printed instructions were provided on exterior of container.

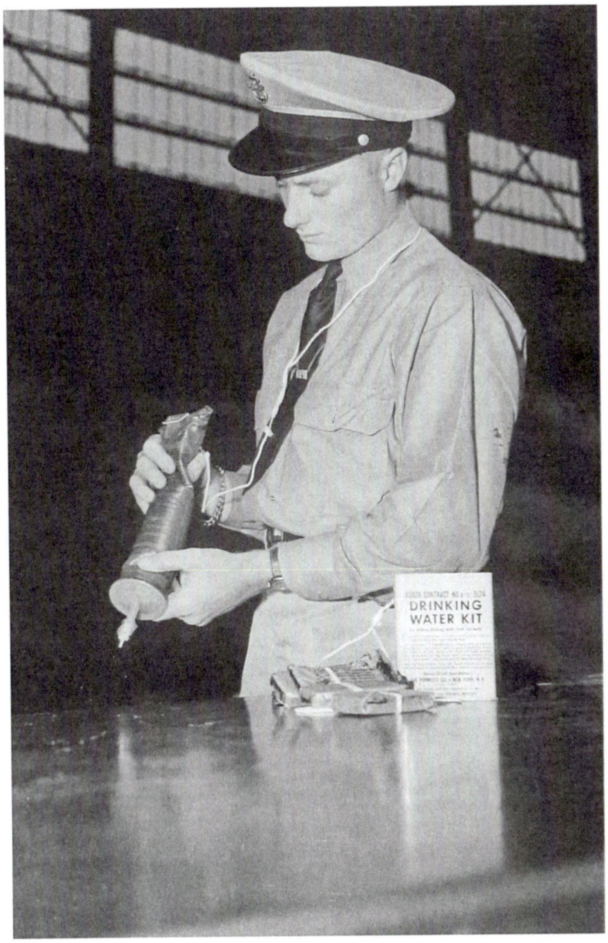

US Navy petty officer examining desalting kit at NAS Lakehurst, September 1944. Desalting kit with its six briquets and vinylite filtered bag.

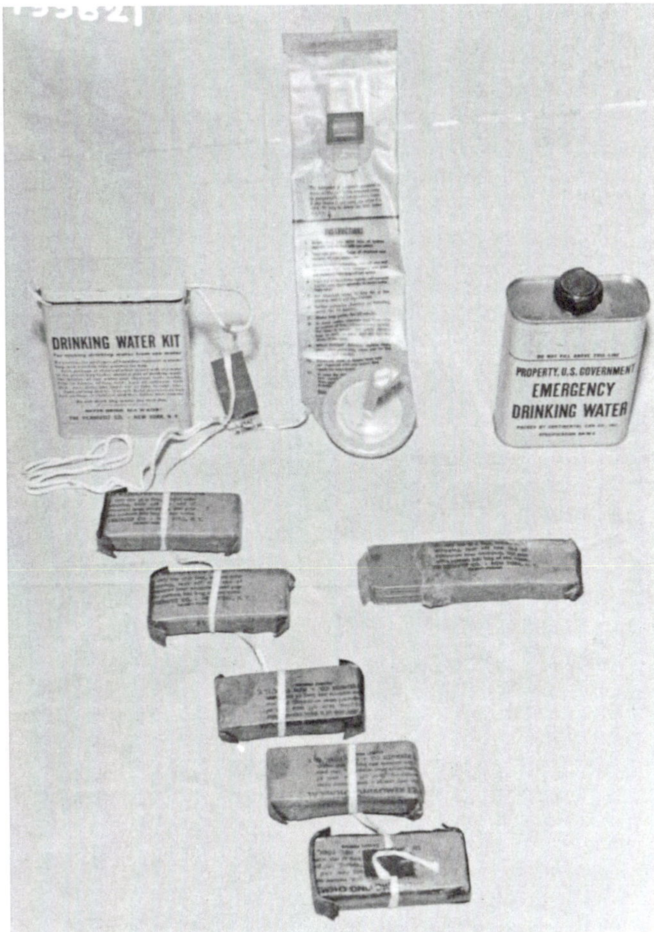

The components were chained together before packaging to prevent loss. An AN-W-5A flat water can is included for size reference. Image also illustrates what would be standard for US Navy emergency equipment: one desalting kit and one can of drinking water per man. (National Archives)

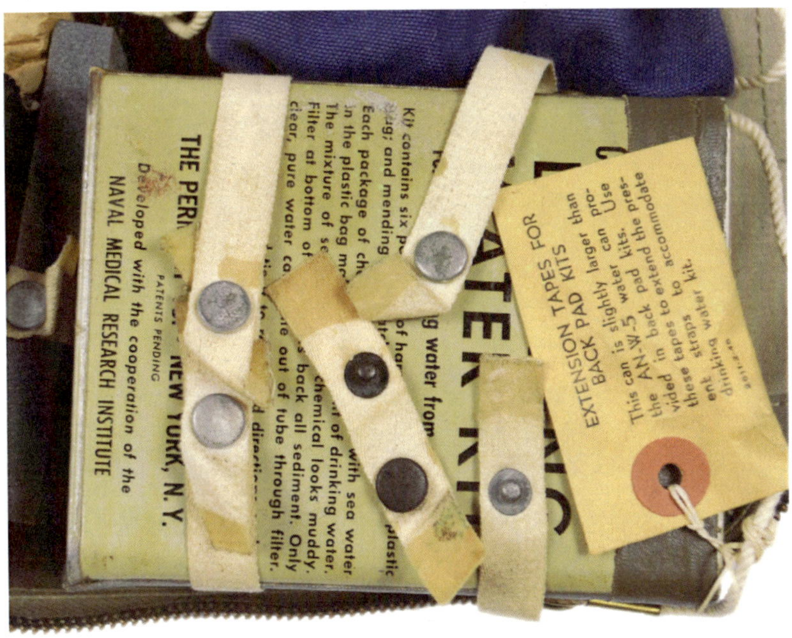

Starting with contract 3124, extension tapes were provided with the M-613 for integration in parachute kits. Tin was slightly larger than water can AN-W-5B. Per Technical Order 100-44 of 30 November 1944 one desalting kit was installed for every two water cans in all emergency equipment, or one per man. M-613 installed in M-592 back-pad kit with extension tapes.

USAAF approved of the desalting kit in March 1944, designating it as Sea Water Desalting Kit Type JJ-1 under Specification No. 94-40725, and standardizing it 6 December 1944. Authority for procurement was released May 1944 under contract 30053-AC-1285 with an immediate additional award in June under contract 33038-AC-3440, available under stock number 4500-526700. By July 1944, over 1 million drinking water kits were on order between the BuAer and USAAF. Due to delays in manufacturing only a limited supply for both services was available by the close of 1944.

(Above) USAAF 7th AF member demonstrated desalting kit by pouring the now-potable water into an empty can during life raft field tests in Hawaii, August 1944. (National Archives)

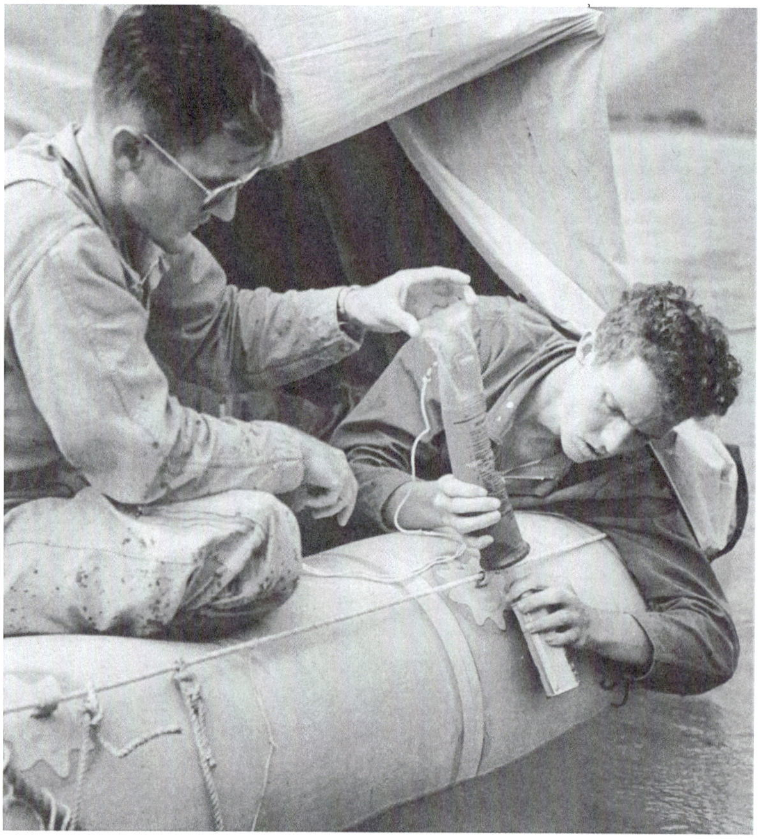

USAAF Type JJ-1 desalting kit. Tin color and printed literature were altered on USAAF procurements. USAAF desalting kit and its complete complement. (National Archives)

The USAAF was less patient in the development process for water-producing devices; it immediately went into the procurement of solar stills. In March 1943, the authority to procure 30,000 Delano solar stills was released based on an initial ratio to supply three per life raft; this was later revised to one per life raft. The Delano still was a private invention of an employee of the Gallowhur Chemical Co, Richard Delano. This solar still was tested by the USAAF and the NMRI in late 1942 but fell from consideration by the US Navy immediately due to its bulk. The procurement was an impromptu decision by General Hap Arnold to meet service requirements and not one by the Materiel Command. Deliveries of the Delano Model-A to the USAAF supply points commenced in December 1943.

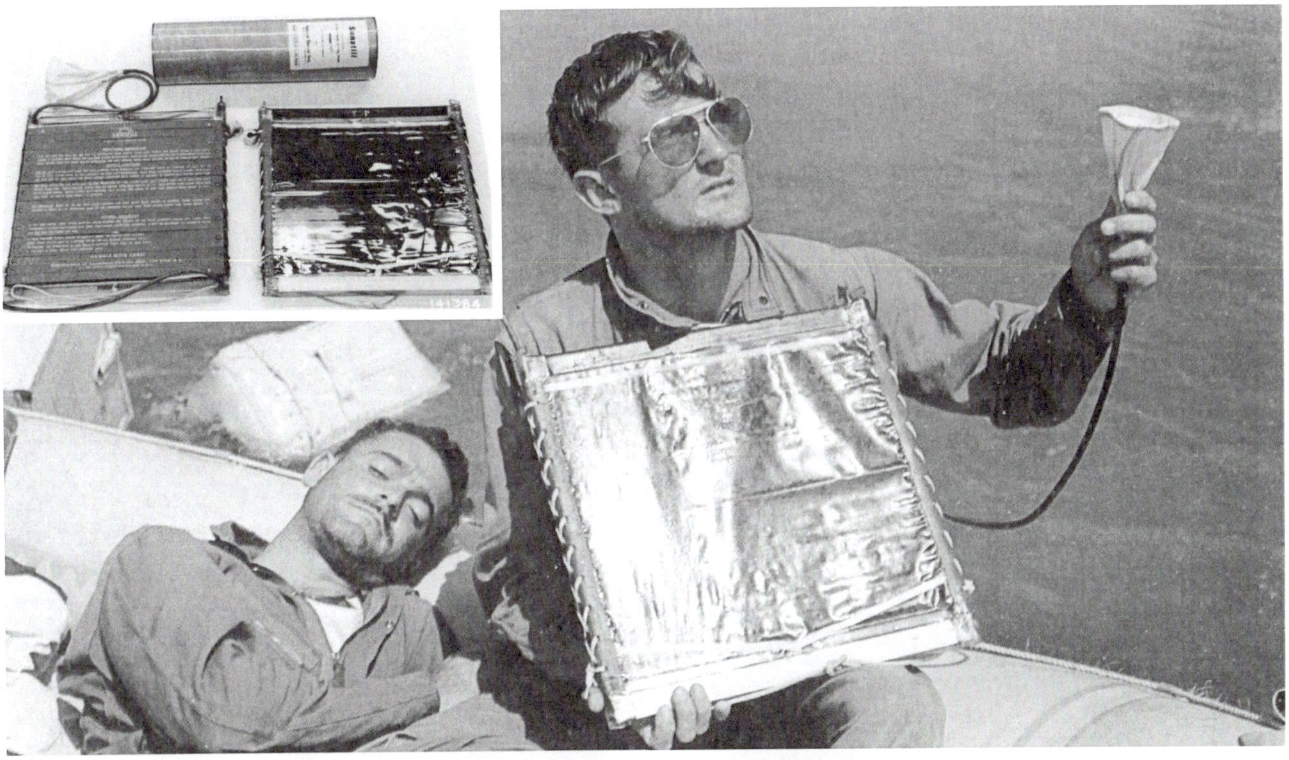

The Delano solar still Model-A consisted of a black filter fabric panel with clear acetate window gapped 1" from the fabric. Seawater was scooped up by a fabric funnel then held in a reservoir at the top. The fabric would filtrate the minerals while sun evaporated seawater, condensing on the acetate sheet then running to a reservoir at the bottom as potable water. The unit was rolled up and stored in a cylindrical fiber container 4 3/4" diameter x 18" L. (National Archives)

With the desalting method all but resolved by the US Navy, the Materiel Command focused efforts on the development of solar stills. In March of 1943, a conference was held with all interested parties to pool ideas. The Navy had no further interest in solar stills as it had found, at best, they can only produce about 16 ounces of water a day compared to the desalting method that had the potential of producing 12-16 ounces in just 30-40 minutes with one treatment and have the capacity to repeat the process multiple times in one compact package. The solar still's other negative feature was that it required the sun for processing; on cloudy days the stills would produce minimal if any potable water. Of the solar stills available at that time, the Delano model proved to be the most effective method, but the packaging was not desirable for the space restrictions on life rafts. The significant breakthrough for solar stills came from Dr. Maria Telkes at the Massachusetts Institute of Technology who worked on solar evaporation processing methods starting in September of 1943. Her contribution was an inflatable still. Building on this concept, Higgins Industries Inc. and the Gallowhur Chemical Corp. developed models with vinylite inflatable bladders that were prepared and ready for field testing by February of 1944.

Upon conclusion of the tests, both stills showed signs of considerable improvements, but further development was required; the best feature was the compatibility of the units. Both companies continued to work on their stills and submitted new designs for field tests scheduled for July of 1944. Upon conclusion of those tests, the improved Gallowhur model was accepted and approved for procurement in August of that year. The USAAF initially directed that the allotment of the still would be at the ratio of two stills per five-man rafts and three per seven-man.

Two styles developed by Higgins and Gallowhur worked under the same principle as the Delano model but as a bladder. (left) Gallowhur Model B-3 solar still. (right) model presented by Higgins. These designs can be best described as the shape of a pancake. (National Archives)

(Below and right) Improved Gallowhur solar still in three states: inflated, deflated, and packaged. Each still was an individually packaged unit. To operate, stills were inflated then tethered together and secured to side of raft resting on the water's surface. (National Archives)

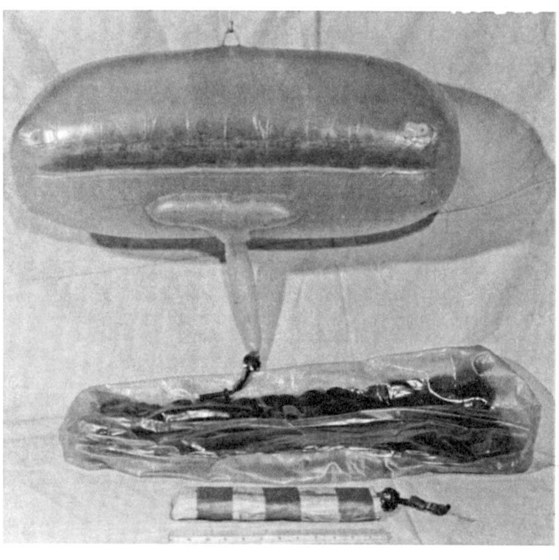

(Above and below) The improved Higgins solar still was tested by the Materiel Command in January 1945. It produced twice as much water as the Gallowhur. The Higgins still went into production as Type LL-2 under Specification No. 94-40957 in June 1945, adopted as standard 24 April 1945. The difference between the two companies' models was the shape, which ultimately affected performance. Both types incorporated a black sponge which retained minerals. Evaporated water condensing along the bladder ran down to a reservoir, from which potable water was obtained. The Gallowhur required orientation towards the sun whereas the spherical shape of the Higgins still did not. (National Archives)

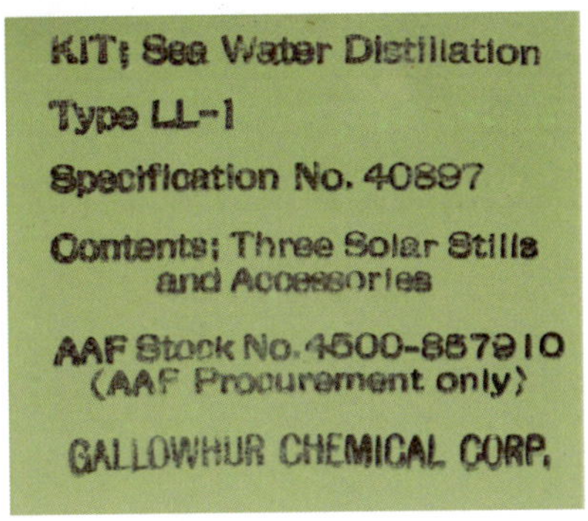
KIT; Sea Water Distillation

Type LL-1

Specification No. 40897

Contents; Three Solar Stills
and Accessories

AAF Stock No. 4500-857910
(AAF Procurement only)

GALLOWHUR CHEMICAL CORP.

(Above) The USAAF awarded a contract for 350,000 Gallowhur solar stills in December 1944 as Type LL-1 Sea Water Distillation Kit under Specification No. 94-40897. Upon award, it was decided to package the stills three to a tin container under stock number 4500-857910. Packaged unit measured 6-1/2" x 6" x 4-1/8". Allotment was one per life raft.

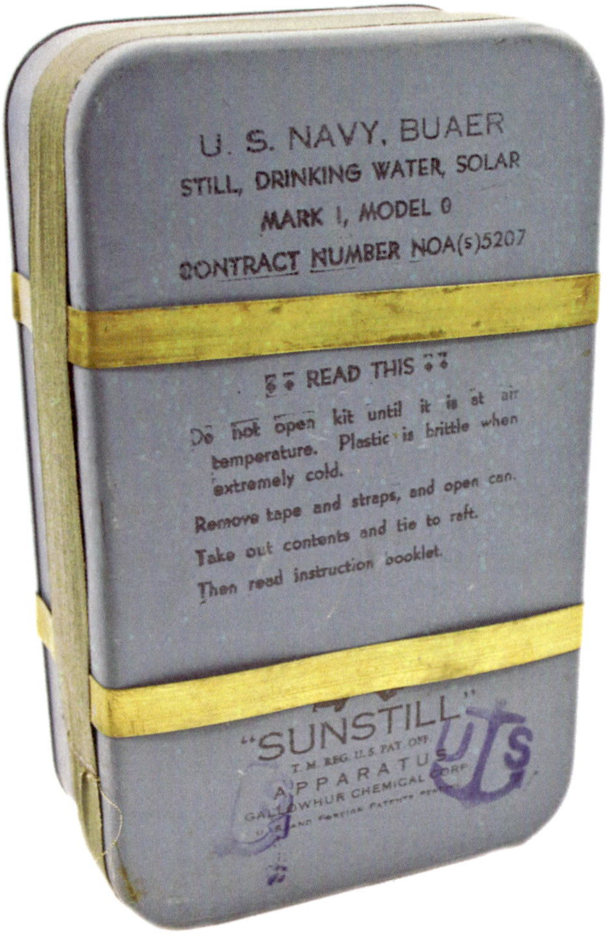

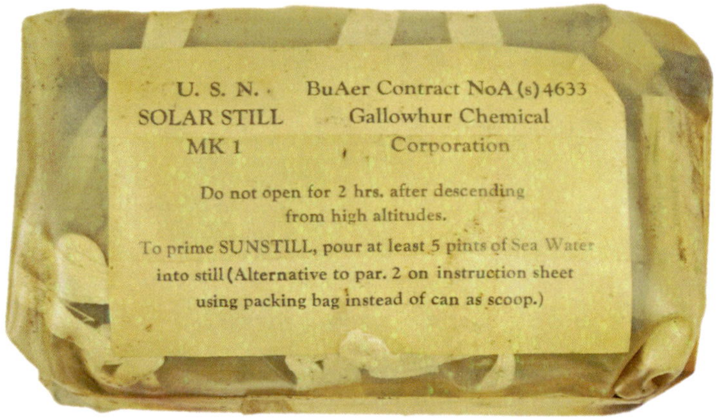

(Left and above) The BuAer procured the Gallowhur solar still under expedited procurement in September 1944, contract NOA-4633. It was titled Solar Still MK-I. Packaging included one still. Primary method for producing fresh water for the US Navy was use of desalting kit; solar still was auxiliary. Unit was standardized 10 January 1945 as the MK-I MOD-0 under NavAer Specification M-663, packaged in more durable tin container. Contracts were awarded for M-663 the same month under contract NOA-5207. When available, solar stills were to be provided one per man on life rafts under stock number R83-S-793050.

(Right) The solar still was designed to be inflated and secured to the side of the raft oriented to the sun. The BuAer adopted the Higgins spherical still as NavAer Specification M-708 in January 1946 as the MK-2 MOD-0. (National Archives)

(Below) In July and August 1944, the BuAer field-tested Gallowhur solar stills at NAS Lakehurst. Naval officer demonstrates how to inflate and use the still. (National Archives)

The third method of distillation evaluated was the age-old fuel type: apply fire to boil water, create evaporation, collect clean water as a distilled liquid. It was impractical for life raft use. The BuAer adopted and standardized a solid fuel still in April of 1938 under NavAer Specification M-362. These were available in three sizes: 17.5, 8, and 4 lb. units, with the 8 lb. unit being most popular. 24 cans of fuel were provided and stored in four tarred tubes. The ignited fuel can was inserted at the bottom of the still. Seawater was placed in a reservoir at the top then boiled. A condensation tube connected to a valve at the top then draped overboard, rising back up into the raft connecting to a collection device, either a vinylite bladder or aluminum pail. Alternate fuels could be used, including a porous block that absorbed aviation fuel or other combustible fluids. Also included were drinking cup and rubber bladder used to store water or keep items dry. M-326J 8 lb. solid fuel still with its complement of components minus three tubes of fuel cans. These were procured and stored on large aircraft but were not standard equipment for life rafts. Available from the ASO under stock number R83-S-793000. These stills were the most efficient, producing potable water at an average of 57 oz./hr., but their bulk rendered them impractical for airborne life rafts. Still in a life raft and equipment display with the addition of the 4 lb. unit. (National Archives)

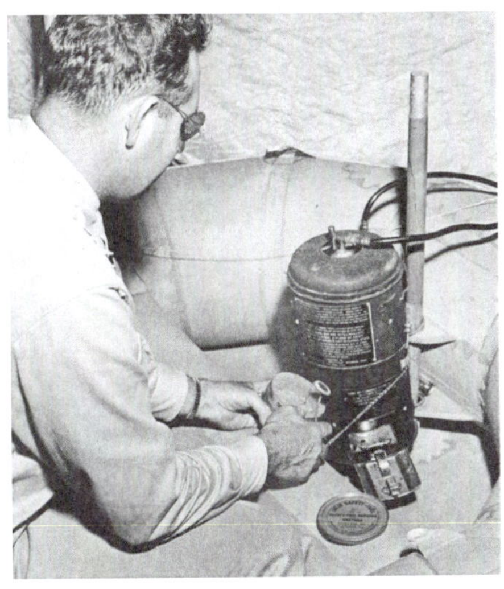

Two other aspects relating to rations were water collection and water stowage. Water was collected by any means available: sails, paulins, and oilskin sheets. Once collected, the problem of water stowage emerged. It was decreed that any container related to the life raft could be used as a water-storing apparatus, such as the outer case, accessories or equipment containers. The standard M-1910 canteen was available but not universally worn; in early USAAF life raft development a pocket for a canteen was added. The necessity to develop and adopt a water-storing device that would be a permanent component for emergency rescue equipment became obvious. Upon development of new emergency parachute kits in 1941 for the USAAF, it recommended that a water storage bag be included for the Type B-2 jungle kit. Companies submitted samples, at the request of the Materiel Command, of rubber-lined fabric bags or some form of a rubber water storage container.

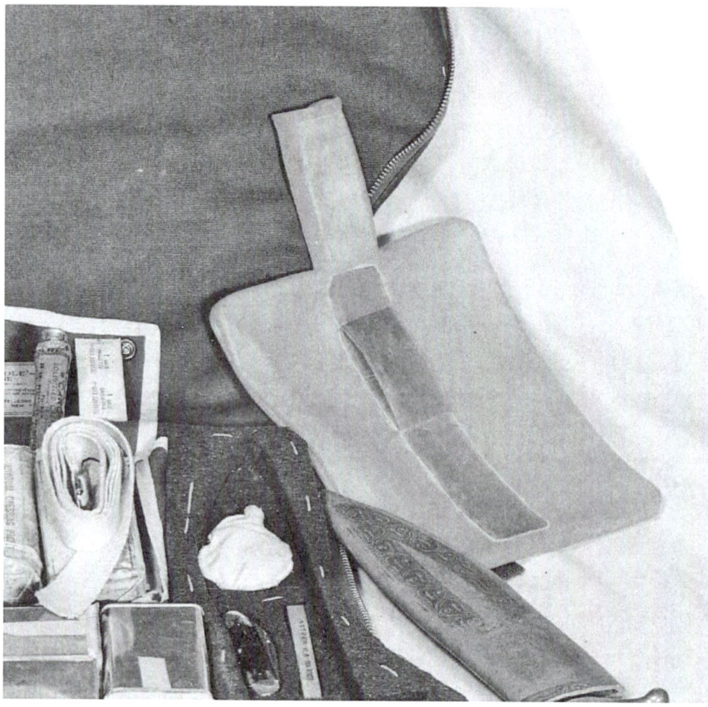

(Left)) Rubber bladder provided and tested in 1941 for use in the proposed Type B-2 jungle kit. After field tests, it proved undurable and was never adopted. (National Archives)

(Below and below left) Multiple companies submitted designs deemed unsatisfactory. To meet immediate demand and as an interim, commercially available fabric water bags were pressed into service until a suitable model could be adopted. 2 gal. water bag was used in sustenance kits like the Type E-11. The 1 gal. water bag was supplied for parachute kits.

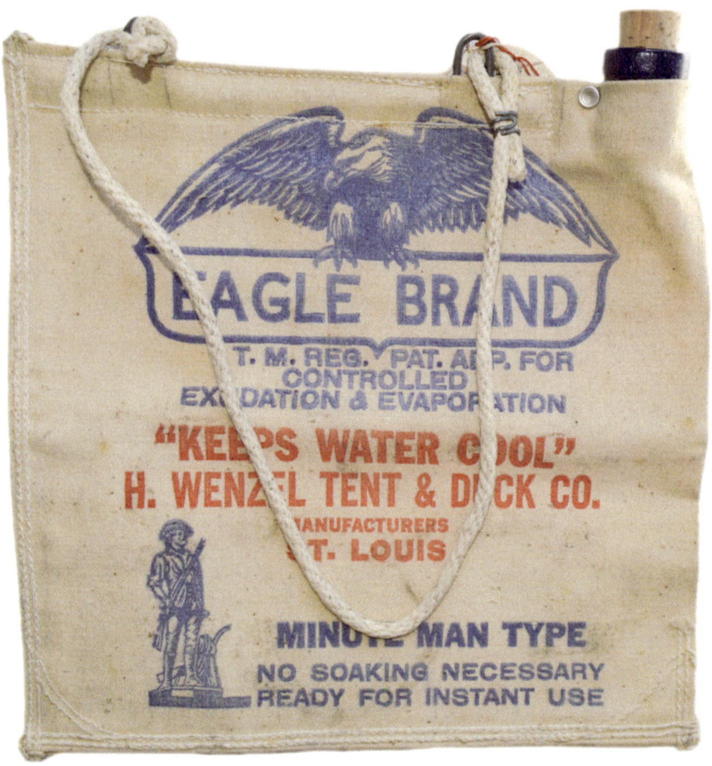

The Materiel Command evaluated a variety of designs and materials for water storage, ultimately approving vinylite bags. Procurement was authorized in December 1942 under Dwg. No. 43B9963. Bags have a 1 qt. capacity and were sealed by rolling the throat down and securing with a Bakelite buckle. S. Buchsbaum & Co. and National Carbon Co. were the approved manufacturers. Stamped on the bladders were the companies' part numbers. Part No. SB 210 was S. Buchsbaum & Co.'s and Part No. 233 was National Carbon Co.'s.

These bladders were not assigned a stock number; they were instead CFE or requisitioned by nomenclature and part number. The BuAer did not adopt the 1 qt. bladder. These bladders were also distibuted as pre-captiure aids by MIS-X.

(Left) National Carbon Co. bladders had multiple variations: 1st production bladders just include the part number, 2nd production added the company name, and 3rd production included instructions.

(Right) S. Buchsbaum & Co. SB 210 has two known variations. 1st prodcution included the part number, only. 2nd production included instructions.

For the BuAer, water storage was part of the complete package of research and development relating to the tablet rations and demineralization methods. As part of the series of field tests conducted in the spring of 1943, water bladders and collection devices were included. Vinylite bladders were tested and found suitable for storing water and other perishable items. The BuAer emphasized collection and capacity, which led to the addition of a funnel to its new tarpaulin, standardized under Specification M-616, and the adoption of a five-quart vinylite water bladder.

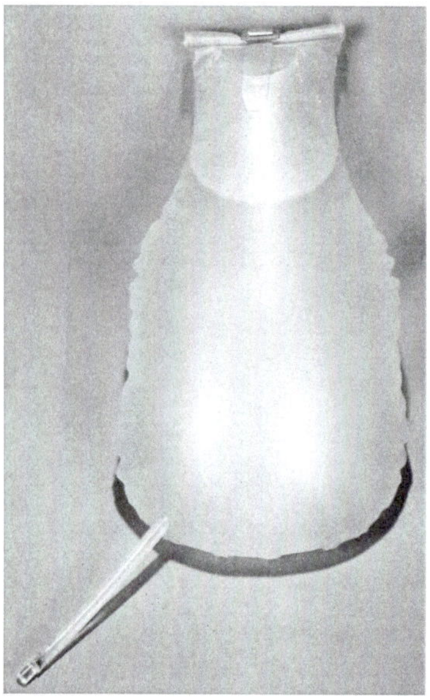

The adopted 5 qt. water bladders were provided by National Carbon Co. under Part No. 283-A as detailed under BuAer Dwg. No. 8985. Original drawing and test samples included drinking tube, but omitted in production. In January 1944 the BuAer procured large quantities of the amended 5 qt. water bladders to supply emergency equipment kits and life rafts at the ratio of one per man, available from ASO under stock number R83-B-30175. No formal specifications were drafted for bladder, manufacturer part number 283-A was referenced. Each man would have one water can, one desalting kit, and one water bladder, with the later addition of one solar still, providing water to an individual for an indefinite period. The 5 qt. water bladder would remain standard for several decades.

The USAAF also adopted the 5 qt. water bladder for its life rafts in 1944, referenced as stock number 4500-240000. The 5 qt. bladder first appears in Technical Order 04-15-1 dated 15 August 1944, which indicated four were to be included in the accessories for life rafts. Those provided were identical to those delivered to the BuAer. The bladder was adopted in FY1946 under Dwg. No. 46G3007-1 & -2. The dash-1 bladder incorporated a canvas shoulder strap with the dash-2 being like the US Navy.

The importance of a canteen for survival was stressed but not always followed. A canteen was issued with personal aids kits Type E-3B and E-17. Here, two B-29 aircrew members of the 677th BS in China, Lt. Fred Bender and Capt. Arthur Hyde (back row, left), carried canteens as a complement to their personal equipment. The canteen's use was optional in some environments; jungle and desert terrain made it mandatory. (National Archives)

(Above) M-1910 canteen was available to all USAAF flying personnel, issued one per individual as detailed by T/O & E for fighter and bombardment squadrons. However, its use was up to personal preference. They were requisitioned through standard QMC supply channels. Here, USAAF B-29 aircrew member recently arrived on Saipan, c.1944. (National Archives)

(Left) Space restrictions aboard airplanes dictated the canteen's location but it was used in all aircraft. P-38 fighter Ace, Lt. Richard Campbell, of the 37th FS wears an M-1910 canteen, North Africa, c.1943. (National Archives)

Recommendation to carry canteens rapidly grew within the intelligence community as part of evasion and escape, and survival indoctrination by medical officers and the ASEO. It was stressed that a canteen should be worn on all flights over water and over arid regions. USAAF, a branch of the US Army, had the canteen as standard issue, whereas it was not a standard item for the US Navy. The BuAer classified the canteen as an item of emergency equipment, which became optional personal equipment for pilots and aircrew requisitioned through the Navy Yard.

(Above) Navy pilot uses M-1910 canteen. Lt. Ralph Wines of VF-5 aboard USS Yorktown, June 1944. (National Archives)

(Above right) Pilot of VF-17 aboard USS Hornet, February 1945. (National Archives)

(Right) US Navy pilot of Air Group-16 aboard USS Lexington, July 1943. Canteen covers, bottles and cups shared the same stock number whether in US Army supply or US Navy supply as Class-74 equipment. (National Archives)

(Left) USMC aviators were no exception to the utilization of the canteen especially in the humid-thirst provoking environment of the SWPA. It was carried both as a convenience and a survival item. Canteens were provided to pilots and aircrew under the same guidelines as the US Army: as standard issue. Both officers and enlisted men were supplied with the standard complement of 782 gear in their training courses, which included canteen cover, bottle, and cup. Here, VMTB pilot wears his partial allotment as a commissioned officer. He would also have been provided pistol belt, compass, and canteen set. Sidearms were issued upon preparing for overseas deployment. The canteen and cup (if present) were US Army M-1910 type. The cover was a late WWII cross-flap type manufactured at USMC Philadelphia Quartermaster Depot. (Courtesy Gregory Pons)

(Below left) 1st Lt. John Hundley of VMF-211 wears a variation of a USMC Philadelphia Quartermaster Depot cover with Lift-the-dot fasteners, c.1943, somewhere in the SWPA. (National Archives)

(Below) 2nd Lt. Dick Morgan wears the same type of canteen, Bougainville, May 1944. (National Archives)

Afterword

Of the accumulated data, the most frequent cause of aircraft loss or damage in the combat zone was enemy anti-aircraft fire, followed by engine failure in close second place. Running out of gas during combat was the third most common cause. These three factors combined were by far the biggest cause of pilots and aircrews being separated from their aircraft during combat. Interestingly, attacks by enemy aircraft barely register in the calculation, hovering around 2%.

Regardless of the cause, individuals could find themselves in a variety of situations: forced landings, water landings (ditching) and bail-outs. Each presents its own unique conditions that needed to be evaluated. So many factors had to be calculated such as: knowledge of the area, dangers from the enemy, personal health and whether to stay or go. This would ultimately be relative to the operational area. Emergency equipment was designed to help lend assistance based largely on the region being flown over.

Injuries were reported in most incidents, the majority being burns, cuts or sprains. Food and water were necessary to continue life. Though most kits only carried a limited supply, they were only intended to provide supplement for a short duration of 2 or 3 days. Rations contributed a feeling of comfort and boosted morale, but most importantly they helped maintain strength. Any extended time after three days, food had to be foraged, living off the land. Basic implements for this task were provided. Maps and compasses were required to navigate back to friendly areas. The most important variable in any of the survival situations was health. Protective clothing and the ability to erect shelters contributed to success. Personal emergency kits were continually revised and adapted to cater to all individuals, in any territory, to assist in preservation of life.

Acknowledgments

This work could not have reached its potential without contributions and support from Robin Bartel and Paul Faulkner. Robin Bartel has been a friend and a peer for more than 25 years, and has contributed many artifacts for reference as well as solid support for the project, consistently urging me on. Paul Faulkner is a long-time friend who shares the same interests, and his knowledge of Naval aviation is unequalled. I would also like to thank Maryjane Stout, Kriss Usherwood, Charlie Flick, Frank Trzaska, Bob Baldwin, Mathieu Bianchi, Remi Bouche, Jeff Klein, Jerry Kolakowski, Ben Major, Steve Milnthorpe, George Petersen, Gregory Pons, Geoff Pringle and Mick Prodger for their contributions and graciously shared information. Significant assistance was provided by Brett Stolle of the National Museum of the United States Air Force, Bob Thomas of the National Naval Aviation Museum and Maranda Gilmore of the Air Force Historical Research Agency.

Author's Note

I have maintained an interest in WWII for as far back as I can recall, playing army in the woods in my home state of Oregon. It was on my 13th birthday, back in 1987, that I established myself as a collector. My gift was an M1 Carbine. Rummaging through my trunk of field gear, I separated the WWII items from everything else, and those, together with the rifle, formed the basis of my collection.

My interest only grew, accumulating more and more items until today, 30 years later, all aspects of United States WWII memorabilia remain a passion of mine. Some 20 years ago I obtained an empty US Navy M-592 emergency back pad kit – my first piece of WWII survival equipment – and so began my transition solely into that area. This initiated a quest to find out and obtain what went in the kit. I quickly found myself frustrated at the lack of information available, gathering more questions than answers. As the investigation moved forward I became more and more interested in the overall subject, accumulating artifacts but still operating with a very limited reference library. Influenced by peers in my local collecting community, I eagerly joined in conversations about their adventures across the country, networking and traveling to various archival holdings. It soon became apparent that I would need make those pilgrimages in order to find the answers to questions that I, and many others, shared. After a relatively short time I decided that if I was going to make that effort, I should put my findings in a reference book for anyone who may be interested.

My initial research efforts were at local university libraries then inter-library loans, until the environment allowed me to visit the National Archives. I have since been back to the NARA multiple times, accumulating documents and images. Not every answer to every question was found, though it was certainly through no lack of effort. Those answers simply may not exist.

It wasn't mandatory for government agencies or facilities to retain all documents, once they were determined obsolete or irrelevant, often ending in their destruction. What survived post-war downsizing and finished up in repositories was at the grace and courtesy of individuals at these government installations, for which I am eternally grateful! However, many records are incomplete, and it can be viewed as a stroke of luck that they still even exist. Information from various sources was sometimes contradictory but through continued accumulation of data, a compilation of redundancy pointed towards the facts. I have attempted to detail each genealogy as clearly as possible.

The overall experience has been very educational for me, and hopefully those who read these studies will find them helpful.

Dustin Clingenpeel
May, 2019

Index

A

Albrook Field 101-102, 104, 106, 118

B

Bag, Sleeping, Type A-2 30
Bag, Sleeping, Type A-3 32
Bag, Water 37, 80, 94, 102, 292
Banks, W.H. 188
Bardo, James 180
Bender, Fred 295
Boder, C.N. 233
Bradford, C.R. 85
Bridges, T.H. 95
Bowman, J.H. 100
Bureau of Aeronautics 6, 46, 50, 52-53, 56, 63, 69, 84-85 92-94, 172, 177-178, 192-193, 197, 211-212, 218, 225, 229, 232. 234, 236, 255, 269, 272-273, 276-277, 280-281, 284, 286, 290-291, 294, 296
Bureau of Medicine and Surgery 172, 174-175, 192, 197, 208, 210, 213, 215, 212, 225, 230, 245, 276-277

C

Cannon, George 170
Campbell, Richard 295
Consolazio, William 282-283
Container, Aerial Delivery, Type A-3 44
Container, Aerial Delivery, Type A-4 27, 30-31
Container, Cooking, 42B2059 71, 74, 90, 92
Container, Cooking, Asbestos 117
Container, Food, Type A-1 269
Container, Sustenance, Type A-13 42-43
Canteen, M-1910 59, 64, 69, 144, 295-297
Compass 21, 25, 29, 59, 86, 92 130, 136
Compass, Liquid Filled, Engineer 32, 35, 37, 41
Compass, Magnetic Card 50
Compass, Match Safe, AN-C-101 47, 68, 141, 145, 156
Compass, Match Safe, M-554 131-132, 136, 139
Compass, Pocket 59, 93 113, 115, 118-119, 121, 123, 131,137
Compass, Prismatic 59-61, 65, 68
Cover, Pistol 114

D

Dixon, Harold 5-9

F

F4F Wildcat 129
F6F Hellcat 53
Fairfield Air Dept 59
Findley, William 130
Fishing Kit 29, 32, 65-66, 69, 71, 74-77, 81, 86-88, 90, 93, 137, 155
Fishing Kit, AN-K-2 35, 37, 47
Fishing Kit, M552 6
Fishing Kit, M552A 138
Fishing Kit, M-627 50, 163
Fishing Kit, Type-5 30, 68
First Aid Kit, Battle Dressing 241
First Aid Kit, Camouflaged 120, 222, 225-227
First Aid Kit, Aeronautic 18, 25, 32, 35, 37,59, 64, 173, 175-176, 181, 183-198, 211-212, 219-220
First Aid Kit, Aircraft Package 6, 174-175, 177, 208
First Aid Kit, Aircrew Member 219-223
First Adi Kit, Ambulance 194, 242-243
First Aid Kit, Arctic 18, 194, 236, 238-239, 243
First Aid Kit, Flight Chest 236, 242
First Aid Kit, Frying Pan Insert 68, 82-83, 90, 93, 205-207, 219
First Aid Kit, Jungle 18, 194, 236, 239-240, 243
First Aid Kit, M-2 229-231
First Aid Kit, M-448 6, 177-180, 192, 208
First Aid Kit, Parachute Packet 187, 194, 198-204, 219
First Aid Kit, RAF, Large Dinghy 211
First Aid Kit, RAF, MK-II 217, 222
First Aid Kit, RAF, Type-3 211
First Aid Kit, Life Raft, Pneumatic 30, 37, 47, 50, 130, 132, 145, 158, 192, 208-217, 225
First Aid Kit, TT-1 240-241
Flare, Fusee 22-23, 35, 37, 68, 77, 80, 90, 134, 137, 161
Flare, MK-I Light 130, 132
Flashlight 21, 25, 35
Flashlight, 17-L-16 164
Flashlight, 17F7 50, 52
Flashlight, Daco 18, 23
Flashlight, M-567 145
Flashlight, TL-122 69
Flashlight, Type A-9 22, 32
Flashlight, Vizlight 18, 23
Fleming, William 200

G

Galley 269-271
Galley, Type C-1 265-268
Galley, Type C-2 265-268
Gloves, Flying, Type A-12 32
Gloves, Mechanics, Type D-2 71, 74-75, 77, 80, 90, 93, 114
Gloves, Mechanics, Type D-3 32, 37, 114, 123
Gloves, Mechanics, Type D-3A 114, 127
Gloves, Work 156
Goggles 91, 115
Green, Thomas 167
Greenhaigh, Selbie 144
Grenade, Smoke, HC-M8 21
Grumman G-21 Goose 46

H

Hat, Sun 35, 37, 41, 113
Hauser, E.J. 170
Headnet, M-1942 113
Headnet, M-1944 159
Headnet, Mosquito 60, 63, 65, 68, 71, 77, 80
Headnet, Mosquito, Type M-565 6, 138, 156
Headnet, Mosquito, Type M-565A 6
Headnet, Mosquito 43G3868 13, 35, 37, 74-75, 124, 130, 132, 134, 137, 156
Held,Charles 129
Helmet, Flying, Type B-9 32
Houston, Charles 238
Hyde, Arthur 295

J

Jacket, Mechanics, Type D-2 37
Jacobs, Howard 128

K

Kit, Back Pad, 69233 130-137
Kit, Back Pad, Original 132, 138-145, 149, 154, 161-162, 164
Kit, Back Pad, M-592 132, 145-161, 164-166, 170
Kit, Parachute, Basic 59-60, 69
Kit, Parachute, Jungle, 35D1751 59-62, 64-65, 69, 248
Kit, Parachute, Jungle, 38J5271 65-69, 71-73, 76-77
Kit, Parachute, Type B-1 70-74, 76-78, 82, 88-89, 90-93, 104, 198, 205
Kit, Parachute, Type B-2 68, 70, 76-83, 88-89, 92, 104, 108, 198, 205, 292
Kit, Parachute, Type B-3 88-89, 95
Kit, Parachute, Type B-4 26, 33, 76, 88-96, 98-99, 104, 108, 119, 205, 258, 293
Kit, Parachute, Type B-5 96-98, 124
Kit, Shipwreck, Type M-594 47-50, 55
Kit, Shipwreck, Type M-594A 49-50
Kit, Shipwreck, Type M-594B 50-51
Kit, Signaling, ADS-1 52, 55
Kit, Sustenance, 39G2156 69-70
Kit, Sustenance, ADR-1 53-55
Kit, Sustenance, Type C-1 33, 39, 41, 96, 102-115, 119, 121-127, 207, 259, 293
Kit, Sustenance, Type C-2 125
Kit, Sustenance, Type E-1 13-16, 25, 27, 30, 32-33, 40, 43, 70
Kit, Sustenance, Type E-2 13, 16-18, 25, 27, 30, 32, 70

Kit, Sustenance, Type E-3 12, 68, 88, 121, 257
Kit, Sustenance, Type E-4 19-20, 268
Kit, Sustenance, Type E-5 21, 23-26, 37, 121
Kit, Sustenance, Type E-6 12, 33
Kit, Sustenance, Type E-7 12, 33
Kit, Sustenance, Type E-8 25-27
Kit, Sustenance, Type E-9 25-27, 33, 40, 43
Kit, Sustenance, Type E-9A 33-34, 40-43
Kit, Sustenance, Type E-10 27-28, 30
Kit, Sustenance, Type E-11 29-30, 33, 37, 39
Kit, Sustenance, Type E-12 30-33, 40-42, 57
Kit, Sustenance, Type E-13 12
Kit, Sustenance, Type E-14 33, 35-36, 40-42, 255
Kit, Sustenance, Type E-15 33, 37-42
Kit, Sustenance, Type E-16 32, 40-43
Kit, Sustenance, Type E-17 12, 121
Kit, Sustenance, Type E-18 39, 42, 127, 241
Kit, Sustenance, Type E-19 40, 43
Kit, Sustenance, Type F-1 69-71
Kelly, W.C. 77
Knife, Boy Scout 29, 37, 47-48, 50, 53, 76-77, 79, 87, 130-132, 134, 138138, 145, 153, 156
Knie, Emergency, Curved 245
Knife, Hunting 13, 21, 30, 32, 59, 67, 69-70, 74, 84, 118

L

Ladd Field 71-72, 102
Lanreth, W.C. 100
Laughling, Joseph 87
Lifeboat, Airborne, Type AR-8 50
Life Preserver, Type B-4 101
Life Raft, Specification M-3 46-47, 50, 53, 129, 131, 208
Life Raft, Type A-2 211, 250
Life Raft, Type A-3 24, 38, 211, 250
Life Raft, Type C-2 77, 80, 95
Life Raft, Type AN-R-2 138, 155, 144
Life Raft, Type Mark-I 129
Life Raft, Type Mark-IV 9, 46
Life Raft, Type Mark-VII 46
Life Raft, Type Model A 145, 161,
Life Raft, Type PK-1 145, 161
Liner, Type A-3 32

M

Machete, 9" 59-61, 65-68, 92
Machete, 10" 17, 41, 77-78, 90
Machete, 43E1068 18, 25, 27, 32, 35
Machete, Bolo 59
Machete, M-1942 25, 32, 36
Machete, V-44 156, 166
Manual, Castaway's Baedeker 130-132, 140
Manual, Emergencies 26, 35, 37, 82-83
Manual, Survival, 21W 32, 37, 41, 82, 108, 119
March Field 69
Martin PBM Mariner 54
Match Safe 29, 59-60, 65-66, 69, 71, 74-75, 115
Match Safe, 42A6930 22, 37, 41, 77, 79, 131, 134, 137
Materiel Command 12, 17, 19, 29, 30, 33, 40-41, 59, 62-64, 71,120, 132, 96, 112, 117, 173, 175, 181, 198, 205, 219-220
Mattress, Type A-2 30
Mattress, Type A-3 32
Middletown Air Depot 60, 65-66, 249
Millet, Lloyd 84
Minot, Henry 142
Mirror, Reflector 21, 29, 132, 141
Mirror, Signaling 23, 27, 29-30, 32, 63, 177
Mirror, Signaling, Type 40653 22, 25, 27, 32, 35, 37, 41, 68, 82-83, 118
Mirror, Signaling, Type M-580 47, 143, 156-157
Mirror, Signaling, Type M-580A 50, 52
Morgan, Dick 297
Mukluks, Type A-14 37

N

Nichols Field 106

O

Obenshain, Roy 68
Oil, M-592 157

P

Panama Air Depot 60
Pastula, Tony 5-9
Parachute, Type G-7 29, 35, 37, 39
Parachute, Type G-8 34, 39, 42
Penney, W.M. 111
Peterson, George 99
Pfeiler, John 162
Philippine Air Depot 60
Phillips, E.A. 149
Pinchot, Gifford 7
Pistol, Caliber, .22 76
Pistol. Caliber, .45 59-60, 64, 76, 100
Poncho, B-4 92, 94
Poncho, P61 50, 84
Poncho, M-549 130, 132

R

Ration, 38D2666 64-65, 68-69, 71, 249-250
Rartion, 41D4293 71, 74, 77, 249-250
Ration, 42B10959 74, 77-78, 134, 250-251, 260
Ration, Abandon Ship 277
Ration, Airborne Lifeboat 263
Ration, Air Corps 59-60, 246, 278
Ration, Air Corps, Cross Country 59, 248
Ration, Aircrew 266
Ration, Armour 247
Ration, Bail-Out 87, 251, 257-258
Ration, Lifeboat, USCG 278-279
Ration, Life Raft, Type-A 30, 33, 37, 43, 255, 260-263
Ration, Lunch, Combat 265
Ration, Parachute 68, 93, 251, 258
Ration, Parachute, Emergency 115, 259-260
Ration, Reserve 59, 247-248
Ration, Mountain 15
Ration, Trench 247
Ration, Type-A 247, 263-265
Ration, Type-C 245, 251-253
Ration, Type-D 13, 21-22, 86, 88, 137, 143, 249-251, 253, 257, 260, 275
Ration, Type-K 22, 26-27, 29-30, 33, 35-37, 251-258, 260, 268, 273
Ration, Type M-539 48, 273-276
Ration, Type M-539B 47, 50, 53,145, 157, 163, 276-277
Repellent, Insect 74-77, 80, 115
Rescue Assembly, Type AR-1 50
Rescue Assembly, Type AR-2 50-51
Rescue Assembly, Type AR-4 50-51
Rescue Assembly, Type AR-7 50-51
Rescue Assembly, Type AR-10 50
Rifle, Carbine 32, 57
Rifle, Stevens 17, 25, 27, 32, 35, 41, 57

S

Sail, Type M-551 138, 166
Sail, Type Size-18 22, 26
Sail, Type Size-77 22, 24, 27, 29-30, 32, 35, 37
San Antonio Air Depot 60
SBD Dauntless 6, 142
Scott Field Air Depot 60, 62
Schofield, Theodore 70
Sea Marker, Type Aluminum 21
Sea Marker, Type AN-S-10 22, 29, 37, 49
Sea Marker, Type M-528 24, 27, 32, 42, 106, 113, 131-132, 134, 168
Sea Marker, Type M-566 49, 52
Shelter, Field, Type A-2 27-28, 30, 32
Shonkin, Courtney 203
Shonkin, Earl 203
Shot, Ammunition, M15 82-83, 103, 118
Sienkiewicz, Stanley 202

Signal, Distress, Parachute, Type M11 29
Signal, Distress, Type T-49/M75 41, 125
Signal, Pistol, Pyrotechnic, Type M8 29-30, 37
Signal, Pistol, Pyrotechnic, Type M9 30, 37
Signal, Projector, Ground M4 21
Signal, Projector, Hand, Type Mark-IV 47, 50, 52,138, 156
Signal, Smoke, Hand, Type MK-I MOD-0 41, 50, 52, 168
Simrill, Francis 100
Spealman, Claire 282-283

T

Tarney, Lucille 242
TBD Devastator 5-6
TBF Avenger 50
Telkes, Maria 286
Transmitter, Type SCR-578 23-24, 55
Trumbull, Robert 5

U

US Army Air Force 12, 20, 41, 46, 57, 70, 59-60, 66, 76, 84, 86-87, 92-93, 95,137-138, 173-174, 177, 181-182, 185, 188-189, 192-193, 204-205, 211-212, 219-220, 222, 226, 229, 236, 241, 242-245, 248, 250-251, 255, 257-260, 263-269, 271-273, 278, 280, 286-289, 292, 294-296
US Army Medical Department 80, 172-176, 181, 184, 189-190, 193-194, 196-198, 204-205, 207, 210-212, 214, 222, 229, 236, 238-239, 242

W

Water, Bladder 76, 100, 292, 294
Water, Bladder, 43B9963 80, 90, 116, 293
Water, Drinking, Demineralization Type JJ-1 43, 286
Water, Drinking, Demineralization Type M-613 50, 53, 138, 146, 148, 152, 282-285
Water, Drinking, Distillation, LL-1 288-289
Water, Drinking, Distillation, LL-2 289
Water, Drinking, Distillation, M-362 291
Water, Drinking, Distillation, M-663 50, 53-54, 290
Water, Drinking, Distillation, Delano 24, 287
Water, Drinking, Type AN-W-5 13-14, 21-22, 26-27, 29, 33, 35, 37, 43, 47, 50, 53-54, 112, 146, 157, 273, 280-282, 284,-285
Water, Drinking, Type M-504 24, 45,111, 168 280
Welby, Bryce 238
Whistle, Police 29, 35, 41, 47, 50, 52, 130-132, 157
Whitmore, James 82

Y

Yost, Coyd 86

Z

Zender, Ralph 169

www.ingramcontent.com/pod-product-compliance
Lightning Source LLC
LaVergne TN
LVRC101546220926
848486LV00041B/386

* 9 7 8 1 9 4 3 4 9 2 9 5 4 *